Climatology

Climatology

Second Edition

Robert V. Rohli
Louisiana State University

Anthony J. Vega
Clarion University of Pennsylvania

JONES & BARTLETT
L E A R N I N G

World Headquarters

Jones & Bartlett Learning
40 Tall Pine Drive
Sudbury, MA 01776
978-443-5000
info@jblearning.com
www.jblearning.com

Jones & Bartlett Learning Canada
6339 Ormindale Way
Mississauga, Ontario L5V 1J2
Canada

Jones & Bartlett Learning
International
Barb House, Barb Mews
London W6 7PA
United Kingdom

Jones & Bartlett Learning books and products are available through most bookstores and online booksellers. To contact Jones & Bartlett Learning directly, call 800-832-0034, fax 978-443-8000, or visit our website, www.jblearning.com.

Substantial discounts on bulk quantities of Jones & Bartlett Learning publications are available to corporations, professional associations, and other qualified organizations. For details and specific discount information, contact the special sales department at Jones & Bartlett Learning via the above contact information or send an email to specialsales@jblearning.com.

Jones & Bartlett Learning Commitment to the Environment

As a book publisher, Jones & Bartlett Learning is committed to reducing its impact on the environment. Many Jones & Bartlett Learning titles are printed using recycled, post-consumer paper. We purchase our paper from manufacturers committed to sustainable, environmentally sensitive processes. We employ the Internet and office computer network technology in our effort toward sustainable solutions. New communication technology provides opportunities to reduce our use of paper and other resources through online delivery of instructors' materials and educational information—and, of course, we recycle in the office.

Production Credits

Chief Executive Officer: Ty Field
President: James Homer
SVP, Chief Operating Officer: Don Jones, Jr.
SVP, Chief Technology Officer: Dean Fossella
SVP, Chief Marketing Officer: Alison M. Pendergast
SVP, Chief Financial Officer: Ruth Siporin
Publisher, Higher Education: Cathleen Sether
Acquisitions Editor: Molly Steinbach
Senior Associate Editor: Megan R. Turner
Production Manager: Louis C. Bruno, Jr.
Senior Marketing Manager: Andrea DeFronzo
V.P., Manufacturing and Inventory Control: Therese Connell
Cover Design: Kate Ternullo
Associate Photo Researcher: Carolyn Arcabascio
Illustrations: Westchester Book Group; Mary Lee Eggart, Cartographic Section, Department of Geography & Anthropology, Louisiana State University
Composition: Westchester Book Group
Cover Image: Courtesy of NOAA
Printing and Binding: Malloy, Inc.
Cover Printing: Malloy, Inc.

Library of Congress Cataloging-in-Publication Data

Rohli, Robert V.
  Climatology / Robert V. Rohli and Anthony J. Vega. — 2nd ed.
    p. cm.
  ISBN 978-0-7637-9101-8 (alk. paper)
  1. Climatology. I. Vega, Anthony J. II. Title.
  QC981.R649 2011
  551.5—dc22
                                                    2010011822
6048

Printed in the United States of America
15 14 13 12 11   10 9 8 7 6 5 4 3 2

Cover image: A land-surface temperature map, generated in real time on March 3, 2010—early spring in most of the northern hemisphere. Orange areas indicate warmer regions; blue represents the lingering cold.

B R I E F

Contents

PART 1   **THE BASICS**   1
Chapter 1   **Introduction to Climatology**   2
Chapter 2   **Introduction to the Atmosphere**   11

PART 2   **CLIMATOLOGICAL PROCESSES**   24
Chapter 3   **The Climate System: Controls on Climate**   26
Chapter 4   **Effects on the Climate System**   54
Chapter 5   **Energy, Matter, and Momentum Exchanges Near the Surface**   78
Chapter 6   **The Global Hydrologic Cycle and Surface Water Balance**   104
Chapter 7   **General Circulation and Secondary Circulations**   126

PART 3   **CLIMATES ACROSS SPACE**   151
Chapter 8   **Climatic Classification**   152
Chapter 9   **Extratropical Northern Hemisphere Climates**   174
Chapter 10   **Tropical and Southern Hemisphere Climates**   220

PART 4   **CLIMATES THROUGH TIME**   259
Chapter 11   **Climatic Change and Variability**   260
Chapter 12   **Anthropogenic Climatic Changes**   286
Chapter 13   **Linking Spatial and Temporal Aspects of Climate Through Quantitative Methods**   309

PART 5   **RELATIONSHIPS BETWEEN CLIMATE AND OTHER ENDEAVORS**   331
Chapter 14   **Applied Climatology, Climate Impacts, and Climatic Data**   332
Chapter 15   **The Future of Climatology**   349

Glossary   354
Index   394

v

Contents

Preface xiii
Acknowledgments xv

PART 1	**THE BASICS** 1	
Chapter 1	**Introduction to Climatology** 2	
	Meteorology and Climatology 3	
	Scales in Climatology 4	
	Subfields of Climatology 5	
	Climatic Records and Statistics 7	
	Summary 9	

Chapter 2 **Introduction to the Atmosphere** 11
 Origin of the Earth and Atmosphere 11
 Atmospheric Composition 12
  The Carbon Cycle 13
  Constant and Variable Gases 16
 Faint Young Sun Paradox 17
 Atmospheric Structure 18
 Summary 22

PART 2 **CLIMATOLOGICAL PROCESSES** 24
Chapter 3 **The Climate System: Controls on Climate** 26
 Latitude 26
 Earth-Sun Relationships 29
  Revolution 29
  Rotation 29
  Axial Tilt and Parallelism 29
  Combined Effect of Revolution, Rotation, and Tilt 30
 Distance to Large Bodies of Water 33
 Circulation 35
  Pressure 36
  Wind 38
  Surface Versus Upper-Level Winds 43
  Vertical Motion 43
  Cyclones and Anticyclones 43
  Oceanic Circulation 46

Topography 47
Local Features 48
Putting It All Together: Spatial and Seasonal Variations in Energy 49
Summary. 51

Chapter 4   **Effects on the Climate System** **54**
Ocean Circulation 54
    Surface Currents 54
    Deep Ocean Thermohaline Circulations 58
El Niño–Southern Oscillation Events 61
    Walker Circulation 62
    Historical Observations of ENSO 63
    El Niño Characteristics 64
    La Niña Characteristics. 66
    Global Effects 66
    Effects in the United States 67
    Relationship to Global Warming? 68
Volcanic Activity and Climate 69
    General Effects 69
    Aerosol Indices 70
    Major Volcanic Eruptions 71
Deforestation and Desertification 71
Cryospheric Changes 73
    Ice on the Earth's Surface 73
    Feedbacks in the Cryosphere 74
Summary 75

Chapter 5   **Energy, Matter, and Momentum Exchanges Near the Surface** **78**
Properties of the Troposphere 78
Near-Surface Troposphere 79
Energy in the Climate System 80
    Sun as Energy Source. 80
    Measuring Radiant Energy. 85
    Radiation Balance 85
    Turbulent Fluxes. 88
    Substrate Heat Flux 89
    Energy Balance 90
Local Flux of Matter: Moisture in the Local Atmosphere 90
    Atmospheric Moisture 90
    Moisture in the Surface Boundary Layer 92
    Measuring Evapotranspiration 92
Atmospheric Statics, the Hydrostatic Equation, and Stability 94
    Statics and the Hydrostatic Equation 94
    Atmospheric Stability 95
    Assessing Stability in the Local Atmosphere 97
Momentum Flux 99

Putting It All Together: Thermal and Mechanical Turbulence and
the Richardson Number 99
Summary. 100

Chapter 6   **Global Hydrologic Cycle and Surface Water Balance** **104**
Global Hydrologic Cycle 104
Surface Water Balance 107
Potential Evapotranspiration. 107
Evapotranspiration 110
Precipitation 110
Soil Moisture Storage. 116
Deficit 117
Surplus and Runoff 117
Putting It All Together: A Worked Example of the Surface
Water Balance 119
Types of Surface Water Balance Models 120
Water Balance Diagrams 121
Drought Indices 121
Summary 123

Chapter 7   **General Circulation and Secondary Circulations** **126**
Circulation of a Nonrotating Earth 127
Idealized General Circulation on a Rotating Planet 127
Hadley Cells. 127
Polar Cells 128
Planetary Wind Systems 129
Modifications to the Idealized General Circulation: Observed
Surface Patterns 132
Land–Water Contrasts 132
Locations and Strength of Features in the Hadley Cells 133
Locations and Strength of Features in the Polar Cells 136
Locations and Strength of Surface Midlatitude Features 137
Putting It All Together: Surface Pressure Patterns
and Impacts 137
Modifications to the Idealized General Circulation: Upper-Level
Airflow and Secondary Circulations 137
Vorticity 139
Constant Absolute Vorticity Trajectory 141
Flow Over Mountainous Terrain 142
Baroclinicity. 143
Rossby Wave Divergence and Convergence 144
Rossby Wave Diffluence and Confluence 145
The Polar Front Jet Stream 145
Mean Patterns of Rossby Wave Flow 146
Summary. 147

PART 3   **CLIMATES ACROSS SPACE** **151**

Chapter 8   **Climatic Classification** **152**

Early Attempts at Global Climatic Classification 153

Classical Age of Climatic Classifications 153

     Modified Köppen Climatic Classification System 154

     Thornthwaite Climatic Classification System..................... 157

     Other Global Classification Systems 159

Genetic Classifications 159

     Air Masses and Fronts 162

Local and Regional Classifications 167

Quantitative Analysis to Derive Climatic Types 168

     Eigenvector Analysis 168

     Cluster Analysis 170

     Hybrid Techniques 171

Summary.. 171

Chapter 9   **Extratropical Northern Hemisphere Climates** **174**

The Climatic Setting of North America 174

     General Characteristics 174

     Severe Weather 175

     Role of the Gulf of Mexico and the Low-Level Jet 176

     Effect of Mountain Ranges..................................... 177

     Effect of the Great Lakes 178

     Ocean Currents and Land–Water Contrast 178

Climatic Setting of Europe 179

     General Characteristics 179

     Effect of Ocean Currents 180

     Effect of Mountain Ranges..................................... 180

     Blocking Anticyclones 181

Climatic Setting of Asia 181

     General Characteristics 181

     Monsoonal Effects.. 182

     Effect of Mountain Ranges 183

     Effect of Coastal Zones on Climate 183

Regional Climatology.. 184

     B—Arid Climates 184

     C—Mesothermal Climates 191

     D—Microthermal Midlatitude Climates 204

     E—Polar Climates 212

     H—Highland Climates 215

Summary 217

Chapter 10   **Tropical and Southern Hemisphere Climates** **220**

Contrasts Between Extratropical and Tropical Atmospheric Behavior 220

Contrasts Between Northern and Southern Hemisphere Atmospheric Behavior... 222

Climatic Setting of Africa. 222
  General Characteristics 222
  Intertropical Convergence Zone 222
  Air Mass and General Circulation Influences 223
Climatic Setting of Australia and Oceania 225
  General Characteristics 225
  El Niño–Southern Oscillation Influences. 226
  South Pacific Convergence Zone 227
  Madden-Julian Oscillation 228
  Quasi-Biennial Oscillation 229
Climatic Setting of Latin America 230
  ENSO Contributions 231
Climatic Setting of Antarctica 232
Regional Climates 233
  A—Tropical Climates 233
  B—Arid Climates 242
  C—Mesothermal Climates 247
  E—Polar Climates 253
  EF—Ice Cap 253
  H—Highland Climates. 254
Summary. 255

PART 4   **CLIMATES THROUGH TIME** **259**
Chapter 11   **Climatic Change and Variability** **260**
                  Climatic Changes in Geologic History 262
                    Temperature 262
                    Ice Ages and Sea Level 266
                    Recent Trends 268
                  How Do We Know What We Know about Past Climatic Changes? 269
                    Basic Principles 269
                    Radiometric Dating 270
                    Lithospheric and Cryospheric Evidence 271
                    Biological Evidence 274
                    Historical Data 276
                    Converging Proxy Evidence 276
                  Natural Causes of Climatic Change and Variability 276
                    Continental Drift and Landforms 277
                    Milankovitch Cycles 277
                    Volcanic Activity 280
                    Variations in Solar Output 281
                    El Niño–Southern Oscilation Events 282
                  Summary 283

Chapter 12   **Anthropogenic Climatic Changes** **286**
                  Global Warming 286
                    The Greenhouse Effect 286
                    Greenhouse Gases 287

The Urban Heat Island 292
Global Warming: The Great Debate 294
Atmospheric Pollution 297
Global Dimming 297
Atmospheric Factors Affecting Pollution Concentrations 298
Air Quality Legislation in the United States 299
Classifying Air Pollutants 299
By Response 299
By Source 300
Reactions and Attitudes to Climatic Change 304
Prevention 304
Mitigation 304
Adaptation 304
Continued Research 305
Summary 305

Chapter 13   **Linking Spatial and Temporal Aspects of Climate Through Quantitative Methods** **309**
Computerized Climate Models 309
Types of GCMs 310
Representing the Earth–Ocean–Atmosphere System in GCMs 311
Data for GCMs 312
The Seven Basic Equations 314
Navier-Stokes Equations of Motion 314
Thermodynamic Energy Equation 316
Moisture Conservation Equation 317
Continuity Equation 317
Equation of State 318
Similarities and Differences Between GCMs and Weather
Forecasting Models 319
Statistical Techniques 320
Atmospheric Teleconnections 321
Extratropical Teleconnections in the Pacific: The Pacific Decadal
Oscillation 322
Extratropical Teleconnections in the Atlantic Ocean 322
Teleconnections over North America 326
Summary 328

PART 5   **RELATIONSHIPS BETWEEN CLIMATE AND
OTHER ENDEAVORS** **331**
Chapter 14   **Applied Climatology, Climate Impacts, and Climatic Data** **332**
Climate Impacts 333
Impacts on Natural Systems 333
Impacts on Societal Systems 334
Impacts on Human Health and Comfort 335
Climatological Data Sources 335
Data Collection Agencies 335

Primary Versus Secondary Data 337
Secondary Data Sources for Applied Synoptic and
Dynamic Climatological Studies 337
Secondary Sources for Applied Studies of the Climate System 345
Secondary Sources Well Suited to Studies in Paleoclimatology
and Climate Change 346
Secondary Sources Well Suited to Studies in
Physical Climatology 346
Summary 346

Chapter 15   The Future of Climatology 349
Relationship Between Climatology and Meteorology 349
Interdisciplinary Work with Other Scientists 349
Interaction Between Climate Scientists and
Nonscientific Professionals 350
Improved Atmospheric Data Availability and Display. 350
Recognition of the Possibility for Rapid Climate Change 351
Climatology as Part of the Ultimate Goal of Sustainability 351
Summary 353

Glossary   354
Index   394
Photo Credits   426

Preface

In the four years since the completion of the first edition of *Climatology*, climate science has taken on an even more prominent role in public discourse, ranging from informal conversations about the weather and the news to serious discussions among policymakers on the local, national, and international scales. Likewise, with so much information available to the current generation of students—the "Millennials"—it has become increasingly important for them to become critical consumers of information pertaining to the global climatic system. Weighing the advantages and potential risks of cloud seeding at the Beijing Olympics, the rationale for instituting "cap and trade" policies, and the concerns that warming of ocean surfaces might reduce coastal ecosystem productivity all require knowledge of basic climate science. It is not unreasonable to suggest that in 2012, an educated person should be able to converse intelligently about issues related to climatology. The value of appreciating and understanding the *contemporary* applications of climate science necessitates our revision of the first edition.

Constructive comments from so many of our colleagues and students, as well as developments in the science and policy, have guided our revision. The process has involved substantial editing of both the text and art package to promote simplicity, clarity, and timeliness. Accordingly, text, figures, and Web links incorporate the most up-to-date information at the time of publication. Most notably, we have updated all maps and graphics pertaining to anthropogenic emissions of pollutants to reflect the most recent data available.

Structurally, the text remains similar to the first edition. The book continues to serve the upper-level undergraduate or introductory-level graduate student, assuming that the basics of introductory atmospheric science have been mastered. For those students without any background in atmospheric science, the basics are covered, particularly in Chapters 1 (Introduction to Climatology), 2 (Introduction to the Atmosphere), 3 (Climate System: Controls on Climate), and 4 (Effects on the Climate System). Such students, however, may need to spend time beyond that normally required for reading in a course at this level. The intent is that the book remain somewhat flexible without sacrificing the rigor that should be expected of upper-level undergraduates preparing for a career in the earth or environmental sciences.

Although all chapters can be covered in a three-hour-per-week, one-semester course, individual instructors may wish to concentrate more on some chapters than others. We designed each chapter, therefore, to be as self-contained as possible, while avoiding too much duplication. For example, Chapters 5, 13, and 14 form the "ABCs" of boundary layer climatology, climate modeling, and applied climatology, respectively, and can be reserved for students with interest in more advanced courses. Some instructors may choose to deemphasize these chapters and can do so without missing key prerequisite concepts.

Chapters 6 (The Global Hydrologic Cycle and Surface Water Balance) and 7 (General Circulation and Secondary Circulations) are critical for understanding the processes that characterize climates across space and time. We examine the spatial variation of climate in Chapters 8 (Climatic Classification), 9 (Extratropical Northern Hemisphere Climates), and 10 (Tropical and Southern Hemisphere Climates). Instructors who do not see the need for so much detail on the various climatological classification systems may skip Chapter 8 without sacrificing understanding of the climatic types that are described in Chapters 9 and 10.

We have received many positive comments about our unique approach in covering the northern hemisphere extratropics together in one chapter and the tropics and southern hemisphere in another. The impacts of the theoretical extratropical synoptic-to-global-scale circulation and energy features described in Chapters 3, 4, 6, and 7 appear in Chapter 9 as the characteristic climatic properties across space. The concepts are further reinforced within Chapter 9 as discussion proceeds from one part of the northern hemisphere extratropics to another and again in Chapter 10 when discussed in the context of the climates of the (small) landmasses in the southern hemisphere extratropics. Most of Chapter 10 applies the properties of the tropical and maritime Earth introduced in Chapters 3, 4, and 7 to tropical and subtropical climates. Our organization of Chapters 9 and 10 capitalizes on the advantage of synthesis afforded by a regional approach and the advantage of repetition of examples afforded by a process-oriented approach.

Concepts and issues presented in chapters 11 (Climatic Change and Variability) and 12 (Anthropogenic Climatic Changes) remain critical toward an educated citizen's understanding of climate. Although the student hears the "whole story" behind these key issues only by the middle of the book, concepts relating to specific concerns about climatic change and variability are peppered throughout other chapters as well, as these notions have perhaps made climate the key environmental concern in the 21st century.

We added text in Chapter 15 (Future of Climatology) on the concept of sustainability from a climatologist's perspective. The book ends with a "call to action" tone and gives it a more contemporary approach that ties the climatic system more tightly to the rest of the earth–ocean–atmosphere system.

Pedagogy

Beginning a new course in climatology can be a daunting experience for many students. *Climatology, Second Edition*, therefore, incorporates several pedagogical tools to help with comprehension and retention. Each chapter has a consistent structure, starting with the *Chapter at a Glance* outline of topics and ending with a narrative *Summary*. *Key terms* within the text are presented in bold and defined in the Glossary at the back of the book. A list of key terms also appears at the end of each chapter to aid in review. For further help with memorizing the vocabulary, students are encouraged to use the interactive glossary, flashcards, and crosswords on the Student Companion Web Site (see following page).

For the benefit of both students and instructors, each chapter ends with a set of *Review Questions* and *Questions for Thought*. The review questions are designed to ensure that students have assimilated the most important concepts of the chapter. These are not short-answer questions that merely ask students to regurgitate information; students must *describe, discuss, explain, compare,* and *contrast*. Answers to the even-numbered review questions can be found on the Student Companion Web Site. The questions for thought take learning a step further, asking students to think critically and apply theory to reality. These questions can be used for independent study, in homework assignments, or to stimulate class discussion.

Ancillaries

For Instructors

An Instructor's Media CD-ROM, compatible with Windows® and Macintosh® platforms, provides instructors with the following traditional ancillaries:

- The PowerPoint® ImageBank contains all of the illustrations, photographs, and tables (to which Jones & Bartlett Learning holds the copyright or has permission to reproduce electronically). These images are inserted into PowerPoint slides. Instructors can quickly and easily copy individual images into existing lecture slides.
- The PowerPoint Lecture Outline Slides presentation package, prepared by Jean Parker of Boise State University, provides lecture notes and images for each chapter of *Climatology*. Instructors with the Microsoft® PowerPoint software can customize the outlines, art, and order of presentation.

To receive a copy of the Instructor's Media CD-ROM, please contact your sales representative.

Online Instructor's Resources

- **Test Bank** This resource, authored by Jean Parker of Boise State University, consists of over 700 exam questions and is available online. Qualified instructors can download these text files from www. jblearning .com/science/geosciences.

For Students

Jones & Bartlett Learning hosts a free *Student Companion Web Site* that provides content exclusively designed to accompany *Climatology, Second Edition*. The site, updated by Matthew Zorn of Carthage College, features an array of study tools including chapter outlines, answers to the even-numbered, end-of-chapter review questions, an interactive glossary, animated flashcards, crossword puzzles, and suggested readings in the primary literature and on the Web for further exploration of the topics discussed in this book. To access the site, please visit http://physicalscience.jbpub.com/climatology.

Acknowledgments

Once again, we thank our families and friends for their support and encouragement. Our colleagues and students have been very supportive in our attempt to improve and update the text and graphics. We also appreciate all of the suggestions and input from the following reviewers of this text:

> Matthew F. Bekker, Brigham Young University;
> Andrew Carleton, Pennsylvania State University;
> Dorothy Freidel, Sonoma State University;
> Chad Kauffman, California University of Pennsylvania;
> Steve LaDochy, California State University, Los Angeles;
> Charles W. Lafon, Texas A&M University;
> Bryan G. Mark, The Ohio State University;
> Corene Matyas, University of Florida;
> Melvin L. Northup, Grand Valley State University;
> Robert Mark Simpson, University of Tennessee—Martin;
> Lensyl Urbano, University of Memphis;
> Lin Wu, California State Polytechnic University, Pomona; and
> Hengchun Ye, California State University, Los Angeles.

We are particularly indebted to the staff of Jones & Bartlett Learning who assisted us in this edition, particularly Caroline Perry, Lou Bruno, Carolyn Arcabascio, Cathy Sether, Molly Steinbach, and Kimberly Potvin. Any errors in the text are our own.

Robert V. Rohli and
Anthony J. Vega

1

The Basics

Chapter 1—Introduction to Climatology

Meteorology and Climatology
Scales in Climatology
Subfields of Climatology
Climatic Records and Statistics
Summary
Review

Chapter 2—Introduction to the Atmosphere

Origin of the Earth and Atmosphere
Atmospheric Composition
Faint Young Sun Paradox
Atmospheric Structure
Summary
Review

1 | Introduction to Climatology

Chapter at a Glance
Meteorology and Climatology
Scales in Climatology
Subfields of Climatology
Climatic Records and Statistics
Summary
Review

Climatology may be described as the scientific study of the behavior of the **atmosphere**—the thin gaseous layer surrounding Earth's surface—integrated over time. Although this definition is certainly acceptable, it fails to capture fully the scope of climatology. Climatology is a holistic science that incorporates theories, ideas, and data from all parts of the Earth–ocean–atmosphere system, including those influenced by humans, into an integrated whole to explain atmospheric properties.

The Earth–ocean–atmosphere system may be divided into a number of zones, with each traditionally studied by a separate scientific discipline. The part of the solid Earth nearest to the surface (to a depth of perhaps 100 km) is called the **lithosphere** and is studied by geologists, geophysicists, geomorphologists, soil scientists, vulcanologists, and other practitioners of the environmental and agricultural sciences. The part of the system that is covered by liquid water is termed the **hydrosphere**; it is considered by those in the fields of oceanography, hydrology, and limnology (the study of lakes). The region comprising frozen water in all its forms (glaciers, sea ice, surface and subsurface ice, and snow) is known as the **cryosphere** and is studied by those specializing in glaciology, as well as

specialized physical geographers, geologists, and oceanographers. The **biosphere**, which crosscuts the lithosphere, hydrosphere, cryosphere, and atmosphere, includes the zone containing all life forms on the planet, including humans. The biosphere is examined by specialists in the wide array of life sciences, along with physical geographers, geologists, and other environmental scientists.

The atmosphere is the component of the system studied by climatologists and meteorologists. Holistic interactions between the atmosphere and each combination of the "spheres" are important contributors to the climate (**Table 1.1**), at scales from local to planetary. Thus, climatologists must draw on knowledge generated in several natural and sometimes social scientific disciplines to understand the processes at work in the atmosphere. Because of its holistic nature of atmospheric properties over time and space, climatology naturally falls into the broader discipline of geography.

Over the course of this book we shall see that these processes can be complex. The effects of some of these interactions cascade up from local to planetary scales, and the effects of others tend to cascade down the various scales to ultimately affect individual locations over time. The processes are so interrelated with other spheres and with other scales that it is often difficult to generalize by saying that any particular impact begins at one component of the system or side of the scale and proceeds to another.

We can state that the scope of climatology is broad. It has also expanded widely from its roots in ancient Greece. The term "climatology" is derived from the Greek term *klima*, which means "slope," and reflects the early idea that distance from the equator alone (which causes differences in the angle or slope of the Sun in the sky) drove

Table 1.1   Examples of Interactions Between the Atmosphere and the Other "Spheres" and Impacts on Thermal Receipt/Climate

Sphere Interacting with Atmosphere	Example of a Potential Impact
Lithosphere	Large volcanic eruptions can create a dust and soot cloud that can reduce the receipt of solar radiation, cooling the global atmosphere for months or years.
Hydrosphere	Changes in ocean circulation can cause global atmospheric circulation shifts that produce warming in some regions and cooling in others.
Cryosphere	Melting of polar ice caps can cause extra heating at the surface where ice was located because bare ground reflects less of the solar energy incident upon the surface than ice.
Biosphere	Deforestation increases the amount of solar energy received at the surface and alters atmospheric chemistry by returning carbon dioxide stored in living plant matter to the atmosphere.

climate. The second part of the word is derived from *logos*, defined as "study" or "discourse." Modern climatology seeks not only to describe the nature of the atmosphere from location to location over many different time scales but also to explain why particular attributes occur and change over time and to assess the potential impacts of those changes on natural and social systems.

■ Meteorology and Climatology

The two atmospheric sciences, meteorology and climatology, are inherently linked. **Meteorology is the study of weather—the overall instantaneous condition of the atmosphere at a certain place and time. Weather** is described through the direct measurement of particular atmospheric properties such as temperature, precipitation, humidity, wind direction, wind speed, cloud cover, and cloud type. The term "weather" refers to tangible aspects of the atmosphere. A quick look or walk outside may be all that is needed to describe the weather of your location. Of course, these observations may be compared with the state of the atmosphere at other locations, which in most cases is different.

Because meteorology deals with direct and specific measurements of atmospheric properties, discussion of weather centers on short-duration time intervals. Weather is generally discussed over time spans of a few days at most. How is the weather today? How does this compare with the weather we had yesterday? What will the weather be like tomorrow or toward the end of the week? All of these questions involve short-term analysis of atmospheric properties for a given time and place. So meteorology involves only the present, the immediate past, and the very near future.

But a much more important component of meteorology is the examination of the forces that create the atmospheric properties being measured. Changes in the magnitude or direction of these forces over time and changes in the internal properties of the matter being affected by these forces create differences in weather conditions over time. Although many meteorologists are not directly involved with forecasting these changes, meteorology is the only natural science in which a primary goal is to predict future conditions. Weather forecasting has improved greatly with recent technological enhancements that allow for improved understanding of these forces, along with improved observation, data collection, and modeling of the atmosphere. Currently, weather forecasts produced by the **National Weather Service** in the United States are accurate for most locations over a period of approximately 72 hours.

By contrast, **climate** refers to the state of the atmosphere for a given place over time. It is important to note that climatologists are indeed concerned with the same atmospheric processes that meteorologists study, but the scope is different. Meteorologists may study the processes for their own sake, while climatologists study the processes to understand the long-term consequences of those processes. Climatology, therefore, allows us to study atmospheric processes and their impacts far beyond present-day weather.

There are three properties of climate data to consider: normals, extremes, and frequencies. These are used to gauge the state of the atmosphere over a particular time period as compared with atmospheric conditions over a similar time period in the past. **Normals** refer to average weather conditions at a place. Climatic normals are typically calculated for 30-year periods and give a view of the type of expected weather conditions for a location through the course of a year. For example, climatologically normal conditions in Crestview, Florida, are hot and humid during the summer and cool but not cold in winter.

Two places could have the same average conditions but with different ranges of those conditions, in the same way that two students who both have an average of 85 percent in a class may not have acquired that average by earning the same score on each graded assignment. Extremes are, therefore, used to describe the maximum and minimum measurements of atmospheric variables that can be expected to occur at a certain place and time, based on a long period of observations. For example, a temperature of 0°C (32°F) at Crestview in April would fall outside of the range of expected temperatures.

Finally, frequencies refer to the rate of incidence of a particular phenomenon at a particular place over a long period of time. Frequency data are often important for risk assessment, engineering, or agricultural applications. For instance, the frequency of hailstorms in a city is a factor in determining a homeowner's insurance premium. Or if an engineer designs a culvert to accommodate 8 cm (3 in.) of rain in a 5-hour period but that frequency is exceeded an average of two times per year, this rate of failure may or may not be acceptable to the citizens affected by the culvert. A farmer may want to know how many days on average have more than 1.5 cm (0.6 in.) of rain in October because October rains are problematic for any crop harvested during that month.

We can say then that both meteorologists and climatologists study the same atmospheric processes but with three primary and important differences. First, the time scales involved are different. Meteorologists are primarily concerned with features of the atmosphere at a particular time and place—the "weather"—whereas climatologists study the long-term patterns and trends of those short-term features—the "climate." Second, meteorologists are more concerned with the processes for their own sake, while climatologists consider the long-term implications of those processes. Third, climatology is inherently more intertwined with processes happening not only in the atmosphere but also in the other "spheres" because the interactions between the atmosphere and the other

spheres are more likely to have important consequences over longer, rather than shorter, time scales. This is particularly true if those processes occur over large areas, because the impacts usually take longer to develop in such cases. For instance, if the Great Lakes were to totally evaporate, such a process would necessarily take place over a long time period. The difference in water level in the Great Lakes between today and tomorrow would not cause much impact on tomorrow's weather as compared with today's. A meteorologist would, therefore, not really need to take this atmosphere–hydrosphere interaction into account when considering tomorrow's weather. However, the difference in water content between the Great Lakes over centuries is more likely to have a noticeable and dramatic impact on climate during that time period. Therefore, interactions between the atmosphere and other spheres, such as in this example, must be considered when evaluating climate.

Regardless of the differences between meteorology and climatology, it is important to recognize that the distinction between the two is becoming increasingly blurred over time. A successful climatologist should have a firm grounding in the laws of atmospheric physics and chemistry that dictate the instantaneous behavior of the atmosphere. An effective meteorologist should recognize the importance of patterns over time and the impacts of those patterns and the influence of other components of the Earth–ocean–atmosphere system.

The holistic perspective of climatology also carries over to include interactions between the atmosphere and social systems. The impact of people on their environment is a theme in climatology that is becoming more prevalent in recent years. It is being increasingly recognized that many features of the human condition are related to climate. This is especially true of climatic "extremes" and "frequencies," because it is the "abnormal" events, and conditions exceeding certain thresholds, that generally cause the greatest impact on individuals and society.

■ Scales in Climatology

We have already stated the importance of temporal scale in climatology. It is also important to emphasize that climatology involves the study of atmospheric phenomena along many different spatial scales. There is usually a direct relationship between the size of individual atmospheric phenomena and the time scale in which that phenomenon occurs (**Figure 1.1**). The **microscale** represents the smallest of all atmospheric scales. Phenomena that operate along this spatial scale are smaller than 0.5 km (0.3 mi) and typically last from a few seconds to a few hours. A tiny circulation between the underside and the top of an individual leaf falls into this category, as does a tornado funnel cloud, and everything between. A larger scale is the **local scale**, which operates over areas between about 0.5 and 5 km (0.3 to 3 mi)—about the size of a small town. A typical thunderstorm falls into this spatial scale.

The next spatial scale is the **mesoscale**, which involves systems that operate over areas between about 5 and 100 km (3 to 60 mi) and typically last from a few hours to a few days. Such systems include those you may have encountered in earlier coursework, such as the mountain/valley breeze and land/sea breeze circulation systems, clusters of interacting thunderstorms known as mesoscale convective complexes, a related phenomenon associated with cold fronts termed mesoscale convective systems, and the central region of a hurricane.

Moving toward larger phenomena, we come to the **synoptic scale**, a spatial scale of analysis that functions over areas between 100 and 10,000 km (60 to 6000 mi). Systems of this size typically operate over periods of days to weeks. Entire tropical cyclone systems and midlatitude (frontal) cyclones with their associated fronts fall into the synoptic scale. Because these phenomena are quite frequent and directly affect many people, the synoptic scale is perhaps the most studied spatial scale in the atmospheric sciences.

Finally, we can also study and view climate over an entire hemisphere or even the entire globe. This represents the largest spatial scale possible and is termed the **planetary scale**, because it encompasses atmospheric phenomena on the order of 10,000 to 40,000 km (6000 to 24,000 mi). Because in general the largest spatial systems operate over the longest time scales, it is no surprise then that planetary-scale systems operate over temporal scales that span weeks to months. Examples of planetary-scale systems include the broad wavelike flow in the upper atmosphere and the major latitudinal **pressure** and **wind** belts that encircle the planet.

■ Subfields of Climatology

Climatology can be divided into several subfields, some of which correspond to certain scales of analysis. For instance, the study of the microscale processes involving interactions between the lower atmosphere and the local surface falls into the realm of **boundary-layer climatology**. This subfield is primarily concerned with exchanges in energy, matter (especially water), and momentum near the surface. Physical processes can become very complex in the near-surface "boundary layer" for two reasons. First, the decreasing effect

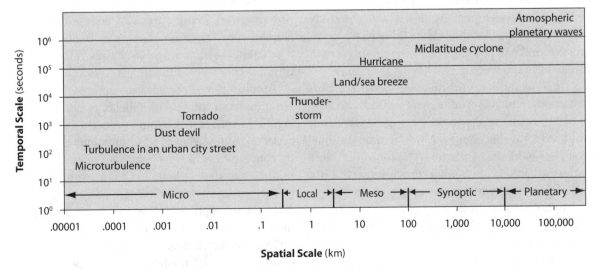

Figure 1.1   Spatial–temporal relationships for selected atmospheric features.

of friction from the surface upward complicates the motion of the atmosphere and involves significant transfer of momentum downward to the surface. Second, the most vigorous exchanges of energy and moisture occur in this layer because solar radiant energy striking the ground warms it greatly and rapidly compared with the atmosphere above it and because the source of water for evaporation is at the surface. Boundary-layer climatology may be further subdivided into topics that examine surface–atmosphere interactions in mountain/alpine regions, urban landscapes, or various vegetated land covers.

Physical climatology is related to boundary-layer climatology in that it studies energy and matter. However, it differs in that it emphasizes the nature of atmospheric energy and matter themselves at climatic time scales, rather than the processes involving energy, matter, and momentum exchanges only in the near-surface atmosphere. Some examples include studies on the causes of lightning, atmospheric optical effects, microphysics of cloud formation, and air pollution. Although meteorology has traditionally emphasized this type of work to a greater extent than climatology, climatologists have contributed to our understanding of these phenomena. Furthermore, the convergence of meteorology and climatology as disciplines will likely lead to more overlap in these topics of research in the future.

Hydroclimatology involves the processes (at all spacial scales) of interaction between the atmosphere and near-surface water in solid, liquid, and gaseous forms. This subfield analyzes all components of the global hydrologic cycle. Hydroclimatology interfaces especially closely with the study of other "spheres," including the lithosphere, cryosphere, and biosphere, because water is present in all of these spheres.

Another subfield of climatology is **dynamic climatology**, which is primarily concerned with general atmospheric dynamics—the processes that induce atmospheric motion. Most dynamic climatologists work at the planetary scale. This differs from the subfield of synoptic climatology, which is also concerned with the processes of circulation but is more regionally focused and usually involves more practical and specific applications than those described in the more theoretical area of dynamic climatology. According to climatologist Brent Yarnal, synoptic climatology "studies the relationships between the atmospheric circulation and the surface environment of a region." He goes on to state that, "because synoptic climatology seeks to explain key interactions between the atmosphere and surface environment, it has great potential for basic and applied research in the environmental sciences." Synoptic climatology may act as a keystone that links studies of atmospheric dynamics with applications in various other disciplines.

Synoptic climatology is similar in some ways to regional climatology, a description of the climate of a particular region of the surface. However, synoptic climatology necessarily involves the explanation of process, whereas regional climatology may not.

The study of climate can extend to times before the advent of the instrumental weather record. This subfield of climatology is termed **paleoclimatology** and involves the extraction of climatic data from indirect sources. This **proxy evidence** may include human sources such as books, journals, diaries, newspapers, and artwork to gain information about preinstrumental climates. However, the field primarily focuses on biological, geological, geochemical, and geophysical proxy sources, such as the analysis of tree rings, fossils, corals, pollen, ice cores, striations in rocks, and sediment deposited annually on the bottoms of lakes (**varves**).

Bioclimatology is a very diverse subfield that includes the interaction of living things with their atmospheric environment. **Agricultural climatology** is the branch of bioclimatology that deals with the impact of atmospheric properties and processes on living things of economic value. **Human bioclimatology** is closely related to the life sciences, including biophysics and human physiology.

Applied climatology is very different in its orientation from the other subfields of climatology. While the others seek to uncover causes of various aspects of climate, applied climatology is primarily concerned with the effects of climate on other natural and social phenomena. This subfield may be further subdivided. One area of focus involves attempts to improve the environment. Examples include using climatic data to create more efficient architectural and engineering design, generating improvements in medicine, and understanding the impact of urban landscapes on the natural and human environment. Other examples involve the possibility of modifying the physical atmosphere to suit particular human needs, such as with the practice of **cloud seeding**, which attempts to extract the maximum amount of precipitation from clouds in water-scarce regions.

In general, each subfield overlaps with others. We cannot fully understand processes and impacts relevant to any subfield without touching on aspects important for others and at least one other nonclimatology field. For example, an agricultural climatologist interested in the effect of windbreaks to reduce evaporation rates in an irrigated field must understand the near-surface wind profile and turbulent transfer of moisture, along with soil and vegetation properties.

■ Climatic Records and Statistics

Because climatology deals with aggregates of weather properties, statistics are used to reduce a vast array of recorded properties into one or a few understandable numbers. For instance, we could calculate the **daily mean temperature**—the average temperature for the entire day—for yesterday at a particular location through a number of methods. First, we could take all recorded temperatures throughout the day, add them together, and then divide by the total number of observations.

A much simpler (but less accurate) method of calculating the daily mean temperature is actually the one that is used: A simple average is calculated from the maximum and minimum temperatures recorded for the day. This method is the most common because in the days before computers were used to measure and record temperature, special thermometers that operated on the principle of a "bathtub ring" were able to leave a mark at the highest and lowest temperature experienced since the last time that the thermometer was reset. Each day, human observers could determine the maximum and minimum temperature for the previous 24 hours, but they would not know any of the other temperatures that occurred over that time span. Thus for most of the period of weather records, we knew only the maximum and minimum daily temperatures.

Of course, the numerical average calculated by the maximum–minimum method differs some-

what from the one obtained by taking all hourly temperatures and dividing by 24. Even though we have automated systems now that can measure and record temperatures every second, we do not calculate mean daily temperatures using this more accurate method because we do not want to change the method of calculating the means in the middle of our long-term weather records. What would happen if the temperatures began to rise abruptly at the same point in the period of record that the method of calculating the mean temperature changed? We would not know whether the "change" represented an actual change in climate or was just an artifact of a change in the method of calculating the mean temperature.

But what about that average temperature? Is it actually meaningful? Let's say that yesterday we recorded a high temperature of 32°C (90°F) and a low of 21°C (70°F). Our calculated average daily temperature would be about 27°C (80°F). This number would be used to simply describe and represent the temperature of the day for our location. But the temperature was likely to have been 27°C (80°F) only during two very short periods in the day, once during the mid-day hours when climbing toward the maximum and again as temperatures decreased through the late afternoon. So the term "average temperature" is actually a rather abstract notion. Most averages or climatic "normals" are abstract notions, but the advantage from a long-term (climatic) perspective is that they provide a "mechanism" for analyzing long-term changes and variability.

"Extremes" are somewhat different. As we saw earlier in this chapter, climatic extremes represent the most unusual conditions recorded for a location. For example, these may represent the highest or lowest temperatures during a particular time period. Extremes are often given on the nightly news to give a reference point to the daily recorded temperatures. We might hear that the high temperature for the day was 33°C (92°F), but that was still 5 C°(8 F°) lower than the "record high" of 38°C (100°F) recorded on the same date in 1963. As long as our recorded atmospheric properties are within the extremes, we know that the atmosphere is operating within the expected range of conditions. When extremes are exceeded or nearly exceeded, then the atmosphere may be considered to be behaving in an "anomalous" manner. The frequency with which extreme events occur is also important. Specifically, if extreme events occur with increasing frequency, the environmental, agricultural, epidemiological, and economic impacts will undoubtedly increase.

Why are climatic records important? During the 1980s and 1990s the rather elementary notion that climate changes over time was absorbed by the general public. Before that time many people thought that climate remained static even though weather properties varied considerably around the normals (averages). With heightened understanding of weather processes came the realization that climate varies considerably as well. Climatic calculations and the representation of climate for a given place over time became exceedingly important and precise. The problems associated with the calculation of various atmospheric properties still existed, however, and the methods of calculating these properties could have far-reaching implications on such endeavors as environmental planning, hazard assessment, and governmental policy.

With today's technology we would assume that calculating a simple average temperature for Earth, for instance, would be easy. However, data biases and methodological differences complicate matters. Many of these issues have been mathematically corrected in recorded data. Given the corrections, it is generally accepted that Earth's average annual temperature has risen by about 0.4 C° (0.7 F°) over the past century.

Another factor that complicates the interpretation of the observed warming is the increasingly urban location of many weather stations as urban sprawl infringes on formerly rural weather stations. Early in the twentieth century many weather stations in the United States and elsewhere were located on the fringe of major cities. This was especially true toward the middle part of the century with the construction of major airports far from the urban core. Weather observations could be recorded at the airport in a relatively rural, undisturbed location. As cities grew, however, these locations became swallowed up by urban areas. This instituted considerable bias into long-term records as artificial heat from urban sources, known as the **urban heat island**, became part of the climatic record. Although the urban heat island is discussed in more detail in Chapter 12, we can say for now that various properties, such as

the abundance of concrete that absorbs solar energy effectively, the absence of vegetation and water surfaces, and the generation of waste heat by human activities contribute to the heat island. The urban heat island provides an excellent example of how humans can modify natural climates and can complicate the calculation and analysis of "natural" climatic changes.

In addition, the long-term recordings themselves may be plagued by other problems. Consider that most weather records for the world are confined to more-developed countries and tend to be collected in, or near, population centers. Developing countries, rural areas, and especially the oceans are poorly represented in the global weather database, particularly in the earlier part of the record (**Figure 1.2**). Oceans comprise over 70 percent of the planet's surface, yet relatively few long-term weather records exist for these locations. Most atmospheric recordings over oceans are collected from ships, and these recordings are biased by inconsistencies in the height of the ship-mounted weather station, the type of station used, the time of observation, and the composition of ship materials. Furthermore, ocean surface temperatures are derived in a variety of ways, from inserting a thermometer into a bucket of collected ocean water to recording the temperature of water passing through the bilge of the ship (with the heat generated by the ship included in the recording). Vast tracts of ocean were largely ignored until the recent arrival of satellite monitoring and recording technology, because the representation of surface and atmospheric properties was greatly limited to shipping lanes.

Even records taken with rather sophisticated weather stations may be biased and complicated to some degree by rather simple issues. Foremost among these are station moves. Moving a station even a few meters may ultimately bias long-term recordings as factors such as differing surface materials and solar exposure occur. Also of note is a different time in the day at which measurements were taken at different stations. Finally, systematic biases and changes in the instrumentation may cause inaccuracies in measurements. The result of these, and a host of other biases, is that considerable data "correction" is required. Both the biases and the correction methods fuel debate concerning the occurrence of actual atmospheric trends.

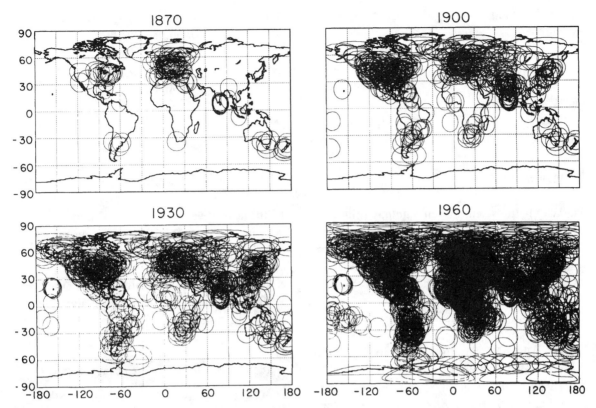

Figure 1.2 Location of global surface observations at various years. *Source:* J.E. Hansen and S. Lebedeff, 1987. Global trends of measured surface air temperature. *Journal of Geophysical Research,* 92 D11, 13345–13372.

■ Summary

This chapter introduced the field of climatology. It described the scope of climatology, the inherent differences between meteorology and climatology, and the associated notions of weather and climate. Meteorology studies changes in weather, the state of atmospheric properties for a given location over a relatively short period of time, while climatology examines weather properties over time for a location. Climatology is a holistic science in that it involves understanding the interaction of the atmosphere with other aspects of the Earth–ocean–atmosphere system using many different spatial and temporal scales. Each scale partially defines the many interlocking subfields of climatology, including boundary-layer, physical, hydro-, dynamic, synoptic, regional, paleo-, bio-, and applied climatology. Interactions occur between the atmosphere, lithosphere, hydrosphere, cryosphere, and biosphere. All are important to the establishment of global, hemispheric, and regional climates.

Climatic data and calculations also were described, with particular emphasis on climatic normals, extremes, and frequencies. Some causes of spurious climatic data, such as the urban heat island, were also introduced.

▶ Key Terms

Agricultural climatology
Applied climatology
Atmosphere
Bioclimatology
Biosphere
Boundary-layer climatology

Climate
Climatology
Cloud seeding
Cryosphere
Daily mean temperature
Dynamic climatology

Evaporation
Extremes
Frequencies
Friction
Global hydrologic cycle
Human bioclimatology

Hydroclimatology
Hydrosphere
Lithosphere
Local scale
Mesoscale
Meteorology
Microscale
Momentum

National Weather Service
Normals
Paleoclimatology
Physical climatology
Planetary scale
Pressure
Proxy evidence
Regional climatology

Synoptic climatology
Synoptic scale
Urban heat island
Varve
Weather
Wind

▶ Review Questions

1. Why is the science of climatology inherently holistic?
2. Briefly describe Earth's "spheres." Give examples of how each of the spheres is connected.
3. Compare and contrast the notions of weather and climate.
4. Compare and contrast the sciences of meteorology and climatology.
5. Describe the various spatial and temporal scales of climatology.
6. Discuss the different subdisciplines within climatology. How are they different from/similar to one another?
7. How are mean temperatures calculated? Discuss the problems inherent in the calculation methods.
8. What is the urban heat island and why is it relevant to temperature assessment?

▶ Questions for Thought

1. Think of several examples of how advancements in science and technology may have helped climatology to evolve as a science since 1950.
2. In today's age of specialization, is climatology's interdisciplinary nature an advantage or a disadvantage?

http://physicalscience.jbpub.com/climatology

Connect to this book's Web site: http://physical science.jbpub.com/climatology. The site provides chapter outlines, further readings, and other tools to help you study for your class. You can also follow useful links for additional information on the topics covered in this chapter.

2 | Introduction to the Atmosphere

Chapter at a Glance
Origin of the Earth and Atmosphere
Atmospheric Composition
    Carbon Cycle
    Constant and Variable Gases
Faint Young Sun Paradox
Atmospheric Structure
Summary
Review

The *atmosphere* is a collection of gases held near the Earth by gravity. It is also pulled away from the Earth because a vacuum exists in the harsh conditions of space. One of the most fundamental properties of the universe is the **second law of thermodynamics**. One form of the second law states that energy (and, therefore, mass, because energy and mass are related by Einstein's theory of relativity) moves from areas of higher concentration (in this case, Earth's lower atmosphere) to areas of lower concentration (outer space). Thus, the atmosphere represents a place where a balance is generally achieved between the downward-directed gravitational force and the upward-directed force of buoyancy. This balance is termed **hydrostatic equilibrium**. It is this extremely thin and delicate zone known as the atmosphere that makes life as we know it possible on the planet. When you look in the sky the atmosphere appears to continue infinitely, but if Earth were the size of an apple, the atmosphere would have the thickness of the apple's skin.

Technically, the atmosphere is a subset of the air because it is composed solely of gases. By contrast, air contains not only gases but also **aerosols**—solid and liquid particles suspended above the surface that are too tiny for gravity to pull downward. Solid aerosols include ice crystals, volcanic soot particles, salt crystals from the ocean, and soil particles; liquid aerosols include clouds and fog droplets.

Because the atmosphere is composed of the lightest elements gravitationally attracted to the Earth, many assume that it has little or no mass. Compared with the mass of the solid Earth (6×10^{24} kg; or 6×10^{21} metric tons) and oceans (1.4×10^{21} kg; or 1.4×10^{18} metric tons) the atmosphere is indeed light. But the atmosphere has a substantial total mass of 5×10^{18} kg (5×10^{15} metric tons)!

The mass of the air is in constant motion, giving considerable impact to the surface environment. For example, a tornado can cause catastrophic devastation to a location. In the case of a tornado, the mass of air has considerable acceleration. According to **Newton's second law of motion**, force is the product of mass and acceleration. The two factors combine, in this situation, to produce a force capable of devastation.

The atmosphere is an extremely complex entity that must be viewed simultaneously on many levels, both temporally and in three spatial dimensions (west–east, north–south, and vertical). Atmospheric processes can be difficult to understand. To appreciate the nature of the atmosphere properly, we must first understand the origins of the atmosphere and its changes since the origin of the planet.

■ Origin of the Earth and Atmosphere

According to the best scientific information, the universe is believed to have begun approximately 15 billion years ago. At that time all matter in the universe was confined to a single space. An explosion of unimaginable proportion sent this matter, mostly hydrogen, outward in all directions. Over time, gravity caused matter to collect in various areas of space to form galaxies. Within the galaxies smaller amounts of matter gravitationally condensed into stars. Star formation began to occur when hydrogen was compressed under its own gravitational weight. If the mass involved was sufficiently large, **nuclear fusion**—a process that converted lighter elements (principally hydrogen) into heavier elements (primarily helium)—began. Such reactions released amazing quantities of radiant energy while creating (fusing) all matter heavier than hydrogen.

Occasionally, a star exploded, sending heavier elements outward through the galaxy. Vast clouds of dust, called **nebula** (**Figure 2.1**), formed, and the original gravitational accumulation process began anew. Our solar system is believed to have formed from a nebula approximately 5 billion years ago. The Sun gravitationally attracted the bulk of the elements that composed the nebula. Planets formed as balls of dust gravitationally collected over various orbits about the primitive Sun. Earth was one such ball of dust. As it grew the elements fused together and collapsed under their own weight and gravity. As Earth grew in size, its gravity increased proportionately. Friction caused the Earth materials to melt. Melting was also encouraged by frequent impacts with large **planetesi-**

Figure 2.1 The Orion Nebula. (See color plate 2.1)

mals, which were essentially very small planets of condensed debris moving over wildly eccentric orbits about the Sun. These planetesimals contributed heavier elements and mass to the growing Earth while shattering and melting its hot surface. A collision between Earth and a planetesimal is thought to have created the Moon. Remnants of early solar system planetesimals are present today in the vast asteroid field between Mars and Jupiter. The **Oort cloud**—a collection of icy comets and dust that surrounds the outer edges of our solar system—also acts as a relic of conditions present in the early stages of solar system formation.

In these early times Earth's atmosphere consisted of light and inert (noble) gases such as hydrogen, helium, neon, and argon. These gases were effectively swept away as the **solar wind**—radioactive particles from the Sun moving through space at nearly the speed of light—developed. Today, Earth is largely devoid of noble gases as a result. So how did the atmosphere that we know today form?

The composition of the atmosphere can be explained by looking at volcanic activity, which is rather limited over Earth's surface today but was apparently widespread billions of years ago as the early Earth cooled slowly from its primordial molten state. As volcanic material cooled, gases were released through the process of **outgassing**, which primarily involved diatomic nitrogen (N_2) and carbon dioxide (CO_2), with lesser amounts of water vapor, **methane (CH_4)**, and sulfur. The **condensation** of water vapor into liquid water in the cool atmosphere formed clouds and precipitation. Precipitation collected in low-elevation areas of the planet and over time built up to form the oceans.

The conditions of our planet are unique in that Earth is the only planet in the solar system known to support the presence of water in liquid form. This is a consequence of many related, and interacting, factors—some of which include distance to the Sun and atmospheric composition. Because water is essential to life, it is not surprising that Earth is the only planet known to support life.

■ Atmospheric Composition

Today, the dry atmosphere consists primarily of N_2 and diatomic oxygen (O_2). Nitrogen is a highly stable gas that comprises 78 percent of the present-day atmospheric volume (**Table 2.1**). The

Table 2.1 Composition of the Dry Atmosphere

Gas	Percentage of Air	
Nitrogen (N$_2$)	78.08	99.03
Diatomic oxygen (O$_2$)	20.95	
Argon (Ar)	0.93	
Carbon dioxide (CO$_2$)	0.039	
All others	0.003	

abundance of N$_2$ has increased as a percentage of the total atmospheric volume primarily because it is not removed as effectively from the atmosphere as are most other atmospheric gases. The **residence time**—the mean length of time that an individual molecule remains in the atmosphere—of N$_2$ is believed to be approximately 16.25 million years.

The next most abundant gas in the present-day dry atmosphere is oxygen, comprising approximately 21 percent of the atmospheric volume today. Added to the quantity of nitrogen, these two gases constitute 99 percent of the dry atmosphere. About 0.93 percent of the remaining 1 percent is composed mostly of argon (A), and a wide array of atmospheric trace gases constitute the remainder.

Of these, CO$_2$ is the fourth most abundant gas in the dry atmosphere, representing 0.038 percent of the dry atmosphere, or 390 parts per million (ppm). It plays an especially important role in maintaining the temperature of the planet at a level comfortable for life in its present form. Earth's early atmosphere apparently contained far more CO$_2$ than today's and little or no O$_2$. So where did most of the CO$_2$ go after outgassing in the primitive atmosphere, and how did O$_2$ come to replace it?

The evolution of Earth's atmospheric composition (including O$_2$) involves significant interactions with the *biosphere*, *hydrosphere*, and *lithosphere*. About 3.5 billion years ago an interesting development occurred in the extensive waters of primordial Earth that profoundly affected the evolution of the atmosphere. Single-celled organisms, called **prokaryotes**, began to appear. These simple ancestors of bacteria and green algae absorbed nutrients directly from the surrounding environment. Prokaryotes released CO$_2$ to the atmosphere as a byproduct of **fermentation**, the process by which simple organisms acquire energy through the breakdown of food. The evolution of prokaryotes

led to more complex, often multicellular, organisms called **eukaryotes**, which contain more complex internal structures and release even more CO$_2$ into the atmosphere. Most life on Earth evolved from the further development of eukaryotes. However, prokaryotes and eukaryotes would have had to develop in the oceans because without oxygen in the atmosphere the protective **ozone (O$_3$)** layer could not have formed to protect terrestrial life from the harmful **ultraviolet (UV) radiation** emitted by the Sun. Over time, CO$_2$ continued to accumulate, as it became a larger and larger component of the atmospheric volume.

By about 3 billion years ago another major development in the history of life on Earth apparently caused another major change to the atmospheric composition. The early evolution of aquatic green plants led to a significant extraction of CO$_2$ from the atmosphere in **photosynthesis**—the process of deriving energy through the breakdown of food—and in the accumulating biomass of those plants. O$_2$ is released into the atmosphere as a byproduct of photosynthesis. As green plants, therefore, began to populate Earth, first in the oceans and later on land—after the presence of O$_2$ gradually led to the formation of ozone (O$_3$) and the O$_3$ layer—atmospheric CO$_2$ decreased in concentration while atmospheric O$_2$ simultaneously increased. Today most of the atmospheric CO$_2$ is stored in vast quantities of sedimentary rock, originally extracted from the atmosphere by living things. The amount of atmospheric O$_2$ present today probably represents a similar percentage to that of CO$_2$ in the early atmosphere.

Carbon Cycle

The process described above essentially represents the atmospheric component of the **carbon cycle**, the continuous movement of carbon through the Earth-ocean-atmosphere system. Carbon can exist in various Earth-ocean-atmosphere **reservoirs**—the components of a system that effectively store matter and/or energy for a certain period of time, after which they allow for the movement (**flux**) of that matter and/or energy to another component of the system. Carbon reservoirs include the atmosphere, which houses CO$_2$; the biosphere, which comprises all living matter; and the oceans, which include dissolved carbonates (**Figure 2.2**). Over time carbon cycles among

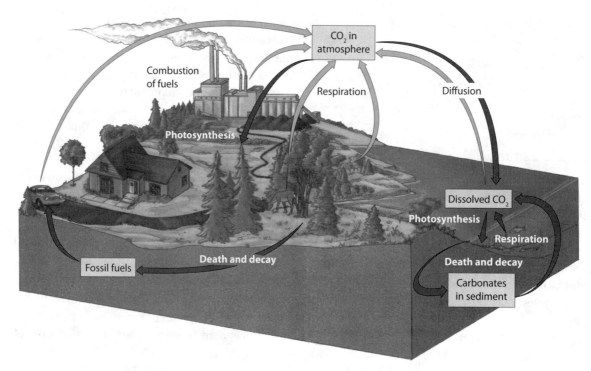

Figure 2.2 The global carbon cycle.

these various reservoirs. The carbon residence time is different for each reservoir, with the rates of exchange directly related to the size of the reservoir. The largest carbon reservoir, comprising the vast majority of carbon in the Earth–ocean–atmosphere system, is sedimentary rock. Carbon may be stored in sedimentary rock layers for billions of years.

The second-largest reservoir is the oceans, which act as a **sink** for atmospheric CO_2 by absorbing many gigatons of CO_2 from the atmosphere each year. Over time CO_2 is transported from the atmosphere into the deep ocean layers in a very slow process. Once the CO_2 is transported into the deep ocean, it may remain in this reservoir for many thousands of years.

The biosphere (including soil) is a reservoir that acts as a sink for most CO_2 from the atmosphere. It is then transported and stored elsewhere in the system. For example, the vast majority of carbon in the rock reservoir was extracted from the atmosphere through biological processes. Residence time associated with the biosphere may be examined on a number of levels. Carbon is directly held in the biosphere as long as the living organism remains alive. Once the organism dies, carbon exits this sink gradually over time as the remains of the organism decay. Some of this carbon may become

buried naturally and transported into the rock reservoir. This process is very effective in marine environments where an abundance of organic matter filters to the ocean floor, building huge layers of organic matter over vast time spans. Some of the decaying carbon dissolves directly into water, becoming part of the oceanic carbon reservoir, and some reenters the atmosphere through **diffusion**. However, much oceanic biomass eventually solidifies into sedimentary rock. Finally, the atmosphere represents the smallest carbon sink, and the carbon exchange rate to this reservoir is rapid—only about 100 years.

Today, many people are concerned that the natural carbon cycle is being disrupted by human activities. Since the dawn of the Industrial Revolution in the late 1700s, people have been burning ever-increasing quantities of **fossil fuels**—deposits of carbon primarily in the form of coal, oil, and natural gas. Fossil fuels represent carbon that was taken out of the atmosphere long ago through natural processes and stored in the vast rock reservoir. Normally, this carbon would be re-released back to the atmosphere over millions of years. However, humans are extracting and releasing this material back to the atmosphere in very short periods of time. The atmospheric quantity of CO_2 increased from 270 ppm in 1800 to 390 ppm today, with the bulk of the increase occurring since 1950 (**Figure 2.3**).

Whereas clues of the atmospheric concentration of CO_2 in the distant past come from chemical analysis of air bubbles trapped in ice, CO_2 concentration has been measured directly since 1957 atop Mauna Loa in Hawaii—a site as far removed as possible from local sources of pollution. The time series of atmospheric CO_2 since 1957 is known as the **Keeling curve** (**Figure 2.4**), named for Charles Keeling, the climatologist who showed that CO_2 released from fossil fuel combustion would accumulate significantly in the atmosphere.

The Keeling curve not only verifies the rapid increase since 1957, but it also reveals the seasonal cycle of CO_2. Maximum concentrations occur in early spring in the northern hemisphere, where most of the world's middle- and high-latitude forests are located. The relative lack of photosynthetic activity during the dormant northern winter months causes a buildup of atmospheric CO_2 into March. Likewise, minimum atmospheric CO_2 in northern hemisphere autumn results from the buildup of biomass throughout the northern hemisphere's spring and summer months.

The long-term exponential growth in the amount of atmospheric CO_2 concerns most climatologists and environmental scientists. Carbon dioxide is an integral component of Earth's energy balance because it absorbs energy that is radiated

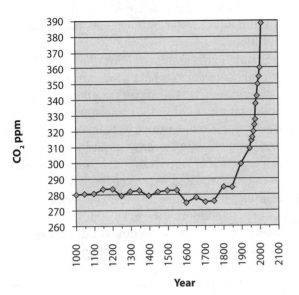

Figure 2.3 Exponential rise in atmospheric carbon dioxide over the past 1000 years.

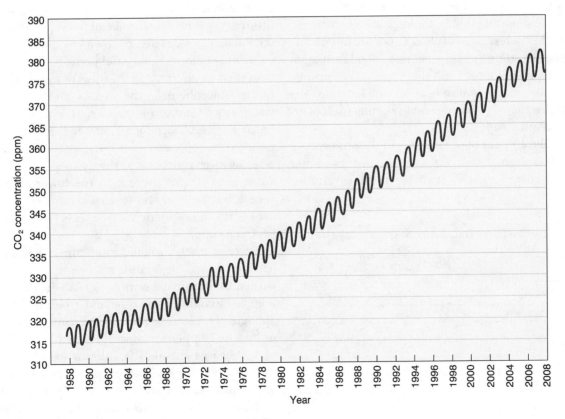

Figure 2.4 The Keeling curve. *Data from*: Scripps CO_2 Program.

from the Earth and then reemits energy back downward to the Earth, thereby keeping the surface warmer than it would be if the CO_2 were not present. This phenomenon is known as the **greenhouse effect**. The rapid increase in the quantity of atmospheric CO_2 is a likely culprit for the observed increases in temperature of Earth's surface in the last several decades. The slow carbon cycle and unknown capacity for the biosphere and oceans to absorb excess atmospheric CO_2 has concerned most scientists about the ultimate impacts of fossil fuel consumption.

Constant and Variable Gases

Constant gases are those that have relatively long residence times in the atmosphere and that occur in uniform proportions across the globe and upward through the bulk of the atmosphere. These gases include nitrogen, oxygen, argon, neon, helium, krypton, and xenon. **Variable gases** are those that change in quantity from place to place or over time (**Table 2.2**). They generally have shorter residence times than constant gases, as various processes combine to cycle the gases through reservoirs. The most notable variable gas is water vapor, which can occupy as much as 4 percent of the lower atmosphere by volume. Higher percentages of water vapor are impossible because atmospheric processes (cloud formation and precipitation) limit the amount that may be present in the atmosphere for any location. The atmosphere is very efficient in ridding itself of excess water vapor.

The amount of water vapor in the atmosphere varies dramatically across space because water and energy must be available at the surface for

Table 2.2   Concentrations of Variable Gases of the Atmosphere

Gas	ppm of Air
Water vapor	0.1–40,000
Carbon dioxide	~390
Methane	~1.8
Hydrogen	~0.6
Nitrous oxide	~0.31
Carbon monoxide	~0.09
Ozone	~0.4
Fluorocarbon 12	~0.0005

evaporation to occur. Water vapor content is maximized over locations with abundant energy and surface water, so the wettest atmospheres occur over tropical waters and rain forest regions. In addition, water vapor is largely limited to the lower atmosphere because as height increases, atmospheric water vapor is increasingly likely to condense to liquid water in the cooler, high-altitude conditions.

Surprisingly, some locations that experience little precipitation may have abundant water vapor in the atmosphere. The maximum amount of water vapor that may exist in the atmosphere is directly related to air temperature. When high temperatures combine with a nearby surface water source, high amounts of water vapor are usually present. For example, the Red Sea region tends to have high quantities of atmospheric water vapor despite the lack of precipitation. This region is dry not because water vapor is unavailable but because the region lacks a means by which the precipitation process can occur easily.

As implied by the Red Sea example, deserts are generally not the regions of lowest atmospheric water vapor content. Instead, polar regions are normally the driest locations on Earth from an atmospheric moisture perspective. This is because little energy is present in cold air to evaporate water. Furthermore, as air cools, water vapor readily condenses to form clouds and perhaps precipitation, thereby minimizing the mass of water vapor in the atmosphere. So the regions with the least water vapor tend to be the coldest locations on Earth. Over such locations in winter, the water vapor content approaches zero. The total will never actually reach zero because there is always at least some water vapor present in the lower atmosphere, but the total may reach about 0.00001 percent of the atmospheric volume, as is the case in central Antarctica.

Several other variable gases are important. Among these, CO_2 is most abundant. As stated earlier, the variable nature of CO_2 stems from its increasing quantity over time, since the late 1700s. The rate of increase is about 0.4 percent per year. or by about 35 percent since 1800. Other notable variable gases include CH_4, nitrous oxide (N_2O), carbon monoxide (CO), tropospheric ozone (O_3), and a family of chemicals known as **chlorofluorocarbons (CFCs)**. Humans have had at least some influence in the concentration of all these gases, and CFCs are entirely human derived. Collectively, these gases make up only a small amount of the atmosphere, but they can have important implications for some processes.

■ Faint Young Sun Paradox

Evidence contained in sedimentary rocks and sediments, ice sheets, and fossils reveals that Earth's average temperature has remained within a range of perhaps 15 C° (27 F°) for most, if not all, of its geological history. This implies that even global-scale shifts in mean environmental conditions, from ice ages to ice-free conditions on Earth, have occurred within a range of temperature variability that is smaller than the summer to winter temperature difference at most locations outside the tropics.

This fact has caused considerable consternation for many climate scientists because of an apparent contradiction between what is known about energy released during the evolution of stars, such as our Sun, and evidence of Earth's temperature through geological time. Stars obtain energy through the constant nuclear fusion of hydrogen into heavier elements. These reactions cause stars to gradually expand and grow hotter and brighter over time. Eventually, stars use up all their sources of energy and burn out. We can assume that energy emitted from the early Sun was about 25 to 30 percent less than that emitted today, because this pattern is observed throughout the life cycle of other yellow dwarf stars like the Sun. We also know that even small changes in solar output can induce drastic climatic changes on Earth. If solar output today were decreased by 25 to 30 percent, temperatures would quickly plummet to a point whereby Earth would be entirely frozen. Because the young Sun must have been weak, at first glance it would appear that Earth must have been frozen for the first 3 billion years of its history. However, no credible evidence has been found to support the notion that Earth was ever below freezing on a global annual average basis. Furthermore, little credible evidence has been found for widespread glaciation during the first half of Earth's existence. This apparent contradiction between a weak Sun but relatively warm global conditions is the **faint young Sun paradox**.

How could temperatures have been above freezing during the early times of geological history? How could Earth maintain a small variance in temperature over time if the Sun were much weaker in the early history of the planet? How could temperatures maintain themselves over time as the Sun grew hotter?

The most logical answer to these questions is that the Earth-ocean-atmosphere system must have some type of internal regulator that keeps temperatures within a reasonable range regardless of changes to solar output over time. This regulator must have been present in the early atmosphere. How would this have happened? Examining the radiation balance of Earth today reveals that surface temperatures are only indirectly caused by **insolation**—incoming solar radiation—being absorbed at the surface. If surface receipt of solar radiation alone determined temperatures, average Earth temperatures would be about −18°C (0°F). Instead, the transfer of energy (either from the Sun or Earth) absorbed in the atmosphere down to the surface augments radiation received at the surface directly from the Sun. Certain gases in the atmosphere, such as H_2O and CO_2, are known to efficiently absorb energy escaping the surface. Much of this energy is then reemitted back down to reheat the surface. This process—the greenhouse effect—is responsible for the life-supporting temperatures we enjoy today. The net result is that the average temperature of Earth is raised from 18°C (0°F) to a more comfortable 15°C (59°F).

So which **greenhouse gases** could have helped the early Earth to remain relatively warm? The two most abundant greenhouse gases in today's atmosphere are water vapor and CO_2. As far as we can tell, the quantity of water vapor has remained relatively constant since primordial times. Until very recently it was assumed that the wide fluctuations of CO_2 over time caused Earth's temperature to remain relatively stable via the greenhouse effect. That argument suggested that the high concentrations of CO_2 in Earth's early atmosphere may have effectively stored large amounts of radiation emitted from Earth and reemitted much of that radiation back downward in the greenhouse effect process during times when the Sun was relatively weak. As the Sun grew in strength, the energy levels, and temperatures on Earth, would have remained fairly constant as atmospheric CO_2 concentrations were decreasing as a result of plant evolution and proliferation.

But if excessive levels of CO_2 indeed caused Earth to remain warm despite a weak Sun, such concentrations probably would have been too high to allow the generation of organic molecules, so life could not have existed easily. Furthermore, no geological evidence has been found to suggest that CO_2 concentrations were ever large enough to have created such a strong greenhouse effect. Specifically, in an oxygen-free atmosphere such as early Earth would have had, CO_2 levels of about eight times today's concentrations would have produced the mineral siderite ($FeCO_3$) in the top layers of the soil as iron reacted with the CO_2, but no traces of $FeCO_3$ have ever been found in ancient soils. Better explanations for the faint young Sun paradox were sought.

A second explanation is that ammonia (NH_3) caused the early greenhouse effect. This hypothesis was proposed by Carl Sagan and George Mullen of Cornell University in the late 1970s and was based on the observation that NH_3 behaves as a very strong greenhouse gas. The problem with this argument is that experiments have shown that NH_3 is easily broken up by UV radiation in oxygen-free conditions, which would have resembled the atmosphere before the arrival of photosynthesis. Nevertheless, the NH_3 explanation is still plausible because shielding by other gases may have caused NH_3 to accumulate in the lower atmosphere, thereby allowing it to be an important greenhouse gas.

A third explanation, proposed by Harvard scientists in the early 2000s, is that in Earth's early history, before photosynthesis, the planet's oxygen-devoid atmosphere made conditions ideal for oxygen-intolerant microbes called **methanogens**. These methanogens (named because they release CH_4 as a waste product) may have allowed CH_4 to produce a very strong greenhouse effect, which would have warmed Earth even though the Sun emitted less radiation at the time. It is well-known that CH_4 is a very effective greenhouse gas. Once oxygen entered the atmosphere from photosynthesis, the methanogens became less dominant, and CH_4 became less important as a greenhouse gas. In today's oxygen-rich atmosphere the concentration of CH_4 is extremely minute—an average of only 1.8 ppm in the atmosphere. Some geoscientists believe that the demise of the methanogens caused Earth's first global ice age and perhaps contributed to subsequent ice ages.

Regardless of the explanation, most of Earth's history is believed to have been dominated by somewhat higher temperatures than exist today, or at least temperatures that are not far below those now. The debate over the explanation of the faint young Sun paradox lingers on.

■ Atmospheric Structure

The atmosphere may be divided into a series of layers based on thermal qualities (**Figure 2.5**). The lowest layer of the atmosphere is called the **troposphere**. This is a very thin zone confined to the first 8 to 20 km (5 to 12 mi) above Earth's surface, yet this atmospheric layer contains approximately 75 percent of the mass of the atmosphere. The compressibility of air allows its weight to exert a downward force on, and compress, the lower atmosphere. This layer, therefore, also contains air of the greatest **density**—the amount of mass per unit volume.

The term "troposphere" is derived from the Greek word meaning "to turn." This indicates that the troposphere is a region in which mass is constantly overturning, largely as a result of thermodynamic (heat-driven) processes. Most heat is absorbed by the atmosphere from Earth's surface after having passed through the atmosphere on its way downward from the Sun. The surface heats the air directly above it through the process of conduction. This gives the lowest layers of the atmosphere buoyancy and causes the air to rise in a process known as convection. Eventually this air cools and sinks. Because this vertical movement is integral to the development of most weather-related processes, the troposphere is sometimes referred to as the "weather sphere."

Because the atmosphere is heated primarily from Earth's surface and because the compression of atmospheric gas decreases with height (as the weight of the atmosphere above it decreases), a decrease of temperature usually occurs with increasing height through the troposphere. This decrease is known as the **environmental lapse rate**. Through the troposphere, air cools at an average rate of 6.5 C°/km (or 3.5 F°/1000 ft), although this value may vary widely from place to place and from day to day, and even on an hourly basis.

According to one form of **Charles' law**, in an ideal gas (which the atmosphere approximates) density decreases as temperature increases, if pressure remains constant. Thus, hot air rises. Because Earth is not heated equally and hot air rises, the troposphere does not have a uniform depth. Instead, the troposphere is thicker near the equator than near

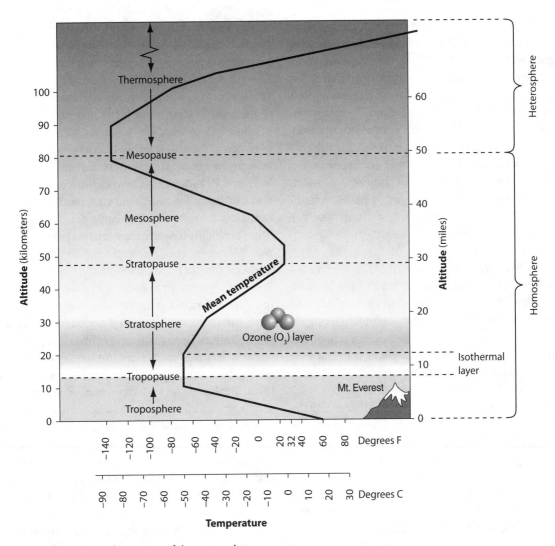

Figure 2.5   The vertical structure of the atmosphere.

the poles. Near the equator the layer is approximately 20 km (12 mi) thick, whereas near the poles in winter the thickness is only about 8 km (5 mi). This is because of thermal expansion of the atmosphere near the equator and thermal contraction near the poles. Roughly the same amount of atmospheric mass exists over the two locations, but the density of that air is less over the equator and greater over the poles.

The top of the troposphere is called the **tropopause**. This feature represents the boundary between the troposphere below and the next layer of the atmosphere above (see Figure 2.5). The average temperature at the tropopause is about −57°C (−70°F), which represents quite a decrease from the 15°C (59°F) average temperature of the surface.

The layer above the troposphere is the **stratosphere**. Temperatures remain somewhat constant

from the tropopause upward into the stratosphere for about 10 km (6 mi). Any zone of relatively constant temperature with height, such as this one, is called an **isothermal layer**. Above the isothermal layer temperatures actually increase with height through the rest of the stratosphere. This **temperature inversion**—any increase of temperature with height—is caused by the absorption of UV radiation by the triatomic form of oxygen (O_3), or ozone.

The so-called ozone layer in the stratosphere involves several processes that have important implications for terrestrial life on the planet. To understand the workings of the ozone layer we must first review the nature of radiation reaching Earth from the Sun. Energy from the Sun arrives in Earth's atmosphere in a wide range of **wavelengths**, which are measured in **micrometers** (millionths of a meter [μm]). Shorter wavelengths are associated with more intense (and, therefore, more harmful to living things) insolation than energy with longer wavelengths. The Sun emits more energy at a wavelength of about 0.5 μm than at any other wavelength, with successively less energy emitted at successively shorter and longer wavelengths (**Figure 2.6**). By convention, energy from solar origin shorter than 4.0 μm is usually referred to as **shortwave radiation** in the atmospheric sciences. Wavelengths of the peak amounts of shortwave radiation occur in the visible part (0.4 to 0.7 μm) of the **electromagnetic spectrum**—the full assemblage of all possible wavelengths of electromagnetic energy.

Any insolation with wavelengths less than 0.4 μm is too intense to allow terrestrial life to exist. UV radiation falls between wavelengths of 0.01 and 0.40 μm, making UV radiation harmful to living organisms. Fortunately, N_2 absorbs electromagnetic radiation of wavelengths below 0.12 μm. But how does the ozone layer protect us from UV radiation at wavelengths between 0.12 and 0.40 μm?

Some of the diatomic oxygen (O_2) that enters the atmosphere from photosynthesis near the surface may reach the stratosphere over time. Because O_2 molecules effectively absorb UV radiation at wavelengths between 0.12 and 0.18 μm, the O_2 reaching the stratosphere is exposed to incoming harmful radiation. When this radiant energy strikes O_2 molecules, a chemical reaction that splits the molecular bonds is triggered. The molecule undergoes **photodissociation**—the process of splitting into two monatomic oxygen (O) atoms caused by exposure to light. Because O is inherently unstable, it bonds quickly and easily with other atoms and molecules. Some of these atoms chemically bond with an O_2 molecule to form an O_3 molecule that effec-

tively absorbs UV radiation at wavelengths between 0.18 and 0.34 μm. But in the absorption process the O_3 becomes photodissociated into O and O_2, and the O then bonds with another O_2 to form O_3. The process then repeats endlessly, ensuring that oxygen is continuously being reworked into O_3 in the stratosphere. UV radiation at wavelengths between 0.18 and 0.34 μm is effectively "absorbed"— actually used in chemical processes—such that only the UV radiation at wavelengths between 0.34 and 0.40 μm filters to Earth's surface. This harmful radiation can cause skin cancer, cataracts, and other problems if we are exposed to it in large doses, but at least we are protected from the much more harmful shorter UV wavelengths.

Early in geological history, the first organisms must have formed in murky waters because no O_2 (and, therefore, no O_3) existed to protect them from UV radiation. By the time life evolved into shallow water areas and onto land surfaces, O_2 released from photosynthesis had built a stratospheric O_3 layer. Life has never known an excessive amount of this type of radiation and has never adapted to it. Introduction of excessive amounts of UV radiation is damaging to virtually all terrestrial life forms.

Humans have contributed to thinning the very fragile O_3 layer over the past half-century by producing CFCs. Most general uses for CFCs involved refrigeration both as a gas (Freon) and as an insulating substance (foam and Styrofoam). The product was also used as a propellant for aerosol sprays. When chlorine from CFCs and bromine are released to the atmosphere, they can make their way upward to the stratosphere where they readily bond with monatomic oxygen atoms. Such a bond does not allow the O to bond with O_2 to produce the O_3 that would absorb UV radiation. The result is that increased amounts of UV radiation reach the surface, where adverse effects on organisms occur. Although O_3 is found throughout the stratosphere, if all stratospheric ozone were compressed to the surface it would create a layer only 3 mm thick. Since the U.S. ban on CFC production took effect on January 1, 1996, the ozone layer has shown some signs of recovery.

The ozone formation process is responsible for the temperature inversion in the stratosphere. As O_3 absorbs UV energy, heat is captured by the molecules. The O_3 at the top of the stratosphere has the first opportunity to gain heat (and, therefore, temperature, because temperature is a measure of

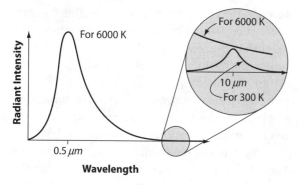

Figure 2.6 Emitted energy by wavelength for the Sun and Earth, assuming a surface temperature of 6000 K for the Sun and 300 K for Earth.

the heat or energy content of matter) from incoming UV radiation. Its temperature is, therefore, higher than for molecules lower in the stratosphere. The process of O_3 production and dissociation happens in the stratosphere because this is the uppermost layer for which atmospheric density is high enough to allow O and O_2 to meet and bond quickly enough so that incoming UV radiation is absorbed effectively. Temperatures rise to approximately −18°C (0°F) at the **stratopause**, which is about 48 km (29 mi) above the surface. The stratopause is the boundary between the stratosphere and the layer above it.

The layer above the stratosphere is the **mesosphere**, from the Greek prefix *meso-*, which means "middle." Although this layer does sit near the middle of the atmosphere from an altitude perspective—in the region between the stratopause and about 80 km (50 mi) above the surface—the low-density mesosphere does not represent the middle of the atmosphere by density or volume. Because of the compressibility of gases, the middle of the atmosphere by density and volume is only about 5.5 km (3.4 mi) above the surface—well within the troposphere.

Similar to the troposphere, temperatures in the mesosphere decrease with height. The temperature inversion characteristic of the stratosphere is not present in the mesosphere because it is too high for photodissociated O_2 to find monatomic oxygen atoms to bond with quickly enough to absorb the incoming UV radiation. Instead, the increased density and proximity to the surface and stratospheric heat sources below the mesosphere make the lower mesosphere warmer than the top of this layer. Temperatures at the **mesopause** average approximately −84°C (−120°F).

Very few processes of consequence to weather and climate are known to occur in the mesosphere because so little atmospheric mass exists in this zone. Charged particles from the Sun that are captured by Earth's magnetic field do create problems in the mesosphere, because they can disrupt telecommunications during their release of energy. These same charged particles are also responsible for some impressive light shows, most notably the northern lights (*aurora borealis*) (**Figure 2.7**) and southern lights (*aurora australis*). But even these processes have minimal effect on Earth's weather and climate.

Figure 2.7   The aurora borealis. (See color plate 2.7)

From the surface up to the mesosphere, the ratio of atmospheric gases is about the same as that at the surface, except for the greater concentration of O_3 in the stratosphere. Thus, the first three "spheres" of the atmosphere are sometimes collectively known as the **homosphere**, which means "same sphere." Above the mesosphere gases stratify into layers according to their atomic weights because there is so little mass to "stir them up." That region is termed the **heterosphere**.

The heterosphere corresponds to the final thermal layer of the atmosphere, the **thermosphere**. Like the stratosphere, the thermosphere is characterized by temperatures that increase with height—a temperature inversion. Unlike the stratosphere, however, where the inversion exists because of O_3 absorption of insolation, the thermospheric temperature inversion occurs because the uppermost N_2 and O_2 molecules have the first opportunity to absorb insolation. Their position allows them to attain extraordinarily high temperatures because Earth's magnetic field captures charged high-energy particles from the Sun.

The number of those molecules with very high temperatures is very small, however, because of the sparseness of the atmosphere at such heights. The total mass of the thermosphere accounts for only about 0.01 percent of the total atmospheric mass. The decrease of density, mass, and volume of the atmosphere can be expressed by the **mean free path** of a molecule—the distance an individual molecule must travel before encountering another molecule. The mean free path at the surface is on the order of a micrometer. By contrast, in the thermosphere the mean free path is on the order of a kilometer or more. Despite the fact that the individual molecules have very high amounts of energy, there are so few molecules to contain the heat that even if you could somehow survive for more than a fraction of a second at those heights, you would freeze to death instantly even at temperatures above 1100°C (2000°F)!

A thermopause does not exist; instead, the atmosphere simply merges slowly into interplanetary space. Individual gas molecules may be gravitationally attracted to the planet for quite a distance into space. However, most agree that the atmosphere extends no higher than about 1000 km (600 mi) above the Earth's surface.

■ Summary

The atmosphere is a fragile and complex collection of gases gravitationally attracted to Earth. It originated early in the history of the planet as volcanic materials cooled and outgassed. The early atmosphere is believed to have been composed primarily of nitrogen and CO_2, but the composition is different today, as O_2 has replaced CO_2 as the second-most abundant gas. This resulted from the evolution and proliferation of simple organisms and green plants, which have stored carbon in their biomass and have output O_2 into the atmosphere over the past 3 billion years.

Carbon follows a cycle in which it may be stored for certain periods of time within a number of reservoirs such as rock layers, the ocean, biomass, and the atmosphere. This cycle of carbon is important in the history of Earth. However, its importance has been challenged recently by evidence suggesting that methane-emitting microbes may have played a greater role than previously believed in the trapping of energy emitted by Earth through the greenhouse effect in Earth's early history. This methane may have kept Earth's temperatures within a narrow range of variability, despite a weaker solar output.

The troposphere is the lowest layer of the atmosphere and is where nearly all weather and climate processes of importance occur. Temperatures in the troposphere usually decrease with height because of the increased density in the most compressed part of the atmosphere—the part nearest to the surface. The stratosphere is the second layer and is characterized by increases in temperature with height because of ozone absorption. In the mesosphere temperatures decrease with height, for the same reason they do in the troposphere. The final thermal layer of the atmosphere is the thermosphere, which is characterized by increases in temperature with height because of the direct absorption of incoming radiation by N_2 and O_2.

▶ Key Terms

Aerosol
Atmosphere
Biosphere
Carbon cycle
Charles' law
Chlorofluorocarbon (CFC)
Condensation
Constant gas
Density
Diffusion
Electromagnetic spectrum
Environmental lapse rate
Eukaryote
Evaporation
Faint young Sun paradox
Fermentation
Flux
Fossil fuel
Greenhouse effect
Greenhouse gas
Heterosphere

Homosphere
Hydrosphere
Hydrostatic equilibrium
Insolation
Isothermal layer
Keeling curve
Lithosphere
Mean free path
Mesopause
Mesosphere
Methane (CH_4)
Methanogen
Micrometer
Nebula
Newton's second law
    of motion
Nuclear fusion
Oort cloud
Outgassing
Ozone (O_3)
Photodissociation

Photosynthesis
Planetesimal
Prokaryote
Reservoir
Residence time
Second law of
    thermodynamics
Shortwave radiation
Sink
Solar wind
Stratopause
Stratosphere
Temperature inversion
Thermosphere
Tropopause
Troposphere
Ultraviolet (UV) radiation
Variable gas
Wavelength
*Terms in italics also appeared in
    Chapter 1.*

▶ Review Questions

1. Explain how Earth and its atmosphere formed.
2. How is today's atmosphere similar to and different from early Earth's atmosphere?
3. Describe how oxygen came to compose almost 21 percent of the atmosphere today.
4. Given that solar output had increased over the past 4.6 billion years, how have Earth's temperatures remained fairly constant over that same time?
5. What is residence time and why is it important?
6. What is the carbon cycle and how does it operate?
7. Describe the thermal structure of the atmosphere.
8. What causes the thermal characteristics associated with each thermal layer of the atmosphere?
9. Compare and contrast the heterosphere and the homosphere.
10. Why is there no defined top to the atmosphere?

▶ Questions for Thought

1. Give as many examples of the second law of thermodynamics as you can think of, both related to and not related to the atmosphere.
2. Why do the *aurora borealis* and *aurora australis* occur in the mesosphere and not elsewhere in the atmosphere?
3. Why are the *auroras* not visible in the tropical parts of the Earth?

---

http://physicalscience.jbpub.com/climatology

Connect to this book's Web site: http://physical science.jbpub.com/climatology. The site provides chapter outlines, further readings, and other tools to help you study for your class. You can also follow useful links for additional information on the topics covered in this chapter.

2 Climatological Processes

Chapter 3—Climate System: Controls on Climate
Latitude
Earth–Sun Relationships
Distance to Large Bodies of Water
Circulation
Topography
Local Features
Putting It All Together: Spatial and Seasonal Variations in Energy
Summary
Review

Chapter 4—Effects on the Climate System
Ocean Circulation
El Niño–Southern Oscillation Events
Volcanic Activity and Climate
Deforestation and Desertification
Cryospheric Changes
Summary
Review

Chapter 5—Energy, Matter, and Momentum Exchanges Near the Surface
Properties of the Troposphere
Near-Surface Troposphere
Energy in the Climate System
Local Flux of Matter: Moisture in the Local Atmosphere
Atmospheric Statics, the Hydrostatic Equation, and Stability
Momentum Flux
Putting It All Together: Thermal and Mechanical Turbulence and the Richardson Number
Summary
Review

Chapter 6—Global Hydrologic Cycle and Surface Water Balance

Global Hydrologic Cycle

Surface Water Balance

Putting It All Together: A Worked Example of the Surface Water Balance

Types of Surface Water Balance Models

Water Balance Diagrams

Drought Indices

Summary

Review

Chapter 7—General Circulation and Secondary Circulations

Circulation of a Nonrotating Earth

Idealized General Circulation on a Rotating Planet

Modifications to the Idealized General Circulation: Observed Surface Patterns

Modifications to the Idealized General Circulation: Upper-Level Airflow and Secondary Circulations

Summary

Review

Climate System: Controls on Climate

Chapter at a Glance

Latitude
Earth–Sun Relationships
 Revolution
 Rotation
 Axial Tilt and Parallelism
 Combined Effect of Revolution, Rotation, and Tilt
Distance to Large Bodies of Water
Circulation
 Pressure
 Wind
 Surface Versus Upper-Level Winds
 Vertical Motion
 Cyclones and Anticyclones
 Oceanic Circulation
Topography
Local Features
Putting It All Together: Spatial and Seasonal
 Variations in Energy
Summary
Review

Several factors control the general state of the atmosphere for a given location and, therefore, govern weather and climate. For instance, we expect locations in the midlatitudes to experience pronounced differences in temperature throughout the year. Most tropical locations show much less seasonal change in temperature, but they may have seasonal precipitation changes. These differences are due in part to the influence of the various climatic controls. This chapter details how and why the climatic "site" of a place is dictated by its latitude, location relative to solar radiation receipt, and local factors. The climatic "situation" of a place relative to other features, such as water bodies and major atmospheric circulation features, is also presented because it impacts the types and distribution of climates that we see on Earth.

◼ Latitude

To understand the importance of location on Earth relative to the Sun's direct rays, we must first review the concept of **latitude**—a set of imaginary lines that run from west to east around the Earth, paralleling the equator and each other (**Figure 3.1**). They are, therefore, usually referred to as "parallels." Latitude lines are named for the angle they make between their location, the center of the Earth, and the equator. Thus, the latitude of the equator itself must be 0° and the North and South Poles have latitudes of 90°. We always indicate whether a latitude coordinate is north or south of the equator (0°) by using the letter "N" or "S" after the angle. For instance, Kent, Ohio, lies at approximately 41° N latitude, or almost half the angular distance between the equator and the North Pole. Parallels of latitude provide an indicator of the location of a place relative to the equator. More specifically, the location of any place can be expressed in terms of how far north or south of the equator the place lies, with one degree of latitude corresponding roughly to 111 km (69 mi) on Earth's surface.

Longitude is defined as a set of imaginary north–south lines running through the north and south poles perpendicular to the equator and every other parallel of latitude (Figure 3.1). Longitude lines are sometimes referred to as "meridians." Unlike lines of latitude, meridians of longitude are not parallel to each other because they converge toward the poles, even though they may not appear to converge because of the way that certain kinds of maps

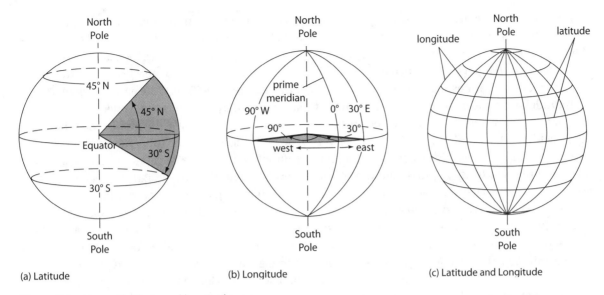

(a) Latitude (b) Longitude (c) Latitude and Longitude

Figure 3.1 Lines of latitude and longitude.

are constructed. The longitude coordinate actually measures the angle created between the point of interest, the center of the Earth at the same parallel of latitude, and a point on that latitude along the arbitrarily defined **prime meridian**—the meridian running through Greenwich, England, which is 0° longitude. Each degree of angle east or west of Greenwich represents one degree east or west longitude (with the direction designated by "E" or "W" after the angle). The maximum longitude coordinate is 180°, or halfway around the world from the prime meridian. The combination of a latitude and longitude coordinate in this grid system allows us to pinpoint any location on the surface of the Earth accurately.

But why does latitude have anything to do with climate? Quite simply, variations in latitude (in addition to variations in time of day and time of year) cause solar energy to strike the surface at some angle varying between 90° (when the ray is directly overhead) and 0° (when the Sun is on the horizon) (**Figure 3.2**). The lower the Sun's angle (i.e., the closer to 0°) above the horizon, the less intense the *insolation*, because the ray can undergo **attenuation**—the depletion of solar rays—more effectively by interacting with atmospheric particles when it reaches Earth from a lower angle. When the Sun is directly overhead (i.e., at a 90° Sun angle), solar rays pass through the atmosphere most efficiently because they strike the surface perpendicular to it and have fewer opportunities to be attenuated. The result is that the rays are more intense when they hit the surface.

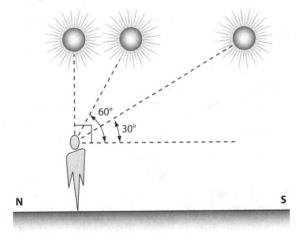

Figure 3.2 Variation in local Sun angle.

Gases, liquid and frozen water, water vapor, dust, and other *aerosols* floating in the air absorb, reflect, and scatter away insolation. Attenuation increases as concentrations of particles in the air and the **path length** of radiant energy increases through the depth of the atmosphere. The more attenuation in the atmosphere, the less intense the insolation when it strikes Earth's surface. Imagine using a flashlight to illuminate the sidewalk at night. Then imagine the same flashlight on a foggy day shining on the sidewalk or the same flashlight atop a skyscraper attempting to illuminate the same sidewalk. The greater number of water molecules on the foggy day or the greater distance atop the skyscraper allows for more attenuation, thereby reducing the amount of light striking the sidewalk. Increased attenuation, therefore, leads to decreased surface heating power. By contrast, a relatively short path length through the atmosphere to the surface allows for the energy to be more concentrated per unit of surface area, with less attenuation by atmospheric aerosols.

To understand the concept and importance of attenuation more completely, imagine you are holding a flashlight that is pointed straight down at the floor. The beam of energy is focused and the illuminated area will likely approximate the actual area of the shining face of the flashlight. Energy is concentrated through a very small area on the floor in that situation. Now imagine you shine the flashlight across the floor some meters away from you. The illuminated region now expands and elongates, even though the same battery is producing the same amount of light from the flashlight. The same amount of energy is now spread across a larger area. In the same way, when the Sun is shining on a location at a low angle, warming is less intense than it if the Sun were shining from a high angle because its energy is dispersed across a wider region of Earth's surface.

Earth's spherical shape causes areas receiving lower solar angles to see higher attenuation rates because of long path lengths through the atmosphere to the surface. To illustrate this concept, imagine two lines, A and B, as shown in **Figure 3.3**. Line A passes from the Sun through the atmosphere to the surface in the equatorial region; we say that this path length equals 1. Line B passes from the Sun through the atmosphere and touches the Earth at the North Pole. For solar energy to reach the surface of a high latitude location, it must travel a much longer distance through the atmosphere. Specifically, the path length increases approximately 2.5 times simply because of Earth's curvature. Because of the great distances between the Sun and Earth, for this and all other discussions of insolation we can assume that solar energy travels in parallel lines to Earth's system. If Earth were shaped like a block rather than a sphere, the path length of the Sun's rays through the atmosphere would be essentially the same for lines A and B. But the spherical nature of Earth makes it curve away from the Sun and gives much greater path lengths to locations at higher latitudes.

The parallel of latitude experiencing the direct rays of the Sun—the rays that fall perpendicular to the surface, or 90° above the horizon—on a given day always has the shortest path length. That latitude is termed the **solar declination**, and it varies only between 23.5°N and 23.5°S latitude. So, for locations north of 23.5°N latitude or south of 23.5°S latitude, the Sun is never directly overhead (i.e., at a 90° angle). The higher the latitude outside the tropics, the lower the Sun's angle. So because of their differences in latitude alone, we would expect Vancouver, British Columbia to have a lower Sun angle, more attenuation, and less intense sunlight than San Diego, California. Latitude affects climate because it affects the Sun angle of a location,

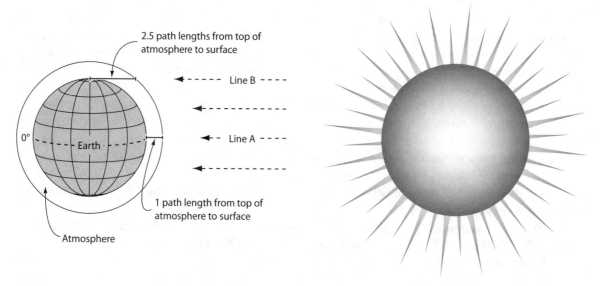

Figure 3.3 Atmospheric path length due to Earth curvature.

which in turn affects the intensity of radiation received at the surface, which in turn affects temperature.

As we see shortly, latitude also impacts day length throughout the year. The greater the number of hours of daylight, the more time exists for the surface to heat up. Variations in both solar declination and day length occur because of the changing relationships between Earth's surface and the Sun through the year. Such relationships are responsible for creating distinct seasons.

■ Earth–Sun Relationships

In combination with latitude, earth–Sun relationships contribute heavily to the characteristics of climatic "sites" on Earth. These influences occur because of Earth's revolution around the Sun, rotation on its axis, and its axial tilt. Together, these factors produce daily and seasonal cycles of energy and temperature.

Revolution

Earth's **revolution**—its orbit around the Sun—creates a distinct impact on temperature on Earth when combined with other factors. The orbit itself is not circular, and its elliptical path is not even perfectly centered about the Sun. The average distance from Earth to the Sun throughout the year is 149.67 million km (92.96 million mi). Given this distance, and the constant speed of light through the vacuum of space (299,338 km s^{-1} or 186,000 mi s^{-1}), it takes approximately 8.3 minutes for light emitted from the Sun to enter Earth's atmosphere. When you see the Sun, you are actually seeing the Sun as it was 8.3 minutes ago!

The earth–Sun distance varies from about 147.09 million km (91.36 million mi) on or about January 3—termed **perihelion**—to about 151.92 million km (94.36 million mi) on or about July 4—termed **aphelion**. Obviously, given the dates at which the perihelion and aphelion occur (during northern hemisphere winter and summer, respectively), the distance from the Sun must play little role in controlling intraannual surface temperature variations. If the earth–Sun distance were the most important factor producing the seasons, January would be summer in both the northern and southern hemispheres, because it is the month when the Earth is closest to the Sun, and July would be winter in both hemispheres.

Rotation

Rotation—the spin of the Earth on its axis—causes differing longitudes to experience the most direct ray of the Sun they receive on a given day at different times. The direction of rotation is from west to east, or counterclockwise when viewed from above the North Pole. One complete rotation (i.e., 360° of longitude) occurs every 24 hours, which is equivalent to motion through 15° of longitude every hour.

Because of rotation, the Earth is divided into **time zones**, 24 longitudinally based regions across the Earth's surface. They are approximately 15° of longitude in width, with the "official" time set 1 hour apart in each adjacent time zone. When one time zone is receiving the most direct ray of sunshine (i.e., is situated in direct sunlight), others are on the fringe of the illuminated part of Earth, while others are in darkness. Exactly one-half of Earth is illuminated at any given time. The **circle of illumination** is the line, as viewed from space, that separates the illuminated and dark halves of Earth. It is constantly changing in position across Earth's surface.

Axial Tilt and Parallelism

To understand the cause of seasonality, we must examine the concepts of axial tilt and parallelism. The **plane of the ecliptic** represents the plane that bisects both Earth and the Sun (**Figure 3.4**).

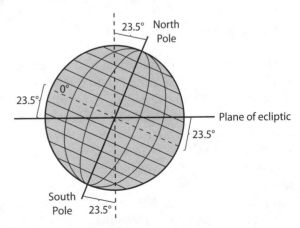

Figure 3.4 Earth's tilt is 23.5° from a perpendicular line through the plane of the ecliptic.

It is the plane on which Earth revolves around the Sun. Imagine a perpendicular line passing through Earth, through the plane of the ecliptic. This line does not pass through the poles, but instead the angle between this line and Earth's axis represents Earth's **axial tilt**. The axis of rotation (which passes through both poles) lies 23.5° from the perpendicular line through the plane of the ecliptic. We could also state that the equator is tilted 23.5° from the plane of the ecliptic. If the Earth had no axial tilt, the equator would align along the plane of the ecliptic and the poles would align along the perpendicular line through the plane of the ecliptic.

This 23.5° axial tilt can be thought of as being "fixed." No matter where Earth is in its revolution around the Sun, the axis still points in the same direction and remains at the same angle. This principle is known as **parallelism**. The northern hemisphere's axis of rotation has a convenient reference point in space—Polaris, or the North Star. If you were standing on the North Pole through the course of the year, you would always see Polaris at a 90° angle, or directly over your head. The other stars would rotate around you throughout the year as Earth orbits the Sun, but Polaris would appear as a fixed point. The axis of Earth is always parallel to itself for every position in its orbit (**Figure 3.5**). In reality, Earth's axial orientation and angle of tilt do vary slightly through geological time, and this variance is believed to contribute to global climatic changes. However, on time scales important in the context of contemporary weather and climate, we can think of the axis as remaining unchanged in orientation and angle of tilt.

Combined Effect of Revolution, Rotation, and Tilt

With the exception of the equator, the proportion of each line of latitude inside the circle of illumination changes throughout the course of the year, as demonstrated in Figure 3.5. In June more than half of any parallel of latitude in the northern hemisphere is in the daylight sector and more than half of any parallel of latitude in the southern hemisphere is in darkness. But for Earth as a whole exactly half is illuminated at any time. In December more than half of any parallel of latitude in the northern hemisphere is in darkness, while more than half is in daylight in the southern hemisphere. The closer to the pole the line of latitude is (in either hemisphere), the greater the difference in day length from June to December.

The number of hours each day during which a given parallel of latitude is inside the circle of illumination is the same as the day length at that latitude on that day. Day length changes throughout the year as the proportion of a parallel of latitude inside the circle of illumination changes. These changes have profound impact on the heating of Earth throughout the year and cause the seasons to exist. To understand how these processes work, we must put together what we have learned about revolution, rotation, and tilt.

The changing orientation of Earth's axis of rotation with respect to the Sun throughout the year causes these relationships to exist. Orientation changes because of Earth's revolution. Remember that Earth's axis maintains parallelism throughout its orbit (**Figure 3.6**). So at one point in Earth's orbit, on or about June 21,[1] the northern hemisphere axis is pointed maximally toward the Sun (this is termed the **June solstice**—or **summer solstice** in the northern hemisphere). This concept is shown on the far left side of Figure 3.6, where we see evidence for the northern hemisphere's long summer days and short nights and generally warmer conditions in the northern hemisphere and colder conditions in the southern hemisphere (which faces away from the Sun) at the corresponding latitude. On all days from March 21 through September 22 the northern hemisphere is pointed toward the Sun, but the pointing is most direct on June 21. On that day the tilt of Earth causes the solar declination to be at 23.5°N latitude—the **Tropic of Cancer**.

It is also clear from Figure 3.6 that on June 21 more than half of any line of latitude in the northern hemisphere is in daylight and less than half of that parallel of latitude is in darkness. Furthermore, the farther north the line of latitude, the greater the percentage of that parallel is in daylight on June 21. Days are longer than nights in the northern hemisphere from March 21 to September 22 (centered on June 21), because the pointing of the axis toward the Sun in the northern hemisphere during that period would leave any line of latitude in daylight for a longer period of time than it would be in darkness as Earth rotates on its axis.

---

[1]Solstice dates may fluctuate, depending primarily on leap years. For the purpose of this text the most common dates are used.

Figure 3.5 Unequal division of lines of latitude into day and night by the circle of illumination with change in orbital position.

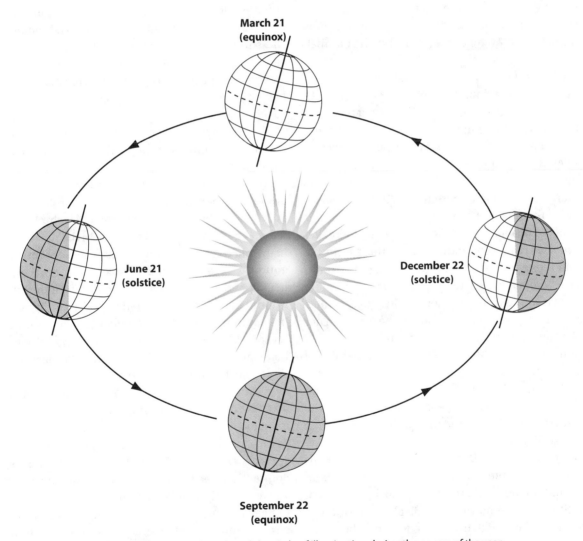

Figure 3.6 Relationships between Earth's axis and the circle of illumination during the course of the year.

In the northern hemisphere, day length increases from the equator to the **Arctic Circle** (66.5°N) on any given day from March 21 through September 22 as a higher percentage of each line of latitude sits within the illuminated half of Earth as latitude increases.

On June 21 day length increases from 12 hours at the equator to 24 hours at the Arctic Circle.

Every location north of the Arctic Circle receives constant daylight through this 24-hour solstice period, and for longer periods of time as one continues north, until day length reaches 6 months (from March 21 through September 22) at the North Pole! The position of the Sun does little to indicate when it is time to sleep or eat—it merely circles around the horizon rather than dipping beneath it.

In the southern hemisphere, nights are longer than days from March 21 to September 22 (centered on June 21), because the South Pole is tilted away from the Sun during that interval, with the most direct tilt away from the Sun occurring on June 21 (the winter solstice in that hemisphere). On any given day from March 21 through September 22, day length decreases from the equator to the **Antarctic Circle** (66.5°S) as the parallels of latitude become less and less included within the circle of illumination as latitude increases. Again, the maximum effect occurs on June 21, when day length ranges from 12 hours at the equator to zero at the Antarctic Circle. A total absence of daylight exists for even longer periods successively south from the Antarctic Circle to a maximum of 6 consecutive months of darkness (from March 21 through September 22) at the South Pole. The effect of the tilt away from the Sun is cold conditions for the southern hemisphere. Because Earth is simultaneously farthest from the Sun (aphelion) near the June solstice, the southern hemisphere's winter season is even colder than it would be because of the effect of tilt alone. Because the equator is neither in the northern nor the southern hemisphere, days and nights are of equal length on June 21 (as on every other day of the year).

On the right side of Figure 3.6 we can see that December 22 marks the opposite point in the Earth's revolution around the Sun. That day, termed the **winter solstice** in the northern hemisphere, or, generically, the **December solstice**, sees conditions that are the same as those that occur on June 21, but the hemispheres are reversed. Parallels of latitude in the northern hemisphere have a decreasing day length from the equator to the Arctic Circle (12 hours to 0 hours, respectively) on that day, whiles in the southern hemisphere day length increases from the equator to the Antarctic Circle (12 hours to 24 hours, respectively). On all days from September 22 through March 21 the north-

ern hemisphere is pointed away from the Sun and the southern hemisphere is pointed toward the Sun, but the effect is most direct on December 22. On that day the tilt of the Earth causes the solar declination to be at 23.5°S latitude—the **Tropic of Capricorn** (Figure 3.6).

Because of these conditions, significant differences in insolation are received at the surface during the course of a year, except near the equator. There are also significant latitudinal differences in insolation received at any given day of the year. The northern hemisphere as a whole experiences an energy surplus in June, July, and August that accounts for warm conditions throughout the hemisphere. The southern hemisphere is in the opposite situation during those months, with a net energy deficit and resulting cold conditions. In December the situation is reversed as the northern hemisphere axis is pointed away from the Sun and the southern hemisphere axis is pointed toward it. The hemispheres are then in a net energy deficit and surplus, respectively.

Between the extremes represented by the solstices, Earth moves through the transition seasons. At the midpoints between the two solstices, Earth's axis is pointed neither directly toward nor away from the Sun. Instead, the axis of rotation is pointed in a "neutral" direction on March 21 (the **March equinox**, or **vernal equinox** in the northern hemisphere) and September 22 (the **September equinox**, or **autumnal equinox** in the northern hemisphere). In the southern hemisphere the autumnal equinox occurs on March 21 and the vernal equinox occurs on September 22, because the seasons are reversed from those in the northern hemisphere. On the equinoxes, the solar declination is at the equator. So, to see the Sun directly over your head (90°) on March 21 or September 22, you would have to be located along the equator at noon local time.

On the equinoxes, the circle of illumination bisects every parallel of latitude because the axis is pointed neither toward nor away from the Sun. The result is that every location in the world experiences 12 hours of daylight and 12 hours of darkness on those days. Even though every location has virtually the same day length on the equinoxes, not every location receives the same intensity of radiation because of differences in solar angle. The angle at **solar noon**—the time when the Sun is at its highest point in the sky on a given day—

decreases with increasing latitude from 90° at the equator on the equinoxes. This decreasing solar angle causes greater attenuation with increasing path length. Local factors, such as cloudiness, may also complicate the relationship between solar angle and intensity of energy receipt.

The only parallel of latitude that has virtually unchanging day length throughout the year is the equator. It should be noted that day length does fluctuate very slightly during the course of the year even at the equator. This is because day length is determined by the time between sunrise (when the leading edge of the Sun's disk—not the middle of it—reaches above the horizon) and sunset (when the trailing edge of the Sun's disk slips below the horizon). At certain times of the year the solar declination is more distant from the equator (e.g., the solstices). At least part of the solar disk remains above the horizon for slightly longer than 12 hours at the equator. Furthermore, refraction—bending of light as it moves through air of varying density—causes sunlight to appear above the horizon for a few minutes before and after the Sun is actually above the horizon. So, at all locations day length is generally a few minutes longer than would be expected; it exceeds 12 hours by a few minutes even at the equator, with the difference slightly greater at times of the year when solar declination is farthest away.

The equator lies at the middle of the "tropics," the region between the Tropics of Cancer and Capricorn. Tropical locations see the vertical ray of the Sun pass overhead twice a year. Realistically, the vertical rays of the Sun at solar noon would never be very far from a tropical location. As a result, the tropics generally have high solar angles at noon throughout the year. Day length also does not vary much from 12 hours in tropical locations. The absence of short days, along with the short path length through the atmosphere and lack of significant attenuation, both as a result of consistently high noon solar angles, causes energy surpluses and high temperatures throughout the year.

■ Distance to Large Bodies of Water

Another major control on climate is a "situation" variable known as continentality—location relative to large bodies of water (especially oceans). Such large bodies are capable of storing huge quantities of energy during high-energy times (summer)

and releasing this energy slowly to the atmosphere during low-energy times (winter). These energy *fluxes* can have a significant effect on the climate of a location adjacent to the water body. Assuming that all other factors are equal, summers over or near oceans are not as warm as they are in the interiors of continents. Likewise, winters in coastal and oceanic locations are generally not as severe as they are in inland locations. In addition, the onset of seasons is delayed significantly over and near oceans because of the oceanic absorption of the energy through the summer and slow release of it to the atmosphere from autumn through winter.

To determine the relative importance of this climate control, let's negate the impact of latitude and Earth–Sun relationships. To do this we compare the climates of three cities that lie on or about the same line of latitude but in various regions of North America. Choosing cities on the same line of latitude ensures that all three cities are subjected to roughly the same amount of energy because of latitude (and, therefore, the effect of path length, attenuation, and sphericity is similar). Would all the cities have the same climate?

The three cities we use in our example are San Francisco, California; St. Louis, Missouri; and Washington, DC. All three cities lie between 37° and 39 °N (**Figure 3.7**). In winter St. Louis is coldest of the three cities, on average, but St. Louis also has the highest summer temperatures. This wide annual temperature range occurs because of the distance from St. Louis to a large water body. The oceanic storage of radiant energy during summer and release of that energy to the atmosphere during winter do not provide much relief from the summer warmth and winter cold for St. Louis.

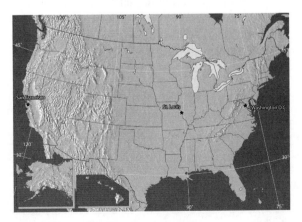

Figure 3.7 Continental position of San Francisco, St. Louis, and Washington, DC.

Instead, interior areas of continents, such as the location of St. Louis, warm significantly during summer and cool dramatically during winter. This creates more seasonal extremes in temperature and a larger seasonal temperature range for affected locations. St. Louis experiences this effect of continentality on its seasonal climate.

On the other hand, the warmest of the three cities in winter and the coolest in summer is San Francisco, which is downwind from the largest body of water on Earth and, therefore, has the most significant **maritime effect**—a moderating influence on temperatures by bodies of water. We will see shortly why Washington, DC has less of a maritime effect than San Francisco, even though both cities are coastal.

The slow response of water bodies to inputs of energy and the slow rate of release of energy that is input through insolation can be represented by a quantity known as **specific heat**—the amount of heat (energy) required to raise the temperature of a 1-g mass by 1 C°. Specific heat for various substances is shown in **Table 3.1**. The specific heat of water is about five times greater than that of land. This means that given the same mass of both water and a sample of solid Earth, the water requires about five times more energy to heat to the same temperature as the solid. Furthermore, energy is easily distributed in a water column through **convection**—the vertical transfer of energy or matter through a fluid—and horizontally through **advection** in the form of ocean currents. This redistribution helps to maintain relatively constant surface temperatures over the seasons. Another factor contributing to consistency of temperature in the oceans is water transparency, which allows insolation to penetrate to a depth of up to about 100 m (330 ft). Because insolation is spread vertically down the water column, the surface temperature does not change appreciably.

Table 3.1 Specific Heat Values for Various Substances at 0°C

Substance	Specific Heat (Joules kg⁻¹ K⁻¹)
Mercury	139
Granite	190
Sand	835
Glass	840
Dry air	1005
Saturated air	1030
Water	4187

Continental versus maritime effects are important contributors to the climate of any location. Globally, the greatest seasonal extremes occur over the largest land masses. Not surprisingly, Asia experiences the greatest continentality on Earth. But the effect of continentality may even manifest itself over relatively short distances from a large water body.

Smaller water bodies such as lakes, swamps, and marshes can also cause important temperature variations. Locations that are **windward** (upwind) of a lake tend to have more extreme variations in temperature than those that are **leeward** (downwind). A good example is Toronto, Ontario, which lies on the windward side of Lake Ontario for much of the year. Toronto has lower winter temperatures and higher summer temperatures than Watertown, New York, which is located on the downwind side of the lake.

An additional factor contributing to the continentality and maritime effects involves the use of insolation at the surface. When water is present, some insolation is required to evaporate water rather than to heat surfaces. To evaporate a bucket of water, a sufficient amount of insolation must be present, and this **radiant energy** is converted to **latent energy** during *evaporation*. This latent energy cannot simultaneously be used for heating air (**sensible energy**). Swimmers may feel cold after getting out of a swimming pool even on a hot summer day because of latent heating at the expense of sensible heating on their skin during the evaporation process. This also explains why rubbing alcohol feels cold even though the alcohol is stored at room temperature. Alcohol is an easily evaporated substance, which means that latent heating is occurring at the expense of sensible heating on the skin as the alcohol evaporates. We can think of evaporation as a "cooling" process; evaporation involves the absorption of insolation and conversion of it to latent form. It is important to note that at least some evaporation occurs in any unfrozen water body, regardless of the overlying air temperature. As such, the water body surface experiences some amount of evaporative cooling.

When the atmosphere is nearly saturated with water vapor (as is often the case in hot, humid climates), little additional evaporation (and, therefore, little cooling) can occur. On a hot day in a rain forest environment, for instance, perspiration cannot evaporate easily if the atmosphere is nearly saturated, and little evaporative cooling can occur through the evaporation of perspiration. This "stickiness," created by high humidity, increases the **heat index**—a contrived "temperature" based on how the temperature "feels" for people. This creates rather unpleasant conditions. By contrast, hot, dry places such as Arizona generally do not feel as uncomfortable in high temperatures because the low humidity allows freer evaporation (and subsequent cooling) through perspiration.

Water in the form of clouds also plays a significant role in daily temperature variations. This is especially true for most tropical locations and in midlatitude locations during the summer months. Typically, these clouds are generated through convection. Thus, they often become most prominent during the late afternoon when warm, humid air near the surface rises and its water vapor condenses into liquid water. Once formed, the clouds absorb, scatter, and reflect significant amounts of insolation, which may lead to a respite from the high temperatures. Often, the clouds generate precipitation, and this precipitation also cools the surface both directly (by adding cool water to the surface) and indirectly (through the cooling effect of evaporation of the surface water after the precipitation ends).

Given the accumulated properties of water, a large water body sees relatively little temperature variance over the course of the year, particularly if it is in the tropical part of Earth where day length and solar angles remain similar throughout the year. Because water temperatures remain nearly constant through the year, associated coastal locations such as San Francisco also see more constant temperatures through the year than inland locations such as St. Louis.

■ Circulation

Climate is also affected by location relative to atmospheric circulation features. The mission of the atmospheric circulation is to balance energy inequalities across the latitudes in accordance with the *second law of thermodynamics*. However, this mission is never accomplished because of the constant inequalities in the amount of insolation being received across the latitudes. Most of this imbalance is initiated by the Earth–Sun relationships and position on the continent relative to oceans, as discussed previously. But other effects, including atmospheric circulation characteristics, also play important roles.

Even if two locations have the same latitude, the same distance to the ocean, and Earth–Sun relationships affect them in the same way (i.e., they are in the same hemisphere), they still may not have the same climate. One location may be affected by atmospheric circulation from a certain direction more often than another. To clarify this idea let's go back to an earlier example. Which of the three cities mentioned previously would experience the second-coldest winter?

On initial glance it may seem that Washington, DC and San Francisco would have the same moderating maritime effect because each city lies adjacent to a large ocean. However, the maritime effect for Washington is much less than for San Francisco because of the direction of the prevailing winds. Because the prevailing wind direction in the midlatitude part of Earth is from the west, Washington is windward rather than leeward of the moderating influence of the ocean, unlike San Francisco. We say that the prevailing wind direction in the midlatitudes is westerly because in the atmospheric sciences wind direction is always named for the direction from which the wind is blowing.

In winter the westerly prevailing winds chill considerably over the cold North American continent. In summer the prevailing winds tend to have a southwesterly component, which adds low-latitude energy. Further heating occurs as the air blows over the warmed continental surface. Moisture originating from the Gulf of Mexico and/or the southwestern North Atlantic Ocean evaporates into the air, creating hot and humid conditions for Washington. The proximity of the North Atlantic does impart a maritime effect on Washington, DC; the maritime effect provides Washington with conditions that are a bit warmer in winter and a bit cooler in summer than they should be, given the energy it receives for its latitude. Nevertheless, Washington, DC is significantly colder in winter and warmer in summer than San Francisco because of the effect of circulation. By contrast, prevailing circulation advects moderate Pacific air from the westerly winds to San Francisco.

The relatively high winter temperatures and low summer temperatures of San Francisco caused by the maritime effect led Mark Twain to his famous quip that "the coldest winter I ever spent was the summer I spent in San Francisco." The pronounced maritime effect is caused by circulation features not only in the atmosphere but also in the ocean. In addition to the westerly winds discussed above, the nearby cold ocean current provides a circulation regime that prevents the summer temperatures from rising too much. The prevailing westerly summer winds are chilled over the cold ocean waters before blowing onshore. Because the temperature of the ocean surface changes relatively little over the course of the year, San Francisco enjoys a relatively constant temperature through the year. High temperatures are often in the low 20s°C (70s°F), while lows dip into the teens (10s°C; ~50s°F). The air is also relatively moist, giving it an added chilling effect, even during summer.

Pressure

Circulations that result in climates differing across space are caused by horizontal inequalities in atmospheric pressure. To understand this process of wind generation fully, we first need to understand the concept of pressure and pressure imbalances (pressure gradients).

Pressure is simply a measure of the amount of force applied to a given surface area. If we imagine a closed volume of air, the amount of air pressure is proportional to the frequency of molecular collisions of the matter composing the air. In such a case there are two ways that pressure could increase. The first is to decrease the volume. Putting the same number of molecules in a smaller volume increases the frequency of collisions. Because density is mass divided by volume, decreasing the volume is the same as increasing the density if mass is held constant. Thus, pressure and density are directly proportional to each other.

A second way to increase pressure is to increase the speed at which the molecules move in the closed container. Faster-moving molecules inherently collide with one another more readily. Temperature is a measure of the **kinetic energy**, or energy of motion, of molecules in a substance. Kinetic energy is present in all matter. According to kinetic theory, motion may cease only if the theoretical **absolute zero** temperature is reached. Although scientists at the Massachusetts Institute of Technology recently created conditions in a laboratory at which absolute zero was approached to a few billionths of a degree, all naturally occurring substances have kinetic energy, and increasing the temperature of matter increases the speed of molecules comprising that matter.

Thus, we can also say that pressure and temperature are directly proportional, if we hold vol-

ume (and, therefore, density) constant. Likewise, a decrease in pressure is associated with an increase in volume, decrease in density, and/or decrease in temperature, as shown in **Figure 3.8**. These relationships are expressed quantitatively in the **equation of state** for an ideal gas (also known as the **ideal gas law**). For perfectly dry air (which approximates an ideal gas), the equation of state is

$$P = \rho R_d T$$

where P represents pressure, ρ is density, R_d is the dry gas constant (287 Joules kg⁻¹ K⁻¹), and T is the temperature using the **Kelvin temperature scale**—a means of representing temperature such that it is proportional to the amount of energy contained by the molecules composing the matter. The **Joule** is the metric unit of energy or work, with units of kg m²s⁻². If these units seem to be cumbersome, keep in mind Einstein's theory of relativity, $E = mc^2$, where E represents energy, m is mass, and c is the speed of light. Mass has units of kilograms and speed squared must be in units of m²s⁻². The equation above also suggests that if pressure remains constant, warmer air is less dense than colder air.

Air pressure is sometimes expressed as the number of pounds of force exerted on a square inch of Earth's surface. Expressed this way the average sea level pressure of Earth is 14.7 lb in⁻². This means that for every square inch of surface area, the "weight" of the atmosphere on it is about the same as the weight of a bowling ball. If we were to add up all the square inches of surface area on the human head, the total atmospheric weight pushing downward would be about 600 lbs. If we were to lie down, we would experience over 5 tons of atmospheric pressure bearing down on us at sea level. So why are we not crushed by this pressure? It is because the pressure force is exerted in all directions—not just downward. Our bodies (and those of other creatures) also have internal fluid pressures that exert themselves outward against this pressure. We are in equilibrium with our physical surroundings.

If we were to increase pressure on our bodies, we would feel the negative effects of this "excess" pressure. For instance, divers readily experience an increase of pressure because of the increasing mass of water pushing downward onto their bodies as they dive deeper and deeper into the water. Eventually, they would reach a point where the fluids and gases of their bodies could not offset the

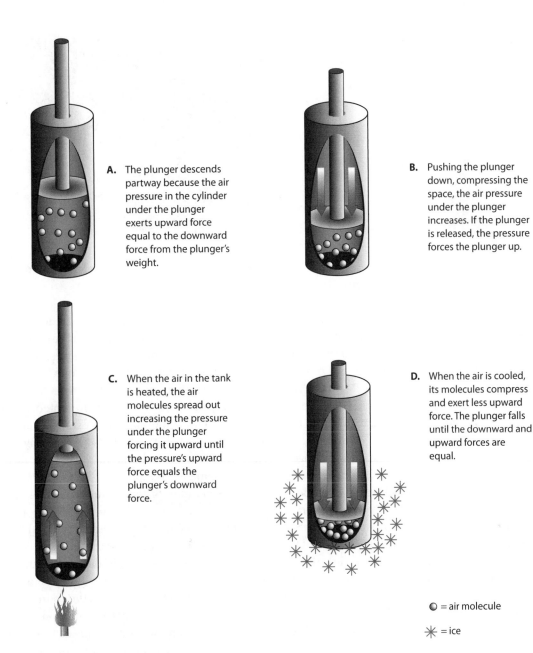

A. The plunger descends partway because the air pressure in the cylinder under the plunger exerts upward force equal to the downward force from the plunger's weight.

B. Pushing the plunger down, compressing the space, the air pressure under the plunger increases. If the plunger is released, the pressure forces the plunger up.

C. When the air in the tank is heated, the air molecules spread out increasing the pressure under the plunger forcing it upward until the pressure's upward force equals the plunger's downward force.

D. When the air is cooled, its molecules compress and exert less upward force. The plunger falls until the downward and upward forces are equal.

⬤ = air molecule

✳ = ice

Figure 3.8   The concept of pressure in a hypothetical pump with no outlet.

increasing pressures and they would implode. Although this example is rather extreme, it does illustrate the notion that pressure increases with depth within a fluid. The atmosphere is a fluid, and we lie at the bottom of a gaseous ocean of air. The weight of the atmosphere represented by its pressure is, therefore, maximized at the surface or, more generally, at sea level.

Even though pressure may be expressed in terms of pounds per square inch, this notation is rarely used in the atmospheric sciences. Other pressure notations are preferred. The first measurements of pressure used the height of a fluid in a vacuum tube.

The downward pressure (force) of the atmosphere actually pushed the fluid up the vacuum tube. This experiment was initially done with water, and huge tubes were needed. It was found that a more conveniently sized tube containing mercury was more useful, and the invention of the mercury barometer in 1643 was credited to Evangelista Torricelli. To determine atmospheric pressure, one needed only to note the number of "inches of mercury" in the tube (**Figure 3.9**).

Today, atmospheric scientists seldom use the number of inches of mercury to represent pressure. Instead, even in the United States they represent

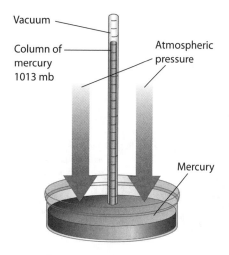

Figure 3.9 Schematic representing a mercury barometer.

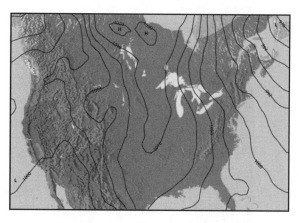

Figure 3.10 Surface weather map of the United States on August 9, 2005 at noon CDT. *Data from*: U.S. Weather Service.

pressure using the **Newton (N)**—a metric unit that measures force per square meter. One Newton of force per square meter is termed a **Pascal (Pa)** and 100 Pascals equal 1 **millibar (mb)**. The top of the atmosphere has a pressure reading of 0 Pa (or 0 mb), because no overlying mass exists to exert pressure, whereas sea level has an average pressure of 101,325 Pa or 1013.25 mb.

Meteorologists and climatologists use average sea level pressure as a benchmark for assessing pressure fluctuations for any place at any time. Because pressure varies greatly in the vertical, as one ascends through the atmosphere there is less mass above and, therefore, a reduction in pressure. Of course, this makes comparisons between locations of different elevations meaningless. To alleviate this problem the actual recorded pressure at each weather station is mathematically "corrected" to that which would occur at a particular spot if it were at sea level. This reduction to sea level equivalent allows for the role of elevation to be effectively eliminated. With pressures undergoing a reduction to sea level equivalent, it is possible to compare pressures at any stations directly, regardless of elevation differences.

Comparison of pressures on a map is usually done by connecting locations that have the same sea level–corrected pressure using **isobars**—lines of equal pressure. An isobaric map of the United States is shown in **Figure 3.10**. The interpretation of isobars is intuitive. All locations on one side of a given isobar have pressure values above that represented by the isobar, and all areas to the other side of an isobar have pressure values below that which is represented by the isobar. For example,

in Figure 3.10 the highest pressures are located in northwestern North America and off the east coast of North America. The lowest pressures on that day are found in the northeast, around Hudson Bay. By convention, on all surface weather maps only every fourth millibar is represented by an isobar. This convention allows a user to compare how closely packed the isobars are on any two maps; the closer the isobars, the greater the variation in pressure across space.

In general, "high pressure" indicates pressures above that of mean sea level, while "low pressures" are considered to be those below that of mean sea level. However, pressure systems can also be considered to be "high" or "low" relative to one another. For instance, we might note a low-pressure system migrating across the country, but that "low" pressure could actually have a central pressure reading in excess of mean sea level pressure if this pressure is lower than other pressure readings surrounding it.

Wind

Now that we have an understanding of atmospheric pressure, we can begin to understand winds and circulation. *Wind* is simply air in motion, or, stated another way, it is the transfer of atmospheric mass from one location to another. The transfer occurs because there is an excess of atmospheric mass in one location and a lack of mass in another. Places where there is "extra" air have higher pressure, and places that have less air have lower pressure. Air moves in relation to pressure imbalances in either the vertical or the horizontal.

Fluid always moves from areas of higher concentration to areas of lower concentration, in accordance with the second law of thermodynamics, unless acted on by some other force.

Imagine a cup of coffee. If the level of coffee happens to be higher on one side of the cup than the other (i.e., the mass or pressure of a "column" of coffee from top to bottom is not the same across the whole cup), you would expect the coffee to slosh back across the cup to even out the level, unless some other force is maintaining the imbalance (such as your hand tilting the cup). Similarly, the atmosphere also always attempts to equal out imbalances of mass or pressure.

Because an area with low pressure actually has less atmospheric mass than surrounding locations, it may be compared with a hole. The word "depression" is often used to describe an area of low pressure. If a fluid exists near a hole in the ground, gravity pulls the fluid into the hole. If the hole is entirely filled, then the fluid no longer moves into the cavity. This is very similar to the way atmospheric mass reacts to a low-pressure area. Atmospheric mass from surrounding higher-pressure areas is forced to move into the low-pressure area. Atmospheric scientists refer to the "filling of a low" when describing the end of a low-pressure system's life cycle. This denotes the point when the fluid pressure equalizes.

Pressure imbalances in the atmosphere are caused directly by thermal inequalities occurring at all spatial scales. Even micrometeorological thermal inequalities can induce and affect circulation. In general, surface air that is warmer than the air surrounding it rises because of a decrease in density. Such a situation creates a low surface pressure. Likewise, cooler air tends to sink as molecular motion decreases, and density and surface pressure increase.

Pressure Gradient Force The pressure gradient force (PGF) initially causes air motion and is generated from pressure inequalities across space. Although pressure inequalities occur both in the vertical (lower pressure always exists at higher elevations) and the horizontal, we concentrate on the horizontal first. When pressure differences occur across any distance, air moves in an attempt to equalize the pressure inequality.

The PGF induces air to move laterally from one location to another. If it could act alone, the PGF

would always produce motion from an area of higher pressure toward an area of lower pressure, as shown in **Figure 3.11**, at least initially. The PGF not only determines wind direction, but it also dictates initial wind speed. The speed depends on the pressure gradient.

As a fluid, air responds to certain forces as other fluids would. Envision water running down a slope. The water moves across a steep slope more quickly than the same amount of water moves across a gentle slope. The speed of the water is,

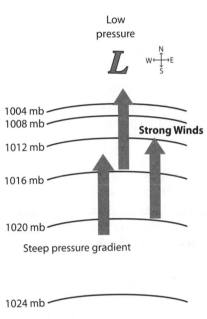
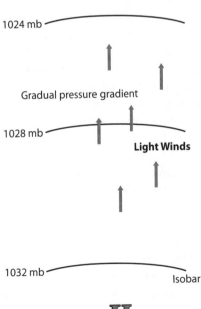

Figure 3.11 The pressure gradient force.

therefore, dictated by the difference between the high-elevation and low-elevation points on the landscape. The larger the elevation difference across a particular distance (or the smaller the distance between points that have the same elevation difference), the steeper the slope and the higher the speed of the moving water.

In the same way, wind speed is proportional to the pressure difference between the high- and low-pressure areas. If there is a large difference between two pressure centers across a given distance, air will move faster than if the pressure differences were smaller. Likewise, if the pressure difference is the same across two pairs of points but the distance between one pair of points is less than the distance between another pair, air will move faster between the first pair of points, and we would say that the pressure gradient is greater. If there is no difference in pressure across a distance (i.e., no pressure gradient), the air will not move.

The Coriolis Effect   Once air is in motion, other forces may change its trajectory and speed. One "force" that causes considerable change on the trajectory of air in motion is the **Coriolis effect (CE)**—a force caused by differences in Earth's angular velocity about the local vertical. To understand the concept of the local vertical, imagine a person standing at the equator and another person standing at the North Pole, as in **Figure 3.12**. As you can see, the person at the equator rotates because she is riding on the Earth, but her rotation is not about her local vertical at all. On the other hand, the person standing at the North Pole is rotating completely about his local vertical.

To make the point clearer, hold a pen in your outstretched hand parallel to the floor and turn yourself around, as if you are the Earth rotating. The pen is rotating because you are rotating, but the pen is not spinning about its local vertical at all because you are not turning the pen itself. Now hold the pen in your hand while reaching up vertically and begin rotating yourself around. If there is writing or some other mark on the pen, notice how it constantly faces a different direction. We can say that the entire component of the pen's rotation is about its local vertical. In reality, most of us do not live either at the equator or at the poles, so our rotation is somewhat about the local verti-

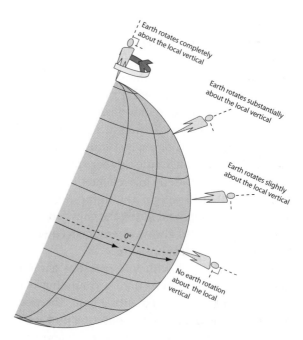

Figure 3.12   Earth rotation and the local vertical.

cal, with the proportion increasing with latitude, as shown in Figure 3.12.

In short, then, the CE is an apparent force that causes deflection of a fluid in motion when it is rotating at least somewhat about the local vertical. If Earth did not rotate, there would be no rotation about any local vertical, and the CE would not exist. Because there is no rotation about the local vertical at the equator, there is no CE there either. The CE increases in strength with latitude because the proportion of the rotation about the local vertical increases with latitude.

The CE is expressed mathematically as

$$CE = 2\Omega(\sin\varphi)v$$

where Ω (omega) equals the angular velocity of Earth (a constant), φ (phi) is the latitude at which the object is moving, and v represents the wind speed. Because Earth rotates through 360° (or 2π radians) in 24 hours (or 86,400 seconds), Ω is actually equal to $\dfrac{2\pi}{86,400\,s}$ or $7.27 \times 10^{-5}\,s^{-1}$.

From the previous equation we can see quantitatively that the Coriolis deflection of a moving object is directly proportional to both the latitude and the wind speed. All other factors being equal, as latitude increases, CE increases. Likewise, the equation shows that a greater speed produces an increased CE. If the latitude is zero (i.e., at the equator), we can see that CE is zero, as suggested previously.

But which way will a moving object appear to be deflected? The west-to-east direction of Earth's rotation causes objects in motion in the atmosphere to be deflected to the right in the northern hemisphere and to the left in the southern hemisphere. To understand this, assume that a rocket is launched from the North Pole to a point on the equator. The rocket initially travels in a straight line along, say, 0° longitude. Will the rocket actually hit the equator at that line of longitude? No! The reason is that once it is launched, the rocket travels in a perfectly straight line from north to south. However, as the rocket travels above Earth, Earth actually spins underneath it from west to east because Earth must rotate faster on its axis as latitude decreases, to move through a greater number of kilometers of circumference in the same 24 hours as it needs to rotate at higher latitudes (where Earth has a smaller circumference). Thus the rocket hits to the west of the intended target or, if you faced the direction of initial movement of the rocket in flight, to the *right* of the intended target.

Now imagine that you are launching the same rocket from the South Pole. Again, Earth moves from west to east underneath the rocket. This places the landing west of the intended target once again. However, the deflection in the southern hemisphere is now to the *left* if you are facing the direction of initial movement while standing at the pole.

As implied above, the amount of deflection changes with the latitude of the moving object because of differences in Earth's rotational speed at different latitudes. In 24 hours a person standing at the equator would have to travel the entire circumference of the planet to arrive back at the same point relative to a point in space. So over 24 hours a person at the equator would travel 40,075 km (24,901 mi) in a circle. By contrast, a person at the pole would simply spin around the same point in a circle over 24 hours. So a person at the equator is moving about 1609 km hr^{-1} (1000 mi hr^{-1}), while a person at the poles does not even leave the point at which he or she started! This difference creates a rotational pull on objects and fluids in motion. The increasing amount of curvature of Earth toward the poles also plays a role in deflection. An object's apparent trajectory is deflected to the right of the initial direction of motion in the northern hemisphere and to the left in the southern hemisphere (no matter the direction of motion),

with the amount of deflection proportional to both speed and latitude.

This apparent force is the subject of misunderstandings concerning the flow of water in bathtubs. Such rotating motions are not initiated by Coriolis deflection. An object must be in motion (have **inertia**) for a particular length of time, varying by latitude, for the CE to be important. This **inertial period** is one half of the pendulum day—the time required for the plane of a freely suspended pendulum to complete an apparent rotation about a local vertical. A fluid would have to be in motion for some number of hours (in most cases) before Coriolis deflection would occur. Water movement in a bathtub cannot meet those requirements. The spin induced on water moving into some sort of drain is actually caused and influenced by the dimensions, shape, and configuration of the basin itself, plus other factors including the water input direction, velocity of fluid input, and size and location of the drain. In general, the CE is negligible at small temporal and spatial scales. It would be an exaggeration to blame a missed putt on the rotation of the Earth!

If inertial period is so important, why do the atmosphere and oceans experience CE deflection instantly when set in motion? This is because such fluids are always in motion and have been since their formation. Their inertial periods have been surpassed billions of years ago and the deflective response is continuous.

Centrifugal Force   Centrifugal force (CF) is an apparent outward-directed force on an object moving along a curved trajectory. CF is simply a manifestation of inertia. Although some physicists argue that there really is no such force (there is simply inertia), atmospheric equations express this force quite frequently.

To understand the influence of CF, imagine air moving from a high- to low-pressure area. Initially, the PGF causes the air to move in a straight line toward the low-pressure core at a particular speed as determined by the pressure gradient. But as soon as the air begins moving, the CE causes it to curve to the right (in the northern hemisphere) with the amount proportional to the speed and latitude of the moving air. At that point the trajectory shows curvature, and CF begins altering the trajectory by pulling the air to the outside of the curve. The faster the wind movement and the tighter the curve, the greater the centrifugal pull. An air stream has its initial trajectory altered first by CE and then again by CF. It is this centrifugal pull that allows tornadoes to maintain their spin, particularly because the wind speeds are so high and the circulation is very tight in a tornado.

Mathematically, CF can be expressed as

$$CF = \frac{v^2}{r}$$

where v is the linear wind velocity and r represents the radius of curvature of motion. For perfectly straight flow, r would be infinite and CF would approach zero. For tighter circulations, such as a tornado, r is small and winds are large, and CF is extremely strong.

The concept of CF can be used to provide entertainment. Virtually every amusement park ride uses CF to stimulate sensors in the inner ear. The spinning cage, which plasters riders to the cage with such force they cannot even lift an arm, is a particularly good example. Simple experiments such as placing water in a bucket and spinning it about the body in a vertical circle also demonstrate the principle of CF. The water does not fall out of the bucket even when the bucket is upside down if the bucket is spun quickly enough. In this case CF exceeds the force of gravity. Of course, if the bucket is slowed beyond a critical point, the bucket spinner will be doused in water. Another example occurs in a washing machine. The spin cycle spins the internal basin. CF forces the clothing and water to the outside of the basin. Small holes in the basin allow water to escape the basin and drain from the machine, partially drying the laundry.

Friction *Friction* is another force that affects wind. It is also the most difficult component to quantify mathematically. As a force of opposition, friction works to decrease the speed of moving objects. In the atmosphere, friction is maximized near the surface and decreases with height until it becomes negligible in the **free atmosphere** (also termed the "friction-free zone") that characterizes most of the atmosphere.

The amount of friction applied to overlying moving air depends on the nature of the surface. Air flowing over smooth surfaces, such as water bodies, encounters much less friction than air moving over mountainous surfaces. Likewise, vegetation, hills, pastures, buildings, and roads all impart varying amounts of friction to moving air. With such variety it is impossible to describe and understand the precise effects of friction across even small spaces. This makes it impossible to account fully for friction and all its effects in mathematical equations that are used to forecast future air motions.

Friction must be accounted for in forecasts because it not only slows air, but it also indirectly alters the trajectory. Remember that both the CE and the CF alter wind trajectories and both depend on wind speed. Faster-moving objects experience greater Coriolis and centrifugal deflection. If friction is applied, speed is reduced. If speed is reduced, so is the amount of trajectory deflection from those two forces. Thus, the varying amount of friction with height causes wind direction to shift with height. For this reason many equations used in forecasting future conditions are solved for the free atmosphere where friction is negligible.

So wind is a combination of interactions between four forces:

- PGF, which determines initial air speed and direction
- CE, which alters trajectory by deflecting moving air to the right (in the northern hemisphere) and is a function of speed and latitude
- CF, which pulls the Coriolis-influenced curving air slightly back toward the initial trajectory (to the outside of the curve) as a function of speed and curvature
- Friction, which slows the air, thus affecting the speed and reducing the amount of deflection from both CE and CF

We can simply state that

$$\text{Surface wind} = PGF + CE + CF + F$$

where F is friction. All forces are interacting with each other to produce a wind moving in a particular direction at a particular speed.

Mathematical equations are capable of quantifying all aspects of air motion accurately except friction. The three **Navier-Stokes equations of motion** (for west–east, south–north, and down–up directions) form the ABCs of numerical weather forecasting. Data are collected at various weather observation sites around the world and input into computers that apply the equations to the initial data to forecast motion a few minutes into the future. The equations are reapplied using the "new" data as input and the process is repeated until forecasts of 6 hours, 12 hours, and each 12-hour increment afterward out to a few days are generated.

Surface Versus Upper-Level Winds

The varying effect of friction with height creates great differences in the representation in computer models of winds at the surface compared with those in the free atmosphere. But how are the equations applied if friction cannot be accounted for properly? The answer is rather simple: We greatly oversimplify the friction term in the equations. Because much of the focus in the computer models is on upper-air circulation, where friction is assumed to be negligible, the harm caused by inaccuracies in the oversimplified friction equations is minimized. Automated forecasts also use surface data, but these often are not used to forecast large-scale motion. But if they are, they are more likely to be erroneous because of the poorly handled friction terms.

In the contiguous United States, "official" upper-air weather data are collected from only 74 locations twice a day. By comparison, surface data are available from thousands of stations, with many providing minute-by-minute data. Needless to say, large spatial gaps occur between the upper-air observations, and large temporal gaps (12 hours) exist between readings. Many conditions change within the atmosphere that cannot be detected in the computer-based forecast because of the coarse set of input observations across both space and time. It is rather amazing that from such scant observations any forecast holds any degree of accuracy, yet the *National Weather Service* forecasts are generally accurate out to 3 days.

Vertical Motion

Air also moves vertically in the atmosphere. An upward-directed pressure gradient always exists on Earth because pressure (and atmospheric mass) decreases rapidly with height. The tendency for air to rise from the surface to lower-pressure regions aloft (buoyancy) is balanced by the downward force of gravity. In all, a large-scale equilibrium exists between buoyancy and gravity, and the atmosphere remains permanently adjacent to the surface of Earth, a situation known as *hy-drostatic equilibrium*. However, local imbalances can occur between the gravitational force and upward-directed PGF. Storm systems, particularly thunderstorms, represent situations in which local hydrostatic imbalances occur. In such cases a net transfer of atmospheric mass occurs from the surface upward and of *momentum* from the upper atmosphere downward. **Updrafts** carry air aloft, while **downdrafts** move air toward the surface.

Cyclones and Anticyclones

A **cyclone** is simply an enclosed area of relatively low pressure. If the cyclone exists at the surface, it is characterized by a convergence of winds into that low-pressure center, which initiates rising motions. These rising motions trigger cloud and precipitation-forming processes. Surface cyclones are, therefore, associated with storms. Many in the United States refer to tornadoes as "cyclones," whereas in parts of Asia a "cyclone" indicates a storm of tropical origin such as a hurricane. Any storm system in which surface air spirals about and into a low-pressure core is technically a cyclone. In general, there are three types of cyclonic storms: the **midlatitude (frontal) wave cyclone**, the **tropical cyclone** (tropical storm/hurricane), and the tornado.

The term "cyclone" technically refers to the type of air motion involved. The PGF dictates that air move toward the low-pressure core at a speed proportional to the pressure gradient minus friction. However, the CE deflects that airflow to the right (in the northern hemisphere) at some angle corresponding to the speed of airflow and the latitude of the system. The result is a system that spins counterclockwise in the northern hemisphere (**Figure 3.13**) and clockwise in the southern hemisphere. At some point in the upper *troposphere*, air diverges to compensate for the convergence near the surface.

Because of friction near the surface, air crosses closed isobars at some angle in a surface cyclone. However, in the free atmosphere air flows parallel to the isobars in a counterclockwise motion above the cyclone (**Figure 3.14a**). Thus, the wind direction changes with height, from a counterclockwise (in the northern hemisphere) and inward motion at the surface to a counterclockwise motion that parallels the isobars aloft. Another way of stating this is that the wind direction is deflected more and more to the right (in the northern hemisphere) with height—through a process called **veering**—until it parallels the isobars. Veering can be thought of as a clockwise change in wind direction with height. In the southern hemisphere, the winds around a surface cyclone move clockwise (because CE pulls winds to the left as they move into the low-pressure center) and inward, whereas winds aloft would simply move clockwise. In the southern hemisphere, then, winds turn more and more to the left, or a counterclockwise change in wind direction with height—through a process called **backing**.

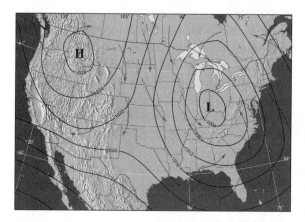

Figure 3.13 Northern hemisphere surface cyclone and anticyclone airflow patterns.

If the flow is sufficiently straight that CF can be ignored and is sufficiently high above the surface that friction can be ignored, we can say that the veering or backing ends when there is a balance between the PGF and the CE, a condition known as **geostrophic balance**. Airflow in geostrophic balance is termed the **geostrophic wind** and flows along constant pressure surfaces instead of across them.

Anticyclones—enclosed areas of high pressure— have opposite circulation patterns from cyclones (Figures 3.13 and 3.14b). These systems are characterized by surface divergence because the PGF initially pulls air outward away from the high-pressure core in an effort to eliminate the pressure gradient. But as the air moves from the central core toward lower pressure regions extending in all directions, the airflow is deflected by the CE. Again, deflection is to the right in the northern hemisphere, resulting in a clockwise and outward spin at the surface and clockwise flow aloft. Southern hemisphere anticyclones spin in an opposite manner: counterclockwise and outward at the surface and counterclockwise aloft.

How can air flow outward from a central point near the surface for any sustained length of time? The air must be coming from somewhere. Again, an anticyclone shows opposite conditions from a cyclone. An anticyclone has air moving toward a central location in the upper atmosphere to

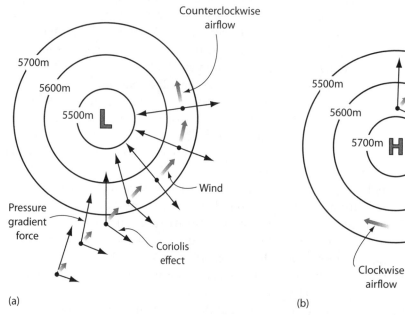

Figure 3.14 Northern hemisphere upper airflow: (a) cyclone and (b) anticyclone.

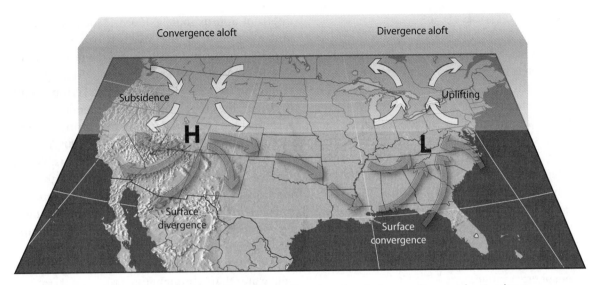

Figure 3.15 Surface and upper airflow characteristics of northern hemisphere cyclones and anticyclones.

compensate for the divergence of air from the central core at the surface. This air converges at the upper levels, forcing sinking motions at the central core and then outward motions from a central surface region. Sinking motions reduce or eliminate the possibility for cloud formation. Without clouds, the chances of precipitation are obviously nonexistent. Therefore, anticyclones are generally characterized by relatively clear conditions (**Figure 3.15**).

As with cyclones, the upper-air characteristics of anticyclones are opposite those at the surface. In anticyclones, upper-air convergence is associated with the surface divergence below it. As we have seen, in cyclones surface convergence is linked to upper-air divergence. So where does rising air go in the cyclones and where does sinking air come from in the case of the anticyclones? If viewed independently, these questions are difficult to answer. We must instead view the atmosphere as an integrated whole in a three-dimensional sense to understand interactions between seemingly disparate systems.

After surface air rushing toward the low-pressure core of a cyclone is deflected en route to the core, it converges with air streams entering from other directions and spirals upward. This air expands and cools on its ascent, forming clouds and possibly precipitation. At some level aloft, the air diverges and continues to move laterally across the upper troposphere until convergence once again occurs elsewhere, in a location associated with the upper-level features of a surface anti-

Figure 3.16 Synergistic relationship between circulation around cyclones and anticyclones (northern hemisphere example).

cyclone. There, the air is forced to sink back toward the surface. Once the air reaches the surface, it spirals outward toward surrounding lower pressure and the process continues with air again rising in those locations.

The motion associated with cyclones and anticyclones is highly integrated. Imagine looking down at the surface from a point in space. Now imagine that you see two large balls, each spinning clockwise over an area of the northern hemisphere. Now imagine another ball inserted between the two larger ones. That ball would spin in a counterclockwise direction, like gears in a machine, in opposition to the surrounding larger ones (**Figure 3.16**). This situation describes the surface interactions between two anticyclones with a cyclone sandwiched between them. The anticyclones feed the cyclone with surface air streams that converge and rise. The cyclone then feeds the anticyclones with upper-airflow that converges and sinks. Such interactions occur continuously.

Oceanic Circulation

We have seen that the purpose of the atmospheric circulation is to balance inequalities of energy incident on the surface, in fulfillment of the second law of thermodynamics. Besides atmospheric circulation, another mechanism for mitigating some of Earth's energy imbalances is the "situation" created by oceanic circulation. Although the ocean is a denser fluid than the atmosphere, it nevertheless has the same "mission" as the atmosphere. Most oceanic circulation characteristics are created by the same energy inequalities present in the atmosphere. Ocean movement follows **Newton's laws of motion**—the same laws that govern atmospheric motions.

Surface ocean currents are considered to be those that extend to approximately 100 m (330 ft) below the surface; they are driven primarily by surface winds. Frictional drag occurs between air moving over a water surface and the water at the surface. This energy causes motion in the water surface. The momentum is then transferred downward into the water column. Surface-generated energy below about 100 m (330 ft) approaches zero. Deeper ocean motions are initiated by processes outside of the atmosphere's direct influence.

Water motion does not correspond as closely to surface airflow as one might expect. Instead, the water surface may be viewed as being a stationary fluid when air begins to move above it. Frictional drag transfers to the ocean, and the water begins to move. Once the surface water begins moving, it is subject to Coriolis deflection. Surface water immediately deflects to the right of the airflow (in the northern hemisphere). Again, deflection is proportional to the speed of motion and the latitude. At the surface a deflection of about 45° to the right (in the northern hemisphere) of the airflow trajectory occurs, but this deflection only affects a column of water a few meters in thickness. The momentum involved in moving that surface column of water is then transferred downward to the next few meters of water. This water is set into motion and again is subjected to Coriolis deflection. The resulting motion is again an offset to the trajectory of the water column moving above it. This motion then initiates motion in the next column of water below it, with additional Coriolis deflection, and so on.

The result is the **Ekman spiral**—a spiral of water motion extending from the surface downward to a depth of approximately 100 m (330 ft). Identified and quantified by the Swedish oceanographer, Walfrid Ekman, in 1902, it causes the bulk of water flow to move at a right angle to the initiating wind stress (**Figure 3.17**). The Ekman spiral is important in climatology because this phenomenon is partly responsible for **upwelling**—an ascending current of cold water from depths of about 1000 meters (3300 ft) to the surface in certain locations. When surface waters move offshore because of the Ekman spiral, the cold water from the depths moves upward to replace it. In addition to providing a mechanism for nutrients to be returned to the surface, upwelling greatly contributes to the climatology of the coastal zones affected because of the cold water that is brought upward.

Deeper ocean currents are initiated by temperature and salinity inequalities across space. Because cold water is denser than warmer water, this density characteristic affects three-dimensional

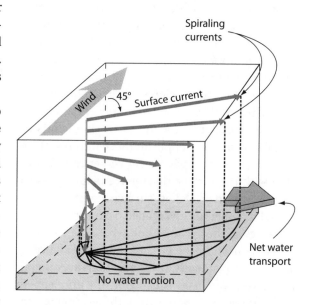

(a) Ekman spiral in the northern hemisphere

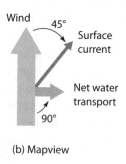

(b) Mapview

Figure 3.17   Ekman spiral.

motion of water in an attempt to achieve an ocean of homogeneous density. Because salty water is denser than fresh water, differences in salinity also set deep waters into three-dimensional motion to accomplish the mission of the oceans. Areas with cold, salty water, such as the northern North Atlantic near Iceland, see sinking motions where near-surface water is carried to the ocean floor. This northern Atlantic water moves equatorward as it sinks, eventually crossing the equator and moving into the Indian Ocean. The water is forced to rise in the western North Pacific near Indonesia where a surface current is initiated that carries the water back to the North Atlantic. This resembles a deep ocean conveyor belt that results in **thermohaline circulation**, a very slow energy transportation system that greatly affects global climatic conditions. "Thermo" refers to differences in thermal characteristics in the deep ocean; "haline" refers to salinity characteristics.

In general, the most important and immediate oceanic circulation effects on climates are the surface currents. These currents generally follow the winds produced by semipermanent pressure zones that exist in the atmosphere. Clockwise (in the northern hemisphere) or counterclockwise (in the southern hemisphere) circulation associated with semipermanent subtropical anticyclones exists in each ocean basin around 30° latitude. These features impart the most influence on oceanic surface circulation, partially because the oceans are so vast at the subtropical latitudes. These circulations cause cold ocean currents on the eastern sides of the ocean basins as cold water migrates from higher latitudes toward lower latitudes. The western ocean basins are dominated by warm surface currents moving from the equatorial areas poleward and transferring excess energy from low latitudes toward higher ones. Cold and warm surface currents greatly affect climates located near the eastern and western coastlines of continents.

Europe provides an excellent example of the influence of surface ocean currents on climates. Rome is positioned at about the same parallel of latitude as New York City (about 42°N), but Rome is climatologically warmer than New York. Likewise, much of Norway exists at latitudes equivalent to northern Canada and Siberia, but, again, the temperatures in Norway are much higher in winter, on average, than those locations. The strong maritime effect in Europe and prevalence of the warm

Gulf Stream current keep much of Europe warm during winter. That ocean current originates in the Caribbean and travels northward past the Atlantic seaboard of the United States before moving away from North America toward Europe. As it traverses the North Atlantic, it is renamed the **North Atlantic Drift**. This warm current brings large supplies of energy to Europe. As the current wraps northward around Scandinavia, it keeps Murmansk, a port city in Russia, ice free despite being north of the Arctic Circle.

■ Topography

Topography also affects climates. Temperature in the troposphere generally decreases with height, at an average rate of 6.5 C°/km. If a location is mountainous, the climate varies widely across short distances because of rapid changes in altitude (and, therefore, temperature) and the amount of exposure to the Sun through the year. If a location is relatively flat, that area responds to energy changes more uniformly.

The rapid decrease in atmospheric mass (and, therefore, density) with height also affects conditions in rugged terrain. One-half of the mass of the atmosphere lies in a layer only 5 km (3 mi) thick. In high elevations the decreased mass of the atmosphere can cause human health implications such as a shortage of oxygen, resulting in shortness of breath, headaches, nosebleeds, and even death.

Because of decreases in mass and density it is common to see high diurnal temperature ranges at high elevations compared with other locations that are lower in elevation but along the same parallel of latitude. This contrast is caused by differences in the transfer of energy through a "thin" versus a "thick" atmosphere. The thicker atmosphere at low elevations stores the heat energy far longer and more effectively than one that contains less mass because the matter composing the dense, thick atmosphere absorbs a relatively large amount of insolation. By contrast, at high elevations less matter is available in the thin, sparse atmosphere to absorb radiation. The result is that mountain peaks have warmer daytime surfaces than one might expect because of differences in pressure alone. At night the sparse high-altitude atmosphere allows the surface to radiate its energy back to space very quickly and effectively, causing temperatures to plummet.

Mass and density changes also affect the behavior of fluids such that at high elevations water evaporates more quickly than in a denser atmosphere. This occurs because individual water molecules can "break free" from the surrounding water more easily under lower pressure and convert into a vapor more easily. Increased pressure associated with a denser atmosphere tends to hold the individual water molecules more firmly together as a liquid. We can see a manifestation of this concept by examining the boiling point of water at various elevations. As elevation increases, the temperature at which water boils decreases. At sea level water boils at a temperature of 100°C (212°F). But at approximately 2.75 km (9000 ft) above sea level, the boiling point of water is roughly 9.2 C° (16.5 F°) lower.

Finally, topography also influences the distribution of precipitation. When air is forced up the side of a slope, it is likely to become saturated, because air saturates more easily as it cools moving uphill. This **orographic effect** may cause a cloud to be created and perhaps produce precipitation. The windward side of a slope often experiences significantly more precipitation than the leeward side. The leeward side is often quite dry, because as the air sinks downhill it warms, which stymies cloud and precipitation processes.

■ Local Features

The last climatic control is collectively described as local features. This catch-all term refers to isolated influences on the climate of a particular small area adjacent to those influences. Such factors include the presence of lakes, wetlands, forests, agricultural areas, and urban areas.

A notable example of a feature's influence on local climate involves the Great Lakes. During winter, **lake effect snows** occur as the relatively warm lake waters evaporate into much colder, drier air masses blowing over them from the northwest. The moisture quickly condenses in the cold air, creating clouds and snowfall for the downwind locations (**Figure 3.18**).

Another notable example of a local influence on climate involves the extra warmth associated with cities, a phenomenon known as the *urban heat island* (also discussed in Chapter 2). Urban areas

Figure 3.18   Average annual snowfall totals (inches) along the Great Lakes. *Adapted from*: Morgan, M.D., and J.M. Moran. 1997. *Weather and People.* Upper Saddle River, NJ: Prentice Hall.

heat more than the surrounding rural regions because of a combination of impervious surfaces, lack of vegetation, building geometry, the presence of glass and steel structures, and machines (including air conditioners and cars) that emit heat into the local environment. In some cases urban areas may be as much as 6 to 8 C° (10 to 15 F°) warmer than the surrounding rural environment.

Pollution can also contribute to local energy and temperature variations because pollution interacts with some radiation streaming toward the surface. However, pollution needs to be very severe for one to notice an appreciable decrease in surface temperatures. Recent conflicting research calls into question the net effect of pollution, as the absorption of insolation and subsequent extra radiation emitted downward to Earth by some types of pollutants may create a net warming effect. We simply do not have a deep enough understanding of the effects of pollution on temperature to reach a firm conclusion on its impact.

■ Putting It All Together: Spatial and Seasonal Variations in Energy

A location's climate results from the culmination of all climatic controls working both independently and cumulatively. Some factors contribute to wide energy variations over the globe through the course of the year, despite the fact that circulation is constantly working to even out these energy imbalances. The result is that temperature varies from one location to another across Earth and at any particular location through the course of the year.

Isotherms—lines connecting points that have the same temperature—can be used to map the influence of the various controls on climate. **Figure 3.19** shows that July isotherms extend poleward over the warm continents of the northern hemisphere as a direct result of continentality influences. The most significant warming occurs over Asia, where continental influences are most pronounced. By comparison, ocean temperatures show a somewhat even and gradual temperature decline with increasing latitude. Figure 3.19 also shows that the longer July days in the high latitudes of the northern hemisphere lead to relatively mild temperatures, while the southern hemisphere's high latitudes (over Antarctica) show extremely low temperatures. In general, a decline in temperatures with increasing latitude is discernible for both hemispheres. Nevertheless, significant alterations occur because of maritime and continentality effects.

In January the northern hemisphere shows very pronounced continental impacts as isotherms

Figure 3.19   July surface air temperatures (°C). *Data from*: NOAA.

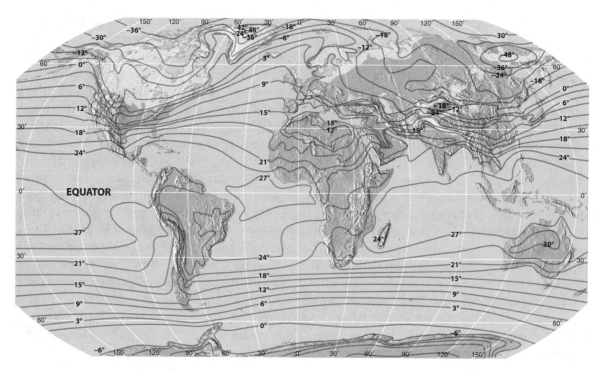

Figure 3.20 January surface air temperatures (°C). *Data from*: NOAA.

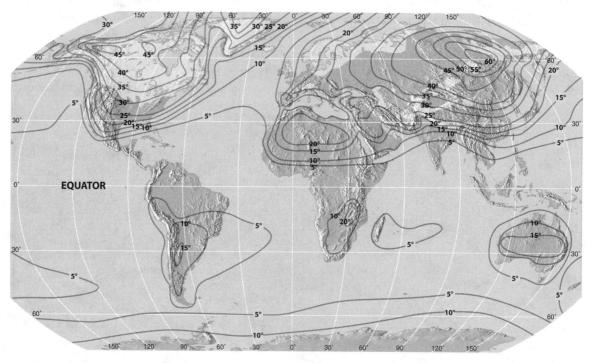

Figure 3.21 Temperature (°C) differences between January and July. *Data from*: NOAA.

dip dramatically toward the equator over the continents (**Figure 3.20**). The pattern is caused by rapid and intense cooling of the land surfaces during this time of net energy deficits. Again, Asia has especially strong effects because of the size of that land mass. The southern hemisphere shows a rather simplistic, latitudinally driven temperature pattern because of the lack of continental effects in that hemisphere. But some continentality is still evident as isotherms shift poleward over the continents in the southern hemisphere summer.

The effects of ocean currents may be seen as well. Isotherms dip equatorward along the eastern ocean edges as cold, high-latitude water is pulled toward the equator, in July (Figure 3.19), in January (Figure 3.20), and in all other months. A further influence stems from coastal upwelling because of the net transport of surface waters by the Ekman spiral away from the coast. Likewise, the western ocean basins show poleward bumps in the isotherm patterns as warm, low-latitude waters migrate poleward. The Gulf Stream is a prime example.

Continental and maritime effects are easily seen on a map depicting the difference between mean January and July temperatures. **Figure 3.21** shows these differences. Large continental regions exhibit significant differences, whereas maritime locations depict more thermal consistency throughout the year. The largest land masses show the widest range of mean monthly temperature through the year. Central Asia experiences an astonishing 60 C° (108 F°) or more temperature change, while the high latitudes of North America record a maximum change of about 45 C° (81 F°) throughout the year. Because of its low latitude, Africa reports a maximum monthly range of only about 20 C° (36 F°) between the extreme seasons, whereas South America and Australia record peak ranges on the order of 15 C° (27 F°). The continent that sees the least amount of intraannual temperature change is Antarctica. This is caused largely by the continuous ice coverage, which effectively reflects summer's insolation.

◼ Summary

Six factors cause the climate of a location to have its fundamental characteristics: latitude, Earth–Sun relationships, position in the continent, atmospheric and oceanic circulation, topography, and local features. Latitude is the position of a location on Earth relative to the equator. Assuming that all other factors are equal, locations with higher latitudes (i.e., more poleward locations) generally experience lower temperatures than locations at lower latitudes. The Sun's rays strike at a 90° angle only at the meridian of longitude that is experiencing solar noon and the parallel of latitude at the solar declination on a given day. All other lines of latitude (and longitude) experience solar angles that are less than 90°. Thus the solar declination changes with the day of the year, as Earth revolves around the Sun, and the longitude of the Sun's direct rays changes with the time of day, as Earth rotates on its axis.

Earth–Sun relationships play a strong role as controls on climate. Earth's orbit is not circular, and the Sun is not at the center of the orbit. Instead, Earth is closest to the Sun during January and farthest from it during July. Thus, Earth–Sun distance does not cause the seasons, but it does affect the severity of the winter and summer seasons. Instead, the changing orientation of Earth's tilt relative to the position of the Sun as Earth revolves around the Sun is the cause of the seasons. The tilt and revolution cause all lines of latitude to receive differing amounts of solar radiation throughout the year and the direct rays of the Sun to strike at latitudes varying from 23.5°N (the Tropic of Cancer) to 23.5°S (the Tropic of Capricorn). Tilt and revolution also cause day length to vary over the course of the year for each line of latitude, except at the equator.

Continental position is also an important climatic control. Interior continental locations are more affected by variations in energy through the course of the year than coastal locations. This difference occurs largely because large water bodies are capable of storing large quantities of energy during warm periods and releasing that energy during colder periods. This imparts a moderating maritime effect on coastal locations, while interior locations experience a wider temperature range through the year as part of a continentality effect.

Atmospheric and oceanic circulation factors also play a role in the climate of particular locations. The mission of both the atmospheric and the oceanic circulation is to balance inequalities in input energy received from the Sun. Persistence of motion of these fluids leads directly to prevailing winds and ocean currents, thereby affecting climates. Wind is simply a transfer of atmospheric mass between two locations, and it is the atmospheric mechanism behind the circulations that affects climate. Wind speed and direction result from several interacting forces, including the PGF, the CE, CF, and friction. Surface winds are affected most strongly by friction, which changes the speed and trajectory of the initial wind. Upper-level winds are affected by friction only negligibly, so air motion in the free atmosphere is quite different from that near the surface. Pressure differences in the atmosphere lead to the development of cyclones and anticyclones.

Topography and local features also play significant roles in the development of the climate of a lo-cation. All controls on climate cause variations in energy across Earth's surface. The influence of these controls across space is evident on a map of isotherms.

▶ Key Terms

Absolute zero
Advection
Aerosol
Antarctic Circle
Anticyclone
Aphelion
Arctic Circle
Attenuation
Autumnal equinox
Axial tilt
Backing
Centrifugal force (CF)
Circle of illumination
Continentality
Convection
Coriolis effect (CE)
Cyclone
December solstice
Downdraft
Ekman spiral
Equation of state
Evaporation
Flux
Free atmosphere
Friction
Geostrophic balance
Geostrophic wind
Gulf Stream
Heat index
Hydrostatic equilibrium
Ideal gas law

Inertia
Inertial period
Insolation
Isobar
Isotherm
Joule
June solstice
Kelvin temperature scale
Kinetic energy
Lake effect snow
Latent energy
Latitude
Leeward
Longitude
March equinox
Maritime effect
Midlatitude (frontal) wave cyclone
Millibar (mb)
Momentum
National Weather Service
Navier-Stokes equations of motion
Newton (N)
Newton's laws of motion
North Atlantic Drift
Orographic effect
Parallelism
Pascal (Pa)
Path length
Perihelion
Plane of the ecliptic
Pressure

Pressure gradient force (PGF)
Prime meridian
Radiant energy
Refraction
Revolution
Rotation
Second law of
  thermodynamics
Sensible energy
September equinox
Solar declination
Solar noon
Specific heat
Summer solstice
Thermohaline circulation
Time zone
Tropic of Cancer
Tropic of Capricorn
Tropical cyclone
Troposphere
Updraft
Upwelling
Urban heat island
Veering
Vernal equinox
Wind
Windward
Winter solstice
Terms in italics have appeared
  in at least one previous
  chapter.

▶ Review Questions

1. Discuss the role of latitude as a climate control.
2. How does position on a continent affect the climate of a location?
3. Compare and contrast the maritime effect and continentality by examining the climate of various cities located along the same line of latitude.
4. Discuss the important characteristics of Earth's orbit.
5. What is the axial tilt of Earth and why is this important?
6. How does parallelism play a role in seasonality?
7. Compare and contrast Earth–Sun relationships during the equinoxes and solstices.
8. How do Earth–Sun relationships affect the spatial and temporal distribution of energy through the year for each hemisphere?
9. What roles do day length, attenuation, aerosols, and path length play in seasonality?

10. How does specific heat compare between water and land surfaces?

11. What is the surface wind equation and how do the forces involved interact to produce wind?

12. How are winds, and the forces involved, in the free atmosphere similar to/different from those at the surface?

13. Compare and contrast characteristics of cyclones and anticyclones.

14. What role do surface ocean currents play in climate?

15. What is the Ekman spiral and how does it develop?

16. How are thermohaline currents important to climate?

17. What is the role of topography and altitude on the climate of a particular location?

18. What local features may influence the climate of a location? How?

▶ Questions for Thought

1. Given what you have learned in this chapter, can a hurricane track from the northern to the southern hemisphere, or vice versa? Why or why not?

2. What forces are most important in producing a tornado?

http://physicalscience.jbpub.com/climatology

Connect to this book's Web site: http://physical science.jbpub.com/climatology. The site provides chapter outlines, further readings, and other tools to help you study for your class. You can also follow useful links for additional information on the topics covered in this chapter.

4 | Effects on the Climate System

Chapter at a Glance

Ocean Circulation
    Surface Currents
    Deep Ocean Thermohaline Circulations
El Niño–Southern Oscillation Events
    Walker Circulation
    Historical Observations of ENSO
    El Niño Characteristics
    La Niña Characteristics
    Global Effects
    Effects in the United States
    Relationship to Global Warming
Volcanic Activity and Climate
    General Effects
    Aerosol Indices
    Major Volcanic Eruptions
Deforestation and Desertification
Cryospheric Changes
    Ice on the Earth's Surface
    Feedbacks in the Cryosphere
Summary
Review

Chapter 3 examined the major controls on the climate system, including the effects of *latitude*, Earth–Sun relationships, *continentality*, atmospheric and oceanic circulation, elevation, and general local features. This chapter explains the role of other important components on the climate system. These contributors may be thought of as internal components because they affect the system already set in place by the aforementioned climate controls. The features discussed in this chapter work to influence and change the already-established climate system. Even though the oceans were mentioned in Chapter 3, they warrant further

scrutiny in this regard. Because over 70 percent of Earth's surface is covered by oceans, processes occurring in the oceans play an important role in climate and climatic variations.

■ Ocean Circulation

Surface Currents

Atmospheric Effects Oceanic surface currents are driven by overlying winds that on the broadest scale are dictated by the great global atmospheric **subtropical anticyclones**—large semipermanent high-pressure cells that exist in each ocean basin approximately centered on the 20 to 30° parallel of latitude. These subtropical anticyclones wax and wane in strength and position seasonally, but they are always present. The clockwise circulation of the anticyclones in the northern hemisphere and the counterclockwise circulation of the systems in the southern hemisphere initiate corresponding surface water motions in the oceans. Of course, the near-surface waters are also moved by the *Coriolis effect* inducing the *Ekman spiral*, but for a large section of the oceans, near-surface waters move in a general clockwise pattern in the northern hemisphere and counterclockwise in the southern hemisphere because of these subtropical anticyclones (**Figure 4.1**). These circular flows, caused by a coupling of the atmospheric and surface oceanic circulations, are termed **gyres**.

As we saw in Chapter 3, the large-scale motions of the oceanic surface ensure that cold currents occupy the eastern sides of the ocean basins and warm currents are found along the western edges of the basins. This spatial pattern is caused by mass *advection*—lateral movement—of warm waters traveling from relatively low latitudes along the

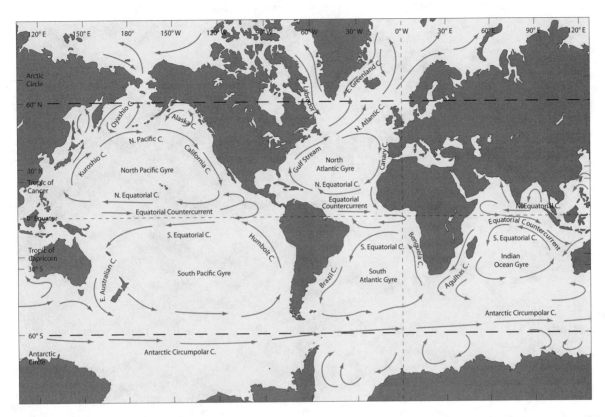

Figure 4.1 Surface currents.

western ocean edge, while cold water from the high latitudes is advected along the eastern edge, in both hemispheres. The effect is that low surface temperatures on the eastern ocean basins (with warmer air above the cold air adjacent to the cold water) tend to stabilize the atmosphere. A **stable atmosphere** is one in which vertical motions are discouraged because the colder, denser air exists below warmer, less dense air and, therefore, it tends to rise above the less dense air. Although this concept is explored in more detail in Chapter 5, we can say now that a stable atmosphere promotes net sinking motions in the atmosphere, which discourages cloud formation processes. For this reason some of the most enjoyable climates on Earth are located along the eastern ocean margins (and, therefore, the western coasts of continents) where pleasant temperatures, abundant sunshine, and little precipitation occur.

By contrast, the western ocean basins are dominated by warm waters that destabilize the overlying atmosphere. In this situation warm, moist air adjacent to the warm ocean currents becomes buoyant and rises. Such an **unstable atmosphere** is generally conducive to cloud formation and precipitation. Therefore, the western ocean basins (and the adjacent eastern coasts of continents) typically have abundant precipitation that is usually distributed fairly evenly throughout the year. During summers the subtropical and tropical midlatitude regions on the east coasts of continents become hot and humid because abundant water vapor usually exists from the *maritime effect.* In winter the effect is less pronounced, at least in the middle and subtropical latitudes, as westerly wind systems dominate the midlatitudes, promoting more continental temperatures along the eastern coasts.

Upwelling, Downwelling, and Mass Advection
Upwelling reinforces the effect of the cold currents along the eastern ocean basins. When offshore surface winds parallel a coastline for some distance, with the coastline to the left of the direction of flow in the northern hemisphere (and to the right of the flow in the southern hemisphere), the net water advection near the surface is directed away from the coastline by the Ekman spiral. Upwelling results from the Ekman spiral's deflection of surface water

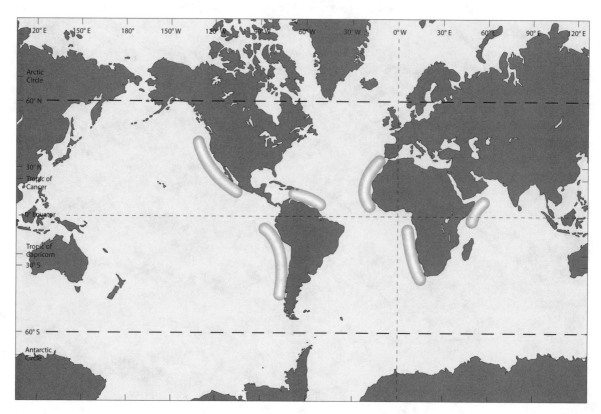

Figure 4.2 Locations where coastal upwelling is important.

to the right (in the northern hemisphere) with depth. Examples of locations around the world where upwelling is particularly effective because of the coastline shape and consistency of the winds are shown in **Figure 4.2**. Upwelling also can occur in areas where two currents diverge, allowing water from the depths to rise to replace the surface water that has been moved away laterally. Regardless of the cause of the upwelling, departing surface waters are replaced by cold deep water from beneath. In such situations an already cold surface current becomes even colder.

Perhaps the best example of upwelling is associated with the **Humboldt** (also known as the **Peru**) **Current** located off the western coast of South America (**Figure 4.3**). The South Pacific subtropical anticyclone is particularly well established and strong, because of the size of the South Pacific Ocean and the lack of interfering land masses. A very strong and persistent counterclockwise circulation results and forces very cold surface waters to move equatorward from the Antarctic. As this water parallels the South American coast, the Ekman spiral forces it westward. The upwelling results in one of the coldest ocean currents on Earth. The low

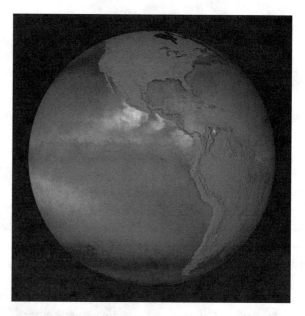

Figure 4.3 Cold water upwelling along the South American coast. (See color plate 4.3)

surface temperatures (even at tropical latitudes) stabilize the overlying atmosphere, causing cool, dry climates to dominate the western edge of the continent. The Atacama Desert of coastal Peru and Chile is the driest desert on Earth.

On the western sides of ocean basins, **downwelling** may occur—an oceanic process that is the reverse situation in which surface airflow pushes water against a coastline where it is forced to sink. In addition, in locations where two surface ocean currents converge, the pile-up of surface water can induce downwelling. Downwelling typically supports high ocean temperatures as sinking warm waters push the cold water to deeper levels. An example occurs along the equatorial western Pacific Ocean where easterly (east to west) winds push waters westward toward Indonesia. This water warms significantly along its westward journey across the tropics. On the western side of the equatorial Pacific, the water is forced to sink. The Ekman spiral does not affect the bulk of this water because the Coriolis effect does not occur along the equator. Without Coriolis deflection the current maintains its straight course. The **Maritime Continent**—the Indonesia–Philippines region—is, therefore, dominated by very high ocean surface temperatures that promote low-level atmospheric instability. Such situations support cloud formation and precipitation. Locations on the western sides of ocean currents (eastern sides of continents) in the subtropics are usually dominated by warm, wet climates.

In the North Atlantic Ocean, for example, the subtropical anticyclone forces a clockwise circulation that supports the cold water **Canary Current** off the coast of southern Europe and northern Africa, the **North Equatorial Current** near the equator, the warm *Gulf Stream* along the eastern coast of North America, and the *North Atlantic Drift* (current), which completes the northern gyre circulation. Again, the circulation regime is initiated by the atmospheric anticyclone and mimicked by the ocean in a corresponding gyre. Similar oceanic circulations occur in all other ocean basins.

Because of the deflection of the Ekman spiral, water tends to pile beneath the subtropical anticyclones in each ocean basin. As the gyre circulates, the Ekman spiral dictates that mass advection occurs 45 to 90° to the right (in the northern hemisphere) of the initial water motion. Equatorward-moving surface water, the Canary Current in this example, is forced toward the central ocean region. Water moving from east to west across the low latitudes in association with the North Equatorial Current is also forced by the Ekman spiral toward the middle of the basin. Sur-

face waters associated with the Gulf Stream and portions of the North Atlantic Drift, which moves essentially west to east, are also forced toward the ocean center. The result is a mound of water located roughly in the middle of the North Atlantic Ocean basin near the center of the atmospheric subtropical anticyclone. The mound supports sea levels exceeding those of surrounding locations.

At the same time, gravity acts on the mound, pulling it downward toward the lower sea level on the periphery. Thus, the water within this mound is constantly affected by offsetting forces. The Coriolis effect (manifested as the Ekman spiral) pushes water inward, whereas gravity pulls water outward (downward). As the water circulates clockwise (in the northern hemisphere) along this mound, it reaches an equilibrium between the two forces. An oceanic version of **geostrophic flow** occurs when these forces are in balance (**Figure 4.4**).

An exactly opposite condition occurs for waters flowing counterclockwise (in the northern hemisphere) in association with semipermanent low atmospheric pressure regions, normally centered around 60° latitude. An oceanic depression occurs at these locations and waters swirl around the depression in geostrophic balance.

The central axes of these mounds and depressions are not perfectly centered beneath the atmospheric highs and lows but instead are offset toward the west of the centers. This is because Coriolis deflection increases with latitude, affecting the amount of water transported via the Ekman spiral along the gyre. Poleward-moving currents have increasing amounts of water advected toward the center of the basin, while equatorward-moving currents have decreasing amounts of net transport along their trajectory. The result is a pile of water with its axis offset to the west of the central ocean basin.

The effect of the offset central axis causes poleward-moving waters (i.e., on the western sides

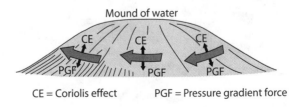

CE = Coriolis effect   PGF = Pressure gradient force

Figure 4.4 Geostrophic balance in the ocean.

of ocean basins in both hemispheres) to pile up between the continent and the mound axis. When a moving fluid converges, as in this case, a speed change is initiated because the fluid is forced through a smaller space over a constant time period. This type of speed change may be seen when placing a finger partially over the opening of a hose. Once the opening constricts, water, under pressure, is forced to increase in velocity. In the ocean the response results in **western-boundary intensification**, a very deep and swift western ocean basin current (**Figure 4.5**). The Gulf Stream is a perfect example of western-boundary intensification, because it is a very deep, fast-moving western ocean basin current. On the opposite side of the ocean basin, the cold water current (in this case, the Canary Current) is typically very shallow and slow moving and is spread across a larger surface area. These properties affect the climates of adjacent locations.

Another way to view western-boundary intensification is relative to the slope of the sea surface. Because the central axis of the water mound is offset to the west, the corresponding slope in sea surface is steeper in that location. This results in water moving along a steeper pressure gradient, which increases the speed of water motion. On the opposite side of the ocean basin, the colder water flows over a much gentler sea surface slope, resulting in a slower rate of flow (Figure 4.5).

Deep Ocean Thermohaline Circulations

Surface currents are not the only currents that influence climate. Deep ocean currents are driven by *thermohaline circulation*, as discussed in Chapter 3. **Thermohaline currents** flow in response to

temperature ("thermo") and/or salinity ("haline") characteristics through the deep ocean. Just as density differences (cold versus warm) in air drive atmospheric motions, deep ocean thermohaline differences establish net pressure gradients. The fluids of the deep ocean respond, because denser waters tend to sink beneath less dense waters. Because colder, saltier water is denser than warmer, fresher water, net downward motions occur where waters become colder and/or saltier. Warmer and/or fresher waters are more buoyant and rise. Any situation in which colder water underlies warmer water represents **stable stratification**, the oceanic equivalent to a stable atmosphere (**Figure 4.6**). Further *convection* (vertical motion) is suppressed unless another force or energy source initiates it.

Interestingly, deep ocean currents are actually driven by surface processes because seawater gains its thermohaline characteristics from surface sources. Intense solar heating in low-latitude regions warms the water surface sufficiently to decrease its density. The reduced solar input and heat loss (to the frigid atmosphere) in high-latitude regions favor a denser ocean surface because of its colder surface characteristics. Oceanic areas with

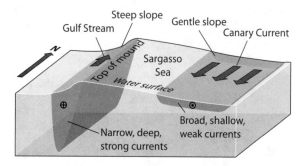

Geostrophic flow around the North Atlantic Ocean

Figure 4.5   Western-boundary intensification.

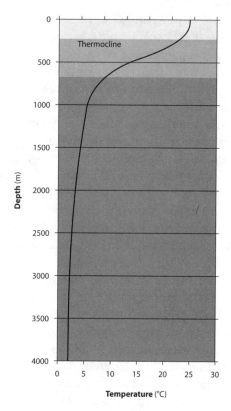

Figure 4.6   Deep-water stratification.

abundant precipitation have lower water densities because of the high input of fresh water. Likewise, oceanic areas near coastlines in humid climates typically have fresher water because of abundant input of stream discharge. By contrast, areas of the ocean that receive less precipitation than *evaporation* and/or have input only from a few small streams typically have dense, salty water. Precipitation characteristics of both the ocean and the surrounding land masses influence the density of ocean waters.

The freshest ocean waters on Earth are typically found near the equator, where high temperatures and abundant precipitation support low-density waters. High salinities usually occur near the centers of the subtropical anticyclones (30° latitude), where abundant sunshine evaporates large quantities of surface water, leaving behind salts. When these salty waters are advected to higher latitudes, the high salt content combines with lower temperatures to create the densest ocean waters. These waters sink to the deep layers of the ocean

where they are forced to move along thermohaline gradients.

Source areas are locations near the surface that provide water to the ocean depths. These are located in relatively high-latitude regions of the Atlantic and Pacific Oceans (**Figure 4.7**). In the North Atlantic, deep water originates at the surface near Iceland, where it sinks and moves equatorward. The South Atlantic also acts as a source area for deep water near Antarctica and in the area east of the tip of South America. In the North Pacific, deep water sinks near the tip of the Aleutian Islands and moves equatorward. The South Pacific deep-water source area is located near 65°S latitude.

There are many deep-water thermohaline currents, but the main one in the Pacific sees deep water moving from the North Pacific southward across the equator and then looping in the extreme South Pacific back toward the equator. In the Atlantic Ocean the main body of deep water also begins in the northern hemisphere near Iceland. It travels

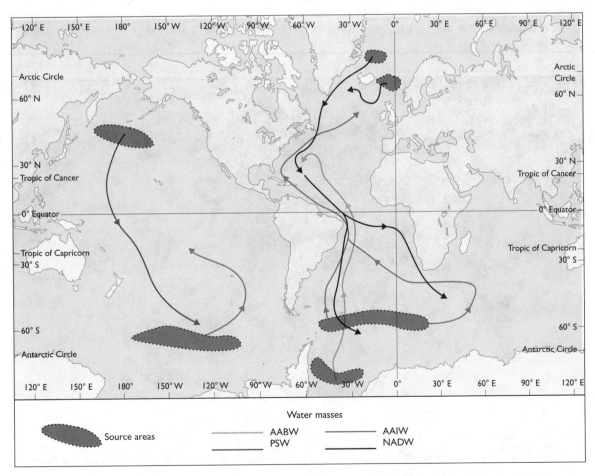

Figure 4.7 Deep ocean currents of the world. *Adapted from*: Gordon, A. L., *Lamont-Doherty Geological Observatory Report,* 1990–1991.

equatorward largely beneath the Gulf Stream, crosses the equator off the eastern coast of South America, and then continues toward Antarctica. Once in the high latitudes the deep water turns eastward before moving back toward the equator. The deep water travels northward near its opposing equatorward-flowing counterpart, before completing the loop near Iceland.

Deep waters travel at varying depths relative to each other. This makes it possible for different deep-water currents to occupy similar regions. The deepest waters in the Atlantic Ocean are those arriving from the **Antarctic Bottom Water**, the coldest ocean current, which forms near Antarctica (Figure 4.7). The **Antarctic Deep Water** moves over that layer as it flows northward and gently upward, eventually flowing beneath the current known as the **North Atlantic Deep Water**. The progression continues upward to the surface.

In the Pacific, two currents—the **Common Water**, initiated again near Antarctica, and the **Pacific Subarctic Water**—occupy the lowest layers of the ocean. These currents are overridden by **Antarctic Intermediate Water** and the **North Pacific**

Intermediate Water. Surface waters override those layers. Thermohaline currents in the Indian Ocean are dominated by Common Water arriving from the region of Antarctica and Antarctic Intermediate Water.

Coexisting with these individual thermohaline currents is a massive conveyor current that connects all the ocean basins on Earth except the Arctic. Much of what is known concerning this deep-water thermohaline conveyor comes from the work of Henry Stommel. In 1958 Stommel devised the simplified model of deep ocean circulation, basing water motions on similar forces that drive western-boundary intensification. The **Stommel model** indicates that net sinking motions are initiated in the North Atlantic near Iceland (**Figure 4.8**). This water sinks to the floor of the North Atlantic as it moves equatorward. After crossing the equator the conveyor moves toward Antarctica, where Antarctic Bottom Water is added. Then, it turns eastward to move between Africa and Antarctica. The deep water continues eastward but forks into two branches, one progressing equatorward through the Indian Ocean and the other moving eastward

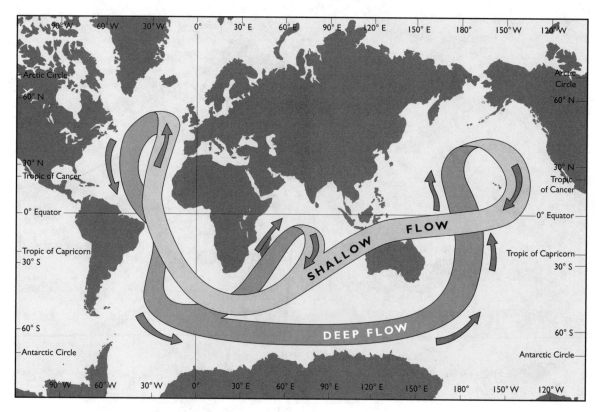

Figure 4.8 Stommel deep water conveyor model.

into the Pacific. The Indian Ocean branch rises toward the surface near the equator south of India. This water then turns southward and moves as a warmer shallow layer back toward the Atlantic. The southern deep-water branch progresses into the South Pacific before moving northward east of Australia. This water stream crosses the equator and rises in the central North Pacific. It then loops back and moves near the Maritime Continent as shallow water. This current ultimately joins the shallow water stream in the Indian Ocean. Both branches then move back around the tip of Africa and into the Atlantic, where the conveyor moves northward, ultimately sinking again near Iceland.

This deep-water conveyor is capable of transporting huge amounts of oxygen and energy from one side of the planet to the other. The oxygen provides a mechanism by which oceanic life is supported. The energy is capable of influencing climate for centuries, as signatures of past climatic variations are present in the slow-moving water itself. Reintroduction of this stored energy into the present or future atmosphere can alter climate, particularly where the conveyor approaches the surface and can transfer the energy back into the atmosphere easily. Likewise, present-day atmospheric changes, most notably **global warming**, can provide additional storage of energy in the conveyor belt and elsewhere in the oceans. This additional energy is likely to cause future changes both to the *hydrosphere* in the form of surface currents, thermohaline circulation, and conveyor belt and to other "spheres" in the climate system. The impacts are unknown.

■ El Niño-Southern Oscillation Events

The **El Niño-Southern Oscillation (ENSO) event** is perhaps one of the most misunderstood of all atmospheric processes. ENSO events affect global weather in a very profound way. Perhaps only Earth–Sun relationships exert more of an influence on the global climate system than ENSO.

The term **El Niño**, translated from Spanish as "the boy," originally referred to the annual warming of the equatorial ocean current off the western coast of South America during the Christmas season (**Figure 4.9**). This sea surface warming is associated with the change of seasons—the beginning

of winter in the northern hemisphere and summer in the southern hemisphere. For most of the year in the tropics strong winds, known as **trade winds,** push the warm surface waters of the tropical Pacific westward, allowing upwelling of cold, deep ocean waters in the eastern tropical Pacific. As the seasonal atmospheric circulation pattern becomes established during the transition of autumn and spring, trade winds along the equator weaken, allowing the warm water in the western equatorial Pacific to migrate back toward the east. This annual migration warms the Pacific's eastern equatorial rim, typically around December.

Over the years the meaning of the term El Niño has changed from its original reference to the normal annual event described above. It is now used to refer only to the unusually extreme increases in sea surface temperatures (SSTs) occurring for several months approximately every 3 to 7 years in the central and eastern equatorial Pacific Ocean. By contrast, **La Niña**—the opposite of El Niño—refers to a strengthened "normal" situation, which typically reinforces cold water conditions in the eastern equatorial Pacific and warm water conditions in the western tropical Pacific near the Maritime Continent. In recent years the neutral situation—one in which neither El Niño nor La Niña conditions occur—has been termed **La Nada**, or "the nothing."

Figure 4.9 Satellite image of water temperatures in the Pacific Ocean during an El Niño event. White represents unusually warm water. (See color plate 4.9)

The **Southern Oscillation**, recognized by the climatologist Sir Gilbert Walker in the 1930s, refers to a seesaw effect of surface atmospheric pressure between the eastern and western equatorial Pacific Ocean in which higher-than-normal pressure in one of these two regions is coincident with lower-than-normal pressure in the other. Climatologist Jacob Bjerknes realized in 1969 that these pressure changes resulted from SST variations over several years in the equatorial Pacific, establishing the link between El Niño and La Niña events and the Southern Oscillation.

Today, most climatologists refer to the entire phenomenon as El Niño-Southern Oscillation, or ENSO. The variations in atmospheric pressure associated with extremes in the Southern Oscillation and the changes in the oceans during extreme phases of ENSO (El Niño or La Niña events) cause unusual global atmospheric circulation features and impacts. Because it is associated with a reversal of the typical Pacific pressure patterns rather than an intensification of them, the El Niño phase usually has wider and stronger impacts than the La Niña phase.

Walker Circulation

In 1969 Bjerknes coined the term **Walker circulation** to describe the connection between the atmospheric pressure centers in the equatorial Pacific associated with the Southern Oscillation, the SSTs, and the tropical trade winds that blow from east to west near the surface across the Pacific Ocean (and other regions of the Earth within the tropics). The Walker circulation exists to balance the normally observed pressure gradients over the tropical Pacific Ocean. The Walker circulation represents "normal" or "La Nada" atmospheric and oceanic conditions in the equatorial Pacific Ocean, characterized by high surface atmospheric pressure in the east and relatively low surface pressure in the west (**Figure 4.10**). A pressure gradient forms in an east-to-west direction at the surface with atmospheric mass (winds) working to "fill the low." Thus, surface winds normally blow strongly from the east toward the west in this region. Known as the **northeast trade winds** (or northeast trades) in the northern hemisphere and the **southeast trade winds** in the southern hemisphere, these winds comprise a major part of the global atmospheric circulation that is discussed more fully in Chapter 7.

The frictional dragging of the sea surface by these winds causes the North and South Equatorial Currents to occur along the equator, approximating the atmospheric circulation pattern, flowing from east to west in both hemispheres. The currents push the surface waters warmed by the Sun

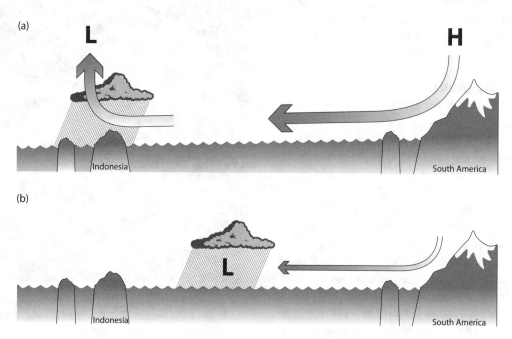

Figure 4.10 "Normal" (or La Nada) Walker circulation (a) and El Niño–related atmospheric circulation (b).

westward. This results in an accumulation of warm water in the western tropical Pacific Ocean near Australia and the Maritime Continent. Average sea level may be as much as 40 cm (16 in) higher and water temperatures 4 to 8 C° (7 to 14 F°) higher in the western tropical Pacific than at the same latitude in the eastern Pacific near South America (**Figure 4.11**). The reduced SSTs in the east are caused both by the cold surface current of the eastern South Pacific and by upwelling of colder, deep water replacing the warmer surface waters that were displaced to the west by the equatorial currents.

These "normal" or La Nada ocean temperatures control the Walker circulation along the equator. Lower SSTs in the east chill the overlying air, which in turn increases the air density, reducing its buoyancy. The sinking air causes high atmospheric pressure and usually restricts precipitation because cool air is less likely than warm air to rise vertically so that its moisture can condense and form vertical clouds. The eastern equatorial Pacific is normally dry as a result, and the extension of this dry area on the land surface is the dry Atacama Desert. Warm waters in the western tropical Pacific cause the density of overlying air to decrease. The rising air leads to lower atmospheric pressure and buoyancy that produces frequently cloudy and wet conditions over the region.

A simple index was devised to describe the atmospheric pressure variations in the tropical Pacific Ocean. **The Southern Oscillation Index (SOI)** is derived from sea level pressure differences between the eastern and western Pacific. The SOI is found by subtracting the air pressure at sea level in Darwin, Australia, from that in Tahiti. This value is often standardized statistically so that a value of zero represents La Nada, values exceeding +1 may be considered to coincide with La Niña in the ocean, and values below −1 may be considered to represent El Niño conditions. Another means of identifying El Niño and La Niña events involves monitoring SSTs in various regions of the equatorial Pacific Ocean. A temperature departure of ±1 C° away from normal conditions for 3 consecutive months constitutes a major event.

Historical Observations of ENSO

Direct information about ENSO events first came from accounts of Spanish explorers in South America during the 1500s. Indirect data sources such as tree rings, flood frequency, sediment cores, and coral reef growth suggest that anomalous weather events associated with ENSO events have been occurring for at least many thousands of years. The earliest written records of impacts believed to be related to extreme El Niño phases extend from the Chimu Dynasty (AD 1100) in the Moche Valley of Peru and indicate periodic extreme flooding, now known as **Chimu Floods**, extending as far back as 2500 BP (before present).

Most accounts are not direct observations of atmospheric or oceanic conditions but suggest indirect effects of ENSO on climate. Sources include ship logs in addition to writings by clergy members who reported unusual natural phenomena in great detail. These records became more detailed between 1600 and the mid-1800s, when South American events were chronicled in European literature. Accounts from historians, explorers, geographers, pirates, and engineers became more

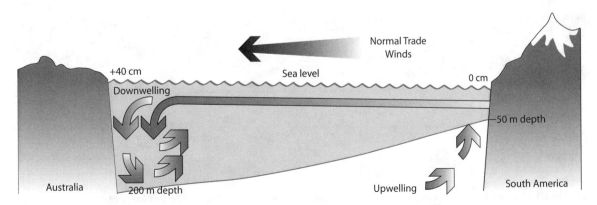

Figure 4.11   Warm water piling in the western equatorial Pacific during "normal" or La Nada conditions.

detailed toward the end of the 1800s. Identification of El Niño events took many forms. Some of these are listed in **Table 4.1**.

El Niño Characteristics

Each extreme of the ENSO cycle (El Niño and La Niña) occurs about every 3 to 7 years. An El Niño event in the ocean coincides with a reversal of the

Table 4.1   Indirect Evidence for El Niño Events in Historical Sources

- Variations in travel times from sailing vessels
- Ship logs noting unusual weather and sea conditions
- Presence of "aguaje" or red tide, a bloom of toxic marine plankton
- Abnormally warm waters along the South American coast
- Severe and unusual weather events, such as heavy rains and flooding
- Property damage caused by floods
- Travel obstructions from washed-out roads or mudslides
- Agricultural destruction
- Increases in sea levels along the South American coast
- Mass mortality of marine sea life caused by a decrease in the upwelling of nutrients
- Death and/or departure of birds
- Reductions in productivity in coastal fisheries

normal Walker circulation. Trade wind flow weakens or may even reverse along the equator, allowing the pool of warm water piled up in the western Pacific to flow back toward the east. Because La Nada and (especially) La Niña conditions pile warm water in the west, the ocean is simply regaining a somewhat uniform sea surface level when the trade wind flow weakens.

As the warm water migrates eastward, the overlying atmospheric low-pressure center follows the warm water migration eastward as well. By the time the warm water pool reaches the eastern boundary of the tropical Pacific Ocean, reduced atmospheric pressures and increased precipitation are well established. At the same time, colder than normal water conditions become established in the western tropical Pacific. Higher pressures build over that area as a result, and precipitation is far below normal in the region around the Maritime Continent. El Niño, then, simply coincides with a reversal of the "normal" or "neutral" equatorial Pacific air/sea conditions set up by the Walker circulation (**Figure 4.12**).

Movement of the warm water pool from west to east during El Niño, usually taking about 4 months, appears in the form of an **equatorial Kelvin wave**—a pool of warm water moving eastward while surface waves propagate westward. The surface waves are initiated by overlying winds that continue to blow from east to west as a part of the normal trade wind flow. However, during El Niño events the trade winds are weaker than normal, which allows the warm water pool to move eastward as an equatorial Kelvin wave.

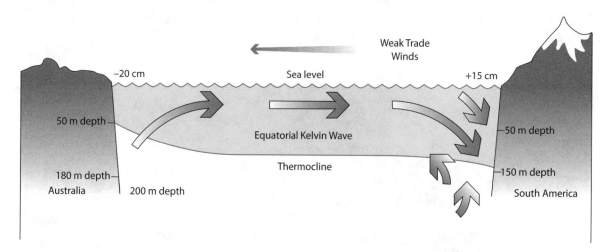

Figure 4.12   Ocean response to El Niño. Sea level values represent change from La Nada conditions.

When the warm water pool reaches the eastern boundary, it splits into three primary components. The main component is the **equatorial Rossby wave**, which sloshes back westward along the equator, typically reaching the starting point within 6 to 8 months, ending the El Niño event. Movement of the return flow in this wave is slower than the original Kelvin wave. These oceanic Rossby waves differ from Kelvin waves in that the bulk of water motion is in the same direction as the surface waves.

The remainder of the original eastward-moving equatorial Kelvin wave splits into **coastal Kelvin waves**—smaller warm water pools that migrate north and south along the North and South American coasts, displacing cold currents off the west coasts of both continents. These waves are responsible for many regional climate abnormalities.

The warm water pool in the eastern equatorial Pacific causes a downward movement of the **thermocline**—the boundary between warm surface waters and colder deep waters. Oceanic upwelling still occurs in the eastern ocean during El

Niño events, but because of the deeper thermocline the cold water involved in upwelling is confined to the deeper layers. Because of these abnormally warm waters near the American coast, El Niño events are sometimes referred to as warm-ENSO events. In its entirety an El Niño event usually lasts between 10 and 14 months.

Because so much energy is transported across such a large distance regardless of the ENSO phase, disruptions to the La Nada situation have worldwide atmospheric consequences. The most direct effects occur in the tropics during the El Niño phase (**Figure 4.13**). The reversal of the usual atmospheric pressure and SST patterns during El Niño events causes atmospheric and oceanic circulation systems on the planetary scale to readjust. These anomalous circulation patterns cause seemingly chaotic weather conditions worldwide. Normally dry regions, such as western North and South America and western Australia, become much wetter than normal, while characteristically wet regions such as northeastern Brazil, eastern Australia, and the Maritime Continent

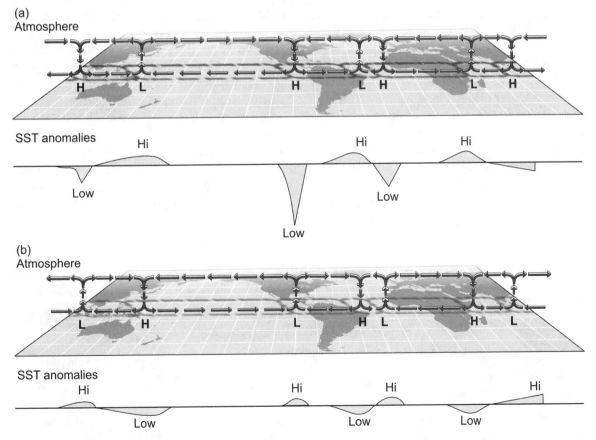

Figure 4.13   Equatorial air/ocean circulation anomalies during (a) La Nada and (b) El Niño periods.

become excessively dry. Such disruption of normal atmospheric and oceanic conditions leads to some devastating ecological and economic problems.

La Niña Characteristics

La Niña may be considered an extreme version of the "normal" or La Nada characteristics in the sense that the Walker circulation does not reverse. During this phase the trade wind flow along the equatorial Pacific Ocean becomes even stronger than normal, which increases warm water accumulation in the western equatorial Pacific and cold water upwelling in the east. This causes a deeper thermocline in the western tropical Pacific and a shallower than normal thermocline in the east (**Figure 4.14**). Sea levels respond accordingly, as the west records higher than normal sea levels and temperatures, while in the east lower than normal sea levels and temperatures occur. The enhanced cold water upwelling near the Americas has given La Niña another name in recent years—"cold-ENSO events."

Global Effects

As suggested above, the consequences of ENSO events do not stop with the equatorial Pacific Ocean. A closer inspection of Figure 4.13 reveals that the effects of warm-ENSO events carry through the downstream circulation patterns of the entire tropics. Because lower pressure builds over the western portion of South America during negative phases of the Southern Oscillation, the circulation regime of eastern South America is altered.

Normally, eastern South America is dominated by low atmospheric pressure associated with the abundant heat energy and humidity from the Amazon rain forest. During the El Niño phase the western South American low disrupts the usual circulation over the eastern portion of the continent. As a result, higher than normal air pressure and subsiding airflow is initiated over tropical eastern South America, bringing drought to much of the rain forest region.

Just as the circulation is altered in South America during warm-ENSO events, the circulation regimes of other tropical locations are affected. The relatively high-pressure regime that normally prevails over southwestern Africa flips to lower atmospheric pressure and the region becomes anomalously wet. This in turn reverses the normally low pressures that occur near southeastern Africa, resulting in higher air pressures and drought near Madagascar. This induces a low-pressure region over the equatorial Indian Ocean, which is normally dominated by higher pressures, and the region becomes wetter than normal. Finally, the circulation is linked to the western side of the Walker circulation over eastern Indonesia and northeastern Australia, which experiences El Niño–induced above-normal air pressures and the drought conditions discussed previously. The effect is that the atmosphere seesaws to an opposite regime from normal.

Energy transfers during warm-ENSO events also alter temperate climates, as indicated in **Table 4.2**. Such interactions originate from increased heat energy and moisture transport from the equatorial Pacific into the midlatitudes. Much of this energy transport is accomplished by the

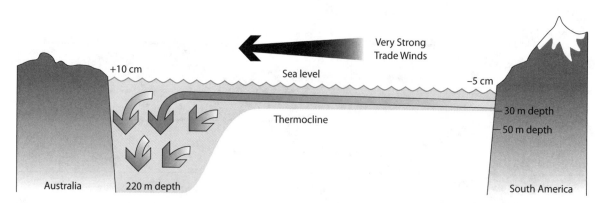

Figure 4.14 Ocean response to La Niña. Sea level values represent the change from La Nada conditions.

Table 4.2   Persistent Global Effects of El Niño

Condition	Areas Affected
Drier than normal	Maritime Continent; southeastern Africa and Madagascar; east central Africa; eastern South America (Brazil)
Wetter than normal	Central equatorial Pacific Ocean; eastern equatorial Pacific Ocean and western South America (Ecuador, Peru); southeastern South America (southern Brazil, Argentina); southeastern United States
Warmer than normal	Japan, eastern Asia (China/Manchuria); northwestern North America (southern Alaska through U.S. Pacific Northwest into central North America); eastern Canada (Labrador, Nova Scotia)

poleward-moving coastal Kelvin waves. These warm water pools displace cold currents along the west coasts of North and South America, overriding the colder, denser currents normally present in those locations. This position allows the abundant stored oceanic heat energy to be transferred easily into the atmosphere, particularly under the conditions of unusually low atmospheric pressure, initiating rising motions in the atmosphere. Such conditions are capable of providing energy for storms that then migrate across the normally dry western continental locations.

As a general rule, La Niña situations strengthen normal atmospheric circulation patterns across the tropics. A normally dry region becomes exceedingly dry; a normally wet region becomes exceedingly wet. Such conditions may have as many negative ecological and economic consequences as an El Niño event in some locations.

Extreme phases of the Southern Oscillation stress ecosystems in several ways. Large-scale animal migrations and die-outs occur in the affected regions. These factors in turn stress humans as food sources are affected and landscapes are degraded. For instance, western South America has seen widespread famine as a result of fish migrations away from the coast during warm-ENSO events. As fish move away from the coast in search of prey, birds that feed on the fish are affected. In addition, human populations are affected directly, because seafood serves as a prime source of nutrition in these coastal cultures affected by El Niño. At the same time, the Atacama region may experience heavy rains that cause widespread floods in a landscape barren of vegetation and without natural stream drainage capable of handling the sudden downbursts. Mudslides become common on hill slopes and occasionally destroy entire villages. Roads become blocked and bridges collapse. In short, widespread ruin may occur during particularly strong ENSO events.

Effects in the United States

Relationships between El Niño and the U.S. climate have been fairly well documented. Because the **jet streams**—fast currents of air in the upper troposphere—are most active during the cool season (November to March), the region sees most ENSO-related changes during that time. In particular, southern and central U.S. precipitation and southwestern U.S. temperatures appear strongly tied to warm-ENSO events. Increased energy and moisture are transported from the tropics to North America as the equatorial Kelvin wave and the smaller coastal Kelvin wave warm pools displace cold currents along North and South America. Thus, increased precipitation and storm activity are also typical effects of El Niño in much of the southern, central, and southwestern United States. The small, northward-migrating coastal Kelvin pool displaces the usual cold **California Current** off the U.S. Pacific coast, destabilizing the overlying atmosphere and causing an adjustment in the North American jet streams.

As expressed more completely in Chapter 7, the two major rivers of air in the high troposphere—the **polar front jet stream** (over midlatitude locations) and the **subtropical jet stream**—transport energy and moisture in a generally west-to-east direction. During warm-ENSO events, the **amplitude**—the north–south and south–north component of motion—of the polar front jet stream tends to increase across the United States. As we see later, jet amplitude dictates where storms form, their intensity, and direction of migration.

During the El Niño phase the polar front jet is altered from normal as a **trough**—an equatorward dip in the flow—overrides the eastern North Pacific Ocean. Midlatitude storm systems form just to the east of trough regions. The trough is accompanied by a **ridge**—a poleward shift in the jet flow—downstream (toward the east), as shown in **Figure 4.15**.

During warm-ENSO events the polar front jet may split into two distinct branches—one north, one south—over the eastern North Pacific Ocean. This situation greatly alters normal jet flow across North America and affects the climate of many associated regions. The northern branch of the polar front jet tends to remain farther to the north than the usual polar front jet and typically prevents the coldest Canadian air from penetrating southward into the United States. Thus warm-ENSO winters are usually warmer than normal in the northern United States and southern Canada. The southern branch of the polar front jet during warm-ENSO events flows across the northern Gulf of Mexico and tends to steer midlatitude storm systems into the Gulf Coast states, after allowing them to acquire moisture over the Gulf of Mexico. The subtropical jet still exists in its normal form, south of the southern branch of the polar front jet, even when the polar front jet bifurcates.

Because El Niño events can also trigger increases in jet amplitude across North America, a trough in the northern branch of the polar jet can cause colder air to infiltrate the eastern United States as far as the Gulf Coast. Northern states are not impacted with extreme events under these conditions because this cold air would have been present whether or not the dip in the jet stream existed. This trough also causes an increase in precipitation for the southern and central United States.

The northeastern United States shows a weaker relationship between El Niño and temperature and precipitation patterns. One explanation for the lack of relationship is that the polar front jet exits the continent over the Northeast regardless of jet flow amplitude. It should be noted, however, that any particular warm-ENSO event can cause drastic changes in northeastern U.S. weather. Each ENSO event changes the jet flow pattern somewhat differently, and strong El Niño events occasionally change the jet pattern substantially. The response to El Niño in the northeastern United States is usually slightly higher temperatures along with slightly drier conditions caused by a net reduction in snowfall.

Recent research indicates that cold-ENSO events have as much influence on precipitation in the southern United States as warm-ENSO events. Recent La Niña years have been accompanied by pronounced drought throughout the southern and south-central United States, even in summer, if the SOI remains positive through the summer. Additionally, cold-ENSO periods are associated with temperature increases throughout the Southwest as the polar front jet stream pushes northward over the western United States. Slight temperature increases also occur in the north-central area of the United States (Dakotas to Wisconsin). La Niña periods also see heightened Atlantic hurricane activity if the event is in force into late summer and autumn, as was evidenced by the very active 2004 and 2005 seasons. Incipient El Niño conditions in late summer and autumn 2006 and early summer 2009 were associated with a much quieter Atlantic hurricane season.

Relationship to Global Warming

Although much research has been done on ENSO events, much work is still needed. Many mechanisms thought to trigger ENSO events must be explained further. One such mechanism that has gained recent publicity is global warming. It is widely assumed that global warming is caused by human emission of pollutants known as *greenhouse gases*. The recent speculation that global warming triggers extreme ENSO events was gener-

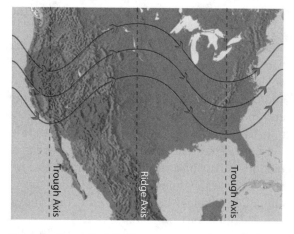

Figure 4.15 A typical ridge–trough configuration in the upper troposphere.

ated by a higher incidence of intense ENSO events during recent years of high global temperatures. Since 1970 warm-ENSO events have occurred with a periodicity of about 2.8 years. This significantly exceeds the frequency of the long-term average of 4.7 years. In time, relationships between a warmer atmosphere and El Niño events may prove valid; however, the reverse may be equally true: that a higher incidence of intense warm-ENSO events may be forcing higher global temperatures. Such a realization may be substantiated given the slow response of oceans to atmospheric forcing in addition to the long-term heat storage capability of oceans.

Recent research supports a more plausible explanation for the increase in extreme ENSO events since 1970. The **Pacific Decadal Oscillation (PDO)**, an oceanic phenomenon described in more detail in Chapter 13, is directly tied to ENSO strength and frequency. The PDO is similar to the Southern Oscillation in that warm and cold phases occur over time. As the name implies, variability associated with the PDO occurs much more slowly than that associated with the Southern Oscillation. A PDO regime may last 20 to 30 years, during which one phase of the oscillation tends to dominate. By contrast, the Southern Oscillation fluctuates over periods of 3 to 7 years. During times when the Southern Oscillation and the PDO are in phase (i.e., both in a "warm phase" or in a "cold phase"), the impacts of the Southern Oscillation tend to be magnified. When the phases offset the Southern Oscillation effects tend to be dampened.

A PDO warm phase occurred from the early 1970s until the late 1990s. This situation supported frequent and strong El Niño events. Beginning in the winter of 1997 and 1998 the PDO shifted to a cold phase. This shift is believed to support and magnify the effects of La Niña over the next few decades.

Clearly, continued research on the causes and effects of extreme ENSO events becomes more important with the increasing demands made by humans upon the natural environment. Given the abnormalities in the temperature and precipitation regimes during El Niño and La Niña events, heightened understanding of these events is important for planning purposes in a wide array of environmental and societal applications.

■ Volcanic Activity and Climate

General Effects

As Earth's crustal plates move very slowly (at about the same rate at which your fingernails grow), volcanic activity occurs in certain locations. This volcanism contributes to both short- and long-term climatic fluctuations. In Chapter 2 we saw that the primordial atmospheric composition resulted from the *outgassing* of molten rock as the Earth cooled and the molten material hardened. This led to an atmosphere rich in N_2 and CO_2. Water vapor was also initially introduced to the atmosphere through outgassing processes. The atmosphere shed excessive water vapor in the form of precipitation, which eventually formed the oceans of today. The composition of the atmosphere is, therefore, linked to volcanic activity.

Evidence indicates that the early landscape of Earth was marked by rather frequent volcanism as the volatile Earth cooled from the surface inward. Over time, volcanic activity decreased. Still, volcanism is an important contributor to the Earth-ocean–atmosphere system. We typically do not hear about active volcanism unless a particular volcanic event directly threatens humans or the event is of such magnitude as to invite attention. Many cases of active volcanism are associated with "cracks" in the oceanic crust that are widening—midoceanic ridges—deep beneath the ocean surface, whereas others are associated with rather remote locations such as the Aleutian Island chain of Alaska. Others, such as Kilauea, Hawaii, are tourist attractions.

Volcanic activity today contributes 1.2 to 2.1×10^{11} kg (130 to 230 million tons) of CO_2 to the atmosphere each year. Volcanoes also add between 3.3 and 6.6×10^{12} kg (3.6 and 7.3 billion tons) of **sulfur dioxide (SO_2)** to the atmosphere annually. Significant amounts of water vapor are also ejected into the atmosphere. Of the gases emitted by volcanoes, the one that produces the most significant climatic signal is SO_2. The amount of CO_2 emitted by volcanic activity is low compared with the amount already present in the atmosphere. The additional amount is quickly absorbed into biomass and/or the oceans. Water, of course, exits the atmosphere very quickly as precipitation. But SO_2 presents a different problem.

SO$_2$ and atmospheric *aerosols* of volcanic origin can trigger either net surface cooling or warming, depending on individual eruption characteristics. SO$_2$ typically combines in the atmosphere with water, dust, and sunlight to produce **vog**, or volcanic smog. Vog can have both local- and planetary-scale impacts. At the local scale vog may form near the surface, where it may hinder loss of longwave radiation back to space, leading to warming. On the island of Hawaii, vog is persistent because of the continuous effusive (gentle) eruption of Kilauea since 1986. The volcano contributes about 1.8 million kg (2000 tons) of SO$_2$ to the local atmosphere daily.

Vog becomes a planetary-scale problem when a major explosive eruption ejects sulfur compounds into the *stratosphere*. Once there, the sulfur compounds can linger up to 4 years, in part because the stratosphere is higher than most precipitation clouds, which would rinse aerosols and SO$_2$ from the atmosphere. SO$_2$ combines with the limited amounts of water found in the stratosphere, dust particles, and sunlight to form vog. In that situation the vog is capable of producing a haze that has a rather high **albedo**—the percentage of incoming energy from the Sun that is reflected off an object. The result is that the haze directly reduces surface air temperatures.

Long-term climatic changes are not initiated by single volcanic events. Instead, volcanoes lead to significant long-term climatic changes only during prolonged periods of above-normal activity. Persistent volcanism is thought to help maintain Earth's rather steady climate state. As we saw in Chapter 2, Earth temperatures have varied by less than 15 C° (27 F°) over its 4.6 billion year history. Volcanic activity is partly responsible for this relatively delicate energy balance that maintains this range of temperatures.

Individual volcanoes are, however, capable of altering hemispheric and global climates over relatively short time periods. When volcanoes erupt, gases and solid aerosols (**particulates**) are ejected into the atmosphere. Most of the aerosols fall back to the surface over short periods of time. Particles that are sand sized or larger fall back to the surface within minutes of an eruption. Smaller aerosols are capable of being suspended in the atmosphere for much longer periods. This is especially true of dust-sized and smaller particles, which may reach the stratosphere if the volcano is high and the eruption is especially explosive. Large eruptions have an average recurrence interval of about 30 years over recorded history (**Table 4.3**).

Aerosol Indices

Two indices have been designed to estimate the amount of aerosols, especially those of volcanic origin, in the atmosphere. The **Dust Veil Index** is based on the amount of material dispersed into the atmosphere. It uses surface air temperatures and the amount of *insolation* reaching the surface, among other variables, to estimate the total amount of particulates in the atmosphere. This index is most useful in the midlatitudes because it has been calibrated in predominantly middle-latitude locations. Its major weakness is that by using temperature to estimate the Dust Veil Index it is impossible to assess the impact of volcanic activity on temperature accurately.

The **Volcanic Explosivity Index** is based only on volcanic criteria. The data used are derived from the magnitude, intensity, dispersion, and destructiveness of individual volcanic events. A scale between 1 and 8 is used for each event, with 8 being the strongest volcanic event. An event with a Volcanic Explosivity Index exceeding 4 is assumed

Table 4.3 Major Volcanic Eruptions of the Past 200 Years

Volcano	Year	Average Resulting Global Temperature Decline (C°)
Tambora	1815	0.4–0.7
Krakatau, Indonesia	1883	0.3
Santa Maria, Guatemala	1902	0.4
Katmai, Alaska	1912	0.2
Agung, Indonesia	1963	0.3
El Chichón, Mexico	1982	0.5
Mount Pinatubo, Philippines	1991	0.5

to produce emissions into the stratosphere, but atmospheric composition is not taken into account in the derivation of the Volcanic Explosivity Index.

Major Volcanic Eruptions

With the possible exception of Eyjafjallajokull in Iceland, which began erupting in Spring 2010, the most noteworthy example of a recent volcanic eruption that led to surface cooling was **Mount Pinatubo**, which erupted in the Philippines on June 15, 1991. Pinatubo's eruption blasted over 1.8×10^{10} kg (20 million tons) of SO_2 and ash into the atmosphere. The vertical column of ejected material was measured at 19 km (12 mi) high during the eruption. The resulting stratospheric SO_2 plume spread rather evenly across the globe over time, leading to an estimated global surface temperature decline of 0.5 C° (0.9 F°) for 2 years after the eruption.

The 1815 eruption of **Tambora** in Sumbawa, Indonesia is regarded as the largest in modern history, as it ejected an estimated 50 km³ (12 mi³) of magma and an astonishing 1.8×10^{11} kg (200 million tons) of SO_2 into the atmosphere. The eruption led to the widely regarded "year without a summer" in 1816. Snow fell in July in Boston, Massachusetts. Global surface air temperatures are believed to have decreased by 0.4 to 0.7 C° (0.7 to 1.3 F°) during the year after the eruption.

The eruption of **Krakatau** in 1883, also in Indonesia, is thought to have exceeded Tambora in explosiveness. The eruption is believed to have killed up to 40,000 people, and it ejected ash and dust as high as 12 km (7 mi) above the surface as most of Rakata Island disappeared. In total, 20 km³ (5 mi³) of material were ejected into the atmosphere, making Krakatau about 20 times as destructive as the Mount Saint Helens eruption. Although global temperatures apparently were not affected as significantly as by Tambora, the eruption caused spectacular sunsets for over 70 percent of the globe over a time period of 3 years.

Another notable eruption was the 1783 **Laki Fissure** eruption in Iceland, which lasted 8 months and was responsible for devastating much of the human and animal population of Iceland. It is estimated that 14 km³ (3.4 mi³) of basalt was ejected. In addition, over 9.1×10^{10} kg (100 million tons) of SO_2 were emitted, leading to an estimated 1 C° (1.8 F°) temperature drop for the northern hemisphere.

The most catastrophic eruption known occurred on **Mount Toba** in Sumatra approximately 71,000 years ago. This eruption was so cataclysmic that it is thought to have accelerated the onset of the last **glacial advance**. Global climates changed so much that massive extinctions and population declines occurred in many plant, animal, and human populations across the globe. In fact, the event is believed to have caused an evolutionary **bottleneck** in humans—a significant drop in population that triggers rapid genetic divergence in surviving individuals. This process occurs as significant genetic differences proliferate at a rapid rate through small populations, as opposed to being washed out of much larger populations.

■ Deforestation and Desertification

Humans play a role in climatic variation through a number of ways. Most **anthropogenic** (human-induced) climatic effects relate to changes in atmospheric composition through the combustion of *fossil fuels* and the manufacture of certain gases and solids. Human land-use activities also can contribute to variation and changes in the climate system. Many of these involve the processes of deforestation and desertification.

Deforestation refers to the systematic and widespread clearing of forested regions. Although most associate deforestation with tropical locations, the most widespread deforestation actually occurred in Europe and North America a few hundred years ago. Entire forests were cleared for fuel and/or building materials during the early part of the Industrial Revolution, beginning in the late 1700s. Today, the most widespread deforestation occurs in tropical rain forests, where large tracts of land are deforested to support either individual subsistence farms or grazing areas.

Because soils in the rain forest regions are leached of their nutrients by the abundant rainfall, they are not well suited to agriculture. As a result, farmers must continually move from location to location because agriculture is usually successful for only a few years before the soil's nutrients become depleted. This type of migratory farming in which forests are continually cleared, often by burning, is called **slash and burn agriculture**. This method became widespread once it was discovered that burning vegetation adds nutrients to the upper soil layers. These nutrients are capable of supporting crops for only a few extra years, after which the soils become too leached to be productive.

If very small plots of land are deforested in this manner, no long-term damage is done, because the surrounding forest is quick to recapture the "slashed and burned" plot. The problem today involves very large deforested plots of land, especially when the activity leaves only small patches of rain forest surrounded by much larger deforested areas. These small plots are not sufficient to maintain their existence, and they become more susceptible to disease and other problems with such a high percentage of the trees near the fringe of the forest. The result is that the isolated plots die off. Human activities then move farther into the forest, taking more and more trees out of existence. In many areas the rain forests have been totally decimated in this manner. Deforestation of old growth forests within the past few centuries in many locations in Central America has left only a tiny percentage of rain forest undisturbed, and total annihilation looms in the near future. Currently, the worst deforestation of tropical rain forests is occurring in South America, especially northeastern Brazil, and in Indonesia. At present rates of clear cutting, it is estimated that all rain forests will be eliminated within the next 100 to 200 years, if not sooner.

This deforestation has dire ecological consequences. Tropical rain forests represent the most diverse ecosystems on Earth's land surface. Elimination of these regions will directly lead to massive extinctions of plants and animals, many of which may be useful for medicinal and economic purposes.

Large-scale deforestation causes numerous impacts on climate. Most of the precipitation water in rain forests is generated locally through the process of **transpiration**—the constant recycling of rainwater through uptake through tree roots and out through the leaves. Tropical rain forests have been referred to as the **rain machine** for that reason. Once the trees are removed, precipitation is more likely to run downslope and leave the area. Temperatures after deforestation increase abruptly, because of the loss of shade as well as increases in **sensible energy** with concurrent decreases in *latent energy* input.

An additional climatic problem is that deforestation involves the removal of one of the planet's primary carbon *sinks*. In 1 year a single acre of trees can uptake the amount of CO_2 released by driving 26,000 miles. The discontinuation of *photosynthesis* when the trees die decreases the rate at which CO_2 can be removed from the atmosphere. This has large-scale consequences because increased atmospheric CO_2 is a primary cause of global warming. Furthermore, the burning of the trees returns CO_2 that was sequestered in the biomass to the atmosphere, directly increasing atmospheric CO_2 concentrations. Finally, photosynthesis in the tropical rain forests produces a large share of the planet's atmospheric oxygen. Obviously, elimination of the planet's "lungs" would be disastrous.

Another human land-use effect that can modify large-scale climate changes is **desertification**—the expansion of deserts into semiarid regions, largely through the impact of human activities such as ranching and overuse of water. The domestication of grazing animals has become very important in many semiarid regions of the world, where grasslands are dominant. When these activities become too concentrated, they can quickly strip a region of vegetation, deplete and contaminate water, and compact the soil, leading to a rapid drying of the landscape.

Semiarid regions typically display wide variability in the precipitation regime. Wet periods may persist for many years or decades, only to be offset by years or decades of exceedingly dry conditions. Humans typically move into the regions during wet periods and establish grazing practices. But during the dry periods that inevitably follow, they are reluctant to leave. Their continued activities quickly compound the problem, leading to further expansion of the surrounding desert.

Perhaps the most prevalent example of desertification is the **Sahel**—a region of Africa that borders the southern rim of the Sahara Desert and has undergone widespread degradation over the recent past (**Figure 4.16**). Much of the problem began in the early twentieth century, when wet conditions brought people into the region. This settlement, combined with natural population growth and the demise of transient herders in favor of more permanent herding establishments, created a situation with far too many grazing animals in the re-

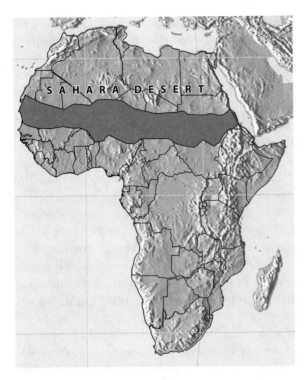

Figure 4.16   Sahel region of Africa.

gion. During the 1970s widespread drought in the region caused the animals to strip far too much vegetation in a short period of time before succumbing to the drought themselves. The concurrent water and food shortages led to tragic famine and death throughout the region. The consequent impact on the landscape was rapid desertification and southern expansion of the Sahara Desert. The region has yet to recover and remains one of the primary regions of food and water shortages on the planet.

Once the desert expands into a semiarid region, it is very difficult for vegetation to reclaim the region because the local water and energy balances are disrupted. The removal of vegetation changes the color and texture of the landscape. The generally dark and rough vegetated surface is replaced by lighter and smoother surfaces, which in turn support higher albedoes, reflecting away a higher percentage of insolation. But this effect is more than compensated by a decrease in water availability after the vegetation is removed. Like a deforested landscape, an unvegetated landscape allows water to run off the surface more quickly, causing less evaporative cooling and a higher percentage of energy devoted to sensible heating

rather than latent heating. The results are increased temperatures, reduced water availability, and larger sections of land being converted to desert.

Several other land cover changes affect climates, particularly at the local scale. For example, irrigation and construction of dams and reservoirs tend to alter the local water and energy budget by producing sudden, periodic, and drastic increases in water availability at the surface. Drainage of swamps tends to have the opposite impact. In all these cases, although local evaporation rates and humidity are likely to be affected, the scale is usually too small to see significant changes in local precipitation totals. And, of course, the most dramatic land cover change on the planet—urbanization—has a distinct and irrefutable global impact. That impact is discussed in detail in Chapter 12.

■ Cryospheric Changes

Ice on the Earth's Surface

The *cryosphere*—the region consisting of all seasonal and "permanent" ice on the planet—is both a direct consequence of and an influence on the climate system. It may exist as a part of semipermanent alpine glaciers, continental ice sheets, seasonal sea ice, and/or seasonal snowpack. During cold periods of the year, all forms of ice accumulate from precipitation and/or **deposition**—the direct conversion of atmospheric water vapor to ice, bypassing the liquid water phase.

In the case of glaciers, the bulk of the ice body remains in below-freezing temperatures throughout the year. However, a portion does exist in areas where temperatures exceed the freezing point for at least part of the year. Parts of glaciers are, therefore, continually melting. Glacial advances occur during cooler and/or wetter periods when the rate of snow/ice **accumulation** in the colder part of the glacier exceeds the rate of **ablation** (which includes melting and **sublimation**—the conversion of ice directly to water vapor, bypassing the liquid water phase) in the warmer part of the glacier. Glacial retreat occurs during times when the rate of ablation exceeds accumulation. In the case of a retreating glacier the ice merely melts or sublimates back from its most equatorward or downhill extent—the glacier does not physically move backward. A net mass balance point exists at the **equilibrium** line—a point within the body of glacier where the rate of accumulation equals the rate of ablation. As this line moves equatorward and/or downhill, the glacier is "healthy," indicating growth, and when it retreats poleward and/or uphill, ablation exceeds accumulation for the system.

Earth has undergone several periods when most glaciers and ice sheets simultaneously advanced over many thousands of years. These glacial advances—intervals of 50,000 to 150,000 years or so—are separated by **interglacial phases**, warmer periods of approximately 8,000 to 12,000 years when net glacial retreat occurs in most glaciers. Summer temperatures generally determine whether glacial advance or retreat is occurring, because even abnormally high winter temperatures are below freezing in most glacial environments. Cool, short summers with snowy transition seasons generally provide the optimal conditions for glacial advance. Long, warm, dry summers cause the high rates of ablation coincident with interglacial phases.

Regardless of whether a glacial or interglacial phase is occurring, any time in geological history when semipermanent ice exists somewhere on Earth's surface is termed an **ice age**. We are in an ice age today, because permanent ice exists on the planet—in Antarctica, Greenland, the north polar ocean, and on the tops of the world's highest mountains at any latitude. The present ice age may have begun as many as 40 to 50 million years ago, but evidence shows that it became more intense some 1.6 million years ago. The last major glacial advance in the present ice age peaked between 12,000 and 18,000 years ago. This advance is known as the **Wisconsin Glacial Phase**, which occurred during the **Pleistocene Epoch**, a generally cold period that occurred from about 1.8 million years ago until about 10,000 years ago. Because planetary temperatures have generally increased for the last 10,000 years or so, most of the ice has been retreating poleward and/or uphill, and we are currently in the **Holocene Interglacial Phase**, an interglacial phase within the present ice age that corresponds to **Holocene Epoch**, the present epoch in geological time.

Feedbacks in the Cryosphere

A very important role in the climate system is played by the **positive feedback system**—an input that creates change to a system in such a way that triggers additional, similar changes in the system.

A small snowball rolling down a hill—picking up more snow on its trek, thereby gaining *momentum* and rolling faster and faster as it gains more mass—is an effective example of a positive feedback. Another example is the suggestion planted by seeing a person in your climatology class yawn; within seconds several classmates yawn, despite the very interesting course material.

By contrast, a **negative feedback system** involves an input to the system that decreases the likelihood of further changes of the same type to the system. An example of a negative feedback system might be your study habits. If you study hard and do well on the first exam, you may then be tempted to study less and then do poorly on the second exam. But then you are likely to study harder and do better on the third exam. The net effect is that your overall grade is average. Negative feedbacks lead to stabilization of a system.

An important positive feedback in the cryosphere is caused by surface albedo changes when the ice- and snow-covered area expands or contracts. The surface of the terrestrial Earth is, for the most part, dark. Expansion of bright white continental ice sheets across continents increases hemispheric and global albedo. In such instances a larger percentage of insolation is reflected from the surface to space. As a result, a decrease in the amount of absorbed insolation occurs at the surface, triggering a reduction in temperature, which in turn supports further growth of the ice sheets.

A similar positive feedback system occurs during periods of slight warming, such as at the beginning of an interglacial phase. Slightly elevated temperatures trigger increased icepack ablation. The albedo then decreases slightly, as the darker surface beneath the ice is exposed. This in turn initiates increased absorption of insolation. The net energy gains then cause increased atmospheric temperatures, which cause further increases in ablation and further albedo decreases.

It is apparent from the discussion above that the cryosphere is intricately linked to the atmosphere over long time scales. But the cryosphere can also influence the atmosphere (and vice versa) on short time scales. For example, a region may be hit with a heavy snowfall event in autumn. In such a case the snow may effectively change the regional albedo. After the event, the surface albedo changes may cause the region to grow colder, which in turn may cause the polar jet stream to shift farther equatorward than normal, because it exists near the boundary between cold and warm air. Because the jet steers midlatitude storm systems, displacement of the jet may initiate even more snowstorms over the affected regions. In this way a positive feedback system is set up, causing even more snowpack and higher albedo rates even farther equatorward, leading to further cooling.

Given such direct relationships between the cryosphere and the atmosphere, it is important that researchers examine and understand processes involving changes in the cryosphere. Such changes may lead to feedback systems that could influence the planetary-scale atmosphere. These atmospheric changes may then cause further feedbacks in the cryosphere and the other "spheres" in the climate system.

Researchers studying regional ice packs have found that the spring melt in western Canada has occurred earlier in more recent years than before, changing by as much as a half-day per year since 1955. These findings support computer modeling studies that suggest widespread reduction in snow cover could occur over the next 50 to 100 years as concentrations of greenhouse gases such as CO_2 increase in the atmosphere. But this trend is not solely dependent on greenhouse warming, as extreme ENSO events, shifts in major circulation patterns, and other factors also play important roles.

Some researchers now estimate that the Greenland ice sheet, the largest in the northern hemisphere, could lose as much as one-half of its mass over the next thousand years, leading to an increase in global sea level of about 2.7 m (9 ft). Adjacent summer sea ice has decreased markedly over the past few decades, and some expect that within the next century the Arctic Ocean could be largely ice free. Alpine (mountain) glaciers are in active retreat worldwide, with only a few regional exceptions.

The largest single planetary ice sheet, in Antarctica, also shows signs of net ablation. In particular, temperatures over the West Antarctic Ice Sheet have increased by approximately 4 C° (7 F°) over the past half century. A decrease in adjacent sea ice extent has also resulted, and two large collapses of major ice shelves have occurred over the past decade. The East Antarctic Ice Sheet appears to have remained stable during the same time period. If all permanent continental ice were to melt, sea levels would rise by approximately 67 m (220 ft).

A common misconception is that the melting of sea ice would not affect sea level because the ice displaces the same volume of seawater as its water equivalent. However, because fresh water is less dense than saltwater, freshwater ice floats higher over the salty seawater and does not displace quite as much seawater as it would if the two had the same density. So when freshwater floating ice melts, it also increases sea level by a small amount because of an increased amount of meltwater compared with the amount of seawater originally displaced.

Both this extra water and (especially) melted continental ice would greatly affect populations, because many of the world's major cities would be inundated by the sea level increases. Fortunately, even in the worst-case scenarios, the entire cryosphere is not expected to melt completely in a short time. But the (unknown) feedback mechanisms involved would determine whether any initial warming and melting would be counteracted (a negative feedback) or accelerated (a positive feedback). Positive feedbacks would further warm the climate system, which itself could have major global consequences.

■ Summary

This chapter explored the effects of several phenomena in the nonatmospheric parts of the climate system on climate. Surface ocean currents derived from atmospheric circulation in turn affect the adjacent climate of affected locations. The subtropical anticyclones centered in each ocean basin play an especially important role in the strength, direction, and variability of these currents. The anticyclones and corresponding semipermanent low-pressure cells—*cyclones*—located poleward of the anticyclones support cold ocean currents in the eastern ocean basins and warm ocean currents in the western basins in each hemisphere. The Ekman spiral, combined with latitudinal differences in Coriolis deflection, causes an accumulation of water to the west of the central ocean basin, which leads to western-boundary intensification—strong, deep, warm ocean currents in the western ocean basins—and shallow, slower cold currents in the eastern basins. Deep ocean currents also influence the broad-scale climate. Deep-water motions were described for each ocean basin, with particular attention paid to the

Stommel model deep-water circulation, which stores and releases large quantities of energy into the climate system over long time periods.

El Niño–Southern Oscillation events represent the greatest single source of variability in the climate, excluding Earth-Sun-related seasonality. El Niño events occur when a warm water pool that normally occurs in the western equatorial Pacific Ocean migrates eastward during times of reduced trade wind flow. The water pool brings changes to typical atmospheric stability patterns. Specifically, clouds and precipitation occur in the normally clear, dry eastern Pacific region. Because of the immense amount of energy carried by the pool, global climatic regimes are affected. The opposite phenomenon—La Niña—is associated with the "normal" (La Nada) directions of oceanic and atmospheric flow but with increased intensities of that flow and effects.

Processes in the lithosphere and biosphere can also impact the climate system. Times of heightened volcanic activity may produce both short-term and even long-term climate variation. Human activities such as deforestation and desertification alter local to regional energy and water balances, which can ultimately affect hemispheric and global climate regimes. Finally, the cryosphere was discussed as an element of the climate system. Positive and negative feedback systems, especially in the cryosphere, are likely to dictate the severity of impacts of climatic change caused by human and natural forces.

► Key Terms

Ablation	Equilibrium line	North Pacific Intermediate Water (NPIW)
Accumulation	*Evaporation*	
Advection	*Fossil fuel*	Northeast trade winds
Aerosol	Geostrophic flow	*Outgassing*
Albedo	Glacial advance	Pacific Decadal Oscillation
Amplitude	Global warming	Pacific Subarctic Water
Antarctic Bottom Water	*Greenhouse gas*	Particulates
Antarctic Deep Water	*Gulf Stream*	Peru Current
Antarctic Intermediate Water	Gyre	*Photosynthesis*
Anthropogenic	Holocene Epoch	Pleistocene Epoch
Bottleneck	Holocene Interglacial Phase	Polar front jet stream
California Current	Humboldt (Peru) Current	Positive feedback system
Canary Current	*Hydrosphere*	Rain machine
Chimu Floods	Ice age	Ridge
Coastal Kelvin wave	*Insolation*	Sahel
Common Water	Interglacial phase	Sensible energy
Continentality	Jet stream	*Sink*
Convection	Krakatau	Slash and burn agriculture
Coriolis effect	La Nada	Source area
Cryosphere	La Niña	Southeast trade winds
Cyclone	Laki Fissure	Southern Oscillation
Deforestation	*Latent energy*	Southern Oscillation Index (SOI)
Deposition	*Latitude*	Stable atmosphere
Desertification	Maritime Continent	Stable stratification
Downwelling	*Maritime effect*	Stommel model
Dust Veil Index	*Momentum*	*Stratosphere*
Ekman spiral	Mount Toba	Sublimation
El Niño	Mount Pinatubo	Subtropical anticyclone
El Niño-Southern Oscillation (ENSO) event	Negative feedback system	Subtropical jet stream
	North Atlantic Deep Water	Sulfur dioxide (SO$_2$)
Equatorial Kelvin wave	*North Atlantic Drift*	Tambora
Equatorial Rossby wave	North Equatorial Current	Thermocline

Thermohaline circulation   Unstable atmosphere   Western-boundary
Thermohaline current   *Upwelling*   intensification
Trade winds   Vog   Wisconsin Glacial Phase
Transpiration   Volcanic Explosivity Index   *Terms in italics have appeared in at*
Trough   Walker circulation   *least one previous chapter.*

▶ Review Questions

1. Discuss the role of atmospheric circulation in creating and maintaining surface currents in the oceans.
2. Discuss the role of surface ocean currents on climate.
3. What is geostrophic balance in the oceans and why is it important?
4. Describe the importance of western-boundary intensification.
5. What are thermohaline currents and how and why are they maintained?
6. How are thermohaline currents tied to the atmosphere and how can these currents affect climatic variations in the future?
7. Discuss the importance of the Stommel model of deep-water motion.
8. What is the Walker circulation?
9. Describe La Nada or "neutral" conditions along the equatorial Pacific Ocean.
10. What is El Niño and how is it initiated?
11. Discuss changes in the sea surface circulation and the thermocline of the Pacific Ocean during an El Niño event.
12. Describe the changes in the overlying atmosphere during an El Niño event.
13. Discuss the global climatic significance of an El Niño event.
14. What is a La Niña event and why is it significant?
15. Discuss the role of volcanism on the primordial atmosphere.
16. Discuss the role of volcanism relative to short-term and long-term climatic change and variability.
17. Compare and contrast deforestation and desertification and their impacts on the climate system.
18. How do atmospheric changes drive processes in the cryosphere?
19. How can changes in the cryosphere lead to changes in the climate system?

▶ Questions for Thought

1. Before 3 million years ago, there was no connection between North America and South America via the Isthmus of Panama. How do you believe the formation of the isthmus may have altered the global oceanic circulation?
2. Think of an example of a possible negative feedback mechanism that might result in a minimization of the global warming problem in the cryosphere.

---

http://physicalscience.jbpub.com/climatology

Connect to this book's Web site: http://physical science.jbpub.com/climatology. The site provides chapter outlines, further readings, and other tools to help you study for your class. You can also follow useful links for additional information on the topics covered in this chapter.

5

Energy, Matter, and Momentum Exchanges Near the Surface

Chapter at a Glance
Properties of the Troposphere
Near-Surface Troposphere
Energy in the Climate System
  Sun as Energy Source
  Measuring Radiant Energy
  Radiation Balance
  Turbulent Fluxes
  Substrate Heat Flux
  Energy Balance
Local Flux of Matter: Moisture in the Local
  Atmosphere
  Atmospheric Moisture
  Moisture in the Surface Boundary Layer
  Measuring Evapotranspiration
Atmospheric Statics, the Hydrostatic Equation,
  and Stability
  Statics and the Hydrostatic Equation
  Atmospheric Stability
  Assessing Stability in the Local Atmosphere
Momentum Flux
Putting It All Together: Thermal and Mechanical
  Turbulence and the Richardson Number
Summary
Review

Chapter 3 described the role of "site" and "situation," including *latitude*, land–sea distribution, circulation, topography, and local features on climates. Chapter 4 examined the effects on the climate system of various features in the *hydrosphere*, *lithosphere*, *biosphere*, and *cryosphere*. These factors and their impacts can be felt on a variety of spatial scales. The next five chapters explain processes that occur in response to these influences on climates at various scales. The scales range

from the **microscale** to the **planetary scale**. This chapter describes the influence of these and other factors on near-surface climates. It also introduces the effect of energy, matter, *momentum*, and *friction* on local climates.

■ Properties of the Troposphere

As we have seen, the atmosphere can be subdivided into various layers based on the expected change in temperature with height. In this chapter we keep the discussion focused on the lower part of the *troposphere*, the layer from the surface up to 8 to 20 km (5 to 13 mi) high (the *tropopause*). We have seen that the troposphere is thickest over the tropics, where warm air tends to expand, stretching the distance vertically, and thinnest over the polar areas in winter, where the intense cold air compresses.

The troposphere is characterized by temperatures that normally decrease with height, both because of the decreasing compression of atmospheric gases with increasing distance from the surface and because of the increasing distance from the (indirect) heat source—the surface. In general, the atmosphere tends to transmit most of the *shortwave radiation*—electromagnetic radiation at *wavelengths* between about 0.4 to 4.0 micrometers (μm)—from the Sun through it. The surface (particularly the land surface) absorbs most of the shortwave energy incident upon it. This causes the surface to emit **longwave radiation** (electromagnetic radiation at wavelengths greater than 4.0 μm) upward. The near-surface atmosphere is more likely to absorb longwave radiation than locations higher in the troposphere, because more atmospheric mass is concentrated near the surface.

Although it varies across space and time, the average rate of cooling with height in the troposphere is 6.5 C°/km. This rate is known as the *environmental lapse rate (ELR, or γ)*. As we see later in this chapter, ELR is a very important determinant of whether vertical motion occurs easily in the troposphere. The root word "tropo" means "turning," and the troposphere is so-named because it is often characterized by *convection*—vertical motions that are caused by hotter air near the surface rising while cooler air aloft sinks, or **turbulence**—vertical motions caused by surface friction on advecting air.

Because about 75 percent of the atmosphere's mass and nearly all its water vapor are located in the troposphere, this layer is sometimes referred to as the "weather and climate layer." The lowest part of this layer, which contains an even more disproportionate share of the atmosphere's mass and water and an even greater tendency for vertical motion, is the site of the most vigorous and intense exchanges of energy, matter, and momentum in the climate system.

■ Near-Surface Troposphere

Four subdivisions within the troposphere can be defined, as shown in **Figure 5.1**: (1) laminar layer, (2) roughness layer, (3) transition layer, and (4) free atmosphere. The **laminar layer** is the part of the atmosphere that is nearest to the surface or elements on the surface, such as a leaf, roof, or wave on the sea. This layer is only a few millimeters thick, at most. By definition, the laminar layer is characterized by smooth flow that parallels the features of the surface over which it moves. Several laminar layers may even exist vertically when the shape of the surface feature is complicated. For example, every leaf on a tree has a laminar layer adjacent to its surface. The features of this layer are most important in plant physiology applications, because the atmospheric flow across the leaves is an important factor affecting the loss of water from the plant's surface.

Above the laminar layer is a zone of strong convection or turbulence, called the **roughness layer**, characterized by a large component of vertical motion compared with horizontal motion. The turbulence in this layer is almost entirely mechanical in origin. **Mechanical turbulence** is convection caused by the friction, irregular flow, and vertical gradients of momentum associated with the roughness elements, such as trees in a forest, buildings, mountains, or crops.

Less important in the roughness layer, particularly within a dense plant canopy, is **thermal turbulence** resulting from unequal heating of the surface, because differences in surface heating are small when shading occurs from the roughness elements. The roughness layer may extend up to one to three times the height of the individual roughness elements (to perhaps 50 to 100 m). It is thicker by day than by night because of the increased differences of surface heating across space during daylight hours. The maximum height of the roughness layer occurs with a medium density of roughness elements. Interestingly, if the roughness elements become too closely packed and uniform in height, the top of the roughness elements may mimic a smooth surface and show relatively little mechanical or thermal turbulence and more laminar-like flow.

The roughness layer is very important from an energy perspective. The vertical motion that dominates this zone transports energy used to drive atmospheric processes between the surface and the atmosphere above this layer. In most cases, especially in daylight hours, the surface is warmer than the atmosphere above it and energy is transported upward via the **turbulent flux** (flow) **of sensible heat**—also known as the **sensible heat flux (Q_H)**.

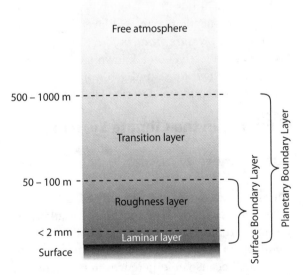

Figure 5.1   The near-surface troposphere.

The most intense turbulence is associated with situations when the temperature difference between the surface and the air above it is very large.

The upward transport of matter is also very important in the roughness layer. The type of matter that is of most concern to climatologists is water. The **turbulent flux of latent heat**—also known as **latent heat flux (Q_E)**—transports water vapor, usually in an upward direction also, from the surface where oceans, soil moisture, streams, and lakes evaporate liquid water to the atmosphere. From there, the turbulent flow may allow clouds to form.

Unlike the fluxes of heat (energy) and matter (i.e., water), the **vertical flux of horizontal momentum** (τ) or **shearing stress** is usually transported downward, from the atmosphere to the surface. This is because winds usually strengthen with increasing height, as friction decreases with height. The net flux of horizontal momentum is downward as turbulence carries air with greater momentum downward and air with lesser momentum upward.

The laminar layer and the roughness layer combined are sometimes referred to as the **surface boundary layer (SBL)**. In the SBL the turbulent fluxes and the momentum flux are generally considered to be constant with height, but not over time. According to **Fick's law**, the flux of any entity is proportional to the gradient of that entity in the fluid (liquid or gas) through which the flux occurs. Thus the gradient of energy or moisture from the surface upward or the gradient of momentum downward must also be constant to the top of the SBL at any given place and time. We can take this logic one step further by saying that the rate at which the energy, water, or momentum is moved vertically is constant from one height to another within the SBL at one moment in time. Because the energy and momentum fluxes are constant with height, wind direction theoretically cannot change within the SBL.

The **transition layer** extends from the top of the SBL to approximately 500 to 1000 m above the surface. Again, this layer extends farther upward by day than by night, as convective processes thrive in the afternoon heat. In the transition layer turbulent features from the SBL remain important. Some of the characteristics of the *free atmosphere* above this layer, however, also have some importance. The shearing stress gradually decreases with height in the transition layer until it becomes zero, which defines the top of the layer, at which a

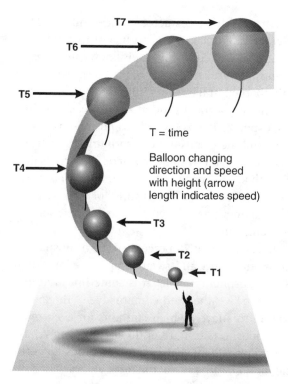

T = time

Balloon changing direction and speed with height (arrow length indicates speed)

Figure 5.2 Change of wind direction with height in the planetary boundary layer.

negligible vertical flux of horizontal momentum occurs. The decrease in friction with height in the transition layer causes winds to change direction in a "spiral staircase" pattern. For instance, a balloon that is released changes direction in a rather predictable pattern (**Figure 5.2**). Collectively, the laminar, roughness, and transition layers comprise the **planetary boundary layer**.

In the free atmosphere friction is generally assumed to be negligible. In that layer flow is governed by the *pressure gradient force*, the apparent horizontal deflective force known as the *Coriolis effect*, *centrifugal force*, and *geostrophic balance*. All these concepts were described in Chapter 3.

■ Energy in the Climate System

Sun as Energy Source

The Sun is the source of essentially all energy that drives atmospheric and oceanic circulations. Without the Sun's unequal heating of Earth, there would be no need for the atmosphere or oceans to circulate, because no differences in heat (energy) would be received from one place to another on

Earth's surface. Because the differences in local energy received from the Sun are so important in the climate system, many applications rely on knowing the position of the Sun in the sky. For example, agricultural models rely on the intensity of *insolation* to determine the rate at which *photosynthesis* allows a crop to grow. Architects design buildings so the overhang over a window is short enough to let the winter Sun (which is relatively low in the sky) into the room yet is long enough to keep the hot summer Sun (which is higher in the sky) out of the room (**Figure 5.3**). Traffic engineers know that at certain times of the day and certain times of the year the Sun will be situated directly behind a traffic light, and the glare will make the signal especially difficult to see. Additional traffic signals should be installed in more visible parts of the intersection.

All these applications require the precise calculation of the position of the Sun. The position of the Sun also largely governs the receipt of shortwave radiation input on a surface. When the Sun is high in the sky, such as near noon near 23.5°N or 23.5°S around the *summer solstice*, its radiation reaches the surface more effectively because it does not have to pass through a great distance of atmosphere (**Figure 5.4**). So we can say that the **optical air mass**—the ratio of the distance that a beam travels through the atmosphere (*path length*) to the minimum distance that it could possibly have to travel to reach the surface—is near 1.0. On the other hand, when the Sun is low in the sky, such as around the *winter solstice* in the high latitudes or near Sunrise and Sunset at any latitude, the long path of radiation through the atmosphere allows a significant amount of the radiation to undergo *attenuation*—which includes the **absorption** of energy by air particles and **scattering** away from the surface—before it reaches the surface. In such a situation the optical air mass is much greater than 1.0.

Sun Path Diagrams   Although the equations used to describe these positions are not presented here, the position of the Sun in the sky at any day of the year and any time of the day, for any latitude, can be plotted on a **Sun path diagram**. Just as a position on a map has coordinates in angular degrees of latitude and *longitude*, we can describe the Sun's location in the sky using the coordinates of **zenith angle** and **azimuth angle**. The zenith angle expresses how many degrees the Sun is located away from the local vertical (**Figure 5.5**). Thus when the Sun is directly overhead at a given location, its zenith angle is 0°. At sunrise and sunset the zenith angle is 90°.

The azimuth angle represents the number of degrees that the Sun is located away from north (**Figure 5.6**). For example, if the Sun is in the northern sky (as it is for locations in the southern hemisphere's extratropical areas near noon), its azimuth angle would be 0°, but if the Sun is in the eastern sky at a given location (as it is near sunrise on some days of the year), it would have a 90° azimuth. Likewise, a 180° azimuth angle suggests

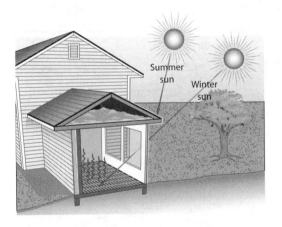

Figure 5.3 Effect of Sun angle on building design: An overhang blocks the hot summer Sun but allows the winter Sun into the window.

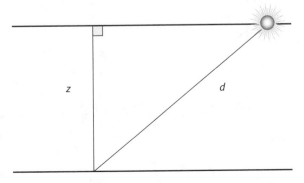

Figure 5.4 The optical air mass would be the ratio of *d* to *z*.

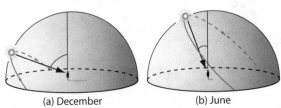

(a) December   (b) June

Figure 5.5 Zenith angle.

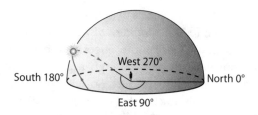

Figure 5.6 Azimuth angle.

that the Sun is in the southern sky (as it is near noon in the extratropical northern hemisphere). And a 270° azimuth angle implies that the Sun is in the western sky.

Figure 5.7 shows Sun path diagrams for locations at various northern hemisphere latitudes. When using Sun path diagrams, the *solar declination* must be known for the day of interest. The dec-

lination is 23° 27′ N latitude on June 21, 0° on the **equinoxes**, and 23° 27′ S latitude on December 22.

Notice that the Sun tracks increasingly toward the southern sky at solar noon as latitude increases in the northern hemisphere. Nevertheless, the position of the Sun at sunrise and sunset at a given day of the year increasingly acquires a northerly component as latitude increases in the northern hemisphere too, until inside the *Arctic Circle*, where sunrise and sunset do not exist for at least 1 day of the year. It is indeed an oversimplification to say that the Sun rises in the east and sets in the west.

Solar Time   **Solar time** is the "true" time at a location based on the location's position with respect to the Sun. Specifically, *solar noon* at a given meridian of longitude occurs when the Sun is at the highest

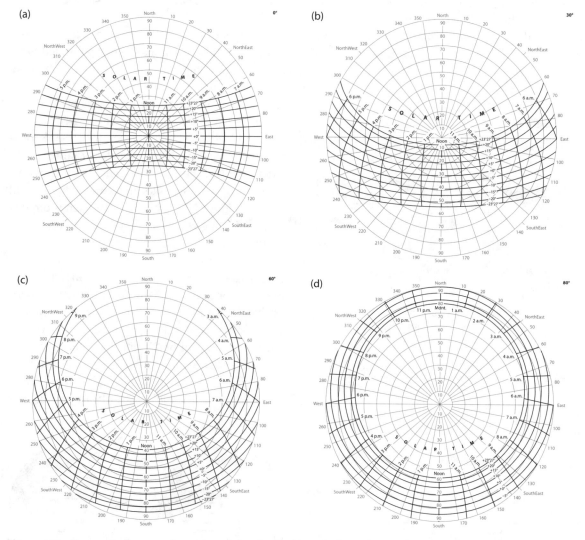

Figure 5.7 Sun path diagrams for various solar declinations, for latitudes (a) 0°, (b) 30°N, (c) 60°N, and (d) 80°N. *Adapted from*: *Smithsonian Meterological Tables,* 6th rev. edition, R. J. List, ed. Washington, DC: Smithsonian Institution, 1958.

position that it will achieve on a given day at that meridian. Put another way, solar noon at any meridian of longitude occurs when the *rotation* of Earth puts that particular meridian directly in front of the Sun. Note, though, that on a given day only one parallel of latitude along that meridian—the solar declination on that day—receives the direct rays of the Sun (and, therefore, would have a zenith angle of zero at solar noon). If we used solar time to represent the time shown on our clocks, locations along every meridian of longitude would have different times on their clocks. This would create undue confusion for everyone when communicating with others across long (or even short) distances.

To remedy this problem we use a simplification called **zone mean time**—a system in which the "clock time" is synchronized across a range of longitudes so that no adjustment is necessary except when crossing from one *time zone* to another. Twenty-four time zones are arranged longitudinally, with the "clock time" set 1 hour apart at each zone (**Figure 5.8**). The zone mean time actually corresponds to the solar time only at the **principal meridians**—24 longitudinal designations around the world. The principal meridians are located at 0° and at 15°, 30°, 45°, 60°, 75°, 90°, 105°, 120°, 135°, 150°, and 165° east and west longitude along with 180° longitude, which roughly corresponds to the **international date line**. For example, because New Orleans, Louisiana is located at 90°W longitude,

the zone mean time is equal to the solar time, and no adjustment is necessary to compute the solar time at that location.

But most locations are not on principal meridians, so an adjustment is required to use a Sun path diagram correctly. Earth rotates through 360° of longitude every 24 hours, so it rotates through 15° every hour (which explains why the principal meridians are 15° apart and have times that are 1 hour apart from adjacent principal meridians). Taken further, we can say that Earth rotates through 1° of longitude every 4 minutes. Because Earth rotates from west to east, locations east of a principal meridian have solar noon before the solar noon of their principal meridian, and locations west of a principal meridian experience solar noon after their principal meridian. In either case the offset is by 4 minutes for every degree of longitude separating a location from the principal meridian in its time zone (**Figure 5.9**).

In actuality, slight variations in Earth's orbital velocity throughout the year cause Earth to have to rotate slightly more or less than 360° to reach the next successive solar noon. This imperfection necessitates another adjustment called the **equation of time**. This adjustment varies not by location (in contrast to the correction for the zone mean time) but by time of year. **Figure 5.10** shows that at the same time each day, the Sun is located in a different position throughout the year, and the path that

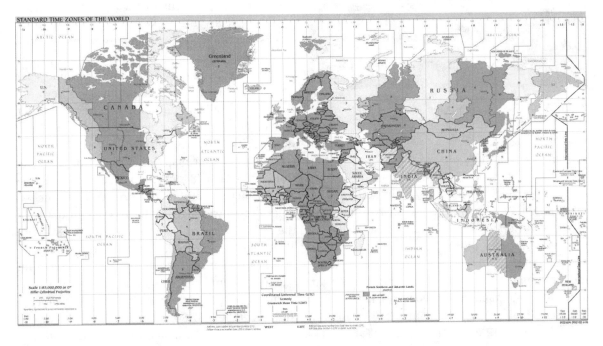

Figure 5.8   Time zones of the world.

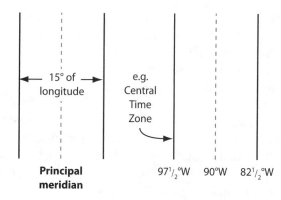

Figure 5.9   Relationship of principal meridian to time zone.

Figure 5.10   The position of the Sun in the sky throughout the year at a given location and time of day.

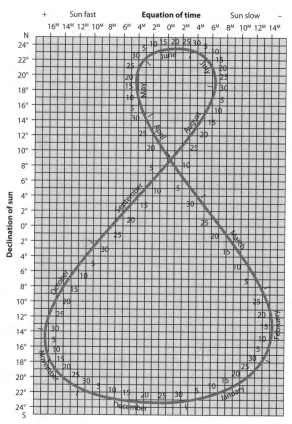

Figure 5.11   An analemma. *Data from*: U.S. Coast and Geodetic Survey.

the Sun takes at the same time of day during the course of a year assumes the form of a figure eight. The equation of time adjustment may add up to 16 minutes and 23 seconds (on or about November 5) or take away up to 14 minutes and 20 seconds (on or about February 13) to the zone mean time to compute the solar time. The adjustment necessary for the equation of time can be shown with an **analemma**—a graph in the general shape of a figure eight (**Figure 5.11**). It can also be shown as a graph (**Figure 5.12**), or in tabular form (**Table 5.1**).

The term **"local apparent time"** is sometimes used as a synonym for solar time to describe the "true" time after the zone mean adjustment and equation of time adjustment are incorporated.

Theoretically then, no location should have clocks that differ from local apparent time by more than about 46 minutes (7.5° of longitude east or west of the principal meridian times 4 minutes per degree of longitude, plus 16 minutes for the equation of time adjustment on November 5). But for convenience and economic advantage some locations are included in time zones beyond 7.5° east or west of a given principal meridian. For example, Cincinnati, Ohio (84.5°W longitude) is closer to the 90°W principal meridian than to the 75°W principal meridian. However, it uses Eastern Time (for which the 75°W meridian is the standard meridian). Thus, Cincinnati actually has solar time that is 38 minutes earlier in the day than clock time, if no equation of time adjustment is necessary (which occurs on or about April 16, June 14, September 1, and December 26). Solar noon on those 4 days occurs at about 12:38 PM on the clock in Cincinnati.

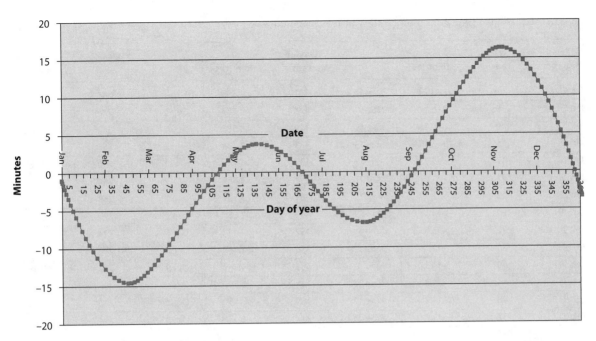

Figure 5.12 An equation of time graph.

Measuring Radiant Energy

Any instrument that measures the flux of radiation over a unit area of the surface (in Watts per square meter) is termed a **radiometer**. Most radiometers operate on the basis of thermoelectricity—that is, the amount of radiation received is proportionate to the temperature increase on the sensor. This temperature increase is then converted back to the intensity of radiation that would have caused such a temperature change.

A **pyranometer** is a radiometer that measures only shortwave (solar) radiation, and a **pyrgeometer** detects only longwave radiation (i.e., radiation emitted from the atmosphere or Earth). Pyranometers and pyrgeometers work by allowing only shortwave radiation (or longwave radiation for the pyrgeometer) to be transmitted through a dome and heat the surface of a sensor. A **pyrradiometer** measures all-wave radiation. In actuality, the use of two of these three types of instruments simultaneously allows the deduction of the third (unmeasured) radiation variable, because the pyrradiometer's reading should be the sum of the shortwave and longwave radiant fluxes.

Any of the above types of instruments with "net" preceding its name implies that the instrument subtracts the energy lost from the energy gained at the surface. For example, a net pyrradiometer measures the amount of all-wave energy incident on a surface less the amount of all-wave energy emitted by the surface. Net radiometers usually require two sensors—one pointing upward and one pointing downward—to measure the net flux of energy both toward and away from the surface.

Radiation Balance

The **net radiation** (Q^*, pronounced "Q-star") received at any given point on Earth varies tremendously over time. Sometimes the net radiant flux is negative. This occurs whenever the local surface is losing more radiant energy than it gains. Such a situation is common over an ice sheet or in any location where a near-surface *temperature inversion*—an increase in temperature with height—is occurring. At other times, strongly positive values of Q^* occur at a place, such as on a hot summer afternoon, when the surface receives insolation faster than it can emit longwave radiation back up to the atmosphere. However, the global amount of shortwave radiation received by Earth's entire surface from the Sun (termed $K\!\downarrow_{sfc}$ and pronounced "K down at the surface") at any one moment in time is essentially constant.

Solar radiation that enters the atmosphere is involved in various processes, including atmospheric absorption (retention by a particle and conversion to internal energy), scattering in all directions by the atmosphere, and **reflection** from the tops of

Table 5.1 Equation of Time Adjustment by Day of Year, in Minutes

	Day of Year	Minutes		Day of Year	Minutes		Day of Year	Minutes
Jan 1	1	−3.23	May 1	121	2.833	Sep 1	244	−0.25
Jan 5	5	−5.1	May 5	125	3.283	Sep 5	248	1.033
Jan 9	9	−6.83	May 9	129	3.583	Sep 9	252	2.367
Jan 13	13	−8.45	May 13	133	3.733	Sep 13	256	3.75
Jan 17	17	−9.9	May 17	137	3.733	Sep 17	260	5.167
Jan 21	21	−11.2	May 21	141	3.567	Sep 21	264	6.583
Jan 25	25	−12.2	May 25	145	3.267	Sep 25	268	8
Jan 29	29	−13.1	May 29	149	2.85	Sep 29	272	9.367
Feb 1	32	−13.6	June 1	152	2.45	Oct 1	274	10.03
Feb 5	36	−14	June 5	156	1.817	Oct 5	278	11.28
Feb 9	40	−14.3	June 9	160	1.1	Oct 9	282	12.45
Feb 13	44	−14.3	June 13	164	0.3	Oct 13	286	13.5
Feb 17	48	−14.2	June 17	168	−0.55	Oct 17	290	14.42
Feb 21	52	−13.8	June 21	172	−1.42	Oct 21	294	15.17
Feb 25	56	−13.3	June 25	176	−2.28	Oct 25	298	15.77
Mar 1	60	−12.6	June 29	180	−3.12	Oct 29	302	16.17
Mar 5	64	−11.8	July 1	182	−3.52	Nov 1	305	16.35
Mar 9	68	−10.9	July 5	186	−4.27	Nov 5	309	16.38
Mar 13	72	−9.82	July 9	190	−4.93	Nov 9	313	16.2
Mar 17	76	−8.7	July 13	194	−5.5	Nov 13	317	15.78
Mar 21	80	−7.53	July 17	198	−5.95	Nov 17	321	15.17
Mar 25	84	−6.33	July 21	202	−6.25	Nov 21	325	14.3
Mar 29	88	−5.12	July 25	206	−6.4	Nov 25	329	13.25
Apr 1	91	−4.2	July 29	210	−6.38	Nov 29	333	11.98
Apr 5	95	−3.02	Aug 1	213	−6.28	Dec 1	335	11.27
Apr 9	99	−1.87	Aug 5	217	−5.98	Dec 5	339	9.717
Apr 13	103	−0.78	Aug 9	221	−5.55	Dec 9	343	8.017
Apr 17	107	0.217	Aug 13	225	−4.95	Dec 13	347	6.2
Apr 21	111	1.1	Aug 17	229	−4.2	Dec 17	351	4.283
Apr 25	115	1.883	Aug 21	233	−3.32	Dec 21	355	2.317
Apr 29	119	2.55	Aug 25	237	−2.3	Dec 25	359	0.333
			Aug 29	241	−1.17	Dec 29	363	−1.65

clouds or the surface. In recent years satellite data have made it possible for us to determine the amount of energy involved in each of the processes of absorption, scattering, and reflection (**Figure 5.13**). On a global annual average basis, approximately 30 percent of K↓ descending to the top of the atmosphere is reflected by cloud tops, scattered by atmospheric particles back to space, or reflected from the surface all the way out to space. Collectively, shortwave energy that exits the atmosphere upward in such a manner is termed K↑ (pronounced "K up"). This 30 percent is sometimes referred to as the **planetary albedo**, but the term "*albedo*" in its strictest sense refers only to energy that is reflected from a surface as a proportion of the total energy incident upon that surface. Of course, cloudy locations tend to experience a greater planetary albedo than places with clear skies, because of increased reflection from cloud tops and scattering by cloud droplets back to space. Typical albedo values for various surfaces are shown in **Table 5.2**.

Insolation that is not part of the planetary albedo may be absorbed in the atmosphere, scattered downward, or transmitted through the air. On a global annual average basis approximately 25 percent of K↓ at the top of the atmosphere is absorbed by atmospheric gases and *aerosols*.

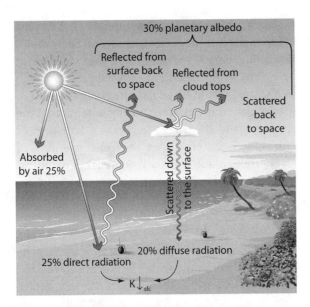

Figure 5.13 Schematic representing the global short-wave radiation budget.

Again, this value varies by several percentage points at any given place and time. On a global annual average basis only the remaining approximately 45 percent of K↓ that reaches the top of the atmosphere is transmitted to the surface.

Insolation that reaches the surface without being scattered downward is termed **direct radiation**, whereas **diffuse radiation** is radiant energy that reaches the surface only after being scattered downward. On a global annual average slightly more of the energy that reaches the surface occurs in the form of direct radiation than diffuse radiation, with a higher percentage of diffuse radiation occurring in cloudy and polluted atmospheres.

Table 5.2 Common Albedo Values for Various Surfaces

Surface	Albedo Value
Fresh snow	75–90
Thick cloud	50–80
Ocean with high zenith angle	50–70
Old snow	50–60
Thin cloud	40–60
Earth–ocean–atmosphere system	30
Dry soil	15–35
Meadow	15–30
Cropland (depending on stage of growth)	10–35
Forest	5–15
Urban land	5–15
Wet soil	5–15
Ocean with low zenith angle	3–5

Diffuse radiation—along with *advection*, the lateral movement of mass or energy—is the reason that the temperature difference between a sunny location and a nearby shady location is not greater than it is, and it also explains why a sunburn is possible even while sitting in the shade.

At a given location even though shortwave energy is being received only between sunrise and sunset, the radiation budget is not "dead" at night. Any object with a temperature above *absolute zero* (which includes every object found so far!) emits radiant energy all the time—day and night. If such an object does not have a temperature that is as hot as the Sun, its emitted radiant energy can be characterized as longwave. Thus Earth and its atmosphere emit longwave energy at all times of the day and night, with the intensity and wavelength of radiant energy emitted depending on the temperature of the emitter. In reality, as we saw in Chapter 2, Earth emits radiation at a wide range of wavelengths but with the vast majority at so-called longwave ranges (see Figure 2.6). Likewise, the Sun's emission spectrum includes a wide range of wavelengths focused on the so-called shortwave range.

For Earth as a whole, on a global annual average most longwave energy emitted by the surface (termed L↑ and pronounced "L up") is absorbed by the air. Only a small fraction (perhaps 7 to 15 percent) of L↑ is transmitted completely through the air and escapes to space unimpeded. The proportion of L↑ that is absorbed by the air is greater in cloudy, humid, and low-elevation environments and less in dry and highland environments because water and water vapor are efficient absorbers of longwave radiation. Any longwave energy from the surface that is absorbed by the air is then reemitted by the air in all directions. Although some of this energy is emitted out to space, more of it is emitted downward (termed L↓ and pronounced "L down") and absorbed near the surface, because most of the atmospheric mass is concentrated closer to the surface rather than near the "top" of the atmosphere. L↓ is sometimes termed **counter-radiation**.

A simple **radiation balance equation** can be created. We can say that

$$Q^* = (K{\downarrow} - K{\uparrow}) + (L{\downarrow} - L{\uparrow})$$

The equation shows that net radiation is the sum of the net shortwave radiant energy (K*) received at the surface and the net longwave radiant energy (L*) received at the surface. K^* is the difference between the shortwave radiant energy that reaches the surface ($K\downarrow$) and that which is reflected from the surface ($K\uparrow$), while L^* represents the difference between the longwave radiant energy that reaches the surface as counter-radiation ($L\downarrow$) and that emitted by the surface ($L\uparrow$).

At night $K\downarrow$ is zero, and $K\uparrow$ must also be zero because no shortwave energy can be reflected if none is incident on the surface. In such a case the first term on the right side of the equation (K^*) is zero. Most of the time, at most locations, ($L\downarrow - L\uparrow$) is slightly negative, because as we saw already the Sun tends to heat the surface more effectively than it heats the air (45 percent of the incident energy absorbed at the surface versus 25 percent absorbed in the air, on a global annual average basis). Thus, most of the time, the longwave energy emission should be greater in the upward direction from the surface than in the downward direction from the air. So at night the typical scenario is for Q^* to be slightly negative. On a sunny summer afternoon, ($K\downarrow - K\uparrow$) is generally a large positive number at the surface, because only a small percentage of the shortwave energy from the Sun is reflected upward. This term is usually more strongly positive than the ($L\downarrow - L\uparrow$) term is negative. Thus, the typical situation on a summer day is for Q^* to be positive until late afternoon.

Q^* is inherently tied to temperature changes at the surface. If Q^* is positive at a location (as it is from morning until at least midafternoon on most days), there is a surplus of energy and the temperature is likely to rise, unless the extra energy is advected away from that location or is sent vertically via turbulent fluxes. If Q^* is negative (as it is from late afternoon until just after sunrise on most days), then there is a deficit of radiant energy at that location and the temperature falls as a result, unless energy from elsewhere is advected into the location.

We might expect that if we computed Q^* at the surface for many locations all around the world at a given instant in time, they should all average out to zero. Some places are experiencing night whereas others are experiencing day, and some are having the opposite season from others. Believe it or not, however, if we calculated surface Q^* for a whole range of locations around the world, we would find a surplus of energy. Likewise, if we calculated Q^* for the air for a large number of locations around the world, we would observe a deficit. The deficit in

the air would be equal in magnitude to the surplus at the surface. If energy transfer occurred by radiation alone, Earth's surface as a whole would warm over time, while the atmosphere above it (for Earth as a whole) would be constantly cooling. Another mechanism must exist to transfer energy from the surface up to the air, to keep the Earth–ocean–atmosphere system in an energetic equilibrium (even though local places on Earth do indeed warm and cool at different times).

Turbulent Fluxes

Convection is the mechanism by which energy is transferred vertically from the surface to the air. This process occurs when more-energized molecules in a heated fluid rise and less-energized molecules in a cooler fluid sink. In the "real" world this flux of energy can take the form of turbulent or convective fluxes introduced earlier in this chapter.

The first convective flux is known as the turbulent (or convective) flux of sensible heat or, more simply, the sensible heat flux (Q_H). This flux is manifested in the atmosphere by the familiar observation that warmer air rises. For Earth as a whole, Q_H transfers only a relatively small percentage of the excess energy from the surface upward because relatively little Earth is covered by land, particularly at tropical latitudes where abundant surface heating occurs.

Nearly all the remainder of the energy is transferred by the turbulent (or convective) flux of latent heat or, more simply, the latent heat flux (Q_E). In Q_E excess radiant energy at the surface is used to evaporate water or to melt or sublimate ice (convert it from solid directly to vapor). In the processes of melting, *sublimation*, and *evaporation* this energy absorbed by the water molecules is stored in latent ("hidden") form. Later, when the evaporated water undergoes *condensation* (changing phase from vapor to liquid form) or when the liquid water freezes or even undergoes *deposition* (changing phase from water vapor directly to ice, bypassing the liquid stage altogether; the opposite of sublimation), the same amount of energy that was required to evaporate, melt, or sublimate the water is then released.

In the tropical part of Earth, where a surplus of radiant energy exists, water is even more abundant than in the rest of Earth. Thus, Q_E is very important. As evaporated water from the tropical oceans is circulated poleward by winds, some of it condenses or deposits. These processes cause the release of latent heat in the cloud-formation process. It is this latent heat (energy) release that can be converted to the *kinetic energy*—energy of motion—that powers winds in storms.

Substrate Heat Flux

A tiny percentage of the excess radiant energy at the surface is also transferred through **conduction**— the transfer of energy by one molecule to another molecule touching it, then to the next molecule, and so on. The flux of energy by conduction is known as the **substrate heat flux** (Q_G). Because molecules are packed closely together in solids, conduction is really only a legitimate energy transfer mechanism in solids, such as a metal rod or Earth itself. Because the atmosphere is not a solid, conduction is negligible in the atmosphere.

Even in solids Q_G is inefficient and slow. For example, the transfer of energy downward from the surface on a hot summer day may occur only down to perhaps 1 m from the surface. If you planted a thermometer at a depth of 1 m and could monitor its readings, you probably would not notice a spike in temperature in the afternoon and a temperature minimum in the early morning shown on that thermometer. Of course, the depth at which the diurnal cycle of temperature becomes negligible does vary depending on the mineral composition and moisture content of the soil. Dogs are well aware that conduction is not an efficient process; on a hot summer day they dig holes in the yard because they know the inefficiency of conduction means the ground is much cooler only a few centimeters below the surface. And after they dig a small hole, the new surface warms quickly, which means they must dig a deeper hole!

At a depth of perhaps 15 m (50 ft), the inefficiency of conduction means that even summer and winter temperatures are indistinguishable. By the time the summer insolation penetrates to such depths via Q_G, the heat gradient moves back upward because the subsurface layer is warmer than the air as summer ends. The temperature beneath such an equilibrium depth remains constant. If you have ever been in a subsurface cave, you know the temperature inside that cave is not dependent on the outside temperature on the day when you visited. Instead, the cave's temperature is always equal to the mean annual temperature at that location.

Despite the fact that conduction is far less important in the global energy budget than convection, Q_G can sometimes play an important role locally (particularly at short time scales) in transferring energy from the surface downward (when the surface is being warmed) or from beneath the surface upward (when the surface is colder than the ground beneath it). For example, conduction allows snow to melt relatively quickly after a relatively warm period when much energy has been stored in the ground. Nevertheless, Q_G is usually small and becomes negligible at relatively shallow depths.

Q_G can be measured relatively easily using **soil heat flux plates**. But because Q_G varies greatly in magnitude over very small distances due to differences in moisture and mineral composition of the soil, measured values should be averaged at several locations in an area for assessing the magnitude of Q_G and its relative importance compared to radiation and convection.

Unlike Q_H and Q_E, by convention Q_G is positive when in the downward direction. All three fluxes can be thought of as positive when directed away from the surface and negative when directed toward the surface. During most of the daytime hours into late afternoon the fluxes are usually positive, and at night they are often negative (**Figure 5.14**).

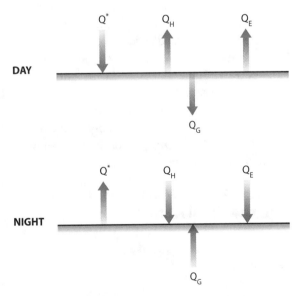

Figure 5.14 Typical direction of the turbulent fluxes by day and by night.

Energy Balance

After including the effects of the **energy balance**, the relationship between the radiant flux received at the surface and the convective and conductive fluxes can be expressed as

$$Q^* = Q_H + Q_E + Q_G$$

Thus, we can say that the net radiation received at a point on the surface must be "used" by some combination of sensible heating, latent heating, and heating of the subsurface layer. Earlier we saw that the radiation balance is

$$Q^* = (K{\downarrow} - K{\uparrow}) + (L{\downarrow} - L{\uparrow})$$

so we can substitute to produce another form of the energy balance equation:

$$(K{\downarrow} - K{\uparrow}) + (L{\downarrow} - L{\uparrow}) = Q_H + Q_E + Q_G$$

The equation above states that the sum of the net shortwave and net longwave radiation receipt at the surface must be balanced by the convective loss of energy from the surface upward through Q_H and Q_E and the conductive loss downward through Q_G. On a hot summer afternoon when $K{\downarrow}$ is high, some combination of a net loss of energy through longwave radiation and convective and/or conductive fluxes must occur to offset this input of energy locally, or the local temperature will increase. Of course, in most cases a high input of insolation cannot be balanced by the radiative, convective, and conductive fluxes, so the temperature does increase. But unlike the radiation balance for Earth's surface as a whole, at any instant in time the total energy balance would theoretically be zero. Some places on Earth gain energy (and, therefore, warming), and others lose energy (and cooling). However, at longer temporal scales (years to millennia) other factors might cause long-term warming or cooling of the entire Earth–ocean–atmosphere system.

■ Local Flux of Matter: Moisture in the Local Atmosphere

Atmospheric Moisture

The second entity of importance in the near-surface atmosphere is matter, and moisture is the most important type of matter in dictating cli-

mate. There are several ways to represent moisture in the atmosphere. **Vapor pressure (e)** is the atmospheric pressure exerted by water vapor. This is divided by **saturation vapor pressure (es)**—the atmospheric pressure exerted by water vapor when the air is at saturation—to calculate **relative humidity**:

$$RH = \frac{e}{e_s}$$

Because e and e_s both have units of pressure—the *Pascal (Pa)* or *millibar (mb)*—relative humidity is simply a ratio. The amount of moisture in the air at saturation depends on temperature. The higher the temperature, the greater the amount of water vapor in the atmosphere when the atmosphere reaches saturation. Thus, we can say that e_s is a function of temperature. The **Clausius-Clapeyron equation,** one of the most fundamental equations in all of atmospheric thermodynamics, shows that e_s can be considered solely a function of temperature at the range of temperatures and pressures experienced on Earth. Mathematically, the Clausius-Clapeyron equation is

$$e_s = 611 \text{ Pa}^{\left[\frac{L_v M_v}{R^*} \left(\frac{1}{273} - \frac{1}{T} \right) \right]}$$

where L_v is the latent heat of vaporization (2.5008×10^6 *Joules* kg^{-1} at 0°C and changing very minimally over the range of Earth temperatures), M_v represents the molecular weight of water vapor (0.018015 kg mol^{-1}), R^* is the universal gas constant (8.314 J mol^{-1} K^{-1}), and T is in the *Kelvin temperature scale*. It should be noted that over an ice surface L_v is substituted by the latent heat of sublimation (L_s, or 2.8345×10^6 J kg^{-1} at 0°C).

If we plot e_s as a function of T using the Clausius-Clapeyron equation, the relationship between these variables is an exponential curve, as shown in **Figure 5.15**. A look at the e_s curve confirms that as the temperature falls (during the course of a "normal" evening, for example), e_s decreases. But a decrease in e_s causes relative humidity to increase, even if the actual amount of vapor in the air (e) remains constant.

To see why, suppose the temperature is 20°C (68°F) and the air has 12 mb of water vapor in it. Figure 5.15 shows that at that temperature the air would be saturated if it had 24 mb of water vapor in it. The relative humidity in this case would be 50 percent. If the temperature began to fall until e equals e_s (as shown by the line from A to B in

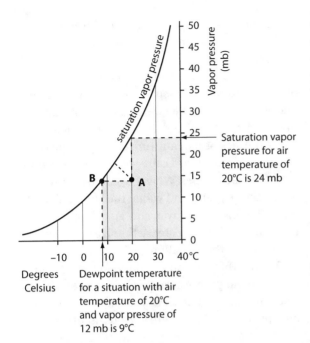

Figure 5.15 Saturation vapor pressure as a function of temperature.

Figure 5.15), saturation will be reached, and the relative humidity will be 100 percent.

The temperature at which saturation occurs at a given place and time is known as the **dewpoint temperature** (T_d). This is shown by the temperature at B in Figure 5.15. To be more precise, whenever T_d is below 0°C it is termed the **frostpoint temperature**. In Figure 5.15 it is evident that if $e = 12\,\text{mb}$ (shown at A), then the air would have to cool from 20°C to about 9°C (48°F) before it would become saturated. Thus T_d in this case would be 9°C.

If T_d is reached, condensation or deposition begins, releasing latent energy during the phase change. The same amount of solar radiant energy that was required to evaporate a certain volume of water (540 calories per gram, at 100°C) would be released when that volume of water then condensed. The amount of energy released during deposition would equal the sum of the energy released when the same volume of water condensed (vapor to liquid) and when it froze (liquid to solid), plus the amount of energy released in cooling the liquid water from the temperature at which it became a liquid to the freezing point.

The energy transformation described above is one manifestation of the **first law of thermodynamics**—a very important principle stating that energy can never be simply created or destroyed in nature; it can only change from one form to an-

other. This concept pervades all the natural sciences. Insolation is converted from a radiant form of energy to a latent form during the evaporation process, and then this energy is released when the vapor returns to a liquid or solid state. This released energy cannot simply vanish; it must be converted to another form. It is this released energy that appears as kinetic energy. Ultimately, then, the energy to drive winds in a storm comes from the Sun.

As we have seen, although T_d is expressed in units of temperature, it is actually a measure of humidity. The closer the dewpoint temperature is to the air temperature, the more humid the air, because in such a case the temperature would have to fall only slightly for it to become saturated. During the course of a day T_d changes very little, unless the actual amount of vapor in the air (e) changes. One scenario that could cause e to change is a shift in wind direction that introduces advection of moisture from an oceanic surface into an area. A small increase in T_d would also occur in the morning if there is dew or frost on the ground. In that case when the Sun rises, its radiation evaporates the dew or melts or sublimates the frost.

In actuality, changes in air temperature do change the dewpoint temperature a bit—about one-sixth of a Celsius degree per Celsius degree of temperature change—even if the amount of water vapor in the air (e) remains constant. But this change is very slight compared with the change in relative humidity with temperature. Dewpoint temperature is actually a more useful and meaningful measure of atmospheric humidity than relative humidity, which only states how close air is to saturation regardless of vapor content.

Another quantity that may be used to represent the amount of vapor in the air is **absolute humidity**—the mass of water vapor per cubic meter (a volume) of air. Because its units are actually those of *density*, absolute humidity is sometimes known as vapor density. This quantity is seldom used in climatological work because a cubic meter of air may contain a drastically different amount of mass at two different vertical heights in the atmosphere. Indeed, as you go up a mountain there is bound to be less water vapor per cubic meter of air because there is less of every type of molecule with height. It is, therefore, usually impractical to use absolute humidity to compare the moisture content of air from one place to another at a different elevation.

Another means of expressing moisture in the atmosphere is by **specific humidity** (usually abbreviated as q)—the ratio of the mass of water vapor to the mass of the air. Because there are no units for q, we say that it is a **dimensionless quantity**—a ratio or constant that has no units associated with it. In this case the units are kilograms per kilogram, which cancel out. Specific humidity is an effective variable for comparing the moisture characteristics across geographic distances because it is not temperature dependent (as are relative humidity and, to a lesser extent, dewpoint temperature) and it does not inherently change because of elevation (as does absolute humidity).

Because specific humidity is so useful, the important question is this: "How is it measured?" The answer is that it is not measured directly for most applications. Instead, it can be computed by a relationship derived from atmospheric thermodynamics:

$$q = \frac{e\varepsilon}{p}$$

where p is atmospheric pressure and e is the ratio of the molecular weight of water vapor to the molecular weight of dry air—a dimensionless constant of 0.622. But how can we measure e? Because relative humidity is the ratio of e to e_s and e_s is a known function of temperature via the Clausius-Clapeyron equation, measurement of temperature (with a thermometer) and relative humidity (with a **hygrometer**) can allow us to compute e, which can then be used to compute q, provided that atmospheric pressure is known.

Moisture in the Surface Boundary Layer

Because of the practicality of specific humidity, it is often the variable used to represent moisture in the SBL. Highly precise and accurate thermoelectric sensors can derive q from surrogate variables. As we will see, this is a convenient means of estimating the flux of water from the surface to the air through the combined processes of evaporation and transpiration.

Less familiar than evaporation, *transpiration* is the process by which vegetation loses water from its leaf surface that had moved from the soil through its roots and stem. The loss of water occurs through the **stomates** (also known as stomata)—tiny pores in the leaf tissue. An analogous process

in humans is perspiration, where water stored in the biomass of humans is lost through pores in the epidermis.

Unlike humans, vegetation can regulate the loss of water through transpiration. During drought and at other times when water is scarce, reductions in this rate reduce water losses. This happens because the lack of soil moisture decreases the water circulated to the **guard cells**—leaf cells that are adjacent to the stomates. When the guard cells receive water, they expand and widen the stomatal openings. The absence of water in guard cells causes them to shrink and decrease the size of the stomatal openings, thereby allowing less water to escape to the air. This natural regulation process allows vegetation to pump excess water to the air during wet times and to conserve water during dry periods, providing them a greater possibility for survival during periods of environmental stress.

The transpiration process is driven by the Sun, in the same way that the Sun provides radiant energy to drive the evaporation process. Transpiration rates are generally greatest on a dry, sunny day when abundant water is available in the soil. This scenario creates a strong gradient of water from the surface to the air and a tendency for the vapor to be moved upward toward drier areas. At night, transpiration decreases to an insignificant rate.

Climatologists are usually not concerned with distinguishing the amount of water that is lost from the surface to the air via the evaporation and transpiration processes individually. Instead, they usually lump the quantities of atmospheric moisture attributable to evaporation and transpiration together to form a variable known as **evapotranspiration (ET)**. Because of its significance in hydrological, biological, agricultural, and economic applications, ET is an important variable in many environmental monitoring applications. Unfortunately, the measurement of ET is often substandard, and high-quality data usually are available only through a sparse network of sites. As a result, simpler but less accurate means of estimating ET are often used and reported in historical data sets as ET.

Measuring Evapotranspiration

The *National Weather Service* determines ET through the use of a **Class A Evaporation Pan**, an instrument that consists of a metal pan 1.2 m (4 ft) in diameter that sits on a short wooden platform and is exposed to the elements (**Figure 5.16**). Inside the pan is a heavy, cylindrical metal tube reaching the bottom of the pan, and inside this cylinder is an analog gauge with the appearance of a fishhook. Water is placed in the pan up to about 8 cm (3 in) from the top, and the analog "fishhook" gauge is positioned vertically with a caliper so that it barely breaks the plane of the water. The next day, after some water has evaporated, the fishhook gauge is repositioned, and the vertical displacement distance read on the caliper represents the centimeters or inches of water that have evaporated since the previous reading. Of course, if precipitation has fallen since the last measurement period, the evaporation reading must take into account the amount of precipitation (measured in an adjacent gauge) that was added to the pan.

In the "real world," water is not always as readily available in the environment as it is in the evaporation pan. The use of an evaporation pan, therefore, provides an estimate of **potential evapotranspiration (PE)**—the maximum amount of ET that would occur if water were not a limiting factor. If this value is less than the amount of precipitation that has fallen over that interval, water is abundant.

However, the pan even overestimates the "true" PE, for three important reasons: (1) the metal pan is heated much more than a natural area is heated, allowing more energy to be available to evaporate water than is present in a natural environment; (2) cohesion between liquid water drops and soil particles in the natural environment slows the evaporation rate as compared with the higher evaporation rates inherent in the pan; and (3) PE is

overestimated in the pan because the presence of wind moving across the pan provides a continual source of relatively dry air into which water readily evaporates. By comparison, if a large field in the real world was saturated with water, some water would evaporate and then move over another part of the field after having already been saturated with evaporated water. It would, therefore, not evaporate any new water from that part of the field. As a result of these factors, a **pan adjustment coefficient** is typically used to estimate PE from pan evaporation. It varies spatially but averages about 0.7, suggesting that PE occurs at a rate of about 70 percent of the evaporation rate in the pan.

Another means of determining ET rates is by means of a **lysimeter**, an instrument installed beneath the surface that weighs the soil and vegetation on top of it (**Figure 5.17**). By knowing the mass of dry soil and vegetation along with the density of water (1000 kg m^{-3}), the water content of the volume can be determined. Although highly accurate, lysimeters are far more expensive to install and maintain than evaporation pans, and the network of lysimetric data is particularly sparse.

Figure 5.17 Installation of a lysimeter at St. Joseph Research Station, St. Joseph, Louisiana.

Figure 5.16 A Class A evaporation pan.

Far more sophisticated methods exist for measuring ET, but because these require precise and expensive equipment they are usually done only experimentally and not on a routine, ongoing basis. An example is **eddy covariance**, where theoretical equations are derived based on the measurement of covariance of fluctuations between vertical wind velocity (for vertical fluxes) and changes in temperature, specific humidity, and horizontal wind speed. These equations allow the estimation of turbulent transfer of energy, moisture, and momentum, respectively. For example, measurement of Q_E, the energy equivalent of evaporation, by eddy covariance or by other means can be used to derive ET. This is based on the premise that the ET rate (in meters of precipitation equivalent per second, which can then be converted to mm hr^{-1}) is a function of Q_E as follows:

$$ET = \frac{Q_E}{L_v \rho_\ell}$$

where L_v is the latent heat of vaporization (2.5008×10^6 J kg^{-1} at 0°C) and ρ_ℓ represents the density of water (1000 kg m^{-3}):

Eddy covariance systems have been deployed in recent years for varying surface types. Unfortunately, despite decreasing costs of equipment in recent years, eddy covariance equipment remains expensive and is not readily available at many locations.

■ Atmospheric Statics, the Hydrostatic Equation, and Stability

The ease with which moisture (along with energy, momentum, and other forms of matter) can move vertically in the air at a given time and place is governed by **stability**—local atmospheric conditions that describe the likelihood of a local mass or "parcel" of air to rise or sink spontaneously. The vertical distribution of absorbed energy in the local near-surface air is extremely influential in determining the stability conditions. If the air parcel has a tendency to rise, the vertically directed force of **buoyancy** pulls it upward in opposition to the downward-directed gravitational force. In other situations the downward-directed gravitational force exceeds the upward-directed buoyancy force and the parcel of air sinks. Stability conditions impact weather and climate greatly, because upward motion is important in enhancing convection and generating clouds and precipitation.

Statics and the Hydrostatic Equation

The **hydrostatic equation** expresses the relationship between the upward-directed buoyancy force and the downward-directed acceleration caused by gravity (9.8 m s^{-2}). It relates the pressure change (Δp) in a static atmosphere to height differences (Δh) over which those pressure changes are observed, as follows:

$$g = -\frac{1}{\rho} \frac{\Delta \rho}{\Delta h}$$

where ρ is atmospheric density. The hydrostatic equation shows that the mean density of a layer of the atmosphere can be calculated if we know the pressure at two known heights—the top and the bottom of that layer. Because the gravitational acceleration is essentially constant near Earth's surface, it is apparent from the equation above that as density increases, the difference in pressure with height must also increase. Pressure decreases quickly with height near the ground (where density is highest), but higher up in the atmosphere (where density is less) pressure does not decrease so quickly with height. We have already seen that the *equation of state* shows that cold air is denser than warm air at a given pressure. So now the hydrostatic equation can be used to verify a simple feature of atmospheric motion: Cold (dense) air has a tendency to sink and create a large vertical difference in pressure (high pressure at the bottom and much lower pressure above—the location from which the air left).

Thus, the hydrostatic equation describes vertical motion in the atmosphere, and it is this vertical motion that is associated with stability. A parcel of air cools as it rises. It is tempting to believe that if it cools it must sink, but this is not necessarily the case. Even as it cools, if it remains warmer than the surrounding air, the parcel continues to rise—positive buoyancy. If its ascent causes the parcel to cool to the dewpoint temperature, the condensation process would begin to create a cloud, which may produce some form of precipitation. In humid locations there is often enough moisture in the air that the parcel does not have to rise far to cool enough to become saturated, so the cloud-formation process happens relatively easily. For this reason the cloud base tends to be relatively near the surface in humid areas. In arid locations a parcel has to rise to great heights before it cools to the dewpoint temperature and forms clouds. This leads to very high cloud bases and less likelihood of precipitation.

Atmospheric Stability

In a *stable atmosphere* if a parcel of air is forced to rise for any reason, it will sink once the lifting force is removed. Thus, the buoyancy force cannot overcome the gravitational force in stable conditions. If the atmosphere is stable at a given place and time, it is difficult for energy and moisture to move vertically from the surface, and little of the moist air can be cooled to the dewpoint temperature, making cloud formation difficult or impossible. Stable conditions are usually associated with fair weather. The downward motion associated with stable conditions allows momentum to be transferred relatively easily from aloft (where friction is relatively low) toward the near-surface atmosphere (where friction is higher). The lack of vertical motion under stable conditions can also have less pleasant effects. For example, pollutants are difficult to disperse from surface point sources under stable conditions.

By contrast, in an *unstable atmosphere* the parcel of air continues to rise even after the lifting force is removed. The buoyancy force exceeds the gravitational force under unstable conditions. Such conditions enhance the upward movement of energy and moisture from the surface and often allow the vertical growth of thunderstorm clouds and the dispersion of local pollutants.

Stability conditions involve a continuum rather than a simple stable/unstable dichotomy. At certain places and times the atmosphere may be extremely unstable, whereas in other situations it may be slightly unstable, slightly stable, or extremely stable. A hypothetical case is the **neutral atmosphere**—an atmosphere where the parcel neither rises nor sinks once the lifting force is removed. It maintains its position no matter where it is placed in the atmosphere because its internal temperature is always equal to the temperature of the ambient (surrounding) atmosphere.

So how do we know the degree of stability (or instability) of a local atmosphere? To understand what determines the stability conditions, it is im-portant to understand the concept of **lapse rate**—the rate at which temperature changes (decreases usually, in the troposphere) with height. By convention, a positive lapse rate is said to occur when the temperature decreases with increasing height. A negative lapse rate is known as a temperature inversion because the temperature profile is "inverted" from what we normally expect in the troposphere; that is to say, in an inversion temperature increases as height increases.

If a parcel of air rises (or sinks) quickly enough that it does not have enough time to exchange any energy with the surrounding air, it is said to have **adiabatic motion**, which implies that the cooling of the parcel with increasing elevation is caused solely from within the parcel itself. If a parcel rises adiabatically, it cools at a totally predictable lapse rate, because it can expand under the decreasing pressure exerted by the (decreasing) column of air above it. It cools as it expands because one implication of the first law of thermodynamics is that energy is used in some combination of either maintaining the parcel's temperature or doing work (i.e., expanding the parcel's area). Because the total amount of energy in the parcel remains constant in an adiabatic process (by definition) and the amount of pressure exerted on the parcel from above decreases as it rises, the use of its energy to expand becomes preferable at the expense of maintaining the temperature of the parcel.

On the other hand, as the parcel sinks adiabatically, it must warm at a totally predictable lapse rate because it must compress under the increasing column of air exerting increasing pressure on it from above. Work is done on the parcel from above in this case, and the energy is used to increase the parcel's temperature. The rate of warming for sinking motion is exactly the same as the rate of cooling for rising motion, in fulfillment of the first law of thermodynamics.

This rate of cooling with ascent and warming with descent is termed the **unsaturated adiabatic lapse rate (UALR)**. The UALR—also represented by Γ_u and pronounced "gamma sub u"—is approximately 9.8 C°/km (or about 1 C°/100 m) and is the rate at which the parcel of air cools with ascent (or warms with descent) at every location on Earth, every day of the year, as long as it remains unsaturated.

If at any point on the parcel's ascent it becomes saturated, the UALR is no longer in effect, even if the motion remains adiabatic. The reason for this difference is that at saturation, latent energy is released to other air molecules inside the parcel in the condensation, freezing, or deposition process. This release of latent energy prevents the parcel from cooling as rapidly as it could cool under unsaturated conditions.

As the parcel continues to rise, less water vapor exists in it at the decreasing temperatures and greater distance from the surface. Consequently, successively less latent energy can be released in the condensation, freezing, or deposition processes, and the parcel can cool more rapidly with its continued ascent than initially upon saturation. Eventually, it cools at nearly the same rate as if it were not saturated. But the net effect of adiabatically rising motion will still always be cooling because the effect of reduction of pressure from above will more than compensate for the effect of the latent heat release. However, the rate of cooling with ascent will never be as great as in unsaturated conditions. With sinking motion, adiabatic warming will always occur at Γ_u because even a minuscule increase in pressure from above will warm the parcel to the extent that it is no longer saturated (because e_s always increases with temperature).

From this discussion, then, it is apparent that unlike the UALR, the **saturated adiabatic lapse rate** (Γ_s) is not a constant value. Instead, it varies according to the amount of moisture in the air to be condensed, frozen, or deposited. In the tropics the high temperatures allow abundant moisture to exist in the air (i.e., high e_s). Under conditions of saturation (such as over the tropical oceans), condensation, freezing, and deposition cause abundant latent energy release, hindering the parcel's cooling on its initial ascent. On the other hand, over polar regions in winter, where e_s is very low under such frigid temperatures, little moisture exists to change phase and release latent energy. The SALR in such conditions is only slightly less than the UALR.

Is the adiabatic assumption valid? The answer is that it is never completely valid, but sometimes it is more valid than at other times. Regardless, we have to make the adiabatic assumption because there is no way to monitor the temperature change of every parcel of air that rises or sinks at all times across the entire Earth. So this is a necessary assumption, but one that can sometimes be invalid.

Weather forecasts are likely to fail when the assumption is not reasonably valid.

The third lapse rate has already been introduced: the ELR (γ)—the temperature profile of the static (ambient) atmosphere surrounding an adiabatically moving parcel of air. The ELR varies daily and across space (**Figure 5.18**). It may be negative (in the case of a temperature inversion), strongly positive (when temperature decreases very rapidly with increasing height), or slightly positive (when temperature only decreases slowly with height). On a global, annual average basis γ is about 6.5 C°/km (3.5 F°/1000 ft). So if we neglect all other factors, we could expect that a location at 1.5 km (5000 ft) in elevation would, on average, be approximately 9.7 C° (17.5 F°) cooler than it would be if it were at sea level.

The ELR can be measured by sending up a **radiosonde**—an instrument package that ascends with weather balloons. At 74 different sites in the contiguous United States, the National Weather Service launches weather balloons synchronously at 0000 UTC and 1200 UTC (i.e., at midnight and noon, respectively, in Greenwich, England). These radiosondes electronically report the temperature, pressure, and dewpoint temperature from many different heights to ground-based computers. The data are then sent to a central computer near Washington, DC where they are used in weather analysis and prediction.

Thus, three different lapse rates are important for assessing atmospheric stability. If a parcel of air is moving adiabatically, its tendency to rise or

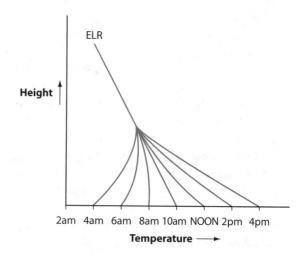

Figure 5.18 Typical variation of the ELR throughout the course of a day.

sink is determined simply by comparing its temperature (which is governed by Γ_u or Γ_s, as the case may be) with the temperature of its surrounding environment (represented by γ). If the parcel of air is warmer than its surrounding environment at a given height, it tends to rise (as in the unstable atmosphere shown in **Figure 5.19** and **Figure 5.20**).

If the air parcel is colder than its surrounding environment at any height, it sinks (as in the stable atmosphere shown in Figures 5.19 and 5.20). However, if an air parcel is in a situation in which it is

forced to rise, as happens when air is moved up a mountain by a wind or when air rises when it encounters colder (denser) air, it rises even if it is statically stable.

It is not uncommon at all for γ to fall between Γ_u and Γ_s. When $\Gamma_s < \gamma < \Gamma_u$ the air parcel in an unsaturated atmosphere is cooler than its surrounding environment and the atmosphere is stable. When $\Gamma_s < \gamma < \Gamma_u$ and the air is saturated, the air parcel is warmer than its surrounding environment at a given height and the atmosphere is unstable. By increasing the amount of moisture in the air to saturation, upward motion is encouraged and the atmosphere is destabilized. This **conditional instability**—with the atmosphere's stability "condition" determined by whether or not the air is saturated—is important because it facilitates the precipitation-formation process because saturated air can rise more easily than unsaturated air.

Assessing Stability in the Local Atmosphere

Thermodynamic Diagrams   The stability conditions of a local atmosphere are identified easily using a **thermodynamic diagram**—a tool that can be used by atmospheric scientists to make a quick determination about whether or not air at a given place and time is likely to rise spontaneously. **Figure 5.21** shows the most common type of thermodynamic diagram used in the United States—the **Skew T–Log P diagram**. The medium zigzagging solid

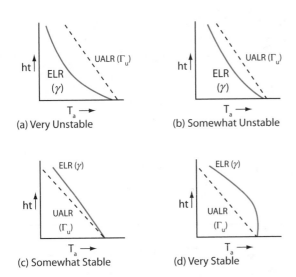

Figure 5.19   Unsaturated adiabatic lapse rates under various atmospheric conditions: (a) very unstable, (b) somewhat unstable, (c) somewhat stable, and (d) very stable.

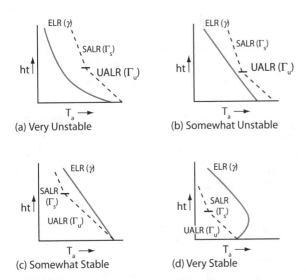

Figure 5.20   Saturated adiabatic lapse rates under various atmospheric conditions: (a) very unstable, (b) somewhat unstable, (c) somewhat stable, and (d) very stable.

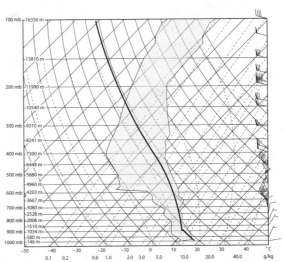

Figure 5.21   Thermodynamic diagram for Washington, DC, at 8:00 AM local time, May 31, 2005. *Data from*: National Weather Service/NOAA.

line on the right represents the temperature profile of the static atmosphere from the surface upward to over 16 km (9.9 mi) high—the ELR. The lines representing the temperature are slanted, or "skewed," upward from left to right. The medium zigzagging solid line on the left side of the diagram represents the dewpoint temperature at any height. Pressure is shown by the horizontal lines, which increase in spacing vertically. The "log P" reminds us that only the logarithm of pressure increases at equal intervals on the graph.

Figure 5.21 shows that at a height of 7300 m, the temperature of the static atmosphere is approximately −30°C. In atmospheric layers where the two medium solid lines are near each other (such as in the lower troposphere in Figure 5.21), the atmosphere is more moist, and in layers where the two medium solid lines are far apart, the air is drier. The thick solid line in Figure 5.21 represents the **adiabat**—the temperature profile of an adiabatically moving parcel of air. Notice how the air parcel's temperature abruptly shifts at an elevation of about 1050 m in Figure 5.21.

This represents the **lifting condensation level**— the level at which the air becomes saturated and the air parcel no longer cools at the UALR but begins cooling instead at the saturated adiabatic lapse rate.

Now look at the example shown in **Figure 5.22**. In this case it is obvious that the air is far more humid than in the previous example, because the ELR and dewpoint plots are near each other.

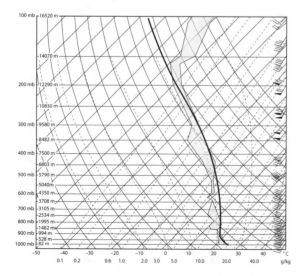

Figure 5.22 Thermodynamic diagram for Tallahassee, Florida, at 8:00 am local time on June 1, 2005. *Data from*: National Weather Service/NOAA.

We can see from comparing the position of the adiabat with that of the ELR that the adiabatically moving air parcel is warmer than its surrounding environment in the lower to middle troposphere. The vertical distance for which the adiabatically moving air parcel trajectory is warmer than γ represents the depth of the unstable layer. The area between the parcel's trajectory and temperature profile is approximately proportional to the intensity of the instability. The same rules hold in cases of stable atmospheres.

Potential Temperature From the discussion above, it follows that in an unsaturated atmosphere, stability is related to the vertical gradient of environmental temperature across some height z as follows:

For a stable atmosphere:

$$\frac{\Delta T}{\Delta z} > -9.8 \text{ C°/km}$$

For an unstable atmosphere:

$$\frac{\Delta T}{\Delta z} < -9.8 \text{ C°/km}$$

For a neutral atmosphere:

$$\frac{\Delta T}{\Delta z} = -9.8 \text{ C°/km}$$

where Δ represents the difference in the value of the variable at two measurement heights. Notice that the negative signs are necessary because temperature would decrease with increasing height at the UALR.

It is convenient to define a new variable at this point to simplify the calculations necessary to determine local stability using the equations above. By definition, **potential temperature** (θ) is the temperature that a parcel of air would have if moved dry adiabatically from its height (z) in the atmosphere to the 1000-mb level. For example, we know that if the parcel were at the 850-mb level, it would have to warm at Γ_u from the height of the 850-mb level down to the height of the 1000-mb level. Mathematically, we can say that

$$\theta = T + \Gamma_u \left(z - z_{1000\,mb} \right)$$

Using differential calculus we could differentiate the equation above and show that a stable atmosphere (or layer of the atmosphere) is one in which θ increases with height. Likewise, unstable conditions are characterized by θ that decreases

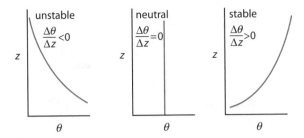

Figure 5.23 A typical gradient of potential temperature with height under unstable, neutral, and stable conditions.

with height, and a neutral atmosphere is one in which θ is constant with height (**Figure 5.23**).

In analyses in the SBL, θ is often replaced by T because the difference between the gradient of temperature with height and the gradient of θ with height is so slight in near-surface, local-scale applications. Recall that because $\Gamma_u = 9.8$ C°/km, this only amounts to approximately 0.01 C°/m. You can see from the previous equation that the difference between θ and T is negligible at such scales. But for synoptic- and global-scale analysis, the difference between θ and T is usually significant.

■ Momentum Flux

In addition to energy, matter (particularly water vapor), and stability, another important feature of the local atmosphere is momentum. For example, when wind speed is much greater at 2 m above the surface than at 0.5 m above the surface, momentum transport is enhanced because the extra momentum associated with the faster winds must be transported downward to satisfy Fick's law. You experience this momentum flux whenever you feel a gust of wind. Whenever winds increase with height, horizontal momentum is transferred downward to locations where less momentum is present. Natural systems always work to eliminate gradients of natural entities, such as pressure, moisture, and momentum. The gust that you feel in a thunderstorm is probably the result of efficient momentum exchange under very unstable conditions.

As we noted at the beginning of this chapter, unlike the gradient of energy and matter (i.e., water vapor), the momentum flux (usually represented by τ) is usually downward rather than upward, because wind speed (and, therefore, momentum) usually increases with height in the troposphere as friction decreases. As was the case for energy and

moisture, the sharpest vertical gradient of momentum is found in the SBL, but in the SBL this flux is constant with height at any given instant. Unlike the effect of friction in the free atmosphere, the vertical difference in momentum is important for any two heights in the SBL separated even by short vertical distances.

■ Putting It All Together: Thermal and Mechanical Turbulence and the Richardson Number

As introduced earlier, both thermal turbulence and mechanical turbulence can produce rising motion in a local atmosphere. Upward motion associated with thermal turbulence occurs when the atmosphere is unstable, which occurs when θ decreases with height (i.e., $(\Delta\theta/\Delta z < 0)$). You may have noticed a bird take off from a locally warm spot, such as a bare rock, because the hot air at the surface of the bare rock provides sufficient thermal turbulence to aid in lift for takeoff. Thermal turbulence is sometimes referred to as **free convection** because vertical motion results spontaneously from differences in energy content between an adiabatically moving air parcel and its surrounding environment.

On the other hand, mechanical turbulence occurs independently of stability conditions. Rather, it is **vertical wind shear**—the change of wind speed with height ($\Delta u/\Delta z$)—that produces the momentum gradient associated with mechanical turbulence. Mechanical turbulence is sometimes referred to as **forced convection** because the vertical motion results from dynamic rather than thermodynamic causes. Even when the atmosphere is stable, mechanical turbulence can cause uplift through forced convection.

The **gradient Richardson number (Ri)** provides a convenient index for describing the relative importance of thermal and mechanical turbulence in the local near-surface atmosphere. It is also useful for assessing local stability conditions. Mathematically, in a vertical layer of the atmosphere,

$$\text{Ri} = \frac{g}{\theta}\frac{\Delta\theta/\Delta z}{(\Delta u/\Delta z)^2}$$

where g is the gravitational acceleration (9.8 m s⁻²) and $[\bar{\theta}]$ represents the mean potential temperature in the layer.

Notice several features of the preceding equation. First, the mechanical turbulence term (in the denominator) is squared, so it cannot be negative (though even before squaring the term it would rarely be negative because we usually expect wind speed to increase with height). The only term that can be negative is the thermal turbulence term (in the numerator). Thus, under unstable local conditions (when, by definition, $\Delta\theta/\Delta z$ is negative), Ri must be negative. As the atmosphere becomes more unstable, $\Delta\theta/\Delta z$ becomes more negative and Ri approaches $-\infty$. By contrast, in a stable local atmosphere (when, by definition, $\Delta\theta/\Delta z$ is positive), Ri must be positive. As the atmosphere becomes more stable, $\Delta\theta/\Delta z$ becomes more positive and Ri approaches ∞. Looking at the equation another way, we can say that at a small vertical wind shear $\Delta u/\Delta z$, |Ri| is very large, and the atmosphere tends to be strongly stable or strongly unstable. In its most extreme case, at $\Delta u/\Delta z=0$ (as when there is no wind or no vertical gradient of wind), |Ri| approaches $\pm\infty$.

In nonneutral conditions both thermal and mechanical turbulence are present unless the winds are calm or constant with height. In an unstable local atmosphere, vertical motion resulting from mechanical turbulence is enhanced by the thermal profile of the atmosphere. In stable conditions the thermal structure dampens vertical motion generated by mechanical turbulence.

The case of a neutral atmosphere is both interesting and important. Under neutral conditions $\Delta\theta/\Delta z$ is zero by definition. In other words, there is no thermally induced turbulence, and Ri must be zero. Thus, for a neutral atmosphere any turbulence must be mechanical in origin, and the source of energy for all turbulence is the kinetic energy of the wind from the free atmosphere. Stated another way, regardless of the sign of Ri, for a given $\Delta\theta/\Delta z$, |Ri| decreases as wind shear (i.e., mechanical turbulence) increases. Neutral conditions are best met under cloudy skies and strong winds in the lowest 1 to 2 m of the atmosphere.

Much theory in boundary layer climatology rests on the assumption that mechanical turbulence dominates thermal turbulence (i.e., that near-neutral conditions exist). For example, if the wind speed at only one height is known, then it is much easier and more accurate to estimate the wind speed at any other height in the SBL under neutral rather than nonneutral conditions. This **logarithmic wind profile** is an important manipulation because climatological data sources usually measure wind at only one height in the atmo-

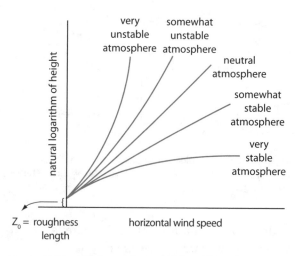

Figure 5.24   The logarithmic wind profile and variations of wind profile with stability. *Adapted from*: Oke, T.R., *Boundary Layer Climates*, 2nd ed. New York: Routledge, 1987.

sphere. **Figure 5.24** shows the plot of wind speed versus the natural logarithm of height under varying stability conditions. Notice that, theoretically, wind speed approaches zero at some height z_0— the **roughness length**—which is approximately 10 percent of the height of the dominant features on the ground (such as blades of grass, trees in a forest, or houses). Thus, in neutral conditions a straight line can be drawn on semilogarithmic graph paper to connect the points represented by measured wind speed at one known height (u_z) and z_0 for a given land cover (where wind speed is zero). This line allows us to estimate the wind speed at any known height. The slope of this line varies; it is inversely related to τ.

Under stable or unstable conditions, measurement of wind speed at one height alone is not as useful for estimating wind speeds at other heights, because as is shown in Figure 5.24, the profile is not linear under such conditions. The validity of making an assumption of neutral stability should always be checked.

■ Summary

The lower troposphere is a very complicated as well as important part of the atmosphere because it lies closest to people and Earth's surface. The laminar layer occupies the lowest few millimeters of the atmosphere, and the roughness layer, where the most vigorous exchanges of energy, matter, and momentum occur through mechanical and thermal turbulence, lies above it. Collectively, these two layers are termed the SBL. In the SBL turbulent fluxes of sensible heat (which generally transport energy upward in the local atmosphere), latent heat (which usually transports moisture upward), and shearing stress (which usually transports momentum downward) are considered to be constant. Above the SBL lies the transition layer, where the influence of friction is present but decreasing with height, and above that layer is the free atmosphere, where friction can be considered to be negligible.

Calculations of local energy transport begin with understanding the position of the Sun in the sky at a location. This depends on the time of day and the time of year. A Sun path diagram can be used to map the Sun for a given latitude and solar declination according to its zenith angle and azimuth angle, once the local apparent time is determined based on the distance of the location from the principal meridian and the time of year.

Radiant energy from the Sun is determined by using radiometers. Special types of radiometers measure only shortwave radiation, only longwave radiation, or all-wave radiation. Shortwave energy from the Sun may be absorbed, scattered, or reflected before it reaches the surface. It may also be scattered down to the surface as diffuse radiation. Direct radiation is transmitted through the atmosphere all the way to the surface. Although some longwave radiant energy emitted from Earth's surface escapes into space, most tends to be absorbed by the atmosphere at most locations and is then largely reradiated back downward in the form of counter-radiation. The net radiation at a given place and time refers to the difference between the amount of all-wave radiant energy absorbed and the sum of shortwave energy reflected and longwave energy emitted.

The transfer of radiant energy alone is not sufficient to make the shortwave energy receipt at the surface balance the loss of energy from the surface through longwave fluxes. Instead, convection is important at all scales to complete the energy balance, and conduction may also be important locally. Convection is accomplished through the turbulent fluxes of sensible heat and latent heat. Conduction occurs through the substrate heat flux. The radiative, convective, and conductive fluxes collectively comprise the energy balance of a location. When the energy balance is positive the local temperature is increasing, and when the energy balance is negative the local temperature is decreasing. But for the whole Earth–ocean–

atmosphere system the energy balance theoretically remains at zero on short time scales. At longer time scales other factors (both natural and human induced) may cause long-term changes to the system's energy receipt and temperature.

Atmospheric moisture is the most important component of mass in the SBL. Moisture can be expressed either by vapor pressure, relative humidity, dewpoint temperature, or absolute humidity, but in many microclimatological studies of atmospheric moisture, specific humidity affords some advantages as the variable of interest. Moisture may be transported upward either directly by evaporation or transpiration through vegetation, but climatologists usually simply consider the combined effect of ET. Most higher-level plants can control the loss of moisture through transpiration to some extent because the stomates on their leaves may partially close during dry periods. Such plants can affect their local atmospheric environments by accelerating the loss of water to the atmosphere during times of waterlogged soils (as long as the atmosphere is not saturated) and slowing the transpiration rates as the soil begins to dry out. ET may be measured using a lysimeter, and potential ET may be estimated with a Class A Evaporation Pan.

The vertical movement of energy and moisture within and beyond the SBL is an important feature of the local circulation. The hydrostatic equation describes the balance of forces between the downward-directed gravitational acceleration and the upward-directed buoyancy effect. Atmospheric stability conditions play a major role in the tendency for energy and moisture to move vertically through the SBL to the free atmosphere.

In an unstable atmosphere vertical motion is enhanced, turbulent fluxes are encouraged, and a rising parcel of air that does not exchange energy with surroundings (i.e., moving adiabatically) may cool to its dewpoint temperature on its ascent whereupon its moisture condenses and forms a cloud. In a stable atmosphere vertical motion is suppressed, dispersion of pollutants is minimized, turbulent fluxes are more likely to be reduced, and an adiabatically moving air parcel will not be likely to rise to the point at which it reaches its dewpoint temperature unless it is forced to rise by some other mechanism.

Stability conditions at a specific place and time can be assessed by comparing the ELR with the unsaturated (or saturated if saturation has been reached) adiabatic lapse rate. If the adiabatically moving parcel of air is warmer than its surrounding environment, the atmosphere is unstable at that height, and if it is cooler than its surrounding environment, the atmosphere is stable at that level. Stability can be assessed relatively easily using a thermodynamic diagram such as the Skew T–Log P diagram.

The replacement of temperature with potential temperature simplifies calculations of stability. Any situation in an unsaturated atmosphere in which the potential temperature increases with height is stable, whereas unstable conditions occur when potential temperature decreases with height. In the lowest few meters of the atmosphere, the difference between the gradient of temperature with height and the gradient of potential temperature with height is negligible.

Unlike the transfer of energy and matter, momentum is typically transferred downward because wind speed usually increases with height.

This momentum *flux* (or shearing stress) is represented by the degree of vertical wind shear or the difference in horizontal wind speed across a vertical length. This flux is much greater in the SBL than in the free atmosphere. The ratio of thermal turbulence (which is a function of potential temperature and indicates stability conditions) to mechanical turbulence (a function of vertical wind shear and representing vertical momentum transfer) is known as the gradient Richardson number. The atmosphere is stable when that number is positive and unstable when negative. The latter case is sometimes termed "free convection." The atmosphere is in a neutral stability condition when the gradient Richardson number approaches zero, and in such a case all vertical motion results from mechanical forcing, also known as "forced" convection. Neutral conditions are often assumed in much boundary layer theory, including the logarithmic wind profile. Thus when such assumptions are made, their validity should be checked.

▶ Key Terms

Absolute humidity	Diffuse radiation	*Joule*
Absolute zero	Dimensionless quantity	*Kelvin temperature scale*
Absorption	Direct radiation	*Kinetic energy*
Adiabat	Eddy covariance	Laminar layer
Adiabatic motion	Energy balance	Lapse rate
Advection	*Environmental lapse rate (ELR)*	*Latitude*
Aerosol	*Equation of state*	Lifting condensation level
Albedo	Equation of time	*Lithosphere*
Analemma	Equinox	Local apparent time
Arctic Circle	*Evaporation*	Logarithmic wind profile
Attenuation	Evapotranspiration (ET)	*Longitude*
Azimuth angle	Fick's law	Longwave radiation
Biosphere	First law of thermodynamics	Lysimeter
Buoyancy	*Flux*	Mechanical turbulence
Centrifugal force	Forced convection	Microscale
Class A Evaporation Pan	*Free atmosphere*	*Millibar (mb)*
Clausius-Clapeyron equation	Free convection	*Momentum*
Condensation	*Friction*	*National Weather Service*
Conditional instability	Frostpoint temperature	Net radiation
Conduction	*Geostrophic balance*	Neutral atmosphere
Convection	Gradient Richardson number (Ri)	Optical air mass
Coriolis effect	Guard cell	Pan adjustment coefficient
Counter-radiation	*Hydrosphere*	*Pascal (Pa)*
Cryosphere	Hydrostatic equation	*Path length*
Deposition	Hygrometer	*Photosynthesis*
Density	*Insolation*	Planetary albedo
Dewpoint temperature (T_d)	International date line	Planetary boundary layer

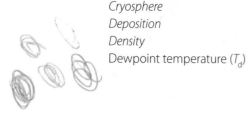

Planetary scale
Potential evapotranspiration (PE)
Potential temperature
Pressure gradient force
Principal meridian
Pyranometer
Pyrgeometer
Pyrradiometer
Radiation balance equation
Radiometer
Radiosonde
Reflection
Relative humidity
Rotation
Roughness layer
Roughness length
Saturated adiabatic lapse rate
Saturation vapor pressure (e_s)
Scattering
Shortwave radiation

Skew T–Log P diagram
Soil heat flux plate
Solar declination
Solar noon
Solar time
Specific humidity
Stability
Stable atmosphere
Stomate
Sublimation
Substrate heat flux (Q_G)
Summer solstice
Sun path diagram
Surface boundary layer (SBL)
Temperature inversion
Thermal turbulence
Thermodynamic diagram
Time zone
Transition layer
Transpiration

Tropopause
Troposphere
Turbulence
Turbulent flux of sensible heat
    (sensible heat flux)
Turbulent flux of latent heat (latent
    heat flux)
Unsaturated adiabatic lapse rate
    (UALR)
Unstable atmosphere
Vapor pressure (e)
Vertical flux of horizontal
    momentum (shearing stress)
Vertical wind shear
Wavelength
Winter solstice
Zenith angle
Zone mean time
*Terms in italics have appeared in at
    least one previous chapter.*

▶ Review Questions

1. Describe the properties of the sublayers in the near-surface part of the troposphere.
2. What is the difference between mechanical turbulence and thermal turbulence?
3. Describe the two convective or turbulent fluxes in the near-surface atmosphere.
4. Describe the conductive flux that is important near the surface.
5. What is shearing stress and how/why does it occur?
6. What does a Sun path diagram tell us and how is it interpreted?
7. How can solar time be calculated from zone mean time?
8. Describe the different types of radiometers and their functions.
9. Describe the variables that can be used to represent atmospheric humidity.
10. How can evapotranspiration be measured?
11. Why is a pan adjustment coefficient necessary when evaluating potential evapotranspiration?
12. How is atmospheric stability evaluated?
13. What is the adiabatic assumption and why is it critical to assessing atmospheric stability?
14. What is potential temperature and how can it be used to simplify the evaluation of stability in the atmosphere?

15. What does the gradient Richardson number tell us about the condition of the near-surface atmosphere?

▶ Questions for Thought

1. Review the Sun path diagrams presented in Figure 5.7. Does day length change linearly across the year? Does the answer to that question depend on latitude?
2. If you have had a course in differential calculus, differentiate the equation shown on page 98 [$\theta = T + \Gamma_u(z - z_{1000\ mb})$] with respect to z to show that in a neutral atmosphere, potential temperature does not change with height.
3. Show that the hydrostatic equation and the gradient Richardson number are dimensionless.

---

http://physicalscience.jbpub.com/climatology

Connect to this book's Web site: http://physical science.jbpub.com/climatology. The site provides chapter outlines, further readings, and other tools to help you study for your class. You can also follow useful links for additional information on the topics covered in this chapter.

6

Global Hydrologic Cycle and Surface Water Balance

Chapter at a Glance

Global Hydrologic Cycle
Surface Water Balance
    Potential Evapotranspiration
    Evapotranspiration
    Precipitation
    Soil Moisture Storage
    Deficit
    Surplus and Runoff
Putting It All Together: A Worked Example of the
    Surface Water Balance
Types of Surface Water Balance Models
Water Balance Diagrams
Drought Indices
Summary
Review

In the last chapter we saw that local moisture is moved from Earth's *hydrosphere* up to the atmosphere through the combined processes of *evaporation* and *transpiration—evapotranspiration (ET)*. We also learned that the *latent heat flux* is the energetic process that moves energy involved in phase changes of water from the surface up into the atmosphere. This chapter focuses on these and related processes in more detail and on a broader scale, as components of the climatic water balance.

The *condensation*, freezing, or *deposition* of every water molecule in the atmosphere and subsequent precipitation of all that water at any moment in time would produce about 2.5 cm (1 in) of **precipitable water**—the depth of water (in centimeters) in a column of atmosphere if all water molecules in that column condensed and precipitated—if it were distributed evenly across Earth's surface. In reality, however, the distribution

of atmospheric moisture is very uneven. Tropical, oceanic locations average about 5 cm (2 in) of precipitable water, while less than 0.5 cm (0.2 in) exists in polar latitudes. The major reason for the difference is that the colder air near the poles has lower *saturation vapor pressure*.

The unevenness of the distribution of precipitable water invites two other important implications. First, because most places on Earth receive well more than 5 cm of precipitation per year, the same water molecules are continually recycled throughout the Earth–ocean–atmosphere system. It has been estimated that the average water vapor molecule spends only approximately 8 days in the atmosphere before it is precipitated back down to the surface. The water you used this morning to brush your teeth may be the same water a dinosaur drank 65 million years ago! Second, the movement of water on Earth must be intricately tied to the planetary atmospheric circulation such that it is accumulated in some places more easily than in others.

The constant cycling of this water (in all its forms) throughout the Earth–ocean–atmosphere system—the *global hydrologic cycle*—is the first major topic of this chapter. Subsequently, the chapter examines the climatic water balance. The global atmospheric circulation that distributes water around Earth is discussed in Chapter 7.

■ Global Hydrologic Cycle

The global hydrologic cycle is illustrated in **Figure 6.1**. The left side of the diagram represents the hydrologic cycle for the terrestrial part of Earth, while the right side depicts the oceanic surface and adjacent atmosphere. Saltwater in the ocean dwarfs all other *reservoirs* of water in the Earth–ocean–atmosphere system.

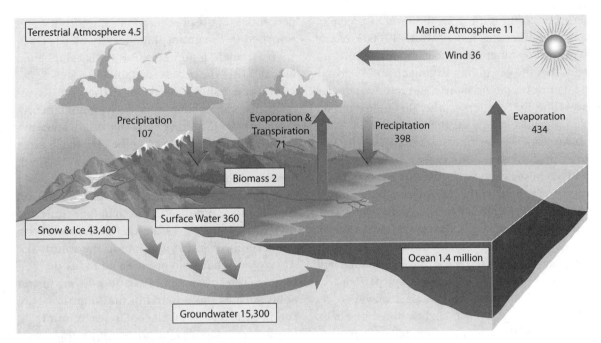

Figure 6.1 Schematic of the global hydrologic cycle. Values represent volume in thousands of km³, with fluxes calculated on an annual basis.

Figure 6.1 also shows that over 73 percent of the total reservoir of fresh water is in the form of solid water in the *cryosphere*. Nearly all the rest, about 26 percent, lies in the form of **groundwater**— water stored beneath the soil (**Figure 6.2**). About 95 percent of the remaining 1 percent of Earth's fresh water lies near the surface in the form of freshwater lakes, soil moisture, and all surface streams. At any given time only a minuscule amount of water—perhaps only 0.05 percent of the fresh water on Earth—is present in the terrestrial and marine atmosphere. It is not surprising that the marine atmosphere contains more water than the terrestrial atmosphere, both because most of the earth is covered by water and because water is more readily available for evaporation over the oceans.

Figure 6.1 also shows the relative magnitudes of the *fluxes* of water in all its forms via precipitation, *advection*, and ET. Once again, because most of the earth is covered by water, the bulk of the fluxes occur over the oceans. Notice also that more water is lost from the ocean through evaporation than is gained by precipitation; the difference is 36,000 km³ yr⁻¹. In the terrestrial atmosphere the opposite occurs. The process of ET moves less water up to the terrestrial atmosphere than is

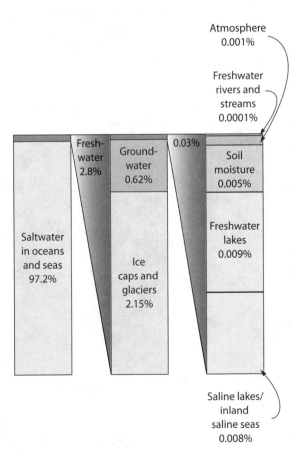

Figure 6.2 Global distribution of water on Earth.

returned to the surface in the form of precipitation. The difference is once again 36,000 km³ yr⁻¹. Without any other fluxes to compensate, this implies that the oceans are drying up and the continents are becoming more and more waterlogged over time. But, of course, other fluxes do compensate. Because nearly all the terrestrial Earth lies above sea level, the excess water (36,000 km³ yr⁻¹) from precipitation flows downhill, into streams, and out to sea via **runoff**. And this 36,000 km³ yr⁻¹ of water is constantly replenished by being advected from the marine atmosphere to the terrestrial atmosphere via winds (Figure 6.1). As a result the global hydrologic cycle results in nearly constant amounts of water on the continental and oceanic parts of Earth over long time periods.

At climatological time scales, however, the global hydrologic cycle is slightly different. For example, meltwater from continental ice sheets and glaciers runs off to the sea, and a new equilibrium is established as a result. At shorter climatological time scales, the loss of biomass in the form of *deforestation* or other removal decreases terrestrial ET. Although this decrease may be slight on a global basis because biomass returns only 2000 km³ yr⁻¹ of water to the atmosphere, it could be very significant on local and regional scales where a higher percentage of the precipitation comes from water returned to the atmosphere locally by ET. Once again, though, a new equilibrium is established globally after the process restabilizes.

Figure 6.3 shows the global distribution of runoff. In general, locations that receive abundant precipitation tend to have large natural waterways to remove the excess water. The classic example of such a situation is the Amazon River in South America, which flows almost entirely through wet climates. Runoff is so significant in the Amazon that this river produces more than four times the discharge (in m⁻³ s⁻¹) than that of the river with the second-highest runoff rate in the world. The Amazon discharges more than 10 times the volume of water than the Mississippi River! In wet areas such as those found in the Amazon basin, runoff carries excess water out to sea, minimizing impacts of flooding and preventing saline water from the ocean from intruding inland. In arid areas runoff may even be seen as wasted fresh water—people are reluctant to "waste" fresh water by letting it reach the sea.

Although other fluxes of water do occur on Earth, these values are relatively small and are not shown in Figure 6.1. Some surface water seeps downward through **percolation**—a very slow

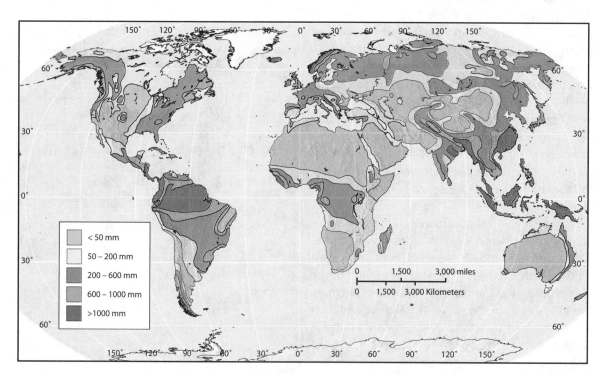

Figure 6.3   Global distribution of mean runoff (mm/year). *Data from*: World Meteorological Organization Global Runoff Data Center.

process through the *lithosphere* that replenishes groundwater supplies. In many locations the rate of removal of this groundwater for industrial, agricultural, and domestic uses far exceeds the rate of replacement by percolation. In addition, a relatively small flux from the groundwater to the ocean also occurs, perhaps offsetting the rate of percolation on a global basis. Another flux not shown in Figure 6.1 because it is subsumed within a larger process is **infiltration**—the downward movement of water from the surface into the soil and plant rooting zone. Infiltration differs from percolation in that it occurs nearer to the surface. On a long-term, global basis much of the infiltrated water is moved from the soil back to the atmosphere through ET.

A simple calculation known as the **global water balance equation** can be used to express the balance between the input and the output of water to and from the surface on a global basis. If we neglect the output fluxes that are small, we can say that

$$P = ET + R,$$

where *P* represents precipitation, *ET* is evapotranspiration, and *R* is runoff.

Two interesting observations of the global water balance equation emerge if we compare it with the energy balance equation introduced in Chapter 5. First, the only term that is shared between the two equations is ET. Recall that the energy equivalent of ET is Q_E, and the two terms are related by

$$ET = \frac{Q_E}{L_v \rho_\ell}$$

where L_v is the latent heat of vaporization $(2.5008 \times 10^6 \, \text{J kg}^{-1})$ and ρ_ℓ represents the density of water $(1000 \, \text{kg m}^{-3})$. Thus, ET affects both the energy balance and the water balance simultaneously. It is no wonder, then, that this is such a critical concept in climatology.

The second observation from the global water balance equation is that, like the energy balance equation, this equation balances on a global, annual-average basis. At any one local place on Earth's surface, however, neither equation may be balanced. Just as the *net radiation* (Q*) may be positive or negative locally, precipitation may exceed, or be exceeded by, the rate of removal of water from the surface via ET and runoff at a given time and place. Local surface moisture content changes constantly, but in the global system the budget must balance.

■ Surface Water Balance

Because the global water balance need not balance locally, it is useful for climatological and environmental monitoring applications to have an idea of how much water is involved in each of the reservoirs and fluxes at a given place and time. For example, if we can model the amount of water in the soil, foresters may have a better idea of whether the poor health of a tree stand is caused by inadequate moisture supplies. The **surface water balance** represents an environmental systems approach to understanding the hydrological cycle on a local level. It emphasizes the input, output, and storage of water and its interactions with the available energy and moisture at Earth's surface. The modern surface water balance was developed by American climatologist C. W. Thornthwaite in the 1940s for the purpose of climatic classification based on the degree of availability of surface water on a climatological basis.

Potential Evapotranspiration

Surface water balance largely depends on the calculation of *potential evapotranspiration (PE)*, an energy term that represents the climatic demand for water from the landscape driven by the amount of *insolation* at the surface. It is theoretically the maximum amount of evaporation and transpiration in centimeters or inches of precipitation equivalent that solar radiant energy could set in motion over a given period of time. It almost always peaks in summer, when the most energy is available to drive the ET process. We can consider PE to represent the maximum amount of ET that would occur over a waterlogged short grass field or the "water need" of the atmosphere. **Figures 6.4** and **6.5** show the annual average PE in the United States and in the world, respectively.

As discussed in Chapter 5, PE can be estimated by using a correction factor to the measured amount of evaporation from a pan of water, or (preferably) by using a *lysimeter*. But evaporation pans and lysimeters are available at few locations, and most have not been in continuous operation for a sufficiently long period of time to develop

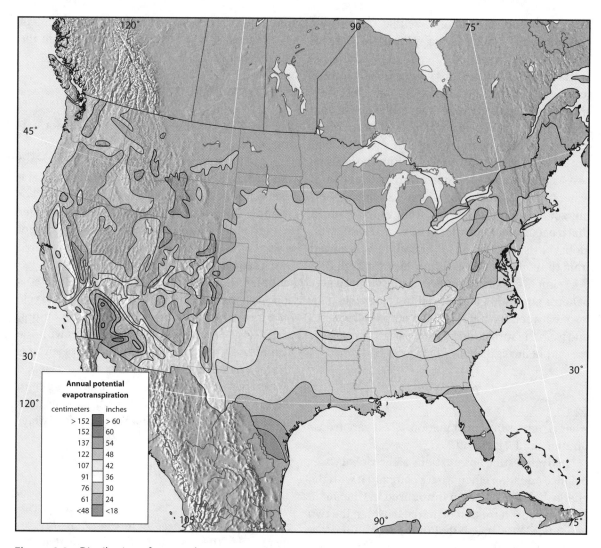

Figure 6.4 Distribution of potential evapotranspiration (PE) in the United States. *Adapted from:* Thornthwaite, C.W., 1948: Toward a rational approach to the classification of climates. *Geographical Review,* 38(1), 55–94.

reliable climatologies of PE. Therefore, it is advantageous to estimate PE based on available data collected both currently and in the historical records.

Although others have devised alternative means of estimating PE, Thornthwaite developed what became known as **Thornthwaite potential evapotranspiration**—a set of empirical equations to estimate PE based only on the temperature and *latitude* of the location of interest. The purpose of including latitude in this calculation is that latitude determines Sun angle and day length, and these, along with temperature, represent the amount of energy availability to drive the ET process. But why did Thornthwaite choose only these variables, when ET rates are also affected by wind speed, atmospheric *stability*, cloud cover, and humidity characteristics? He decided that even though other variables are also important for determining ET,

the lack of data for most locations on Earth would limit the utility of the surface water balance to the few places where data were available for all these variables. For this reason he chose to introduce a simpler but more applicable model.

Cort Willmont and colleagues at the University of Delaware modified Thornthwaite's original PE by first computing an "unadjusted" PE, as shown in **Equation 6.1**, where T_i is the mean monthly air temperature in degrees Celsius and $I = 1$ for January, 2 for February, and so on. In addition,

$$I = \sum_{i=1}^{12} \left(\frac{T_i}{5} \right)^{1.514}$$

and

$$a = 6.75 \times 10^{-7}I^3 - 7.71 \times 10^{-5}I^2 + 1.79 \times 10^{-2}I + 0.49$$

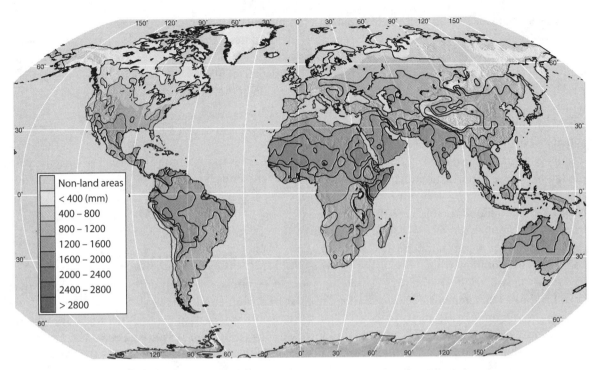

Figure 6.5 World distribution of potential evapotranspiration (PE). *Adapted from*: United Nations Environment Programme Global Resource Information Database—Nairobi.

But, because different months have different numbers of days, PE can be adjusted such that

$$PE_{i\,\text{adjusted}}(\text{mm month}^{-1}) = PE_{i\,\text{unadjusted}}\frac{n}{30}\frac{h}{12}$$

where n is the number of days in the month and h is the number of hours of daylight on the 15th day of the month. This set of statistically derived equations was developed in humid environments of the eastern United States. It tends to overestimate PE in arid, non-midlatitude environments and in arid conditions in normally humid environments.

Over the years other researchers have offered alternate methods of calculating PE. Most notably, the British climatologist H. L. Penman's method and several variants of it frequently appear in the climatology research literature. Although some believe the Penman family of methods—what has become known as the **Penman-Monteith equation**—is

more accurate because the equation is derived from physical laws rather than statistical relationships (as is Thornthwaite's equation), the multitude of input data required often presents a problem for application. For example, to calculate the PE based on this model, net radiation (Q^*), *substrate heat flux* (Q_G), *vapor pressure* (e, or some humidity measurement that can be converted to vapor pressure), temperature, and **resistance** must be measured or estimated. Resistance is a component of force exerted by the air that works in the opposite direction of the flux (so it is generally directed downward in opposition to the latent and **sensible heat fluxes**) and has units of seconds per meter. In general, resistance is higher during times of atmospheric stability and lower during times of instability. This is a cumbersome quantity to measure accurately, and several other quantities required in the Penman-Monteith equation are not measured routinely at weather stations.

Equation 6.1 $PE_{i\,\text{unadjusted}}(\text{mm month}^{-1}) = \begin{cases} 0 & \text{if } T_i < 0^0\text{C} \\ 16\left(\dfrac{10T_i}{I}\right)^a & \text{if } 0^0C \leq T_i < 26.5^0\text{C} \\ -415.85 + 32.24T_i - 0.43T_i^2 & \text{if } T_i \geq 26.5^0\text{C} \end{cases}$

One version of the Penman-Monteith equation is

$$\text{ET} = \frac{S}{S+\gamma}\frac{(Q^* - Q_G)}{L_v\rho_\ell} - \frac{\rho C_p \Delta D}{L_v\rho_\ell r_a}$$

where S represents the slope of the saturation vapor curve as a function of temperature (Pa K⁻¹, calculated using the *Clausius-Clapeyron equation*); γ is the psychrometric constant (65.5 Pa K⁻¹); L_v is the latent heat of vaporization (2.5008×10^6 *Joules* kg⁻¹ at 0°C and changing very minimally over the range of earth temperatures); ΔD is the difference in **wet bulb depression**—the difference between the air temperature and temperature after being cooled by evaporation when measuring *relative humidity*—at two heights in the surface boundary layer (in the *Kelvin temperature scale*); and r_a represents aerodynamic resistance (s m⁻¹).

The most critical assumption of the Penman equation is that S is constant. In other words, the ratio of the difference between the vapor pressure and the saturation vapor pressure at that temperature to the difference between air temperature and *dewpoint temperature* is constant. Because this condition occurs near saturation, Penman's ET is often taken to represent PE.

Evapotranspiration

Another fundamental variable in the calculation of the surface water balance is ET. This variable is distinguished from PE because PE estimates only the maximum possible ET that could occur from a waterlogged grass field. By contrast, ET estimates the actual amount of water transpired through plants or evaporation that does occur. ET does not represent the demand for water but what actually transpires and evaporates. Thus, ET must always be less than or equal to PE. Perhaps a useful analogy is the amount of money you spend shopping compared with the maximum amount you could possibly spend if money were not a limiting factor. For most of us these values are not the same. Although no distinction is made between evaporation and transpiration in the climatic water balance, most water loss from vegetated surfaces is through transpiration.

Although ET can be determined using more complicated methods, several researchers have developed simple equations to estimate ET based only on available data. A relatively simple ET equation that offers some advantages, particularly in the arid western United States, has been the **Blaney-Criddle model** formulated by soil conservation scientists H. F. Blaney and W. D. Criddle. The original version of this model is

$$\text{ET} = kf$$

where ET is in mm/month, k is a crop-specific coefficient, and f is a consumptive use factor given by

$$f = \frac{TP}{100}$$

where T is the mean monthly temperature (in °F) and P represents the monthly percentage of annual daylight hours.

Another technique is the **Turc method**, which was developed for western Europe but has provided reasonable results in some other locations. This method can be expressed as

$$\text{ET} = a_T 0.013 \frac{T_{\text{mean}}}{T_{\text{mean}}+15}\frac{23.8856\text{K}\downarrow + 50}{L_v}$$

where ET is in mm day⁻¹, T_{mean} is the mean daily temperature (in °C), and K↓ is the mean solar radiation (MJ m⁻² day⁻¹, where MJ represents "megajoules," or millions of Joules). The coefficient a_T is a humidity-based value such that if the mean daily relative humidity is greater than or equal to 50 percent, then $a_T = 1.0$. If the mean daily relative humidity is less than 50 percent, then a_T has the value of

$$a_T = 1 + \frac{50 - \text{RH}_{\text{mean}}}{70}$$

In addition, many other techniques for evaporation estimation are available. Some have been shown to be more useful in certain geographical areas, while others have been more appropriate estimators in other regions.

Precipitation

Precipitation (P) is the "input" water to the surface in the surface water balance. It represents the atmospheric delivery of moisture to Earth's surface. This delivery occurs in widely varying totals from place to place (**Figures 6.6** and **6.7**). Precipitation

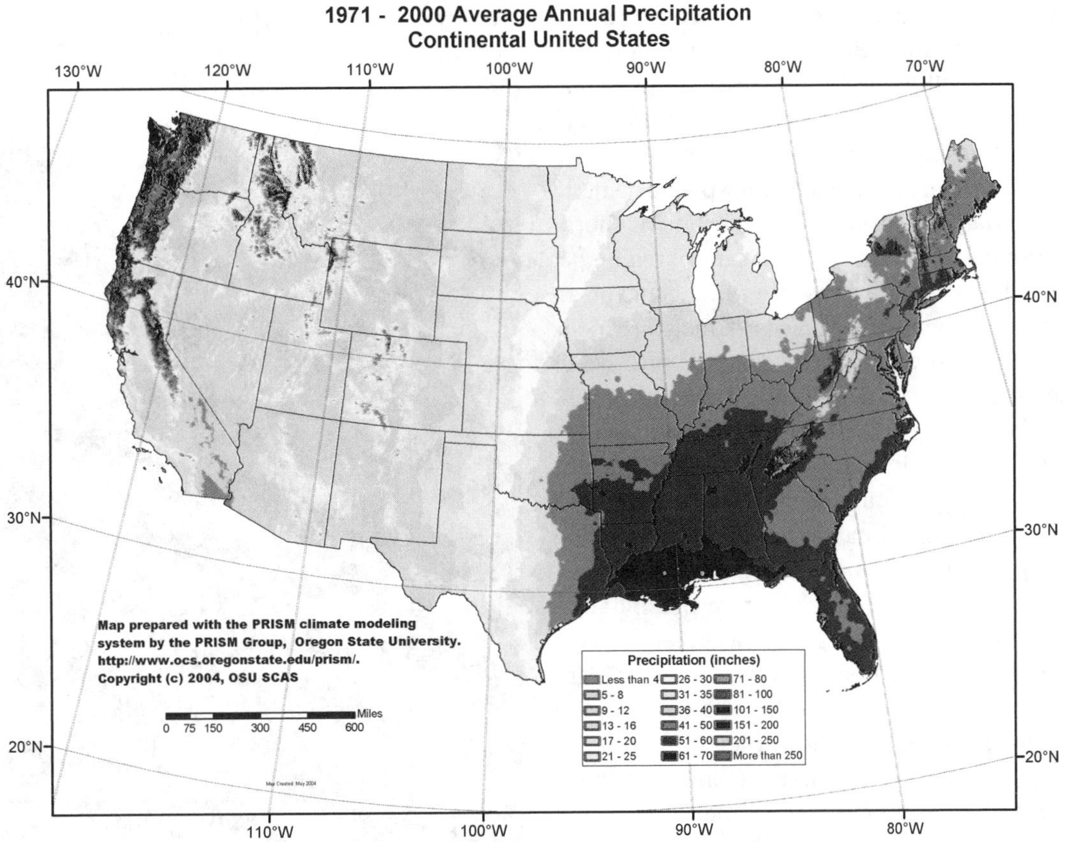

1971 - 2000 Average Annual Precipitation
Continental United States

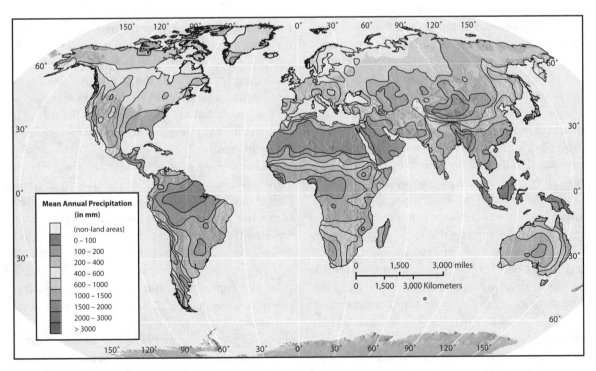

Figure 6.6 Average annual precipitation in the United States. (See color plate 6.6) Copyright © 2004, OSU PRISM Climate Group.

Figure 6.7 Average annual precipitation throughout the world. *Adapted from*: United Nations Environment Programme Global Resource Information Database—Geneva.

is simply compared with PE for the place and time of interest to determine the moisture conditions. In a wet month, P–PE is positive because more precipitation is input to the surface than can be removed from it by the atmosphere. In a dry month, P–PE is negative because there is a greater atmospheric demand for water than is delivered by precipitation. Thornthwaite created a climatic classification system that uses a moisture index based on this concept. In this system, whenever P exceeds PE moisture index values are positive. Magnitudes of the positive or negative moisture indices are used to delineate climatic boundaries in the Thornthwaite system.

Although PE almost always peaks in summer, precipitation may or may not peak in summer at a given location. In climates that have precipitation peaking in summer, floods and droughts may not be as common as in places where precipitation peaks in winter. In the former case precipitation arrives at the same time that the atmospheric demand for water is high so neither too much nor too little water remains on the surface. In the latter case, droughts and wildfires are common in summer and floods and/or mudslides may be common in winter, because the atmosphere's demand for water is coming precisely at the time when moisture is unavailable through precipitation, and the water arrives when insufficient energy exists for the atmosphere to remove it easily.

Although precipitation at some sites is measured using technologically advanced equipment, the vast majority of the 13,000 official precipitation-reporting stations in the continental United States relies on the collection of precipitation from a simple, nonrecording cylindrical container with a diameter of 20 cm (8 in) (**Figure 6.8**). Inside the cylinder sits a funnel that opens into a smaller cylindrical tube. The smaller tube collects up to 5 cm (2 in) of rainfall before the precipitation overflows into the larger tube. Because the inner tube collects precipitation from an opening that is 10 times the area of the opening in the smaller tube, a special stick with "centimeters" or "inches" set apart at increments of 10 cm (or 10 in) is used to measure the precipitation. The purpose is to facilitate the precision of measurement down to a millimeter or hundredth of an inch. When the stick is wet, it leaves a temporary "wetness" mark on it, so that the stick can be placed through the opening in the funnel, removed, and then precipitation can be measured easily. If more than 5 cm (2 in) of precipitation falls,

Figure 6.8   A standard nonrecording rain gauge used in the United States.

the collected precipitation in the larger tube can then be poured back into the smaller tube for measurement. Other nonrecording rain gauges have graduated markings on the inner tube itself that also lengthen each "inch" 10 times; these collect 1 inch of precipitation in the inner tube before spilling into the larger tube that accommodates 5 inches of precipitation.

Measurement of precipitation in the form of snow is only slightly more complicated. Most snow gauges have flanges surrounding the gauge to minimize the impact of blowing snow (**Figure 6.9**). Snow collected in the gauge is then simply melted and converted to centimeters (or inches) of rainfall equivalent. Heavy, wet snow may have two to three times the depth of its rainfall equivalent, and dry, powdery snow may be 25 times as deep as its rainfall equivalent. A reasonable (though not always accurate) rule of thumb is to assume that, on average, 10 to 12 cm of snow equals 1 cm of water equivalent. Average annual snowfall totals for the United States are shown in **Figure 6.10**, and those for the world are shown in **Figure 6.11**. Snow can detract from the water balance model's accuracy, because in some areas snow precipitation may remain on the surface for months before it melts. Thus, snow may be counted as an input of precipitation months before it can actually be used at the surface or runs off from the surface.

Even though it sounds as if 13,000 rain gauges would provide high-quality coverage of precipitation across the United States, this coverage amounts to an average of only one gauge per 723 km² (284 mi²) in the continental United States. This equates to one rain gauge for every 27 × 27 km (17 mi × 17 mi) plot of land across the continental United States, although the spatial distribution is not uniform. You have certainly observed situations where rain and snow events can differ drastically across such a large area. The network of gauges across the United States is especially inadequate in summer, when localized, intense thunderstorms are commonplace. Nevertheless, the United States remains much better served by precipitation data than most other parts of the world. As a result of the lack of high-quality data, current research involves the application of radar- and satellite-derived measurement of precipitation to supplement the existing network of stations.

Aside from the problem of having too few observations of precipitation, data from the stations that do exist are prone to reporting inaccurate results. Because the vast majority of precipitation stations are operated and maintained by humans, errors exist that cause problems with the precipitation data. Some stations report missing observations more frequently than others.

In addition, there are systematic, instrumental errors in the precipitation gauges. According to estimations by climatologist David Legates, several factors constitute an **undercatch** of approximately

Figure 6.9 A snow gauge.

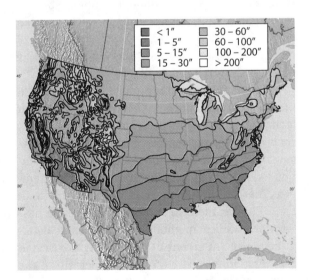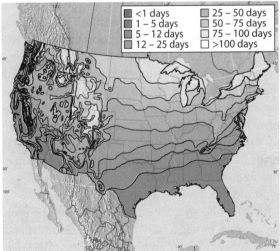

Figure 6.10 Mean annual snowfall in the United States, and mean number of days with 1 inch or more of snow on the ground, 1961–1990. *Adapted from*: Doesken, N.J. and A. Judson, *The Snow Booklet: A Guide to the Science, Climatology, and Measurement of Snow in the United States*. Colorado Climate Center, Colorado State University.

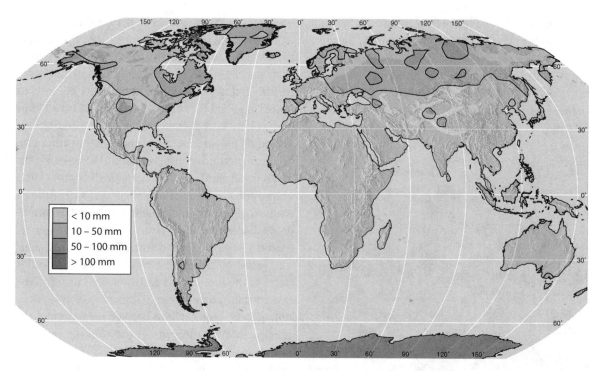

Figure 6.11   Global annual mean snow-cover depth (in millimeters of water equivalent). *Adapted from*: Willmott, C.J., C.M. Rowe, and Y. Mintz, 1985: Climatology of the terrestrial seasonal water cycle. *Journal of Climatology*, 5, 589–606.

11 percent. First, the movement of wind across the gauge creates a **Bernoulli effect**, a phenomenon whereby the increase in wind speed with height over the gauge causes a decrease in air pressure above the gauge. This decreases the chances that a drop of precipitation will fall into the gauge and instead causes precipitation to fall away from the gauge. Winds around rain gauges have been estimated to cause an average undercatch of perhaps 8 percent, with higher values in windy conditions and lower values in calm conditions. In addition, the presence of water "stuck" on the side of the gauge, termed the "wetted perimeter" (which escapes measurement), and evaporation of collected water inside the gauge collectively may amount for another 3 percent of undercatch. Despite these errors, precipitation data in the United States are of relatively high quality, but these potential errors should be considered when using precipitation data to evaluate the "true" amount of precipitation that has fallen.

In the absence of radar-derived precipitation data (which provides continuous precipitation measurement across an area), the precipitation measured at a point must be converted to an area estimate for use in hydrologic modeling, flood analysis, and water conservation planning. For example, four precipitation stations exist in Hancock County, Illinois. The sites measure 5.0, 1.0, 1.5, and 4.0 cm of rainfall for a particular rainfall event (**Figure 6.12a**). Is it fair to simply average these four and say that 2.9 cm of rain fell across the county? This number could be multiplied by the area of the county (2061 km², or 2.061×10^{13} cm²) and then converted to cubic meters to provide an estimated total volume of water supply that fell as precipitation in Hancock County (59.7 million m³). But as you can see in Figure 6.12, the stations are not distributed evenly across the county. So a simple average of the four precipitation totals may not be a fair way to assess the total volume of water that fell.

Spatial analysis methods such as the isohyetal method, Thiessen polygon method, and gridpoint technique have been used to handle problems such as the one described above. The **isohyetal method** uses **isohyets**—lines of constant precipitation totals, analogous to the *isotherms* for temperature. These are drawn across Hancock County, as in Figure 6.12b. Then, **geographic information systems** or graphical methods could be used to determine that 3.9 percent of the county's area had precipitation exceeding 5 cm, 15.6 percent of the county was between 4.0 and 5.0 cm, 8.6 percent was between 3.0 and 4.0 cm, 11.2 percent was between 2.0 and 3.0 cm, 7.8 percent was between 1.0 and 2.0 cm, and 52.8 percent was below 1.0 cm (Figure

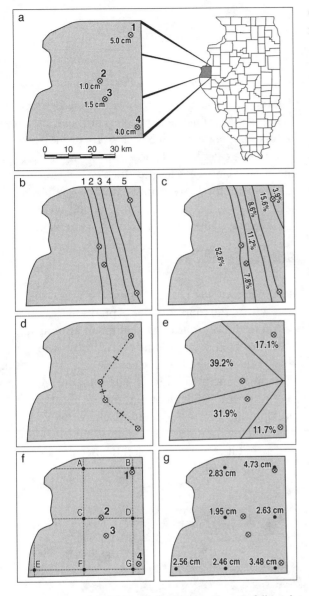

Figure 6.12 Spatial methods of estimating rainfall totals in Hancock County, Illinois. (a) Measured precipitation at the four weather stations; (b) isohyets of precipitation, in centimeters; (c) percentage of the county's area between each isohyet, for calculation of the isohyetal method; (d) connecting the lines that connect each adjacent weather station and drawing perpendicular bisectors for those lines; (e) calculating the area enclosed within each polygon formed by the perpendicular bisectors, for the Thiessen polygon method; (f) superimposing a grid over the county; and (g) estimating the precipitation at each point, for the gridpoint technique.

6.12c). For the part of the county with totals exceeding 5.0 cm, we might assume a uniform total of 5.5 cm. This is reasonable because if any station had measured as much as 6 cm, another isohyet would have been drawn. Likewise, uniform totals of 4.5 cm are assumed for the locations between 4.0 and 5.0 cm, 3.5 cm is used for the part of the county between 3.0 and 4.0 cm, 2.5 cm is used for the area between 2.0 and 3.0 cm, 1.5 cm is assumed for the area between 1.0 and 2.0 cm, and finally, 0.5 cm is assumed for that part of the county with totals below 1.0 cm of rainfall. We then multiply the total by the percentage of area having that assumed total and sum the parts (**Equation 6.2**). This technique produces a county-averaged precipitation of 1.88 cm, a total considerably below that calculated by simply averaging the four measured values.

The **Thiessen polygon method** is also a spatial or graphical method, but it operates differently. In this method a person or a computer draws perpendicular bisectors for the lines connecting each adjacent station in the study area, as shown in Figure 6.12d. These perpendicular bisectors are then extended until they meet, as shown in Figure 6.12e. This procedure ensures that any location in the county falls within the same polygon as the nearest precipitation measuring station. Then, the point-based totals are weighted by these percentages to determine an overall precipitation total for the county (**Equation 6.3**). In this example the Thiessen polygon method gives a result of 2.19 cm across the county, somewhat more than the 1.88 cm computed by the isohyetal method.

A final strategy that can be implemented is the **gridpoint technique**, a method in which an artificial grid is superimposed over the study region, with the coarseness of the grid dependent on the amount of computing power and sophistication necessary. The precipitation totals derived for each of the gridpoints are calculated according to an inverse-distance-weighting function of the precipitation at each of the known stations. For example, in Figure 6.12f, the total distance between gridpoint A and all four stations is 118 km, with 21 km to station 1, 22 km

Equation 6.2 County Average Precipitation = (5.5 cm × 0.039) + (4.5 cm × 0.156) + (3.5 cm × 0.086) + (2.5 cm × 0.112) + (1.5 cm × 0.078) + (0.5 cm × 0.528)

Equation 6.3 County Average Precipitation = (5.0 cm × 0.171) + (1.0 cm × 0.392) + (1.5 cm × 0.319) + (4.0 cm × 0.117)

Equation 6.4

$$\text{Precip. at Point A} = \frac{\frac{118}{21}(\text{Precip. at Station 1}) + \frac{118}{22}(\text{Precip. at Station 2}) + \frac{118}{29}(\text{Precip. at Station 3}) + \frac{118}{46}(\text{Precip. at Station 4})}{\frac{118}{21} + \frac{118}{22} + \frac{118}{29} + \frac{118}{46}}$$

to station 2, 29 km to station 3, and 46 km to station 4. The distance-weighting scheme provides a derived precipitation at gridpoint A of 2.83 cm, calculated from **Equation 6.4**. Once precipitation is estimated in this manner at all gridpoints in the county (Figure 6.12g), a simple average of the derived precipitation totals can be used to estimate the countywide precipitation. In this example, the county-averaged precipitation is 2.95 cm, far more than either the isohyetal or the Theissen polygon methods. Interestingly, in this case the gridpoint method produces a value that is very close to the simple arithmetic mean of the four stations.

Of the three techniques, the isohyetal method is considered by some to be most accurate because the effect of local features, such as topography, water bodies, and land cover, should implicitly influence the weighting scheme. For example, if a mountain is causing a local *orographic effect*, then the region with high precipitation should have a small area. The Thiessen polygon method and the gridpoint technique weigh the stations merely on distance, with no implicit ability to distinguish local features that influence the values at those station points. On the other hand, each storm event would require the calculation of new isohyets and weighting systems, whereas for the Thiessen polygon and gridpoint methods, weightings could be established one time and always be the same, as long as no stations appeared or disappeared from the network.

Soil Moisture Storage

Soil moisture storage represents the water available within the rooting zones of plants for transpiration and evaporation. Global mean annual soil moisture is shown in **Figure 6.13**. Even though this quantity is a ratio of the volume of water to the volume of Earth in which it sits, this number is converted into equivalent centimeters or inches of precipitation, for easy comparison with PE, ET, and precipitation. Just as a sponge has a limit of water that it can store, the soil too becomes saturated if too much water is present.

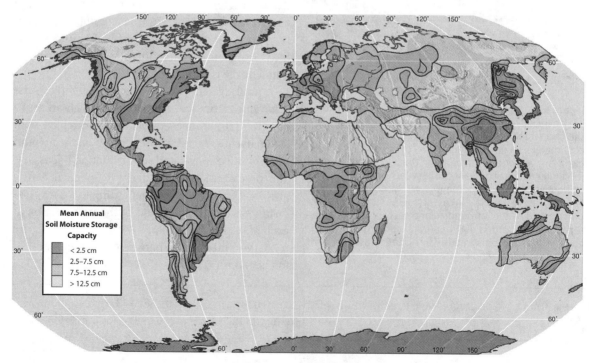

Figure 6.13 Global annual mean soil moisture, in centimeters. *Adapted from*: Willmott, C.J., C.M. Rowe, and Y. Mintz, 1985: Climatology of the terrestrial seasonal water cycle. *Journal of Climatology*, 5, 589–606.

The amount of water (in centimeters or inches of precipitation equivalent) that saturated soil can retain against gravity is known as **field capacity** or **soil moisture storage capacity**. Although this quantity varies locally with the texture, shape, and uniformity of soil particles and the depth of soil, averages in the United States are generally highest in Iowa and the Lower Mississippi Valley and lowest in the Mountain West.

Porosity—the ratio of the volume of all the pore spaces in a soil to the volume of the whole—is related to the texture, shape, and uniformity of the soil and also to soil moisture storage. In general, because grains are bigger in sandy soils than in silty or clayey soils, the pore spaces are usually largest. However, porosity may be greater in silty soils because of the high number of pore spaces. On average, perhaps 25 percent of the volume of a soil may be pore space. Thus, if a soil of 25 percent porosity extends 60 cm (24 in) into the subsurface, 15 cm (6 in) of water can be held in storage.

Just as soils have a threshold of the maximum amount of water that they can hold, there is also a minimum. Although zero moisture is the theoretical minimum, in actuality even when there is some moisture present in the soil, it may be useless because the **surface tension** of water may be high, which could prohibit it from being usable. Small isolated water drops may adhere well enough to the soil particles that plants cannot pull them upward for transpiration, and they cannot be evaporated. This minimum amount of water (in centimeters or inches of precipitation equivalent) that is necessary in the rooting zone to allow extraction by plants is known as the **wilting point**. Water in the soil is considered useless to vegetation unless it exists in quantities within the rooting zone exceeding the wilting point. As the name suggests, any amount of water in the soil below the wilting point is unable to nourish plants. In most cases a farmer wants to provide sufficient irrigation water to avoid situations that decrease the soil water to the wilting point.

Deficit

Moisture is not always available in the soil when it is needed. Anytime the atmospheric demand for moisture from the surface (PE) is not fulfilled by some combination of precipitation and by water withdrawn from storage in the soil, **soil moisture withdrawal** occurs and a **deficit** results. By analogy, if the demand for your money (the bills) cannot be paid either by your paycheck (the input) or by your savings in the bank (storage), you are said to be in deficit as well. Deficit can be considered to represent the part of PE that is not available to be transmitted to the atmosphere. It is the result of a situation in which the atmosphere has the energy to evaporate more water than it can acquire from the surface either through ET of precipitation or water stored in the soil.

Deficit may be considered as an indicator of environmental stress, and, therefore, it is an important concept for climatologists and those who use climatological data. The lack of precipitation alone may not always represent the onset of drought; instead, a drought occurs whenever serious deficit occurs, whether or not precipitation has been falling recently. A week without rainfall may produce deficit conditions in summer (when PE is high) but not in winter (when PE is low).

Surplus and Runoff

At the other extreme, during periods of sufficient precipitation we may have a situation of excess water even after PE is satisfied and soil moisture is at field capacity. Any water that is not needed for ET or **soil moisture recharge**—the process of bringing the soil moisture storage to its field capacity—becomes **surplus** water. The rate of recharge depends on the soil's **permeability**—the property that describes the ability of water to flow between soil particles (or rock, in the case of groundwater recharge). Permeability depends on the shape and orientation of the soil (or rock) grains. If the pore space between individual grains is small and/or poorly connected, the permeability is low and falling precipitation generally will not provide fast recharge rates. Note that despite a low permeability, porosity may still be high if there are many small, poorly connected spaces, so soil moisture storage in such a case would be high even though the recharge rate would be slow. On the other hand, a situation in which there are large and/or well-connected pore spaces allows for high permeability and facilitates the recharge of the soil or groundwater.

Surplus occurs only after precipitation falls on saturated soils. This is the water that runs off into streams and lakes or percolates to the water table and replenishes deep groundwater supplies. On a watershed basis, the surplus term gives a good estimate of streamflow volumes by applying a statistical relationship, often by estimating that 50 to 60 percent of the surplus eventually becomes runoff to streamflow.

What happens to the remaining surplus? Some becomes **interception**—water that is held on surfaces other than the soil, such as tree leaves and rooftops—until it eventually evaporates. Other water becomes **depression storage**—water that is stored in puddles, gutters, and other natural or artificial "depressions" on the landscape. Still other water percolates to replenish groundwater supplies. These sources generally are not considered separately in the surface water balance—a simple ratio is usually chosen that relates the amount of surplus to the amount of runoff.

Runoff is an important output of the water balance model for numerous reasons. This is the water that is supplied to rivers and streams. Too much runoff may cause floods and also makes

river navigation hazardous. Too little runoff can cause other types of navigation hazards, and it can also threaten drinking and irrigation water supplies. In addition, the lack of sufficient runoff near the mouths of rivers can cause **saltwater intrusion** from the open ocean up into normally freshwater areas. During the drought of 1988 in the United States, saltwater intrusion into the Mississippi River penetrated over 160 km (100 mi) upriver from the mouth, endangering coastal habitats and ecosystems and threatening drinking water supplies.

In the United States the 13 River Forecast Centers operated by the **National Oceanic and Atmospheric Administration** are responsible for using the precipitation forecasts from the *National Weather Service* as input to water balance models for predicting the streamflow on the nation's waterways (**Figure 6.14**). This precipitation must be area-weighted using one or more of the schemes described previously. Then, the water balance model is used to predict not only the

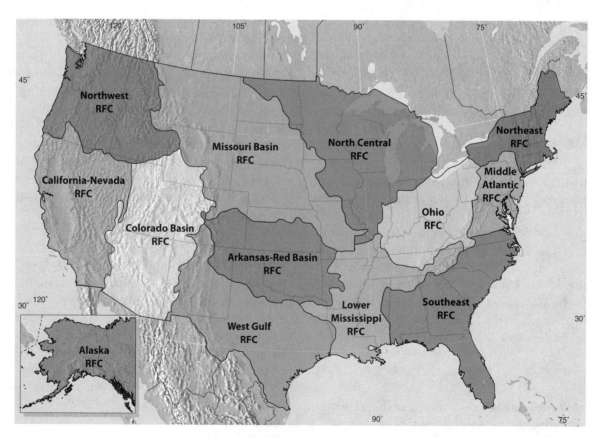

Figure 6.14 The 13 River Forecast Centers in the United States, operated by the National Oceanic and Atmospheric Administration. *Data from*: NOAA.

crest—the peak volume of streamflow—but also the time of peak streamflow after a precipitation event at various points along the stream resulting from either local and/or upstream precipitation. The quality of the forecasts depends on both the accuracy of the National Weather Service precipitation forecasts and the accuracy of the water balance results for upstream regions produced by other river forecast centers. Errors in either step cause errors in the stream forecasts. Accurate stream forecasts can be of great societal and economic benefit, because floods claim more lives and property losses in the United States and the world than any other form of weather-related hazard.

Putting It All Together: A Worked Example of the Surface Water Balance

The best way to see how the surface water balance operates is to work through an example. **Table 6.1** shows a mean monthly water balance for Baton Rouge, Louisiana. The mean monthly PE (in centimeters, based on the modified Thornthwaite calculations) and precipitation values (based on 30-year averages of rain gauge measurements) are given. These represent the data needed to complete the rest of the calculations.

From these data and the assumption that the balance begins with a saturated soil, we can do some simple arithmetic to calculate the water balance. Specifically, we can see that in January mean precipitation exceeds PE by 9.4 cm (3.7 in). But because the soil is already at field capacity at the beginning of January (by assumption), all this extra water goes to surplus. The same situation occurs in February, March, and April, with 9.9, 8.9, and 5.4 cm, respectively (3.9, 3.5, and 2.1 in), going into surplus. May is the first month of the year when mean PE exceeds mean precipitation. The 0.7 cm (0.3 in) that the atmosphere demands but does not receive from precipitation is accommodated by soil moisture withdrawal. This results in a change in soil moisture storage (ΔST) of −0.7 cm (−0.3 in) and a total ST of 14.3 cm (5.7 in). Likewise, in June, July, and August the amount of water that the atmosphere needs but cannot receive through precipitation is also available in storage. So neither surplus nor deficit exists for the period of May through August. Continual soil moisture

Table 6.1 Mean Monthly Water Balance in (a) Centimeters and (b) Inches for Baton Rouge, Louisiana

(a)

	J	F	M	A	M	J	J	A	S	O	N	D	Yr
PE	1.8	2.3	4.1	7.6	11.9	15.7	17.0	16.0	12.4	7.1	3.0	2.0	100.9
P	11.2	12.2	13.0	13.0	11.2	9.7	14.0	11.9	9.7	6.6	9.7	15.2	137.4
P-PE	9.4	9.9	8.9	5.4	−0.7	−6.0	−3.0	−4.1	−2.7	−0.5	6.7	13.2	38.5
ΔST	0	0	0	0	−0.7	−6.0	−3.0	−4.1	−1.2	0	6.7	8.3	0
ST	15	15	15	15	14.3	8.3	5.3	1.2	0	0	6.7	15	0
ET	1.8	2.3	4.1	7.6	11.9	15.7	17.0	16.0	10.9	6.6	3.0	2.0	98.9
D	0	0	0	0	0	0	0	0	1.5	0.5	0	0	2.0
S	9.4	9.9	8.9	5.4	0	0	0	0	0	0	0	4.9	38.5

(b)

	J	F	M	A	M	J	J	A	S	O	N	D	Yr
PE	0.7	0.9	1.6	3.0	4.7	6.2	6.7	6.3	4.9	2.8	1.2	0.8	39.8
P	4.4	4.8	5.1	5.1	4.4	3.8	5.5	4.7	3.8	2.6	3.8	6.0	54.0
P-PE	3.7	3.9	3.5	2.1	−0.3	−2.4	−1.2	−1.6	−1.1	−0.2	2.6	5.2	14.2
ΔST	0	0	0	0	−0.3	−2.4	−1.2	−1.6	−0.5	0	2.6	3.4	0
ST	6.0	6.0	6.0	6.0	5.7	3.3	2.1	0.5	0	0	2.6	6.0	0
ET	0.7	0.9	1.6	3.0	4.7	6.2	6.7	6.3	4.3	2.6	1.2	0.8	39.0
D	0	0	0	0	0	0	0	0	0.6	0.2	0	0	0.8
S	3.7	3.9	3.5	2.1	0	0	0	0	0	0	0	1.8	15.0

withdrawal through these months results in only 1.2 cm (0.5 in) of precipitation in storage by the end of August. This water is not adequate to accommodate the 2.7 cm (1.1 in) demand for water in September that is not provided by precipitation. Therefore, a deficit of 1.5 cm (0.6 in) is accumulated in September. Likewise, a deficit of 0.5 cm (0.2 in) occurs in October. But the decreasing PE values cause precipitation to exceed the atmospheric demand for water in November and December. Soil moisture recharge in Baton Rouge occurs until the soil has returned to field capacity (15 cm, or 6.0 in) by the end of December.

When working through a continuous surface water balance simulation, the calculations should begin at least a year before the time for which data are needed. For example, if the monthly water balance variables are needed for the 1999 through 2009 period, the calculations should begin perhaps in 1997 or 1998. The reason is that an "adjustment" time is needed so that calculations can begin at an accurate initial amount of soil moisture storage. By including a year or 2 before the "actual" period, the chances are more likely that at some point in that initial period (particularly in winter, when PE is lowest), the simulation will correctly indicate that soil moisture is at field capacity. From that point forward, the results will be "correct." Of course, in drier climates it may be necessary to use a much longer initial period to ensure that at some point before the "actual" data are computed the simulation became calibrated correctly.

When performing any water balance analysis, two equations can be used to verify that the calculations are correct:

$$P = ET + S \pm \Delta ST$$
$$PE = ET + D$$

where S is surplus, ST is soil moisture storage, and D is deficit. If either of these equations does not balance, then an error has been made. Conceptually, the first equation suggests that the input of precipitation must be balanced by some combination of the three places that the water can be assumed to go: ET, surplus, and soil moisture recharge/withdrawal. The second equation states that the amount of water that the atmosphere could use must equal the amount of water that the atmosphere receives plus the amount that it could use but does not receive. These equations verify that the calculations shown in Table 6.1 are correct for each of the 12 monthly totals and for the annual sum.

Finally, it should be noted that a surface water balance model may be run for time intervals other than monthly. Some applications require daily water balance conditions, whereby daily estimates of PE and measurements of precipitation are used to simulate moisture conditions in the soil. Agriculturalists may use daily water balance models along with daily weather forecasts to determine the optimal amount and timing of irrigation to use for a crop.

■ Types of Surface Water Balance Models

Some surface water balance models, such as the one just described, are **equal availability water balance models**. These assume that soil water is equally available to be evapotranspired regardless of the amount of storage in the soil. In applications for which more precision is needed, a **decreasing availability water balance model** may be implemented. Such models assume that as soil moisture withdrawal increases, the water becomes increasingly more difficult to be moved upward via ET. This is done through a new variable called **accumulated potential water loss**, which is usually given in a tabular format, based on the field capacity of a given soil, following work done by Thornthwaite and his colleague John Mather. The accumulated potential water loss is a running total of the P − PE value (provided it is a negative number). In principle, the decreasing availability model mimics the "real world" because as the soil begins to dry out, *guard cells* begin to resist the loss of moisture through the *stomates* of plants via ET. Furthermore, as the soil dries an increasingly higher percentage of water in the soil is "stuck" by surface tension to soil and rock particles and cannot be lost to the atmosphere easily.

Another modification that may be made to the surface water balance is to divide the surface zone into two or more layers. The model may be tweaked to allow water to be lost more freely from the top layer but more stringently from successively lower layers. For example, an equal availability assumption may be made in the top layer but a decreasing availability model may be implemented for the bottom layer.

■ Water Balance Diagrams

Because of the difficulty of ascertaining a general impression of the hydroclimatology by simply examining a table of numbers, water balance diagrams provide a helpful way of evaluating the water balance of a location. **Figure 6.15** shows an example of a mean monthly water balance diagram for Dallas, Texas. Note, however, that water balance diagrams need not be for mean hydroclimatic conditions but may also depict actual moisture conditions over any period of time desired.

In the water balance diagram, the PE curve always mimics the temperature curve, because PE is a function of temperature and the number of daylight hours. In Figure 6.15 PE is represented as a dashed line. The precipitation line for Dallas (the solid line in Figure 6.15) shows a fairly even distribution throughout the year, with a slight peak in spring and a secondary peak during autumn. The graph for Dallas shows that for the period of October through May, precipitation exceeds PE, resulting in an abundance of water at the surface.

Soil moisture recharge is represented by the area between the precipitation and the PE curves when precipitation exceeds PE and the soil is not yet at field capacity. This happens climatologically in Dallas from October through late January (Figure 6.15). On the water balance diagram, surplus is represented by the area between the precipitation and the PE curve when precipitation exceeds PE and the soil is at soil moisture capacity. The graph also shows that by the end of January, the soil is at field capacity in Dallas, and all excess water from that point until the end of May becomes surplus.

In June, P – PE is negative in Dallas, and soil moisture withdrawal commences. Soil moisture withdrawal is represented by the area between the precipitation curve and the ET curve (the dotted line in Figure 6.15), when precipitation is less than PE. The ET curve also usually peaks in summer, as it does in Dallas, but unless the soil is so wet that the atmosphere is always getting enough water (which does not happen in Dallas), this curve is not always as high as the PE curve.

Between June and October in Dallas, some of the water demanded by the atmosphere cannot be fulfilled by the amount of water in storage. The graph shows this deficit as the area between the PE and ET curves. Thus, the period of July through September tends to accumulate the largest deficits in Dallas on a climatological basis. By October the return of "wet" (positive P – PE) conditions puts an end to the deficit.

■ Drought Indices

Several drought indices based on the water balance model have been developed to provide a sense of the severity of flood or drought conditions in the United States. Maps of several drought indices are available on the Internet on a near–real-time basis. For example, the **Palmer Drought Severity Index (PDSI)** provides an assessment of long-term moisture conditions on a weekly basis by climate division across the United States (**Figure 6.16**). The PDSI is produced by the Climate Prediction Center and is derived by comparison with normal conditions for a **climate division**, with zero representing average soil moisture conditions at that climate division for that time of year. Likewise, positive values represent above-normal moisture conditions for that location and negative values are used for drier-than-normal-soil moisture. Notice in Figure

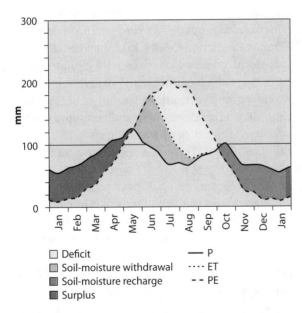

Figure 6.15 Mean water balance diagram for Dallas, Texas. Modified with permission using Web/WIMP version 1.01 implemented by K. Matsuura, C. Willmott, and D. Legates at the University of Delaware in 2003.

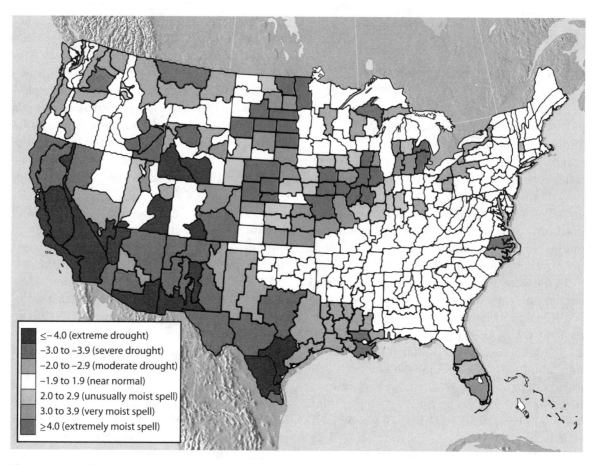

Figure 6.16 Palmer Drought Severity Index for the week ending June 27, 2009. *Data from*: NOAA.

6.16 that in some cases the lack of abundant data in sparsely populated areas may make for interesting spatial patterns. For example, Wyoming's PDSI values contrast with those in adjacent states.

Because the calculations for PDSI involve relatively deep soil moisture conditions, PDSI is often considered to represent long-term moisture. A modest rain event or two is unlikely to end dry conditions at a location, and a few rainy days are unlikely to instigate flood conditions. For some applications the lag between the onset of a wet or dry period and the appearance of a PDSI that corresponds to these moisture conditions is too long to provide adequate warning. Thus, this index should be used only when assessing long-term moisture conditions. For example, water resource planners are likely to use the PDSI to assist in policy related to long-term water budgeting. Similar to the PDSI is the **Palmer Hydrological Drought Index**, a measure used for even longer-term analysis than the PDSI that reflects conditions relating to groundwater availability and reservoir supplies.

Another commonly used tool based on the water balance is the **Crop Moisture Index (CMI)**, which is also produced for the United States by the Climate Prediction Center. Like the PDSI, the CMI is customized for comparison with climatological conditions at a location, with positive values representing moist conditions for that location at that time of year and negative indices representing successively drier conditions (**Figure 6.17**). Unlike the PDSI, the CMI is weighted heavily in the uppermost soil layers, which causes changes in soil moisture conditions to appear relatively quickly compared with the PDSI. Thus, a rain event or 2, or a few dry days, can strongly impact the CMI at a particular location. This index is more useful for short-term agricultural interests because most crops require the presence of moisture in the near-surface soil so that it can be easily accessed by the crops.

A final important index is the **Keetch-Byram Drought Index**—a measure designed by the U.S. Forest Service to assess the risk of fire potential by

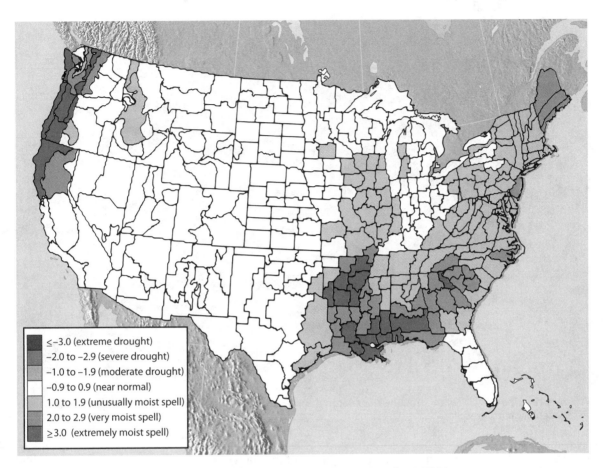

■	≤−3.0 (extreme drought)
	−2.0 to −2.9 (severe drought)
	−1.0 to −1.9 (moderate drought)
	−0.9 to 0.9 (near normal)
	1.0 to 1.9 (unusually moist spell)
	2.0 to 2.9 (very moist spell)
	≥3.0 (extremely moist spell)

Figure 6.17   Crop Moisture Index for the week ending January 2, 2010. *Data from*: NOAA.

examining the relationship between ET and precipitation in the organic matter on a forest floor and in the uppermost soil layers. Like the PDSI and CMI, the Keetch-Byram Drought Index is run continuously and is available on a near–real-time basis. Values range from 0, representing no fire hazard, to 800, representing the most severe hazard potential (**Figure 6.18**).

■ Summary

The global hydrologic cycle involves both reservoirs and fluxes. Reservoirs are dominated by the world ocean and the cryosphere, but the most accessible water for human consumption is in very small and unevenly distributed reservoirs across Earth, including ground and soil moisture, streams, freshwater lakes, and the atmosphere. Net fluxes in the global hydrologic cycle circulate water from the ocean surface to the marine atmosphere, to the terrestrial atmosphere, to the land surface, and back to the ocean via runoff. Although these fluxes seem robust, they are affected by human activities, particularly on climatological time scales. The global water balance equation can be used to describe the major fluxes, but it ignores locally important features such as infiltration, percolation, interception, and depression storage.

Hydroclimatic conditions at a location can be modeled using the surface water balance. In this technique PE is calculated—for example, by using a variation of Thornthwaite's method or by the Penman-Monteith method—and is compared with precipitation during the time period of interest. Precipitation is measured via a network and, like any of the other water balance variables, can be areally weighted to convert point estimates to an areal estimate. If excess water is available at the surface after PE is fulfilled, soil moisture recharge occurs, and if the soil reaches field capacity, surplus and runoff result. On the other hand, if insufficient moisture is available through precipitation to satisfy the atmospheric demand for water (represented by PE), soil moisture withdrawal

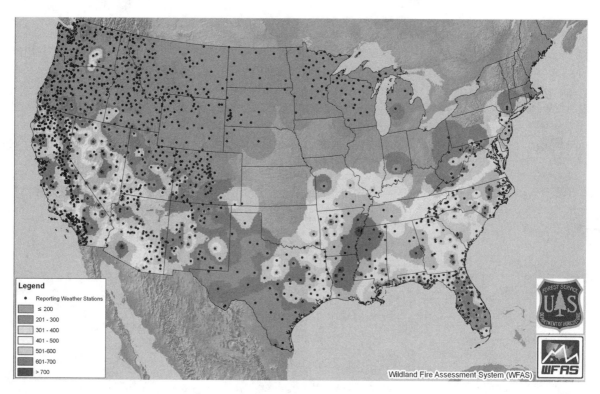

Figure 6.18 The Keetch-Byram Drought Index for January 4, 2010. *Data from*: The U.S. Forest Service/Wildland Fire Assessment System, Missoula, MT.

occurs. If the soil moisture is inadequate to supply the necessary water to the atmosphere, the actual ET is less than the PE and deficit results.

Equal availability water balance models assume that regardless of soil moisture content, water is withdrawn equally easily during wetter and drier periods, whereas decreasing availability water balance models assume that the loss of moisture is hindered when it becomes limiting, because

of resistance from vegetation. Water balance diagrams represent a relatively simple means of assessing moisture conditions at a place graphically, either on a mean climatological basis or for a particular duration of time. Likewise, drought indices based on water balance models, such as the PDSI and the CMI, have been developed to provide a near–real-time assessment of hydroclimatic conditions across the United States.

▶ Key Terms

Accumulated potential water loss
Advection
Bernoulli effect
Blaney-Criddle model
Clausius-Clapeyron equation
Climate division
Condensation
Crest
Crop Moisture Index (CMI)
Cryosphere
Decreasing availability water balance model
Deficit

Deforestation
Deposition
Depression storage
Dewpoint temperature
Equal availability water balance model
Evaporation
Evapotranspiration (ET)
Field capacity
Flux
Geographic information systems
Global hydrologic cycle
Global water balance equation
Gridpoint technique

Groundwater
Guard cell
Hydrosphere
Infiltration
Insolation
Interception
Isohyet
Isohyetal method
Isotherm
Joule
Keetch-Byram Drought Index
Kelvin temperature scale
Latent heat flux
Latitude

Lithosphere
Lysimeter
National Oceanic and Atmospheric
    Administration
National Weather Service
Net radiation
Orographic effect
Palmer Drought Severity Index
    (PDSI)
Palmer Hydrological Drought
    Index
Penman-Monteith equation
Percolation
Permeability
Porosity

Potential evapotranspiration (PE)
Precipitable water
Relative humidity
Reservoir
Resistance
Runoff
Saltwater intrusion
Saturation vapor pressure
Sensible heat flux
Soil moisture recharge
Soil moisture storage capacity
Soil moisture withdrawal
Stability
Stomate
Substrate heat flux

Surface tension
Surface water balance
Surplus
Thiessen polygon method
Thornthwaite potential
    evapotranspiration
Transpiration
Turc method
Undercatch
Vapor pressure
Wet bulb depression
Wilting point
Terms in italic have appeared
    in at least one previous
    chapter.

▶ Review Questions

1. Where are the major reservoirs and fluxes in the global hydrologic cycle? What are their relative sizes?

2. Explain the difference between percolation and infiltration.

3. What are some of the major advantages and criticisms of the Thornthwaite method of calculating PE?

4. How does the Penman-Monteith method of calculating PE differ from Thornthwaite's method?

5. How does the depth of snow in a gauge compare with the equivalent amount of liquid precipitation?

6. What is undercatch? How and why does it occur?

7. How can point-based estimates of rainfall be converted to areal estimates?

8. What advantages does the isohyetal method offer over the Thiessen polygon method and the gridpoint technique?

9. Under what circumstances is ET less than PE in the natural environment?

10. How does porosity relate to soil moisture storage capacity?

11. Why is the wilting point an important measure of moisture content in the soil?

12. Which of the surface water balance variables is most indicative of drought? Why?

13. In the surface water balance, why is runoff only a part of the total surplus value?

14. What is the difference between an equal availability and a decreasing availability water balance model?

15. When should the Palmer Drought Severity Index be used instead of the Crop Moisture Index to assess environmental moisture conditions?

▶ Questions for Thought

1. What effect does a large soil moisture storage capacity have on a location as compared with another location that has the same precipitation and PE totals but a lower soil moisture storage capacity?

2. How do impervious surfaces such as parking lots impact the water balance terms?

3. Why is it so complicated to estimate the potential impacts of global warming on the surface water balance at a location?

http://physicalscience.jbpub.com/climatology

Connect to this book's Web site: http://physical science.jbpub.com/climatology. The site provides chapter outlines, further readings, and other tools to help you study for your class. You can also follow useful links for additional information on the topics covered in this chapter.

Chapter at a Glance

Circulation of a Nonrotating Earth
Idealized General Circulation on a Rotating Planet
  Hadley Cells
  Polar Cells
  Planetary Wind Systems
Modifications to the Idealized General Circulation:
  Observed Surface Patterns
  Land–Water Contrasts
  Locations and Strength of Features in
    the Hadley Cells
  Locations and Strength of Features in the
    Polar Cells
  Locations and Strength of Surface Midlatitude
    Features
  Putting It All Together: Surface Pressure
    Patterns and Impacts
Modifications to the Idealized General Circulation:
  Upper-Level Airflow and Secondary Circulations
  Vorticity
  Constant Absolute Vorticity Trajectory
  Flow Over Mountainous Terrain
  Baroclinicity
  Rossby Wave Divergence and Convergence
  Rossby Wave Diffluence and Confluence
  Polar Front Jet Stream
  Mean Patterns of Rossby Wave Flow
Summary
Review

Processes occurring within the various components of the climate system have been described in previous chapters. Collectively, these processes contribute to the "real-world" atmospheric circulation and its variations. This chapter explains **general circulation**—the overall, prevailing pattern of winds on large spatial and temporal scales. Certainly, some circulations are local and unusual compared with the overriding, prevailing patterns. At broad spatial scales over a long period of time, however, the atmosphere's preferred directions of circulation are particularly noticeable. These patterns steer energy, matter (especially water and solid *aerosols*), and *momentum* from one part of Earth to another.

Circulation results from fulfillment of the *second law of thermodynamics*. One form of this law states that energy tends to be moved from areas of greater concentration to areas of lesser concentration. Because the Sun heats Earth's surface unequally, a circulation must exist in the atmosphere and oceans that attempts to equalize this energy imbalance. But the energy can never be balanced perfectly because as the general circulation of the atmosphere and the oceans continuously attempts to redistribute the energy, the Sun continuously warms different parts of the earth at different rates.

Specifically, the general circulation redistributes heat (energy) that arrives at the earth in greater quantities near the equator than near the poles. Factors such as the unequal distribution of land mass with *latitude* and variation of landforms and land cover types across the terrestrial earth have concomitant implications for the radiation balance, thereby complicating the general circulation. Chapters 8 through 10 describe the specific climate types that result from this general circulation.

In addition to general circulation, this chapter focuses on **secondary circulations**—smaller circulation systems that are characterized by traveling high- and low-pressure systems that affect climate. These circulations are influenced in the midlatitudes by their position relative to planetary-scale waves of motion. This chapter also examines the relationship between these waves and surface *anticyclones* and *cyclones*.

■ Circulation of a Nonrotating Earth

Before we can understand the atmosphere's general circulation on the "real" Earth, let us first consider the properties of the circulation if Earth did not rotate. The British meteorologist Sir George Hadley first postulated the general atmospheric circulation in 1735. His studies centered on a simple model of hemispheric *convection*. He posited that air should warm, become less dense, and rise in the area near the equator (**Figure 7.1**). Following the model of a convection cell, he thought that this rising equatorial air would then travel poleward far above the surface (in both the northern and southern hemispheres), ultimately cooling and sinking in the polar areas. This air stream would then diverge at the surface in the vicinity of the poles and move back equatorward across the surface to begin the process anew. The sinking air at the poles from this circulation would produce a permanent surface anticyclone at each pole. A belt of low pressure would exist in the equatorial region, with the lowest pressure along the *longitude* that was being heated most directly—experiencing *solar noon*—at that particular time of day.

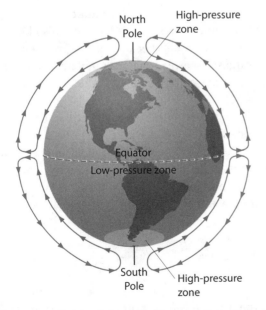

Figure 7.1 The simple hemispheric convection cell model of atmospheric circulation proposed by Sir George Hadley, a British meteorologist.

The major problem with Hadley's view of atmospheric circulation was that it did not match observed data of wind patterns reported from sailing vessels across the globe. The circulation of the planet is more complex than the simple single hemispheric convection cell pattern suggested by Hadley. Still, Hadley's work was the first to include the upper atmosphere in thoughts centered on the general circulation of the atmosphere and for that it gains merit.

There are two main complications that Hadley did not consider. The first is that Earth's surface is not uniform. Rather, it has many different topographic and land/water surface variations that upset the formation of a single hemispheric convection cell in each hemisphere. Because these variations are often local or regional, they tend to contribute even more to local and synoptic circulation patterns than to changes in the "general" circulation. The second complication—the rotation of Earth—exerts a more direct influence on the general circulation. This rotation initiates trajectory changes in moving fluids through the *Coriolis effect*.

■ Idealized General Circulation on a Rotating Planet

Hadley postulated correctly that, in general, air near the equator gains thermal buoyancy because of its warmth. He was also correct to infer that the cold air near the poles has a tendency to sink. We have already seen that rising air is generally associated with atmospheric low-pressure situations, whereas sinking air is typically linked to high-pressure systems. Thus, a broad belt of low surface pressure dominates the equatorial region, whereas higher surface pressures exist over the poles.

Hadley Cells

Hadley also proposed that air rising near the equator begins moving toward the poles in the upper *troposphere*. This upper-level poleward motion occurs because of an upper atmospheric pressure gradient that pushes air away from the equator (where there is an excess of pressure aloft caused by the "extra" air that rose from the surface). **Figure 7.2** shows a model of the general circulation on a rotating planet.

Unlike the model originally proposed by Hadley, most of the air aloft flowing away from the

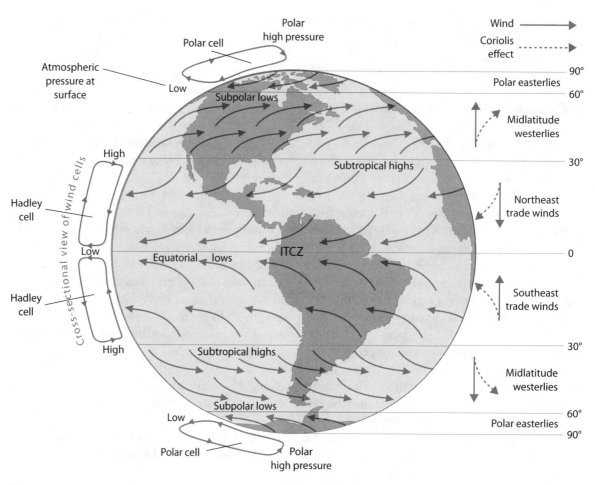

Figure 7.2 Pressure and wind belts of the generalized three-celled general circulation model.

equator does not actually travel all the way to the poles. Instead, much of it cools while it is high above the surface and sinks in the vicinity of 30° latitude (both north and south of the equator). This air diverges from the point of impact at the surface near 30° latitude, forming the *subtropical anticyclones*, or subtropical highs, which are centered at approximately 30° latitude in each hemisphere. As we see later, the highs are situated over each major ocean basin.

The region of the subtropical anticyclones is also known as the **horse latitudes**—a term derived from the days of European exploration when sailors are said to have discarded horses overboard when they found themselves under the stagnant regions beneath the high-pressure zones. The weak descent of air through the atmosphere in these regions provides little in the way of steering currents necessary to push sailboats across vast distances. Thus, sailing vessels frequently stalled in these areas. Water shortages ensued, which sealed the

fate of horses onboard. The subtropical anticyclone region of the North Atlantic Ocean where the **Bermuda-Azores high** is located is also known as the **Sargasso Sea** because of the prevalence of the sargasso seaweed that accumulates seasonally in the rather calm, stagnant waters under sunny skies.

This circulation between the equator and 30° latitude is largely convective, as vertical motions within it are driven solely by heat energy. This convection cell, with rising motion near the equator and sinking motion near 30° latitude, is termed a **Hadley cell**; one Hadley cell exists in each hemisphere.

Polar Cells

Near the poles the net energy deficits reinforced by the high *albedo* of snow and ice in polar areas cause the air to increase in *density*, which initiates sinking motions through the atmosphere. This air diverges at the surface, inducing the permanent high atmospheric pressure at the surface—the **polar highs** (one at each pole). Upon reaching the surface, the diverging air travels equatorward toward lower latitudes, is deflected by the Coriolis effect, and eventually converges at approximately 60° latitude with air that was diverging from the subtropical anticyclones at 30°. The converging air near 60° latitude is forced to rise. As we have seen, rising air motions are associated with low atmospheric pressures. The convergence creates **subpolar lows**—areas of relatively low surface pressure centered on 60° latitude in each hemisphere. These circulation systems are the **polar cells;** one polar cell exists in each hemisphere.

Planetary Wind Systems

As we have seen, starting at the poles and moving toward the equator at 30° intervals of latitude are surface oscillations of high (at 90°), low (at 60°), high (at 30°), and low (at 0°) pressure in each hemisphere (Figure 7.2). These differences in pressure across space set up differences in atmospheric mass, because pressure is a function of force (which is the product of mass and acceleration according to *Newton's second law*). Because mass is transferred from areas of higher concentration to areas of lower concentration, this general surface pressure distribution sets up the planetary wind systems. Winds are simply the transfer of atmospheric mass between an area of excess mass and an area containing less mass. Because pressure differences define winds, let's begin our discussion with an examination of the influence of the major pressure features on surface wind patterns.

Surface Wind Systems   As with all surface anticyclones, air in the subtropical anticyclones at approximately 30° latitude in each hemisphere sinks and diverges across the surface in all directions. The *pressure gradient force* pushes air directly toward lower pressure areas—namely, toward 0° and 60° latitude in each hemisphere, the equatorial and subpolar lows, respectively. Because of the rotation of the planet, the Coriolis effect deflects these surface winds to the right in the northern hemisphere. This causes air diverging from the northern hemisphere's subtropical highs to flow clockwise (anticyclonically) as it moves outward. Likewise, air diverging from the southern hemisphere's sub-

tropical highs flows counterclockwise—also anticyclonically for that hemisphere—as it moves outward away from the high, as the Coriolis effect deflects motion to the left in that hemisphere (Figure 7.2).

Thus, the surface winds moving between 30°N and the equator take on a motion that is from the northeast. As we saw in Chapter 3, winds are always named for the direction from which they blow. These winds are, therefore, appropriately called the *northeast trade winds* (or northeast trades). They are termed "trade winds" because they constituted the primary sailing trade route between southern Europe and Spanish America across the North Atlantic Ocean. The northeast trade winds steered Christopher Columbus from his point of departure in Spain toward his unintended destinations in the Caribbean, Central America, and northern South America. A strong argument can be made that Spanish is the primary language spoken in Latin America and the Caribbean because of the northeast trade winds!

The southern hemisphere also has trade winds, but the leftward deflection of the Coriolis effect causes air flowing out of the subtropical anticyclones to move counterclockwise and surface winds to move primarily from the southeast between 30°S latitude and the equator. These are called the *southeast trade winds* (or southeast trades). It should be easy to remember that the northeast trades occur in the northern hemisphere, whereas the southeast trade winds occur in the southern hemisphere.

Because the northeast and southeast trade winds converge into the low-pressure equatorial low, the equatorial low is sometimes known as the **Intertropical Convergence Zone (ITCZ)**. Although the name appears rather imposing, it is quite explanatory as it describes surface airflow across the tropics that converges in a zone across the low latitudes. The ITCZ also carries another name—the **doldrums**, a belt of relatively stagnant winds that form as air rises from the surface because of trade wind convergence from both hemispheres. Sailing ships venturing into this area would often stagnate for long periods. Being "stuck in the doldrums" achieved colloquial status, signifying a depression of some sort. Interestingly, the doldrums became the region of many mutinies, including the famous mutiny of Captain Bligh on the *HMS Bounty*.

On the poleward sides of the subtropical anticyclones, the surface pressure gradient force pushes air from the high-pressure core toward the subpolar lows. This air is then deflected to the right (in the northern hemisphere) or the left (in the southern hemisphere) on its trek to fill in the subpolar low. In this case, though, the latitude is higher than in the case of the tropical trade winds, so the Coriolis deflection is stronger. Recall that the magnitude of the Coriolis deflection is proportional to latitude. This strong deflection causes the general circulation between 30°N and 60°N latitude to be from west to east near the surface. This surface wind belt is known as the **midlatitude westerlies**. The southern hemisphere also has westerlies between 30°S and 60°S latitude. Even though the Coriolis deflection in that hemisphere is from the opposite direction, the fact that the midlatitude zone is south of the subtropical anticyclone rather than north of the high (as in the northern hemisphere) causes surface westerly winds to exist in the southern hemisphere as well.

Air sinking to the surface and diverging equatorward at the polar highs also undergoes Coriolis deflection. As the air travels from the region of the North Pole toward lower latitudes, it is deflected strongly (because of the high latitude) to the right. This creates the surface wind belt known as the **polar easterlies**, as near-surface prevailing air is generally from the east. Surface air moving from the region of the South Pole is deflected to the left as it moves toward that hemisphere's subpolar low. Coriolis deflection to the left also creates polar easterlies near the surface for that hemisphere between 90° and 60°S latitude.

To summarize, moving from the region of the poles toward the equator, the following surface wind belts occur:

- Polar easterlies (90–60°N latitude)
- Midlatitude westerlies (60–30°N)
- Northeast trades (30°N–0°)
- Southeast trades (0–30°S)
- Midlatitude westerlies (30–60°S)
- Polar easterlies (60–90°S)

Upper-Level Winds in the Hadley and Polar Cells Upper-level winds overlaying the surface wind patterns are also an important component of the general circulation. As is shown in Figure 7.2, the upper-level pressure gradient force pushes air in the opposite direction from that at the surface in both the Hadley cells and the polar cells. At the surface the pressure gradient force is directed from the subtropical anticyclones to the equatorial low in the Hadley cells. But aloft, the upper-level pressure gradient force pushes air from the equatorial region toward the poles, therefore, moving air over the subtropical anticyclones. Over the polar cell the upper-level pressure gradient force pushes the air from the area over the subpolar lows to the area over the polar highs, a direction opposite to that at the surface. Of course, in both the Hadley cells and the polar cells, the Coriolis effect causes a deflection of these winds generated by the pressure gradient force. The deflection is less in the Hadley cell than in the polar cell because of the lower latitudes of the Hadley cell. The result is upper-level southwesterlies over the northern hemisphere Hadley cell, northwesterlies over the southern hemisphere Hadley cell, and westerlies over the polar cells in both hemispheres (**Figure 7.3**). These polar westerlies aloft advected the tephra from the Eyjafjallajökull volcanic eruptions in Iceland eastward toward Europe beginning in the spring of 2010.

As air aloft in the Hadley cell continues to move poleward (the southwesterlies aloft in the northern hemisphere and the northwesterlies aloft in the southern hemisphere), the Coriolis effect increases in strength. The deflection to the right of the flow (or to the left of the flow in the southern hemisphere) increases poleward, and flow becomes increasingly westerly near the poleward edge of the Hadley cell in each hemisphere. Winds at these positions are often very strong, in large part because of **conservation of angular momentum**—an important property of curved motion in a fluid on Earth. Anything that is "conserved"—in this case, angular momentum—remains constant.

Like any type of momentum, angular momentum is the product of a mass and a velocity. Angular momentum simply includes an additional variable—the radius of curvature. The "tighter" the curve, the smaller is the radius of curvature. An infinitely large radius of curvature occurs for straight-line flow. The conservation of angular momentum can be expressed quantitatively as

$$mvr = \text{constant}$$

where m is mass, v represents velocity, and r is the radius of curvature. In the atmosphere mass is virtually constant. Therefore, the equation above

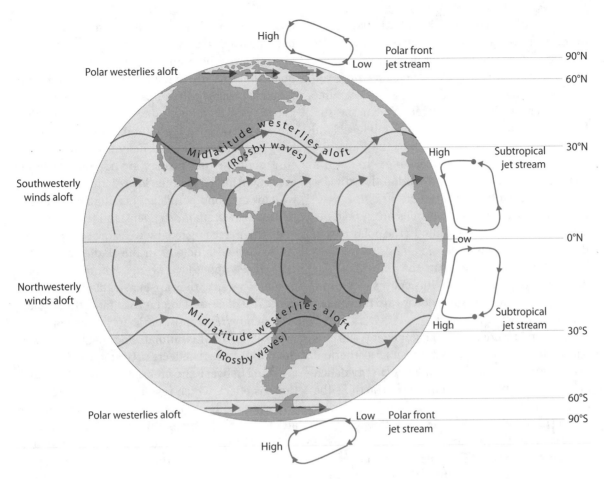

Figure 7.3   Global upper-level wind systems.

suggests that as r decreases, v increases in order to maintain a constant angular momentum. Ice skaters know that to spin faster (increase velocity), they must become "thinner" by moving their arms inward to reduce their radius of curvature. In the atmosphere, westerly winds moving around the hemisphere at higher latitudes have smaller radii of curvature (the distance between the westerly wind and the north Earth's axis) than westerly winds moving around the hemisphere at lower latitudes. As air moves poleward, its radius of curvature decreases (i.e., it goes from the wider part of Earth near the equator to the narrower part of Earth near the poles) and begins to curve increasingly to the right (in the northern hemisphere) because of the increasing Coriolis effect. Its velocity must increase simultaneously as r decreases to ensure that angular momentum is conserved. The result is the westerly *subtropical jet stream* that exists near 30°N and 30°S latitude (Figure 7.3). Although much remains to be discovered about the influence, variability, and impact of this fast-moving river of air in

the upper troposphere, it is known to exert a significant impact on weather and climate in subtropical latitudes.

Upper-Level Winds in the Midlatitudes   To understand the upper-level winds overlying the midlatitude surface westerlies, upper-level pressure gradients can be considered in terms of **geopotential height**, which is simply the altitude (in meters above sea level) of a given pressure surface over a particular location at a certain time. For example, the 500-mb geopotential height represents the altitude at which 500 mb of pressure is being exerted by the weight of the atmosphere in an atmospheric column. Because mean sea level pressure is 1013.25 mb, the 500-mb geopotential height also represents the approximate middle of the atmosphere from a mass perspective; roughly half of the atmospheric mass is above and the other half is below the 500-mb geopotential height level.

In relatively warm air, such as in the equatorial part of Earth, the surplus of heat increases the molecular *kinetic energy* of the gases in the air column. As this energy causes molecules of air to expand upward, the warm air becomes less dense than the colder air as the gases occupy more volume and, therefore, weigh less per unit volume than when pushed closer together near the surface. We would expect that the 500-mb level (and other constant-pressure levels) over the equatorial part of Earth would be relatively high. In a warm column of atmosphere the elevation that divides the total mass of the atmosphere in half is relatively high because the molecules are spread out more vertically in the column. In the polar part of Earth, the high density of cold air occurs because molecular kinetic energy in the individual gas molecules is relatively low. This causes a contraction of the air column and a downward movement of the constant-pressure levels.

In the real world, then, the net energy surplus within the tropics causes tropospheric thermal expansion in that area, and the net energy deficit at the poles causes tropospheric contraction. The equatorial *tropopause* averages approximately 19 km (12 mi) above the surface, but the polar tropopause averages only about 10 km (6 mi) above the surface. A steady decrease in upper-level geopotential heights tends to occur from the equator toward each pole as colder conditions prevail (**Figure 7.4**). Furthermore, the difference in a given geopotential height level between the equator and pole increases with increasing height in the troposphere. For example, it is apparent from Figure 7.4 that the

850-mb geopotential height is somewhat higher over the tropics than over the poles. But above the 850-mb level, the tropics are also warmer than the poles (particularly in winter), so the 700-mb geopotential height difference between the equator and the pole is compounded on the difference that already occurred at the 850-mb level. The latitudinal slope in geopotential heights creates a pressure gradient that pushes air horizontally from areas with higher heights (lower latitudes) to areas with lower geopotential heights (higher latitudes). Because the gradient steepens and friction weakens with increasing height, we expect the winds to strengthen with height. In the Hadley and polar cells, these upper-level winds cannot strengthen as much because they are constrained by the flow in the rest of the cell, including that at the surface.

When upper-tropospheric air moves poleward in response to this midlatitude height gradient, the Coriolis effect deflects it. Ultimately, a balance is gained between the pressure gradient force and Coriolis effect, which results in upper-level westerly winds in both hemispheres from the tropics all the way to the poles. The midlatitude region is the only one of the three zones per hemisphere (Hadley, polar, and midlatitude) in which upper-level winds occur in the same direction as the mean surface winds.

■ Modifications to the Idealized General Circulation: Observed Surface Patterns

Land–Water Contrasts

The model described above is a highly idealized conception of the circulation system of Earth. It assumes that there are no surface variations—the model applies only to a planet composed solely of water. But in the real world, land–water contrasts complicate circulation in the climate system. As we saw in Chapter 3, large water bodies maintain relatively constant temperatures through the course of a year compared with inland areas, assuming that all other factors are equal. The consistency is caused by a combination of water transparency, high *evaporation* rates, vertical and horizontal currents, and a high *specific heat* in water, all of which dampen fluctuations in temperature near the water surface.

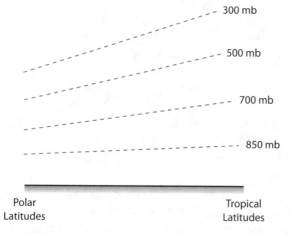

300 mb
500 mb
700 mb
850 mb

Polar
Latitudes

Tropical
Latitudes

Figure 7.4 The slope of geopotential height surfaces from the equator to the North Pole. A similar slope would occur in the southern hemisphere, as heights decrease toward the South Pole.

On the other hand, continents show wide temperature variations across the seasons because their opaque surfaces absorb *insolation* in a tiny surface layer and then readily transfer that energy back to the atmosphere. Inland surfaces also have relatively low evaporation rates (causing a relatively high percentage of energy to be devoted to *sensible heat*), virtually no convective/advective motion to redistribute energy, and a low specific heat. These factors produce surfaces that heat quickly when energy is present and cool quickly during periods of energy deficit.

Given these factors and variations in surface elevations associated with continents, it is no surprise that including continents into the idealized model causes disruptions to the circulation features. But the idealized model does provide a useful starting point to understanding the more complex real-world features more fully.

Locations and Strength of Features in the Hadley Cells

ITCZ and Trade Winds   The ITCZ exists in a similar form in the real world as in the idealized pattern (**Figure 7.5**). It may be thought of as a constantly moving convective thunderstorm chain that follows the migration of the vertical solar ray across varying longitudes through the course of a day and between the tropical latitudes through the course of the year. But the ITCZ is much more physically constrained by surface temperature characteristics than the faster-moving solar rays. It, therefore, lags considerably behind the solar radiation maximum in its latitudinal migration. Also, it never reaches the maximum latitudinal position of the 90° vertical ray, which is at the *Tropic of Cancer* (on June 21) and the *Tropic of Capricorn* (on December 22) (**Figure 7.6**).

(a)

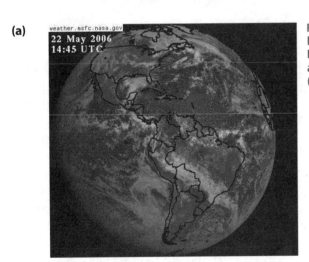

Figure 7.5   Global view of ITCZ represented by (a) the line of low-latitude clouds and (b) the most concentrated lightning flashes. Notice how the ITCZ is most prominent at the longitudes experiencing afternoon at a given time. (See color plate 7.5b).

(b)

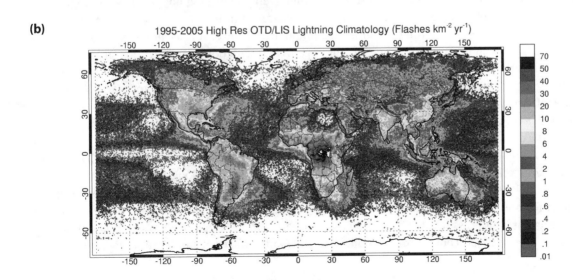

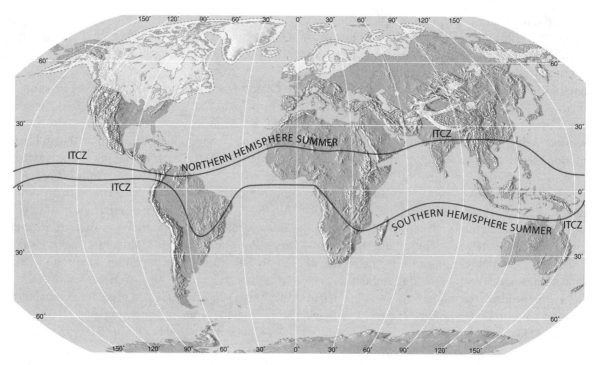

Figure 7.6 Migration of the ITCZ through the year. *Data from*: NOAA/GOES.

Instead, the ITCZ tends to migrate only approximately 10–20° of latitude away from the equator in most locations, lagging behind the *solar declination*. The ITCZ moves poleward into the summer hemisphere in response to the displacement of the vertical solar rays. The ITCZ does not migrate much over places affected by cold ocean currents, such as the tropical west coast of South America, even if these locations are in the path of the Sun's vertical rays.

As the ITCZ migrates seasonally, the trade winds also migrate—to their northernmost extent in northern hemisphere summer and to their southernmost extent in northern hemisphere winter. Thus, some locations on Earth may experience the trade winds for only part of the year. This migration can have a significant influence on the seasonal climate of tropical locations.

Subtropical Anticyclones The idealized model suggests that a continuous belt of surface high pressure should exist across the planet near 30°N and 30°S—the subtropical anticyclones. In reality, however, these anticyclones occur only over the ocean basins near 30° latitude, and they are larger in summer than in winter (**Figure 7.7**). The reason is that the tendency for high pressure at 30°N and 30°S is reinforced by the *relatively* cool summer conditions

(compared with the adjacent land surfaces) over the oceans. The relatively cool ocean surface chills the overlying air, increasing low-level atmospheric *stability*. During that time internal pressure in the subtropical anticyclones reaches a maximum. The subtropical anticyclones are strongest when and where the ocean surface is much colder than the continental surface in the subtropics. The sinking air that results diverges at the surface, spilling toward lower-pressure locations in all directions before experiencing Coriolis deflection.

By contrast, the subtropical land surfaces heat effectively under long day lengths and high solar angles in summer. They tend to encourage rising motion (and lower surface pressure) that offsets the general circulation's tendency for sinking motion and high surface pressure. The subtropical continents are, therefore, typically too warm in summer to support the development and maintenance of semipermanent surface high-pressure cells.

The subtropical anticyclones migrate somewhat from their summer position, latitudinally with solar declination and longitudinally depending on where the coldest water lies, within the subtropical oceans. They also weaken in winter because the ocean surface is *relatively* warm compared with the land surfaces at similar latitudes. This relative

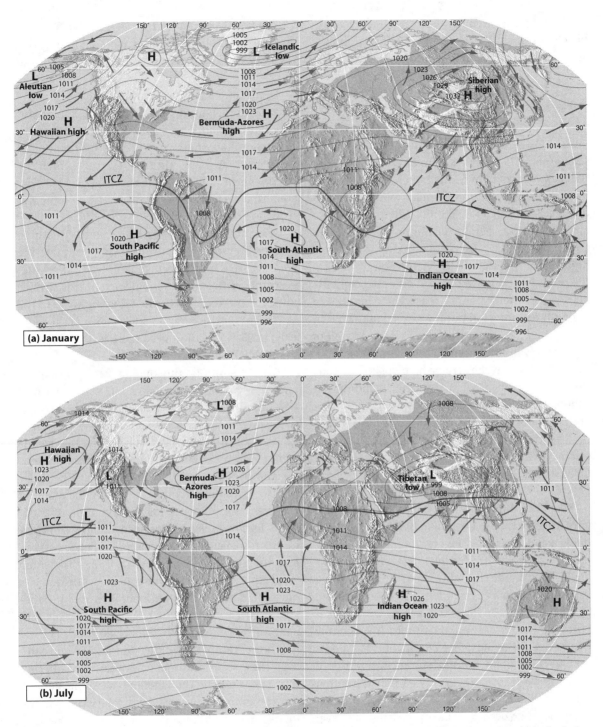

Figure 7.7   Mean surface isobars, positions of the semipermanent surface pressure cells, and associated winds, during (a) January and (b) July. *Adapted from*: January Pressure and Predominant Winds, and July Pressure and Predominant Winds, 1995. *Goode's World Atlas*, 19th ed. Edited by E.B. Espenshade. Skokie, IL: Rand McNally and Co., 1995.

warmth induces rising motion that tends to offset the tendency for sinking air at 30°N and 30°S, as suggested by the idealized general circulation model. Instead, pressure tends to increase over the relatively cold terrestrial atmosphere.

The major surface-based, semipermanent, high-pressure cells located near 30°N latitude are the Bermuda-Azores high in the North Atlantic Ocean and the **Hawaiian high** (also known as the **North Pacific high**) in the North Pacific Ocean

(Figure 7.7). Southern hemisphere subtropical anticyclones include the **South Pacific high**, the **South Atlantic high**, and the **Indian Ocean high**.

Locations and Strength of Features in the Polar Cells

Surface pressure features in the polar cells also strengthen and weaken seasonally. During winter the lack of insolation in the high latitudes causes intense surface chilling, which greatly increases low-level atmospheric stability. The frigid, dense air sinks in the vicinity of the poles and moves equatorward vigorously. The sinking air creates high surface pressure in the vicinity of the poles— the polar highs. These permanent anticyclones— one in each hemisphere—reach their maximum strength during winter and weaken somewhat in summer as the long summer days provide a bit of warming even at the poles. The intensification of the polar high in the winter hemisphere extends the entire polar cell equatorward, including the subpolar low-pressure zones, to approximately 55°N and 55°S latitudes. The weakening of the polar high in the summer hemisphere contributes to the contraction of the polar cell to approximately 65°N and 65°S latitude, coincident with the poleward migration of the subtropical anticyclones.

Like the subtropical anticyclones, the subpolar lows are confined to the ocean basins rather than being a continuous feature across the globe as depicted in the idealized model. The subpolar lows are also similar to the subtropical anticyclones in that their existence is tied to surface temperature variations between land and sea. In winter, oceanic regions at the subpolar latitudes are generally warmer than the adjacent continental locations, both because of the thermal storage properties of water and because of warm ocean currents that transport energy from lower latitudes. The relatively warmer ocean surface promotes low-level atmospheric instability, thereby reinforcing the tendency for rising motion as suggested by the idealized general circulation. This supports the development and maintenance of semipermanent surface cyclonic cells, particularly in winter—the subpolar lows. Buoyant air in the subpolar lows is associated with strong pressure gradients, windiness, and storminess, particularly during winter. The semipermanent North Atlantic subpolar cyclone is known as the **Icelandic low** and that in the

North Pacific is referred to as the **Aleutian low** (Figure 7.7). Southern hemisphere subpolar lows are very weak and exist only in a broad, continuous belt.

Continental areas at approximately 55–60°N latitude are frigid in winter. These cold land surfaces support sinking motion, which counteracts the general circulation's tendency for having rising motion at these latitudes. During winter, the *continentality* over the massive Asian continent sometimes causes the lowest surface temperatures for the northern hemisphere to exist over Siberia rather than at the North Pole. This high surface pressure expands over Asia, even into the zone where the idealized general circulation suggests that rising motion should occur. This surface anticyclone is known as the **Siberian high**. In the southern hemisphere, no such continental areas exist at 55–60°S latitude, so there is no equivalent to the Siberian high.

During summer, the long day lengths in high-latitude locations increase continental surface heating substantially, particularly in the northern hemisphere. Under such conditions the thermal gradient between the land masses and the oceans in the vicinity of 60° latitude decreases, weakening the pressure gradient. With *relatively* cooler oceans in summer (compared with adjacent land surfaces), the tendency for rising motion over the subpolar low regions is reduced, and the semipermanent surface oceanic subpolar lows weaken and shrink, as shown in Figure 7.7. At the same time, the Asian land mass heats considerably, once again because of continentality, and the Siberian high disappears. Instead, a semipermanent zone of low pressure occurs over south-central Asia—the **Tibetan low**. The flip-flop of pressure over the Asian land mass between the Siberian high in winter and the Tibetan low in summer as a result of continentality is the primary cause of the monsoon circulation, discussed in Chapter 9.

Regardless of season, the subpolar lows are generally located where the cold surface air moving equatorward from the polar high meets the much warmer poleward-moving air diverging from the subtropical anticyclones. As we have seen, this convergence is displaced poleward in summer and equatorward in winter. The convergence of these two surface air streams causes rising motion that is enhanced by the temperature differences between the two streams, with the warmer tropical air pushed up vertically over the colder polar air. Thus, this region is the boundary between cold, dense, polar air and warm, less dense, tropical air. Storm formation (or **cyclogenesis**) frequently occurs along this band of latitudes because the steep thermal gradient supports the development and motion of migratory cyclones as secondary circulation features.

Locations and Strength of Surface Midlatitude Features

The seasonal fluctuation in strength of the subpolar lows and subtropical anticyclones works in tandem. In the winter hemisphere the subpolar lows are strengthened and displaced equatorward, while the subtropical anticyclones are relatively weak and displaced equatorward also. In the summer hemisphere the subtropical highs are strengthened and displaced poleward, whereas the subpolar lows simultaneously shrink and retreat poleward. The areas between the subpolar lows and the subtropical highs—the midlatitudes—are affected differently throughout the year by these pressure patterns.

Both the subpolar lows and the subtropical highs are affected more by the changing thermal surface characteristics of the northern hemisphere than by those in the southern hemisphere. The increased concentration of land area in the northern hemisphere promotes stronger land–sea contrasts than in the southern hemisphere. Likewise, the semipermanent subtropical highs and subpolar lows are more consistent in their presence and strength in the southern hemisphere because of the lack of large land masses at the subtropical and (especially) subpolar latitudes.

Putting It All Together: Surface Pressure Patterns and Impacts

To summarize, the summer hemisphere sees weakening and contraction of the polar high and subpolar lows and expansion of the subtropical highs. The ITCZ pushes poleward through the tropical part of the summer hemisphere at the same time. This migration of the ITCZ contributes heavily to tropical precipitation regimes. The winter hemisphere sees expansion and strengthening of the polar high and the oceanic subpolar lows into lower latitudes, whereas the subtropical anticyclones weaken and retreat equatorward.

The waxing and waning of these pressure systems affect and are affected by the changing of the seasons in the high and midlatitudes. The changes in the semipermanent circulation cells trigger direct precipitation regime changes for many high and midlatitude locations. In lower latitudes, precipitation changes directly define the seasons because thermal changes are minimal or nonexistent because of day length consistency through the course of the year. There, the ITCZ is the primary precipitation-forcing mechanism. When the ITCZ migrates over or near a region, it brings thunderstorms and associated precipitation. This occurs during the high-sun period of "summer," which is defined not so much by higher temperatures as by direct rays of sunshine falling in the same hemisphere as the point of interest. During the lower Sun season, the ITCZ migrates to the opposite hemisphere, taking with it the primary precipitation-forcing mechanism. This causes a distinct wet/dry seasonal regime for many locations.

■ Modifications to the Idealized General Circulation: Upper-Level Airflow and Secondary Circulations

As we have seen, flow in the upper-level midlatitudes is primarily westerly. However, these upper-atmospheric air currents typically exhibit flow characteristics that meander northward and southward as they move in the general west-to-east direction. We may characterize the broad-scale, upper-level atmospheric flow at a given time based on the *amplitude* or "waviness" of the flow pattern. If the upper airflow exhibits a relatively deamplified, or nearly straight, west-to-east pattern on a given day, the atmospheric pattern is classified as having a **zonal flow (Figure 7.8a)**. If significant latitudinal amplification occurs in the midlatitude flow pattern on a given day, then the atmospheric pattern is said to have a **meridional flow** (Figure 7.8b).

Virtually all airflow has at least some meridional component, meaning that the flow pattern is at least somewhat wavy. Because some amplitude is present in all flow regimes, any upper-airflow pattern may be broken into wave segments. Any portion of the wave pattern in either hemisphere

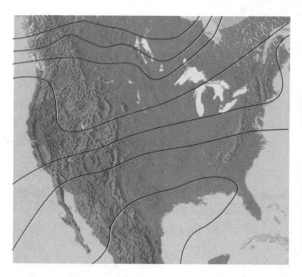

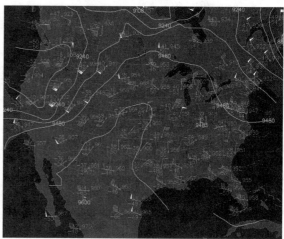

Figure 7.8 (a) The zonal flow over most of the United States, showing a 300-mb geopotential height, recorded at 8 AM EDT on September 18, 2005. *Data from*: The Ohio State University WWW Weather Server, courtesy of the Department of Geography/Atmospheric Sciences Program. (b) A 300-mb map of the United States, recorded on May 28, 2010. On this date, the strongest flow is over the western U.S. and eastern Canada. Reproduced with permission of Unisys Corporation © 2010.

that "bumps" poleward is termed a *ridge*—an elongated zone of high pressure (as opposed to an *enclosed* area of high pressure, which is an anticyclone). The elongation of high pressure refers to the fact that where the midlatitude flow moves poleward, the high pressure of the subtropical anticyclones can be elongated poleward as well. A ridge at the 300-mb level is shown in Figure 7.8b centered over the upper Mississippi Valley.

Any portion of the flow pattern that "dips" equatorward is termed a *trough*—an elongated area of low pressure. By contrast, a cyclone is an *enclosed* zone of low pressure. In Figure 7.8b a 300-mb trough is situated over the western states. When the midlatitude flow dips equatorward, the relatively low pressure associated with the subpolar lows can also be elongated toward the equator.

Interestingly, ridges in the southern hemisphere appear the same as troughs in the northern hemisphere, and vice versa. This is because a dip *southward* of midlatitude flow in the southern hemisphere represents a place where the subtropical anticyclones to the *north* can be elongated, creating a ridge. Likewise, a jog to the north in the southern hemisphere represents a location where the subpolar low-pressure systems can be elongated equatorward, thereby forming a trough.

Like other types of waves, atmospheric waves can be described according to *wavelength*—the

distance from one wave crest (ridge) to an adjacent wave crest (or the distance from one wave trough to an adjacent wave trough). Typically, midlatitude waves that have wavelengths of hundreds of kilometers are simply called **long waves**, or **Rossby waves**, named after the atmospheric physicist Carl Rossby, who was instrumental in explaining upper-atmospheric dynamics during the first half of the twentieth century. Rossby waves are continuous around the hemisphere and circumnavigate each pole.

During winter the northern hemisphere usually has three or four long waves. The latitudinal thermal gradient is maximized during winter when the polar areas receive little to no insolation and the low latitudes still receive about as much energy as it receives in summer. Rossby waves must have some amplification to redistribute this energy imbalance latitudinally in an attempt to equalize the differences. In summer the number of long waves increases to as many as six, but the amplitudes decrease because the latitudinal thermal gradient decreases when the high latitudes warm up under long day lengths. In both winter and summer Rossby waves tend to be less meridional in the southern hemisphere than in the northern hemisphere. Instead, the near-continuous ocean in the southern hemisphere plays a greater role in the redistribution of energy than it does in the northern hemisphere.

Because these upper-air Rossby waves result from latitudinal thermal inequalities, the ridges and troughs are essential in the energy balance of a hemisphere. The trough-to-ridge sides of these waves allow warm, and often moist, low-latitude air masses to move poleward, and the ridge-to-trough sides of Rossby waves allow cold and often drier air to move equatorward. Without these long waves helping to redistribute energy across the midlatitudes, energy differences would continuously increase between the tropics and the poles. This would result in the poles becoming increasingly colder while the equator increasingly warmed, and soon life would be restricted to the very narrow zone where a thermal balance was achieved.

Vorticity

An important feature of upper-atmospheric Rossby wave flow in climatology is **vorticity**—the rotation, or spin, of any object. Although seemingly simplistic, vorticity may be viewed in a number of ways. The vorticity of an object such as a person, a particle in the atmosphere, or a storm system can occur either because the object itself is rotating or because it is situated on or in another object that is rotating (i.e., Earth), or both. **Relative vorticity** (usually denoted by ζ, the lowercase Greek letter zeta) is the spin that occurs because the object itself is turning. If you hold your pen vertically and twirl it between your fingers, you impart relative vorticity on the pen. If you hold your pen stationary but instead *you* begin rotating, your pen is still rotating but for a different reason.

Particles in the atmosphere are always rotating because they possess **planetary vorticity** (f)—the spin they acquire because of the rotation of Earth. The amount of this vorticity is proportional to the Coriolis effect, which as we have seen is itself proportional to latitude and speed of motion. Specifically, in the expression of the Coriolis effect from Chapter 3,

$$CE = 2\,\Omega\,(\sin\varphi)v,$$

the term $2\Omega(\sin\varphi)$, where Ω is the angular velocity of Earth and φ represents latitude (with negative values for southern hemisphere latitudes), is sometimes referred to as the **Coriolis parameter**, but it really represents f.

Earth rotates on its axis in such a way that if we were to look down from space directly on the North Pole we would see counterclockwise rotation (from west to east). But when viewed from below the South Pole, Earth is simultaneously rotating clockwise. By convention, any rotation in the same direction as Earth's rotation in that hemisphere is termed **positive vorticity**. Thus, counterclockwise-spinning objects in the northern hemisphere and clockwise-spinning objects in the southern hemisphere (both of which include cyclones in their respective hemispheres) are said to exhibit positive or "cyclonic" vorticity. Thus, for residents of either hemisphere, "positive vorticity advection" suggests the approach of a storm. Objects spinning in a clockwise manner in the northern hemisphere (or counterclockwise in the southern hemisphere) are said to have **negative vorticity**, because this rotation opposes that of Earth itself. Anticyclones in either hemisphere have negative vorticity.

In the northern hemisphere air flowing around ridges also has negative vorticity, because it is moving clockwise around the ridge, while air moving around troughs has positive vorticity. In the southern hemisphere ridges are also associated with negative vorticity (as the air moves counterclockwise through the ridge) and troughs are linked to positive (clockwise) vorticity.

Usually, vorticity in the middle atmosphere (i.e., the 500-mb geopotential height level) is considered to be most important for analysis. This is because if positive (or negative) vorticity is occurring halfway up the mass of the atmosphere, that same spin is likely to be propagated upward and downward to encompass most of the mass of the atmosphere over that location. If that vorticity is positive, the likelihood for storminess is enhanced. If it is negative, storm development is less likely.

The shearing motions that result from variations in wind speed across a horizontal surface are described as **transverse wind shear**, a force that is responsible for much of the ζ in air and can lead to positive or negative vorticity (**Figure 7.9**). To understand how transverse wind shear can cause ζ to exist in the Rossby waves, imagine a fast-meandering core of air, represented as a place where the pressure gradient is steeper than elsewhere in an upper-level weather map (**Figure 7.10a**). As this jet stream moves across a horizontal

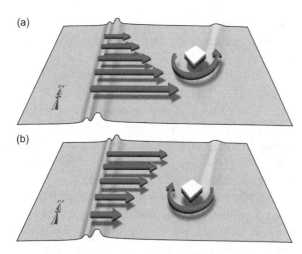

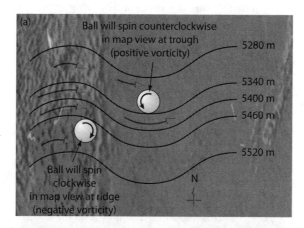

Figure 7.9 Transverse wind shear that produces (a) counterclockwise rotation, or positive ζ, and (b) clockwise rotation, or negative ζ.

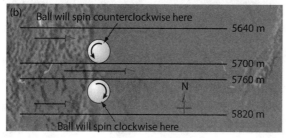

Figure 7.10 An example of the impact of transverse wind shear on relative vorticity (a) in a Rossby wave and (b) in a zonal flow.

surface in the midtroposphere in the area approaching a ridge, the air is pushed poleward. If we place a giant ball on the inside of the curve in the jet stream at the ridge, it would rotate against the spin of Earth's surface—in a clockwise direction in the northern hemisphere. This is because transverse shear is created from the north

side of the ball in the northern hemisphere (where air is moving quickly) to the south side of the ball (where air is moving slower). Thus, the ridge area is characterized by negative ζ (regardless of hemisphere). Now imagine a position downstream from the ridge. There the jet stream meanders such that the belt dips equatorward. If we place a giant ball on the inside of this trough at the jet, the ball would spin in a counterclockwise manner in the northern hemisphere. This causes positive ζ (regardless of hemisphere) as a result of transverse wind shear.

If we track a parcel of air moving from a ridge to an adjacent downstream trough, we find that ζ increases as the air parcel moves downstream, regardless of hemisphere. Air near the ridge axis experiences the strongest negative ζ (in either hemisphere) because the parcel spins maximally in a clockwise (or counterclockwise in the southern hemisphere) manner at that location. Air near the trough axis experiences the maximum amount of positive ζ because a parcel spins maximally in a counterclockwise (or clockwise in the southern hemisphere) manner at that location. Between the ridge and the trough, air parcels show a ζ increase from negative to positive values in either hemisphere.

As the parcel continues toward the next downstream ridge, ζ begins to decrease as the air moves from a positive ζ region (the trough axis) toward a negative ζ region (the ridge axis) in either hemisphere. Again, this ζ change is caused by transverse wind shear. At the midway point between the trough axis and the ridge axis, the parcel of air experiences zero ζ, after which the rotation reverses.

Because ζ is a function of airflow curvature, strongly meridional patterns show the greatest ζ changes between ridges and troughs. Zonal patterns have only small ζ changes between the ridge and trough axes if the pressure gradient is constant. However, changes in the latitudinal pressure gradient cause transverse wind shear even if the flow is zonal (Figure 7.10b). Rossby wave patterns provide both favorable and unfavorable zones for cyclogenesis, which requires counterclockwise rotation (in the northern hemisphere) or positive ζ (in either hemisphere). Deamplified or zonal flow provides neither strong support for nor resistance to cyclogenesis, except when differing wind speeds across space cause transverse shear.

Changes in ζ as air moves also impact the development and life cycles of weather systems. If the air acquires positive ζ, the formation of low pressure and cyclogenesis is encouraged. Surface anticyclones are supported by negative ζ in either hemisphere. This occurs at 500-mb ridges. Surface cyclones, such as those manifested as midlatitude storm systems, are supported by 500-mb troughs. The greater the amount of positive and negative ζ present aloft, the greater the spin of air through the column of air between the middle troposphere and the surface. This is associated with the strength of the individual surface systems and their resulting atmospheric features.

The characteristics of ζ are so closely related to surface system characteristics that most weather forecasters examine a map of 500-mb geopotential heights and ζ to determine initial forecasting scenarios. In the case of midlatitude storminess, low geopotential heights combined with positive vorticity advection (in either hemisphere) indicate the position of the core of a cyclone in the secondary circulation. Numerical models project changes in ζ and positions of the ζ maxima over time. If the ζ in the region shows signs of maintaining itself or strengthening in the computer models, then the surface cyclone is likely to either maintain its strength or increase in strength accordingly. If the midtropospheric ζ in a region shows signs of weakening, then so will the surface system, ultimately to the point of dissipation. The future trajectory of the surface system may be forecasted by using computer-projected 500-mb vorticity characteristics. Similar analyses are conducted at climatological time scales for forecasting potential impacts of long-term changes in the general circulation on precipitation and storminess across space. Analogous modeling endeavors have estimated climatic conditions in the distant past.

The climate of a region is determined by the frequency and distribution of weather systems (and, therefore, vorticity) passing over it. However, ζ is only one component of the total vorticity. Even in the total absence of transverse wind shear (hypothetically a perfectly zonal pattern with no changes in pressure gradients), f is present in the moving atmosphere, except at the equator where the Coriolis effect is zero. **Absolute vorticity** (denoted by N, the uppercase Greek letter nu) is the sum of the relative and planetary vorticity. Expressed quantitatively,

$$N = \zeta + f$$

The most important feature of N is that it is conserved—that is, it remains constant—in the atmosphere. This property has important ramifications for planetary-scale atmospheric flow.

Constant Absolute Vorticity Trajectory

As we have seen, cyclones have converging air motions at the surface but offsetting diverging air motions aloft (regardless of the hemisphere). Anticyclones have diverging air motions at the surface but have offsetting converging air motions aloft. The upper-air level where atmospheric motions oppose those near the surface maximally is typically located just beneath the tropopause. Somewhere between the two zones is the **level of nondivergence**, where neither convergence nor divergence occurs. Although the level of nondivergence may vary, the 500-mb geopotential height surface is often a good approximation of the level of nondivergence, because it is located where about half of the mass of the atmosphere is above it and half is below it.

Carl Rossby showed that at the level of nondivergence air motion follows a **constant absolute vorticity trajectory**. A parcel of moving air simply follows a path that conserves N over time. Imagine a parcel moving laterally in the midlatitude westerlies along the level of nondivergence. If it begins to move higher in latitude at all—say from southwest to northeast in the northern hemisphere (or northwest to southeast in the southern hemisphere)—Coriolis deflection is increasing (**Figure 7.11**). At the same time f is increasing, to conserve N the amount of ζ must decrease as it moves poleward. This causes it to acquire negative ζ (clockwise flow in the northern hemisphere), thereby creating

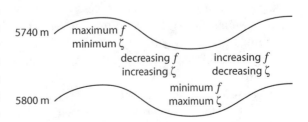

Figure 7.11 Relative and planetary vorticity features associated with midtropospheric Rossby waves in the northern hemisphere.

a ridge (in either hemisphere). When the parcel reaches the area of maximum negative vorticity (and maximum f)—the ridge axis—its clockwise (in the northern hemisphere) flow begins to take it toward the southeast (or northeast in the southern hemisphere). N is conserved.

As the parcel continues to move southeastward (northeastward in the southern hemisphere), its f decreases; so ζ must increase. This increasing ζ manifests itself as counterclockwise flow in the northern hemisphere (clockwise in the southern), and continued counterclockwise flow (clockwise in the southern hemisphere) eventually takes the parcel to a point at which it is as far equatorward as it will go. That point is the one at which f is at a minimum and ζ is at its maximum—the trough axis. Again, N is conserved. Continued motion toward the next downstream ridge once again sees increasingly positive f and increasingly negative ζ in either hemisphere.

Changes in ζ and Coriolis deflection cause adjustment in motion throughout the entire wave train of air around the hemisphere. For instance, as air moves into a ridge axis, Coriolis deflection increases as the parcel moves into higher latitudes. Even though there is an offset of ζ, the Coriolis deflection to the right in the northern hemisphere (left in the southern) causes the air parcel to turn equatorward on exiting the axis. The parcel moves toward lower latitudes along a northwest to southeast trajectory (or southwest to northeast trajectory in the southern hemisphere). At the trough axis f is at its weakest point, while the amount of ζ is maximized. This ζ overshadows the Coriolis deflection such that the parcel now turns to the left in the northern hemisphere (right in the southern hemisphere), causing the air to move along a southwest to northeast trajectory (northwest to southeast in the southern hemisphere). The result is that a ridge of a particular amplitude is associated with downstream adjustments that affect the development of a trough. The process continues with N being conserved as a function of ever-changing Coriolis deflection and ζ values. A series of long waves circumnavigating the entire hemisphere results.

Because N is conserved, any change in airflow at any location along the wave pattern is "felt" by all points upstream and downstream. For instance, if air were to speed up over a trough-to-ridge side of the Rossby wave in a meridional flow pattern, the Coriolis effect would increase because it is proportional to wind speed (and because f has increased). This increased pull to the right (in the northern hemisphere) would be instantly offset by a decrease in ζ to conserve N. Such a situation would amplify the ridge. The flow would then readjust (again) because increasingly meridional motion would cause more extreme decreases in f with equatorward flow, initiating rapid increases in ζ to conserve N. This would cause the downstream trough to deepen, and meridional flow would be further intensified. Adjustments in this area would trigger further readjustments in the next downstream ridge and so on, leading to an increasingly meridional pattern over time. Any adjustment of airflow in one location is propagated downstream until the entire longwave pattern fully readjusts to the new stimulus.

At some point a critical maximum is reached when sufficient energy has been redistributed meridionally. At that time the pattern readjusts to a more zonal pattern. Such a situation often creates closed migratory low- and high-pressure areas, which represent airflow regions very similar to cutoff meanders in streams.

When we view Rossby wave patterns over time, we see preferred geographical areas where flow adjustments occur. Pairs of locations where ridging and troughing or opposite pressure anomalies frequently occur simultaneously are termed **action centers**. Any adjustment in the pressure pattern in an action center triggers an offsetting adjustment at the other action center. Likewise, changes over time occur in preferred locations of ridge–trough pairs along the longwave pattern.

The correlated atmospheric flow patterns resulting from a "see-saw" pattern of pressure and ridging/troughing at action centers are termed **teleconnections**. Because teleconnections are associated with adjustments in midtropospheric patterns, they are responsible for driving climatic variations at the surface for many regions.

Flow Over Mountainous Terrain

Changes in flow pattern characteristics can occur for reasons other than conservation of N. For instance, in mountainous areas the atmosphere constricts vertically. This occurs as the space between the surface and the tropopause (h) decreases as surface elevation increases (**Figure 7.12**). When a moving fluid constricts to a smaller vertical

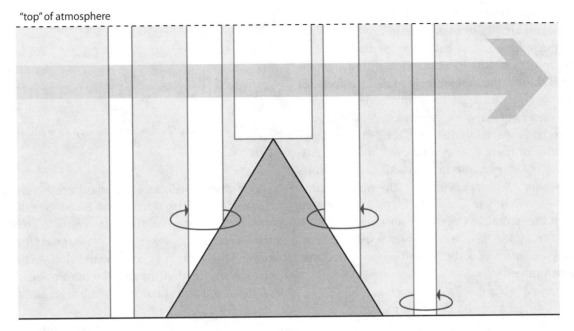

Figure 7.12   Prevailing airflow conserving potential vorticity encountering a mountain range.

space, a speed change occurs because, like N, **potential vorticity** is conserved. Potential vorticity can be expressed as

$$PV = \frac{\zeta + f}{\Delta h} \quad \text{or} \quad PV = \frac{N}{\Delta h}$$

Consider the case of air moving from west to east in the midlatitude westerlies across western North America, when it encounters the Rocky Mountains. As air moves upslope Δh decreases, so N must also decrease. But if the air is flowing from west to east in the midlatitude westerlies, f is unchanging because f is a function of latitude, which does not change in west-to-east flow. For potential vorticity to be conserved, ζ must decrease as the air moves upslope. As we have seen, decreasing ζ is associated with increasingly anticyclonic (clockwise in the northern hemisphere) flow. This encourages a ridge to form immediately west of a high mountain range that is oriented perpendicular to the direction of the flow.

By contrast, as air flows downslope on the *leeward* side of the mountains, Δh increases, so N must also increase (Figure 7.12). But again, f is constant for westerly flow, so for potential vorticity to be conserved, ζ must increase. Increasing ζ is associated with cyclonic (counterclockwise in the northern hemisphere) flow. This encourages a trough to form downstream from the peaks in the midlatitudes in

each hemisphere. Cyclogenesis is common leeward of major midlatitude mountain ranges.

Baroclinicity

Another situation that leads to changes in Rossby wave flow involves **baroclinicity**—the intermixing of steep thermal gradients across small regions. Such gradients often occur at coastlines, where there is a differential heating of land masses versus ocean surfaces at a given line of latitude. The most pronounced thermal gradients occur on the leeward side of midlatitude continents during winter as the continents chill substantially while the ocean temperatures remain relatively mild. This effect is reinforced by the warm ocean currents that dominate the western sides of ocean basins (eastern sides of continents) as low-latitude waters move poleward in the oceanic **gyres**. During winter the thermal gradient across the eastern coasts of continents is maximized.

Any region (at the surface or aloft) where air is being advected into a region with a drastically different temperature is termed a **baroclinic zone**. Baroclinic zones are regions of active and sometimes severe weather in the midlatitudes. Geographical areas where baroclinic zones are preferred have frequently stormy climates. Low-level baroclinicity generally causes wind speed to strengthen with elevation as the heights of constant upper-air pressure surfaces attempt to even out over a small region. These changes manifest themselves through the longwave pattern. In general, if a baroclinic zone occurs in association with an upper-level trough, the development of surface cyclones is strongly encouraged. Baroclinic zones are apparent on a map where isotherms are packed closely together, especially when they are nearly perpendicular to the Rossby wave flow. The latter indicates that efficient thermal advection is occurring. In **Figure 7.13**, notice how the 500-mb baroclinic zone focused over Utah and Idaho is associated with the longwave trough. As air moves through that trough, it acquires counterclockwise rotation (i.e., positive vorticity) and supports cyclonic development.

Rossby Wave Divergence and Convergence

Surface cyclones require diverging air aloft to sustain surface convergence. Such air is found on the trough-to-ridge side of the Rossby waves (**Figure 7.14**). To see why, imagine air moving around the trough aloft (in the northern hemisphere). In such a case it must turn to the left to remain parallel

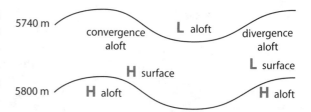

Figure 7.14   Areas of favorable surface cyclone and anticyclone formation relative to the Rossby wave pattern.

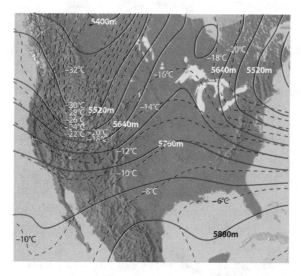

Figure 7.13   A 500-mb weather map for April 18, 2006. Notice how the isotherms (dashed lines) are packed closely together in the Mountain West of North America on this day, coinciding with the closely packed isohypses (solid lines), creating a strong baroclinic zone on that day. By contrast, few isotherms appear in the southern United States where lighter 500-mb winds occur. *Data from*: The Ohio State University WWW Weather Center, courtesy of the Department of Geography/Atmospheric Sciences Program.

to the **isohypses**—lines of constant geopotential height. Flow in the upper-level maps generally parallels the isohypses because *geostrophic balance*—a balance between the pressure gradient force and the Coriolis effect—occurs aloft. To turn against the direction induced by the Coriolis effect, the speed must be slow enough to reduce the Coriolis effect sufficiently.

On the other hand, as air moves through a ridge (in the northern hemisphere) it turns to the right to remain parallel to the isohypses. Because this is the direction that the Coriolis effect deflects the air, the speed increases coincident with the strengthening of the Coriolis effect. Note the Coriolis effect weakens at the trough and strengthens at the ridge (in either hemisphere) because latitude is relatively low at the trough and higher at the ridge.

Thus, air diverges on the trough-to-ridge side of the Rossby wave aloft, moving relatively slowly around the trough and then relatively fast around the downstream ridge (assuming that the isohypse gradient is the same for both). This divergence on the trough-to-ridge side of the upper-level Rossby wave induces rising motion to replace the air that has "spread out" aloft. Air cannot approach laterally because it must remain parallel to the isohypses to ensure the balance of forces suggested by geostrophic flow. The rising motion is supportive of surface low pressure—a cyclone. Therefore, surface cyclones are found on the trough-to-ridge side of the wave in either hemisphere (Figure 7.14).

By contrast, the ridge-to-trough side of the wave is associated with convergence aloft (in either hemisphere) as air slows down and "piles up." This abundance of air aloft induces a sinking motion on the ridge-to-trough side of the wave and surface high pressure. Thus, surface anticyclones are located beneath the ridge-to-trough side of the Rossby wave in either hemisphere.

Rossby Wave Diffluence and Confluence

Additionally, areas of diffluence and confluence in the upper-level Rossby waves can create rising and sinking motion. **Diffluence** refers to the horizontal spreading of air streams at a given height, and **confluence** involves areas where air is converging horizontally (**Figure 7.15**). Upper-level diffluence supports rising air motions by developing an upper-air vacuum as air parcels spread away from each other. Air rushes into the vacuum from below, which initiates and maintains surface low-pressure centers.

Confluence consists of air streams that converge horizontally toward each other. At upper levels the "pile up" of air associated with such motions causes air to slow down. Gravity prevents this "extra" air from moving out to space, so the only place the air can move is downward. Thus, sinking motions result. These motions initiate and support the development of surface anticyclones.

Polar Front Jet Stream

Embedded within the upper-level midlatitude westerlies are cores of extremely fast airflow. These places are indicated on upper-level maps by isohypses that become very closely spaced (but, of course, isohypses can never intersect each other). If we examine a map with the north (or south) pole in the center, as in **Figure 7.16**, an enclosed area of closely spaced isohypses would be apparent some distance from the pole over the midlatitudes. Airflow through these closely spaced isohypses is continuous around the hemisphere. The enclosed region represents the **circumpolar vortex**, an area

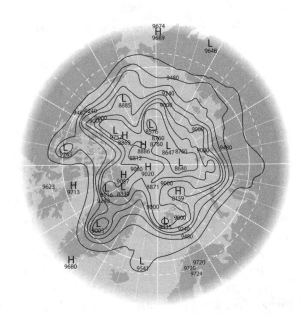

Figure 7.16 The northern hemisphere's circumpolar vortex at the 300-mb level on May 21, 2006. The locations where the isohypses are packed closely together represent the extent of the northern hemisphere's circumpolar vortex. *Data from*: National Weather Service/NOAA.

of strong winds aloft that encircles the surface polar high. The *polar front jet stream*, which marks the periphery of the circumpolar vortex, is a core of fast-moving air that meanders through the upper troposphere. There is one polar front jet in the northern hemisphere and another in the southern hemisphere. The strongest wind speeds occur just beneath the tropopause in the 300-mb to 200-mb height field area where the geopotential height gradient is strongest and friction is minimal.

The region of greatest wind speed within the polar front jet stream—the **jet streak**—occurs over the primary baroclinic zones, which exist at the boundary between cold air of polar origin and equatorially derived air. The speed of the polar front jet stream is directly proportional to the latitudinal thermal gradient, with steeper thermal gradients and baroclinic zones inducing steeper geopotential height gradients and a faster jet. Baroclinicity supports and maintains jet stream characteristics.

Air of tropical and polar origin meets in the midlatitudes to produce active baroclinic zones. The tropical air is usually also humid, because a large geographical area of the tropical Earth is over oceans and because warm air has high *saturation vapor pressure*. The polar air is often also

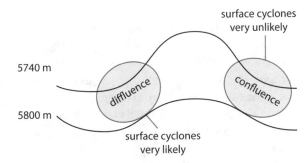

Figure 7.15 Areas of favorable surface cyclone and anticyclone formation relative to the Rossby wave pattern: diffluence and confluence.

dry, because relatively more of the polar Earth is either covered by land or sea ice and because the saturation vapor pressures decrease with decreasing temperatures. The boundary between the tropical and polar air pools produces a distinct thermal–moisture boundary through the vertical profile of the atmosphere. Air temperatures change significantly over very small areas near the boundary. The location of this thermal gradient at a given time determines the location and shape of the circumpolar vortex and the location of the polar front jet stream in the zone between the subpolar lows and the subtropical highs.

The height of geopotential pressure surfaces is directly related to the thermal properties of air, in fulfillment of the *equation of state*, also known as the *ideal gas law*, for dry air:

$$P = \rho R_d T$$

where P represents pressure (in *Pascals*), ρ represents density, R_d is the dry gas constant, and T is in the *Kelvin temperature scale*. If we were to examine the 500-mb geopotential height field (an unchanging pressure value), we would notice that the density of that pressure surface decreases over areas of higher temperature and increases over areas of lower temperature, in fulfillment of the equation of state. Density decreases when air rises, so warmer areas must have higher constant pressure surfaces than colder areas (Figure 7.4). This is a direct result of thermal expansion (warming conditions) and contraction (cooling conditions) of a fluid, as described by the equation of state.

A sharp thermal gradient occurs between warm and cold air at the surface with the boundary extending through a vertical profile of the atmosphere. In the vicinity of the 300-mb geopotential height layer and extending to about the 200-mb layer (just beneath the tropopause), geopotential heights slope sharply across the thermal boundary. Remember that air in the upper atmosphere flows down the geopotential thermal gradient from equatorward to poleward latitudes. The pressure gradient force is always pulling air toward the poles as the geopotential height fields are lower in height (thermally contracted) than over lower latitudes (thermally expanded). This creates a very steep geopotential height slope

above the thermal boundary between warm and cold air.

As air begins to flow across this gradient, it accelerates. This increase in slope provides energy to the jet stream. However, air does not simply flow over the sloping heights toward the pole. As we already know, Coriolis deflection pushes air to the right in the northern hemisphere, allowing for geostrophic balance to be reached with the pressure gradient force, which is pulling air toward the pole. This results in a rapidly moving stream of air flowing generally from west to east above the warm–cold air boundary. The same situation (heights sloping poleward) occurs in the southern hemisphere, which also results in a west-to-east-flowing jet stream as air is pulled to the left by Coriolis deflection.

The polar front jet stream's location varies as the flow of air travels along the area of constant N. The jet stream as identified on upper-air weather maps can be used to identify ridges and troughs, and corresponding Rossby waves, in the upper troposphere. The jet is also important in governing surface weather, because it represents the boundary between cold and warm air and it tends to govern the development and movement of mid-latitude storm systems. Analysis of the polar front jet stream and prediction of its future motions are essential for short-term and long-term atmospheric prediction and analysis.

The thermal boundary associated with the polar front jet stream is generally much more prominent in winter because the latitudinal thermal gradient is much greater during that season than during the summer. Wind speeds typically reach 160 km hr^{-1} (100 mi hr^{-1}) in winter but drop to about 80 km hr^{-1} (50 mi hr^{-1}) in summer.

Mean Patterns of Rossby Wave Flow

Climatological analysis of the polar front jet stream is often done by examining average patterns, even though variability and extremes are important components of climatology. The jet stream's position varies substantially from day to day, but analysis of the position and intensity of the jet over long time scales shows distinct monthly and seasonal patterns.

As explained above, the sharp surface thermal gradients associated with coastal locations produce baroclinic regions that generally correspond to the location of the polar front jet. The constraint of upper airflow by conservation of absolute vorticity results in rather distinct atmospheric teleconnections. The polar front jet is located maximally equatorward during the coldest part of the year. This results from the expansion of the polar anticyclone and subtropical lows and contraction of the subtropical anticyclones during winter. In summer the semipermanent pressure cells reverse from their winter tendencies. This causes the polar jet to retreat maximally poleward.

The presence of topographical barriers and baroclinic zones causes persistence in the long-term jet pattern. For instance, air motion over North America is typically characterized by a ridge over the western mountain cordillera and a trough over the east. This pattern is associated with the conservation of potential vorticity as the height of the air parcel decreases as it moves upslope and then decreases as it moves downslope. The amplitude of the jet changes in association with vorticity and thermal characteristics as described above. A reverse flow—trough in the west and ridge in the east—does occur, but it is rather infrequent. Recent studies show that once the reverse pattern establishes itself, it can be quite persistent.

Climatological averages of middle-tropospheric patterns for the northern hemisphere reveal many of the principles discussed above. **Figure 7.17** shows that 500-mb geopotential heights are generally higher in summer than in winter over any given location and that the 500-mb geopotential height gradients are steeper in winter than in summer. The equatorial areas have very weak 500-mb geopotential height gradients year-round. In contrast, few major seasonal changes occur in the southern hemisphere. The dominant presence of water, resulting in a fairly stable latitudinal thermal gradient in the southern hemisphere, keeps the patterns fairly constant throughout the year.

■ Summary

General circulation exists to redistribute energy that arrives at Earth in greater quantities near the equator than near the poles. Because of the rotation of Earth, this circulation is characterized by a Hadley cell and a polar cell in both the northern and the southern hemispheres. Air in the Hadley cell moves upward near the equator at the Intertropical Convergence Zone, with a poleward motion to about 30°N and 30°S latitude; sinks at the subtropical anticyclones; and then moves equatorward again in the form of the northeast trade winds (northern hemisphere) and the southeast trade winds (southern hemisphere). The subtropical jet stream exists aloft at the poleward edge of the Hadley cell in each hemisphere as angular momentum is conserved, and it exerts significant influences on the subtropical atmosphere, many of which are poorly understood.

The polar cell is characterized by sinking air at the pole, equatorward surface motion to about 60°N or 60°S (deflected by the Coriolis effect to become the polar easterlies), rising air at the subpolar lows, and then poleward motion (deflected to become westerlies) aloft. Between the Hadley and polar cells in each hemisphere are surface westerly winds, with stronger westerlies aloft. Within the midlatitudes atmospheric flow aloft takes the form of Rossby waves, which pull air toward the pole (ridge) and toward the equator (trough) as it moves generally from west to east embedded in the westerlies.

Vorticity is an important feature of atmospheric motion. Air flowing in upper-level Rossby waves conserves its absolute vorticity, thereby ensuring that ridges and troughs remain confined to the middle latitudes. Because potential vorticity is conserved, air moving upslope in the midlatitude westerlies acquires anticyclonic (negative) vorticity, while that moving down the leeward slope acquires positive vorticity. The trough-to-ridge side of the wave and any areas of diffluence along the wave support surface cyclones and cyclogenesis. The ridge-to-trough side of the wave and any areas of confluence along the wave support surface anticyclones and anticyclogenesis.

The circumpolar vortex and in particular the polar front jet stream exist over the zone of the sharpest temperature contrasts and the strongest baroclinic zone. This feature is embedded within the Rossby waves and can exert a significant influence on the weather and climates beneath it.

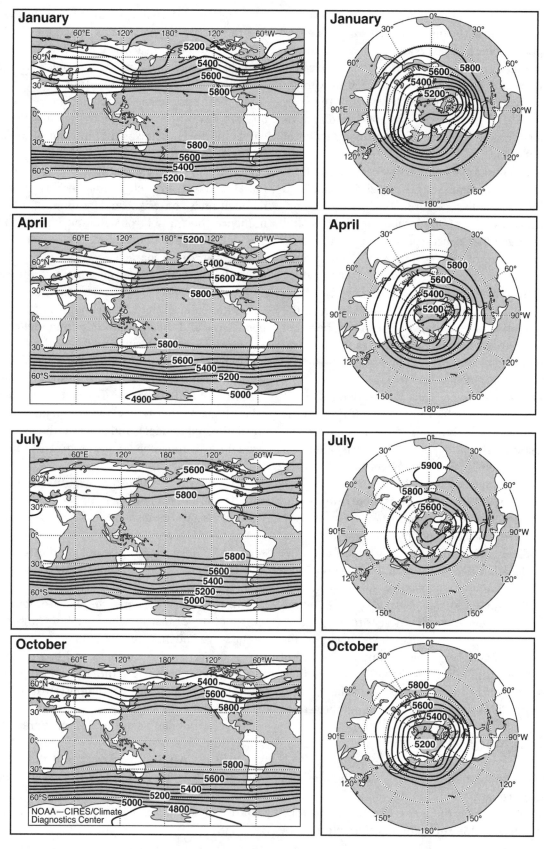

Figure 7.17 Mean 500-mb geopotential heights (1948–2004) in the northern hemisphere. *Data from*: NOAA/CIRES Climate Diagnostics Center and the University of Colorado at Boulder.

▶ Key Terms

Absolute vorticity
Action center
Aerosols
Albedo
Aleutian low
Amplitude
Anticyclone
Baroclinic zone
Baroclinicity
Bermuda-Azores high
Circumpolar vortex
Confluence
Conservation of angular
　momentum
Constant absolute vorticity
　trajectory
Continentality
Convection
Coriolis effect
Coriolis parameter
Cyclogenesis
Cyclone
Density
Diffluence
Doldrums
Equation of state
Evaporation
General circulation
Geopotential height
Geostrophic balance
Gyre
Hadley cell
Hawaiian high

Horse latitudes
Icelandic low
Ideal gas law
Indian Ocean high
Insolation
Intertropical Convergence Zone
　(ITCZ)
Isohypse
Jet streak
Jet stream
Kelvin temperature scale
Kinetic energy
Latitude
Leeward
Level of nondivergence
Long wave
Longitude
Meridional flow
Midlatitude westerlies
Momentum
Negative vorticity
Newton's second law
North Pacific high
Northeast trade winds
Pascal
Planetary vorticity
Polar easterlies
Polar front jet stream
Polar cell
Polar high
Positive vorticity
Potential vorticity
Pressure gradient force

Relative vorticity
Ridge
Rossby wave
Sargasso Sea
Saturation vapor pressure
Second law of
　thermodynamics
Secondary circulation
Sensible heat
Siberian high
Solar declination
Solar noon
South Atlantic high
South Pacific high
Southeast trade winds
Specific heat
Stability
Subpolar low
Subtropical anticyclone
Subtropical jet stream
Teleconnection
Tibetan Low
Transverse wind shear
Tropic of Cancer
Tropic of Capricorn
Tropopause
Troposphere
Trough
Vorticity
Wavelength
Zonal flow
Terms in italics have appeared in at
　least one previous chapter.

▶ Review Questions

1. Discuss the primary atmospheric circulation features present on a nonrotating planet with a continuous, uniform surface.

2. Describe the key pressure and wind features in the idealized circulation model.

3. Compare and contrast the idealized circulation model to reality. Why are they different?

4. Does the idealized global circulation model adequately describe upper-atmospheric motions? If not, why not?

5. Describe the waxing and waning of the semi-permanent pressure cells through the course of the year.

6. Discuss the annual migration of the semipermanent pressure cells.

7. Identify and compare relative and absolute vorticity.

8. Discuss the role of positive and negative vorticity in association with ridges and troughs.

9. Describe the role of vorticity in the development of zonal and meridional Rossby wave patterns.

10. Why is absolute vorticity important? Of what is it a function?

11. Discuss the role of constant absolute vorticity trajectory relative to atmospheric motions aloft.

12. What are teleconnections and teleconnection action centers and why are they important to upper-atmospheric flow patterns?

13. Discuss the role of baroclinicity in the development of upper-atmospheric flow patterns.

14. What is the polar front jet stream, how does it form, and why is it an important feature of midlatitude upper airflow?

▶ **Questions for Thought**

1. The text says that the Rossby waves in the southern hemisphere tend to be less meridional than those in the northern hemisphere. Why do you believe that this might be the case?

2. If the world were to warm up, what do you believe might happen to the various semipermanent pressure features around the world?

3. Assume an earth surface comprised solely of water. Describe midlatitude jet stream characteristics for this earth. Would troughs and ridges develop? Why or why not?

---

http://physicalscience.jbpub.com/climatology

Connect to this book's Web site: http://physicalscience.jbpub.com/climatology. The site provides chapter outlines, further readings, and other tools to help you study for your class. You can also follow useful links for additional information on the topics covered in this chapter.

Climates Across Space

Chapter 8—Climatic Classification
Early Attempts at Global Climatic Classification
Classical Age of Climatic Classifications
Genetic Classifications
Local and Regional Classifications
Quantitative Analysis to Derive Climatic Types
Summary

Chapter 9—Extratropical Northern Hemisphere Climates
Climatic Setting of North America
Climatic Setting of Europe
Climatic Setting of Asia
Regional Climatology
Summary

Chapter 10—Tropical and Southern Hemisphere Climates
Contrasts Between Extratropical and Tropical Atmospheric Behavior
Contrasts Between Northern and Southern Hemisphere Atmospheric Behavior
Climatic Setting of Africa
Climatic Setting of Australia and Oceania
Climatic Setting of Latin America
Climatic Setting of Antarctica
Regional Climates
Summary

Chapter at a Glance

Early Attempts at Global Climatic Classification
Classical Age of Climatic Classifications
    Modified Köppen Climatic Classification System
    Thornthwaite Climatic Classification System
    Other Global Classification Systems
Genetic Classifications
    Air Masses and Fronts
Local and Regional Classifications
Quantitative Analysis to Derive Climatic Types
    Eigenvector Analysis
    Cluster Analysis
    Hybrid Techniques
Summary
Review

If you were given a list of 500 cities around the world and were asked to divide that list into categories based on the climates those cities experience, how would you do it? You would probably not put Seoul, South Korea, in the same category as Bogotá, Colombia, simply because the temperature regime is too different between those cities to be listed in the same category. Similarly, Cairo, Egypt, and Kuala Lumpur, Malaysia, probably do not belong in the same category because the former city is very dry and the latter is very wet. But what categories are the most logical?

Because climate represents the integrated total of weather, and weather represents a great variety of individual atmospheric phenomena such as temperature, precipitation amount, type, seasonality, humidity, wind direction and speed, and cloud type, many different atmospheric variables can be used to categorize climate. All these variables are important contributors to the *overall* cli-

mate of a location. For some applications, however, some variables are more important than others. If you were conducting your climatic classification as an employee of a giant energy conglomerate that was investigating the feasibility of installing hundreds of wind turbines around the world to generate wind power, the most logical atmospheric variables to include might be wind speed and direction and seasonal variability of wind. Perhaps some stations have consistently strong winds from a similar direction year-round, another set of stations may have light winds from variable directions in summer but strong winds from the same direction in winter, and so forth. In short, the criteria used in the classification of climates depend on the use of the classification.

A characteristic of all sciences, classification is used to simplify our conception of a complicated process. Biologists have a very elaborate, hierarchical classification of life, with kingdom at the broadest level and species (and variety) at the most specific. Similarly, geologists classify rocks and minerals, and chemists classify elements. Even in everyday life we use classification as a shortcut to describing things. When you hear the word "chair," you understand that the person speaking with you may be conveying information about anything from a recliner to a beanbag chair, to a throne, to a device for execution.

Effective classifications can be helpful, but ineffective classifications can be more confusing than helpful. The number of classes is also important. Classifications that have too many categories can be very cumbersome and difficult to implement, but those that have too few categories group items together that have little in common. Any effective classification procedure, including those in climatology, minimizes the *within-group vari-*

ability of the variables being considered and max-imizes the *between-group variability*.

Tens of thousands of weather stations around the world record at least some meteorological information on a regular basis. Much of these data are reported to national and/or worldwide climate centers on monthly to annual time scales. Climatic classification systems were devised to summarize the overwhelming amount of atmospheric data that can accumulate even over a short time period and to simplify the climatic characteristics of any location.

Regionalization of climate is useful for several reasons. First, it is an effective tool to gauge the expected pattern of weather for particular locations. Expected weather over the course of the year may be used as an initial forecasting tool. Second, regionalization of weather can identify useful *extremes* and climatic *normals* to which daily meteorological variables may be compared even if no data are available at the exact point of interest. Third, long-term shifts in the boundaries of climatic regions as indicated by climatic classifications may provide an indicator of the degree of climatic variability and change shown over time, which could be manifested as shifts in bioclimatic and agricultural zones, domains of diseases, and weather-related human mortality.

Planetary-scale climatic classification schemes allow for the identification of generalized climatic regions. Although local and regional-scale classifications can be extremely useful in resolving particular climate-related problems, more generalized climatic information at broader scales can also be useful in aiding understanding of climatic variation across space. Generalized planetary-scale climatic classification schemes are usually devised using the most widely needed and readily available atmospheric variables: temperature and precipitation and seasonality of those variables. As noted previously, however, they may require other data, depending on the intended application.

■ Early Attempts at Global Climatic Classification

Early attempts by the ancient Greeks at classifying climate resulted in the identification of three principal climate regions: the Frigid Zone, the Temperate Zone, and the Torrid Zone. The **Frigid Zone**

obviously represented the polar *latitudes* of the planet, an area once thought to be uninhabited due to excessive cold. The **Torrid Zone** was also deemed uninhabitable because of the extreme heat thought to occur within the tropics. The **Temperate Zone**, representing the supposedly only habitable portion of the planet, lay between the other two latitudinal extremes. Not surprisingly, Greece fell in the Temperate Zone. This identification scheme was devised around 500 BC by Paramenides, sometimes known as the Father of Logic. It represents an early attempt to classify climate based solely on logic, in the total absence of climatic data.

Other climatic classification schemes followed, including one by Hipparchus who updated the Paramenides classification by including information on the calculated day length for particular locations. Calculations about day lengths were based on solar angle on the spherical Earth's surface (yes, early Greek scholars knew that Earth was a sphere and were able to calculate its circumference). According to this classification the zones were renamed *klimata*—a word referring to slope, which in this case represents the inclination of the noon Sun at a point on Earth.

Logic-based climatic classification systems ruled until the development and proliferation of weather recording instrumentation. Of particular importance was the development of the thermoscope by Galileo in 1593, which led to the development of the first thermometer created by Santorre in 1612 and the barometer by Evangelista Torricelli, a student of Galileo, in 1643. Once these and updated forms of the instruments were dispersed worldwide, the formal analysis of weather and climate became a realistic possibility.

■ Classical Age of Climatic Classifications

Because of relatively short data records (most weather records persist for less than 100 years) and the massive amount of atmospheric data collected from individual weather stations, the identification of true climatic regions is a relatively recent endeavor. The classical age of climatic analysis began with the mathematical and vegetation-based **Köppen Climatic Classification System** devised by the botanist Vladimir Köppen in 1870. It was Köppen who coined the term "climatic classification." The Köppen system uses monthly temperature and precipitation data in making the calculations.

The **Thornthwaite Climatic Classification System**, formulated by C. Warren Thornthwaite in 1931 and completed in 1948, represents an alternative to the Köppen system. The Thornthwaite system is built on the physical interactions between local precipitation and temperature rather than only the moisture and temperature data. It represents a more sophisticated and precise scheme of classification based on local surface water availability.

The Köppen system is the most widely used system, largely because of its simplicity and its strong correspondence between climatic regions,

natural vegetation communities called **biomes**, and soil types. The Thornthwaite system is preferred by physical scientists in need of precise climate information, particularly when moisture availability is of critical importance, as it is in agricultural, hydrological, and environmental planning applications.

Modified Köppen Climatic Classification System

Not coincidentally, the Köppen system is largely based on dominant vegetation types, because the botanist recognized that most vegetation types respond directly to climatic inputs, especially temperature and moisture variations. Believing that

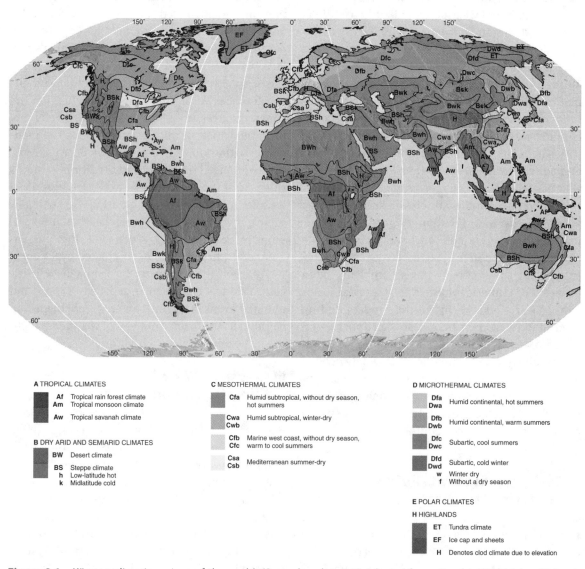

Figure 8.1 Köppen climatic regions of the world. (See color plate 8.1) *Adapted from: Goode's World Atlas,* 19th ed. Edited by E.B. Espenshade. Skokie, IL: Rand McNally and Co., 1995.

natural vegetation could be used as a rough indicator of generalized climatic regions, he determined that the same mean value of the atmospheric variable (e.g., temperature or precipitation) that separates two biomes should also separate two climatic types.

Over the years the Köppen climatic classification has undergone some revisions, most notably by the climatologist Glenn Trewartha, to provide more accurate correspondence between climatic zones and biomes. The modified Köppen climatic regions are shown in **Figure 8.1**. As we shall see, the system does have some inherent drawbacks, but the fundamental attraction lies in the simplicity and accuracy of the system.

Köppen observed and mapped the **ecotone**—the transition zone between two biomes—and then used temperature and precipitation data to derive equations to define the climatic boundary between those two biomes. Using such criteria Köppen identified five main climatic groups: A (tropical), B (arid), C (**mesothermal** or midlatitude mild), D (**microthermal** or midlatitude cold), and E (polar). With some exceptions the A, C, and D climates support the growth of trees, whereas the B and E climates generally do not, being either too dry or too cold, respectively. Later, an additional group, H (highland), was added to account for the extreme climatic variations across short distances in mountainous locations. The main types were further subdivided into additional second- and third-order subdivisions to classify regional climates more precisely. **Tables 8.1**, **8.2**, and **8.3** detail the Köppen climate system.

Specific Köppen climatic types are derived by combining appropriate first-, second-, and third-order subdivisions. In the case of A, C, and D climates the second-order subdivision refers to the precipitation seasonality (with "f" representing climates that are wet year-round, "m" representing tropical monsoon conditions, "s" indicating dry summer climates, and "w" representing dry winter climates). For B climates the second-order subdivision is "W" if the dry climates are true deserts and "S" if the dry climates are only semiarid.

Table 8.2   Major First-Order Köppen Climatic Classification Subdivisions

Subtype	Classification	Criteria Specifics
BW	Desert	True desert
BS	Steppe	Semiarid
ET	Tundra	Warmest month < 10°C (50°F) but > 0°C (32°F)
EF	Ice Cap	Perpetual frost All months < 0°C (32°F)

Table 8.3   Partial Listing of Köppen Second- and Third-Order Subdivisions

Second Order	Third Order	Criteria Specifics
	a	Warmest month > 22°C (72°F)
	b	Warmest month < 22°C (72°F)
	c	Fewer than 4 months > 10°C (50°F)
	d	Same as c but coldest month < −38°C (−36°F)
f		Constantly moist Rainfall through all months of year
m		Monsoon rain, short dry season Total rainfall sufficient to support rain forest
s		Summer dry season
w		Winter dry season
W or S	h	Hot and dry year-round Average annual temperature >18°C (64°F)
W or S	k	Cold and dry year-round Average annual temperature <18°C (64°F)
	n	Frequent fog
	n'	Infrequent fog but high humidity and low rainfall

Table 8.1   First-Order Köppen Climatic Classification Division

Subtype	Classification	Criteria Specifics
A	Tropical	Coolest month > 18°C (64°F)
B	Dry	See specifics
C	Mesothermal	Coldest month > 0°C (32°F) but < 18°C (64°F) Warmest month > 10°C (50°F)
D	Microthermal	Coldest month < 0°C (32°F) Warmest month > 10°C (50°F)
E	Polar	Warmest month < 10°C (50°F)
H	Highland	Undifferentiated highland climates

Notice the capitalization on this and only some of the other second-order letters. Because German nouns are always capitalized, the capital letters represent their equivalent nouns in German. In the case of "E" climates, the second-order subdivision is "T" for **Tundra climate**, a milder polar subtype, while "F" (frozen) is used to represent the Ice Cap subtype.

For the mesothermal and microthermal climates, third-order subdivisions identify the characteristics of summer temperatures, with "a" used for hot summers, "b" for warm summers, "c" for mild summers, and the rare "d" for cool summers. Arid climates have a third-order subdivision of "h" and "k," which denote "hot" and "cold" arid or semiarid regions, respectively.

For example, a Cfa climatic type indicates a mild midlatitude climate (C) featuring an even distribution of precipitation through all months of the year (f) and hot summers (a). The system yields 13 major climatic types with additional subtypes. **Table 8.4** describes the major climate types in the Köppen system. Although many other second- and third-order subdivisions have been identified, those listed in the tables are the most commonly used.

Because natural vegetation boundaries are ambiguous in most locations, they should be viewed as broad transition zones rather than absolute boundaries between very dissimilar climates. This is especially true of tropical climates, which are delineated by natural vegetation boundaries that are governed by annual precipitation variations.

Midlatitude climates tend to be separated by temperature thresholds. Specifically, mesothermal climates have a mean temperature of the coldest month above 0°C (32°F) but below 18°C (64°F), with the mean temperature of the warmest month above 10°C (50°F). Such conditions sustain midlatitude broadleaf and mixed forests, temperate rain forests—including the redwoods of California—and in one subcategory tall grasslands. Microthermal climates are those in which the mean temperature of the coldest month is below 0°C (32°F), while the mean temperature of the warmest month lies above 10°C (50°F). These climates support hardier broadleaf, mixed, and needleleaf forests. In essence, these two may be considered to represent warm extratropical and cool/cold extratropical (but nonpolar) climates, respectively.

The warmest month in polar climates has a mean temperature below 10°C (50°F). The ecotone between subtypes of E climates is delineated by whether any months in the year have an average temperature above freezing. If so, the ET (Tundra) designation is used. Tundra includes various types of shrubs and successively shorter vegetation in the poleward direction, including Arctic grasses, hardy species of mosses, and lichens (a symbiotic combination of fungus and algae that clings to rock), but no trees. The **treeline** separates the D from the E climates in the Köppen system. EF climates are characterized by very limited, highly specialized life forms—if any at all.

Virtually all the climates discussed thus far are derived from only a few input variables. However, dry climates (B) are a bit more involved. A "dry" climate is one in which mean *evapotranspiration* tends to exceed mean precipitation during most months. These two interacting variables must be accounted for and quantified using a simplified water balance approach. A region in which most precipitation occurs during the cool (winter) season shows a markedly different water balance from one in which the majority of precipitation falls during the warm (summer) season. These climatic differences present themselves in the type and distribution of vegetation given that summer evapotranspiration rates are much higher than those during winter. Likewise, a similar annual precipitation total at the two locations may produce

Table 8.4   Major Climatic Types of the Köppen System

Class Type	Description
Af	Tropical rainy
Am	Tropical monsoon
Aw	Tropical wet/dry
BW	Desert (truly arid)
BS	Steppe (semiarid)
Cf	Midlatitude rainy, mild winter
Cw	Midlatitude wet/dry, mild winter
Cs	Mediterranean
Df	Midlatitude rainy, cold winter
Dw	Midlatitude wet/dry, cold winter
ET	Tundra
EF	Ice Cap
H	Highland (variable, mountainous)

completely different ecosystems built around the timing of the precipitation and the resultant water balance.

Dry (B) climates occur if any one of the following three sets of criteria are completely true, where P represents annual precipitation in centimeters and T is mean annual Celsius temperature:

- More than 70 percent of the annual precipitation occurs in the summer half of the year *and* $P < 2(T + 14)$.
- More than 70 percent of the annual precipitation occurs in the winter half of the year *and* $P < 2T$.
- Less than 70 percent of the annual precipitation occurs in the summer half of the year, less than 70 percent occurs in the winter half of the year, *and* $P < 2(T + 7)$.

A given B climate is classified as a true desert climate (BW) if any of the following are completely true:

- More than 70 percent of the annual precipitation occurs in the summer half of the year *and* $P < (T + 14)$.
- More than 70 percent of the annual precipitation occurs in the winter half of the year *and* $P < T$.
- Less than 70 percent of the annual precipitation occurs in the summer half of the year, less than 70 percent occurs in the winter half of the year, *and* $P < (T + 7)$.

Otherwise, the B climate is classified as BS—a semiarid or **steppe climate.** Notice that to distinguish BW from BS climates in each case above, the term on the right is halved from the criteria to distinguish B climates from the humid climates. The BW and BS climates can be further subdivided as shown in **Table 8.5**.

Although the Köppen classification may seem complex, in reality it is quite simple. The system is fairly accurate at presenting generalized climatic regions, but it is not ideal if specific local and short-term climatic influences are important. Regional climatic boundaries may shift over short periods of time given the annual variability of both temperature and precipitation at particular locations. The system is inaccurate in places where climate is not the primary reason for the distribution of particular natural vegetation types; soil type, topographical relief, and other factors

Table 8.5   True Desert (BW) and Steppe (BS) Köppen Climatic Subdivisions

Subdivision	Description
h	Mean annual temperature > 18°C (64°F)
k	Mean annual temperature < 18°C (64°F)
k′	Mean temperature of warmest month < 18°C (64°F)
S	Receives at least 70 percent of precipitation during warm season (summer)
w	Receives at least 70 percent of precipitation during cool season (winter)

can play more significant roles than subtle climate differences in some places. There is, therefore, is not an exact correspondence between climatic type and vegetation distribution. Finally, the system does not explain why the climatic types exist and is instead solely based on the large-scale *results* of climate—the data. This makes it difficult to quantify specific attributes that *cause* particular climatic types. Despite these drawbacks the system is widely regarded as the most useful broad-scale climatic classification system. The Köppen system is, therefore, used in this book to discuss regional climatic types. But first we examine some other commonly used systems.

Thornthwaite Climatic Classification System

In the 1930s C. Warren Thornthwaite devised a very precise climatic classification system. Thornthwaite thought that climate could be ultimately classified by the interactions between the amount of energy present at a location over the course of the year and the amount of available moisture. These interactions between energy and moisture are the keys to identifying individual climatic zones in this system.

To complete the system, Thornthwaite devised a number of specific indices to quantify necessary climatic components. The **moisture index (MI)** is based on the precipitation amount and the *potential evapotranspiration* (PE) rate for a location. As we have seen in Chapter 6, PE refers to the number of centimeters of equivalent precipitation that could be lost from the surface to the atmosphere from both surface *evaporation* and plant *transpiration* from a short grass field given a nonlimiting supply of soil moisture. The Thornthwaite method of calculating PE differs from others, but regardless of the system of estimating PE this variable can be used to calculate a simple water balance for a location. The MI , this water balance, specifically is calculated from

$$MI = \frac{100(S - D)}{PE}$$

where S is the *surplus* water amount and D represents the *deficit* water amount, both in centimeters of precipitation equivalent.

For any period when PE meets or exceeds the precipitation amount, the surplus is zero. In such a scenario the equation above shows that the MI is zero or negative, indicating aridity. The opposite scenario—a precipitation amount that exceeds PE—is associated with $S > 0$ and $D = 0$, and it suggests humid conditions and a positive MI. Using the MI, Thornthwaite calculated nine climatic moisture regimes. This classification is detailed in **Table 8.6**.

Thornthwaite also derived a **thermal efficiency index**—the ratio of temperature to a calculated evapotranspiration value. Again, nine thermal climatic types are identified (Table 8.6). In addition to the MI and temperature-to-evapotranspiration ratio, Thornthwaite calculated a **dryness index (DI)** and a **humidity index (HI)** to identify the times of the year with water deficit or surplus. Calculations for each are as follows:

$$DI = 100 \frac{D}{PE}$$

$$HI = 100 \frac{S}{PE}$$

The DI and HI delineate additional climatic subcategories, as shown in **Table 8.7**. A final subdivision of the system comes from the examination of **summer thermal efficiency concentration**—the ratio of PE in the summer months to total annual PE in the surface water balance at a location. This is shown in **Table 8.8**.

Similar to the Köppen classification system, individual climatic types are denoted by the combination of letters produced by the four indices.

The major Thornthwaite climate regions are shown in **Figure 8.2**. This system provides more detail than the Köppen system. It also expresses climatic regions by inherently incorporating the synergistic impacts of energy and moisture availability for examined locations. The classification system also implicitly links the energy balance to the water balance through the concept of PE. On the other hand, the Thornthwaite classification is complex, perhaps to the point of becoming cumbersome. **Figure 8.3** shows the results of an over-

Table 8.6   Major Moisture and Temperature Divisions of the Thornthwaite Classification System

	Moisture Divisions			Temperature Divisions		
	Climatic Type	Moisture Index		T/ET (cm)	Climate	Type
A	Very humid	≥100		14.2	E'	Ice
B_4	Humid	80–99		28.5	D'	Tundra
B_3	Humid Microthermic	60–79		42.7	C'_1	Microthermic
B_2	Humid Microthermic	40–59		57	C'_2	Microthermic
B_1	Humid Mesothermic	20–39		71.2	B'_1	Mesothermic
C_2	Moist subhumid Mesothermic	0–19		85.5	B'_2	Mesothermic
C_1	Subarid Mesothermic	−33–0		99.7	B'_3	Mesothermic
D	Semiarid Mesothermic	−66 to −32		114	B'_4	Mesothermic
E	Arid Megathermic	−110 to −65		>114	A'	Megathermic

Table 8.7 Thornthwaite Climatic Classification Subcategories

	Humid Climates (A, B, C$_2$)	DI
R	No or very little water deficit	0–16.7
S	Moderate summer deficit	16.7–33.3
W	Moderate winter deficit	16.7–33.3
s$_2$	Large summer deficit	>33.3
w$_2$	Large winter deficit	>33.3
	Dry Climates (C, D, E)	**HI**
D	No or very little water surplus	0–10
S	Moderate winter surplus	10–20
W	Moderate summer surplus	10–20
s$_2$	Large winter surplus	>20
w$_2$	Large summer surplus	>20

Table 8.8 Summer Thermal Efficiency Subdivisions of the Thornthwaite Classification System

Subtype	Summer Thermal Efficiency Concentration (percent)
a'	<48.0
b'$_4$	48.0–51.9
b'$_3$	51.9–56.3
b'$_2$	56.3–61.6
b'$_1$	61.6–68.0
c'$_2$	68.0–76.3
c'$_1$	76.3–88.0
d'	>88.0

lay of the generalized Köppen and Thornthwaite classifications for the contiguous United States.

Other Global Classification Systems

In 1947 American botanist and climatologist L. R. Holdridge devised a similar classification system that came to be known as the **Holdridge Life Zones Climatic Classification System**. Although it was intended to be for global application, it became most widely used in tropical areas, where it has proved useful in ecological and alpine applications. One version of the Holdridge system is shown in **Figure 8.4**. The Holdridge system defines **biotemperature** as an adjusted temperature that considers temperatures below freezing or above 30°C to be noncontributors to the proliferation of life (i.e., to have a biotemperature of zero), because plants are dormant whether the temperature is freezing, far below freezing, or above 30°C.

The Holdridge model also includes annual precipitation and the ratio of PE to precipitation, but Holdridge's method of calculating PE differs from Thornthwaite's. Ratios above 1.0 represent arid climates and ratios below 1.0 suggest humid climates. Holdridge's "triangular" approach to climatic classification results in 36 different possible climatic types (**Figure 8.5**).

Other researchers, such as the Russian climatologist (born in what is now Belarus) M. I. Budyko, also used the concept of evaporation in classifying climate. The **Budyko Climatic Classification System** uses an energy budget approach to classifying climates. In Budyko's system the ratio of *net radiation* (Q*) to the energy required to evaporate a unit volume of local precipitation is calculated. Excess energy results in increasingly dry climates. Thus, the wet–dry climate boundary occurs where the **Budyko ratio** is 1.0.

■ Genetic Classifications

The classification systems discussed above are all driven by data, operating on the premise that locations sharing similar climatic data are grouped into the same climatic type. In the Köppen system the data involve monthly temperature and precipitation and the seasonal variations of each. The Thornthwaite and Holdridge systems rely on the computation of PE based on monthly temperature and precipitation data—a water balance approach. Budyko includes a radiation variable in his system.

A system that requires the input of data to calculate the climatic type based on predetermined class boundaries, such as those discussed above, is termed an **empirical climatic classification system**. Such systems have the advantage of being easy to implement in locations where high-quality and abundant climatic data are available. They also ensure that two locations with similar climatic attributes for the variables in question are categorized in the same group.

A **genetic classification system** is used to categorize climates solely on the basis of the major forcing mechanisms that make climate the way that it is. For example, the proximity of the *subtropical anticyclone* in summer but not winter is a forcing mechanism that is shared by San Diego, California, and Capetown, South Africa. A genetic classification system would, therefore, probably consider these two cities to be in the same category.

Perhumid
Humid
Subhumid
Semiarid
Arid

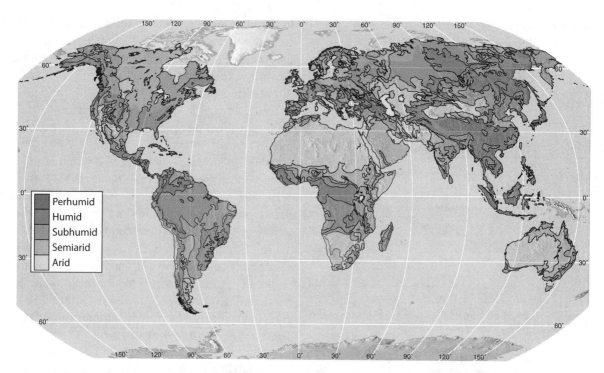

Figure 8.2 Generalized Thornthwaite climatic regions. *Adapted from*: Carter, D.B., and J.R. Mather. 1966. Climatic classification for environmental biology. *Publications in Climatology* 19(4): 304–395.

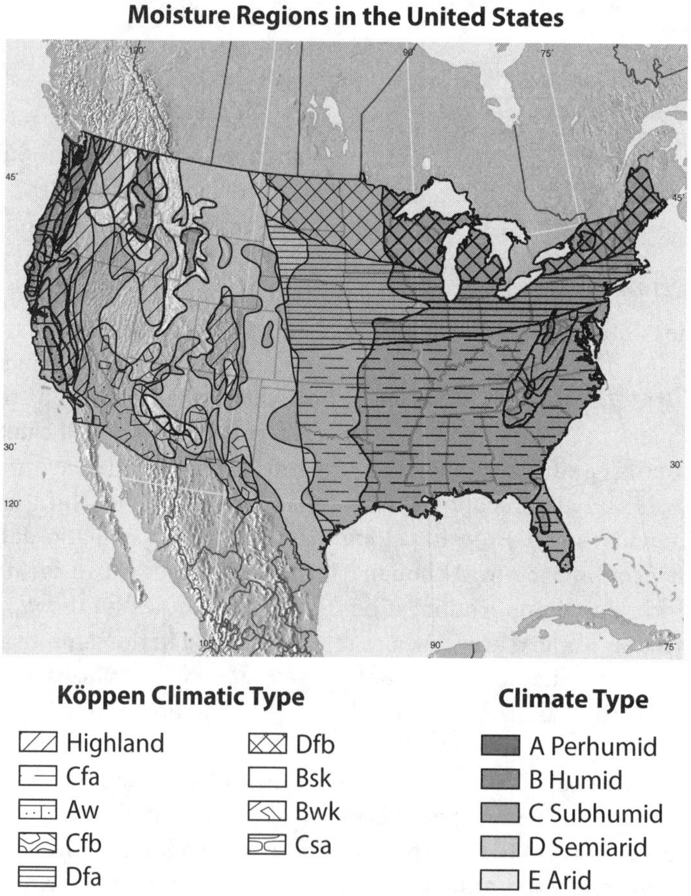

Moisture Regions in the United States
Köppen Climatic Type
Highland
Cfa
Aw
Cfb
Dfa
Dfb
Bsk
Bwk
Csa
Climate Type
A Perhumid
B Humid
C Subhumid
D Semiarid
E Arid

Figure 8.3 Overlay of the Köppen and Thornthwaite climatic classification for the United States. (See color plate 8.3.) Courtesy of Timothy Melancon.

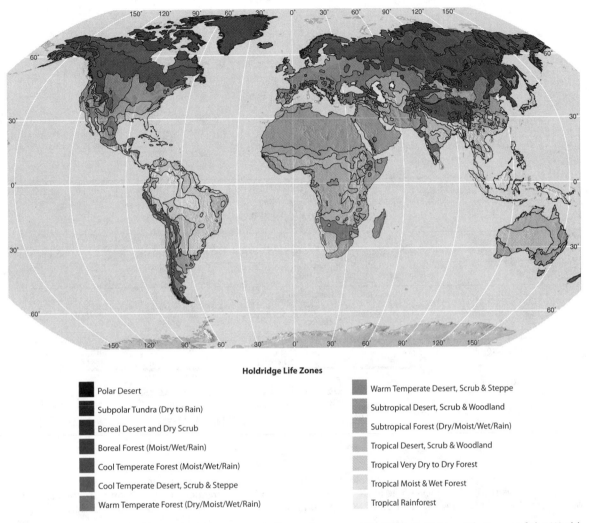

Holdridge Life Zones

- Polar Desert
- Subpolar Tundra (Dry to Rain)
- Boreal Desert and Dry Scrub
- Boreal Forest (Moist/Wet/Rain)
- Cool Temperate Forest (Moist/Wet/Rain)
- Cool Temperate Desert, Scrub & Steppe
- Warm Temperate Forest (Dry/Moist/Wet/Rain)

- Warm Temperate Desert, Scrub & Steppe
- Subtropical Desert, Scrub & Woodland
- Subtropical Forest (Dry/Moist/Wet/Rain)
- Tropical Desert, Scrub & Woodland
- Tropical Very Dry to Dry Forest
- Tropical Moist & Wet Forest
- Tropical Rainforest

Figure 8.4 World map of Holdridge climatic types. *Source*: Leemans, Rik, 1990. Holdridge Life Zones of the World. Global data sets collected and compiled by the Biosphere Project, Working Paper, IIASA-Laxenburg, Austria). Adapted from Oak Ridge National Laboratory Distributed Active Archive Center, Oak Ridge, Tennessee, U.S.A. [http://daac.ornl .gov/NPP/html_docs/hold2_npp.html].

In this case it just so happens that the temperature and precipitation features of San Diego and South Africa are similar enough to be categorized in the same Köppen category also. By contrast, let's consider the case of Reno, Nevada, and Spokane, Washington. The climates of both cities are influenced by their position on the *leeward* side of the major mountain ranges—the Sierra Nevadas in the case of Reno and the Cascades in the case of Spokane (**Figure 8.6a**). The data in Figure 8.6b show that Reno is warmer, drier, and better radiated than Spokane. A position on the leeward side of the slope does not necessarily provide the same conditions to both locations, but the climates of both locations are dictated largely by their positions on leeward slopes.

Genetic classifications do not compare the absolute numbers that represent the effect of the forcing mechanisms. Rather, they simply consider the causes involved. They are inherently more subjective than empirical methods, which can use rigid rules to classify the climates. How can we be sure that being on the leeward side of the mountain is a more important forcing mechanism than location with respect to common storm tracks, elevation, or any other potential forcing mechanism?

The **Bergeron Climatic Classification System**, devised by Swedish meteorologist Tor Bergeron in 1928, is one early-modern genetic classification system. This system categorizes the climate at a location based on the manual and somewhat subjective determination of frequency with which it is

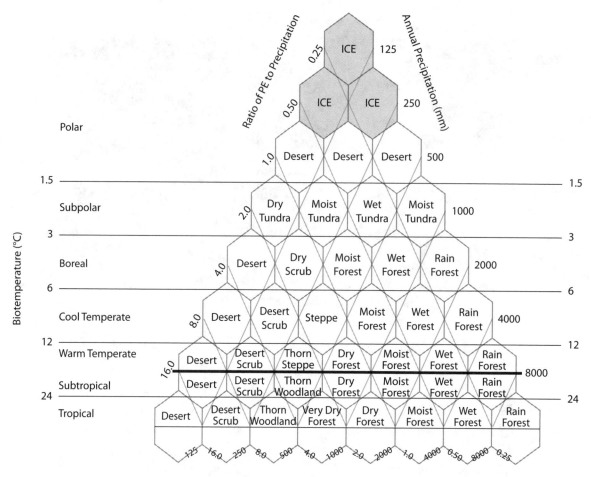

Figure 8.5 The Holdridge Life Zones pyramid. *Source:* The Holdridge Life Zones Pyramid. Cuming, Michael J. and Barbara A. Hawkins, 1981. "TERDAT: The FNOC System for Terrain Data Extraction and Processing." Technical Report Mil Project M-254, 2nd ed. Prepared for USN/FNOC and published by the World Meterological Organization.

dominated by certain types of weather. Improvements on this system were suggested by several climatologists, most notably by the German-American Helmut Landsberg. These genetic classification strategies evolved during the time that the important concepts in the science of meteorology described below were developing in the United States.

Air Masses and Fronts

As discussed above, classification schemes have been used to denote major types of atmospheric conditions. Simple classifications became the focus of weather forecasting early on in the development of meteorology as a separate scientific discipline. These classification schemes were based on absolute and relative temperature and moisture characteristics, and they proved somewhat useful until Jacob Bjerknes showed, in 1922, that

the development of frontal storm systems was of critical forecasting importance.

The Bjerknes model classified areas of a surface midlatitude *cyclone* and associated pressure features into specific sectors with each section exhibiting particular weather characteristics. Forecasting the motion of the migratory cyclone became the new focus, allowing forecasters to anticipate specific weather characteristics over time for a location as a cyclone tracked over a region. This classification scheme improved the accuracy of weather forecasting and is still used today. Knowledge of air masses and the Bjerknes model, which applies only to midlatitude systems, is necessary to fully understand the midlatitude regional climates described in Chapters 9 and 10.

Air Masses An **air mass** is a body of air that is relatively uniform in its characteristics for distances of

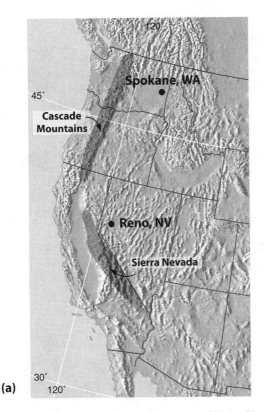

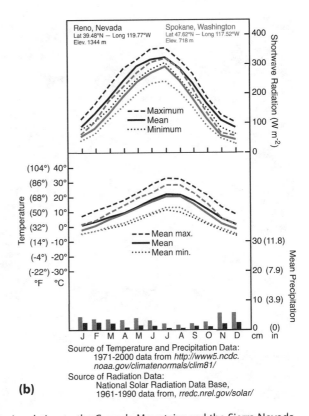

Figure 8.6 (a) Spokane, Washington, and Reno, Nevada, in relation to the Cascade Mountains and the Sierra Nevada, respectively. (b) Climograph for Reno, Nevada (black). Average annual temperature is 10.78°C (51.38°F) and average annual precipitation is 19.0 cm (7.5 in). Climograph for Spokane, Washington (blue). Average annual temperature is 8.58°C (47.38°F) and average annual precipitation is 42.3 cm (16.7 in).

hundreds to thousands of square kilometers. The characteristics of an air mass result largely from the properties of the **source region**—the place where the air mass forms—usually a relatively broad, homogeneous region where air may stagnate. **Figure 8.7** shows the major air mass source regions for the globe. Remember that air gains its temperature and moisture properties directly from the surface through the absorption of terrestrial radiation and the evapotranspiration of moisture, respectively. Source regions that are cold and dry, such as those occurring over high-latitude continental regions, produce cold, dry air masses. By contrast, air stagnating over a large, warm ocean surface likely gains the properties of that surface, resulting in a warm, moist air mass. The transfer of the source region's thermal and moisture characteristics to the air mass overlying it is enhanced through turbulence. The more unstable the atmosphere, the more rapidly the atmospheric properties of the source region can be transferred to the air mass over it. The longer an air mass remains over its source region, the more likely the air mass is to attain the properties of its source region.

Source regions are typically flat because such areas attain fairly uniform energy and moisture characteristics, as opposed to a mountainous terrain where those factors change appreciably across small spaces. Source regions tend to be tropical and high-latitude ocean regions and broad, flat continental regions. Most midlatitude regions are poor source regions because in the midlatitudes different air masses converge and weather changes rapidly.

The air mass classification system typically uses only two variables—moisture and temperature characteristics (**Table 8.9**). The moisture indicator appears first in the designation system. The lowercase letter "c" represents air of continental origin and, therefore, dry atmospheric conditions, and "m" indicates moist air of maritime origin. Temperature characteristics follow with uppercase designators. The uppercase letter "T" denotes warm air of tropical origin, "A" denotes very cold air of Arctic origin, and "P" indicates somewhat less severe cold air of polar origin.

Classifying the various air mass types is simply a matter of combining the moisture and temperature designators. Two cold, dry air mass types

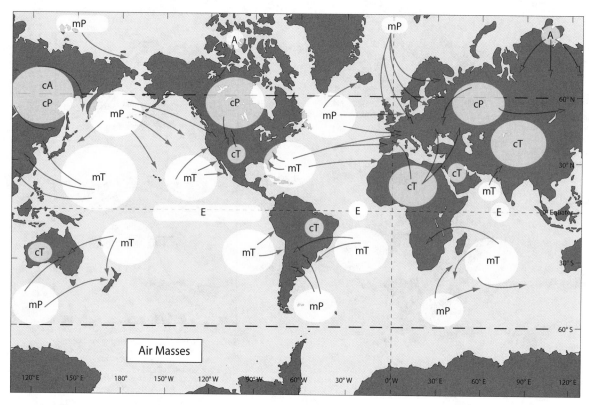

Figure 8.7 Major air masses and their source regions.

Table 8.9 Air Mass Classification System

Temperature Characteristics	Moisture Characteristics	
	Dry	**Humid**
Cool/cold	A, cP	mP
Warm/hot	cT	mT, E

affect the northern sections of the northern hemisphere. These are the intensely cold and dry **Arctic (A) air masses** and the slightly warmer and more humid **continental polar (cP) air masses**. Likewise, high-latitude oceans produce cool, humid air masses called **maritime polar (mP) air masses**. Portions of the northern hemisphere extratropics are affected by tropical air masses as well. Tropical oceanic source regions generate **maritime tropical (mT) air masses**, while inland tropical source regions produce the hot, dry **continental tropical (cT) air masses**. The system also includes a category for the **equatorial (E) air mass**—a more extreme version of the mT air mass

type. It is important to distinguish the *air mass* classification system from the Köppen *climatic* classification system—in the latter "A" represents a tropical climate and "E" represents a polar climate!

The midlatitude, upper-tropospheric *Rossby waves* transfer air masses away from their source regions. Arctic and polar air masses migrate equatorward on the ridge-to-trough side of the wave, whereas tropical air masses undergo *advection* in a poleward direction on the trough-to-ridge side of the wave. These waves exist as a mechanism to even out the imbalance between energy receipt at the poles and equator. These continuous motions of air, which respond to energy inequalities across the latitudes, represent the primary mission of the atmosphere. But because the Sun is, of course, constantly sending more intense energy to the equator than to polar regions, this mission of achieving thermal balance across the latitudes can never be achieved. The result is a warmer polar area and a cooler equatorial area than would exist without the Rossby wave activity, but polar areas remain colder than the tropical parts of Earth.

Classification of Fronts

Cold, Warm, and Stationary Fronts   Because the midlatitude part of Earth is the region where cold air and warm air masses usually converge, this is the area where vigorous exchanges of energy occur. A **front** is a narrow zone of perhaps 100 km (60 mi) where air masses of different temperature and humidity meet. A **cold front** represents a situation in which a colder air mass is pushing a warmer air mass back toward its source region. Cold fronts are represented on surface weather maps by blue lines, with blue triangles pointing in the direction in which the cold air is moving (**Figure 8.8a**). The warm air at a cold front is pushed upward at an angle of perhaps 2° above the horizontal. Although this angle may seem gentle, it is actually steep. The atmosphere is thin enough that if air moves upward at a 2° slope and we assume that the tropopause is 12 km (7.5 mi) above the surface, the horizontal distance required to reach all the way up to the tropopause (x) could be calculated as

$$\tan 2° = \frac{12 \text{ km}}{x}$$

or only 344 km (213 mi)! This relatively narrow zone is sufficient to ensure that cloud cover at cold fronts be vertically oriented. This produces **cumuliform clouds** and includes the severe weather-producing **cumulonimbus clouds**. Cumuliform clouds generally produce intense precipitation of short duration. "Cumuli-" means "piled up" and originates from the same root that produces words such as "accumulation" and "cumulative."

A **warm front** is a region where a warmer air mass is displacing a colder one poleward. These are represented on surface weather maps by red lines with red semicircles pointing in the direction that the warmer air is moving (Figure 8.8b). The ascent of relatively warm air at the warm front proceeds only at perhaps an angle of 1°. At this rate of steepness the air would need 687 km (427 mi) to reach a 12-km-high tropopause. This gentler angle ensures the formation of more horizontally oriented **stratiform clouds**, which are widespread and associated with precipitation of only light or moderate intensity that lasts for long periods of time. Precipitation intensity is directly proportional to cloud thickness and is thus less in stratiform clouds, which are much thinner than cumuliform clouds. Warm fronts may produce convective clouds if the air is particularly unstable, but this is a less common phenomenon.

For periods of up to several hours the cold and warm air masses may temporarily meet at a **stationary front**—a stalemate in which neither air mass is pushing the other backward. Stationary fronts are depicted on a weather map by an alternating red and blue line, with blue triangles pointing in the direction that the colder air would move if it could move and red semicircles pointing in the direction that the warmer air would move if it could move (Figure 8.8c).

When referring to fronts it is important to note that the temperatures are relative ones. The air behind a cold front may be 0°C (32°F) in winter but 30°C (86°F) in summer. It is also colder in higher latitudes than in lower latitudes, as the air masses warm as they move equatorward.

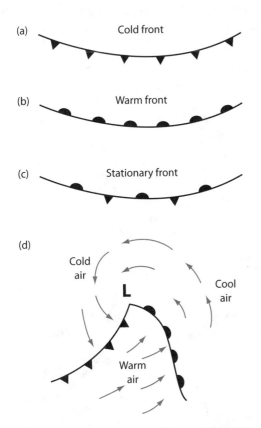

Figure 8.8   The various types of fronts. (a) cold front; (b) warm front; (c) stationary front; and (d) the midlatitude wave cyclone.

Midlatitude Wave Cyclone   Fronts do not typically occur in isolation. Instead, they are usually part of a larger system known as the *midlatitude wave cyclone* (Figure 8.8d). These cyclones are the largest storm systems on the planet, with diameters between 1600 and 3200 km (1000 and 2000 mi). Because weather conditions vary across space depending on location with respect to the midlatitude wave cyclone, such a cyclone provides an ideal model for classification systems. At the core of the midlatitude wave cyclone is a cyclonic (i.e., low-pressure) center. The rotation of air around the cyclone (counterclockwise in the northern hemisphere) causes colder air to be pushed equatorward behind (i.e., to the west of) the cyclone. This is the cold front. Likewise, ahead of (i.e., to the east of) the cyclonic center, the counterclockwise rotation pushes tropical air poleward. This is the warm front.

Midlatitude wave cyclones are sometimes referred to as *secondary circulations* because they are embedded within the larger general circulation. Specifically, air associated with a midlatitude wave cyclone rotates cyclonically (counterclockwise in the northern hemisphere) while it is also moving in a general west-to-east direction as dictated by the general circulation (i.e., the midlatitude westerlies). Such circulation is common in nature; a river has a general circulation downslope toward sea level, with smaller eddies spinning while embedded within that general circulation.

Migratory anticyclones may also act as secondary circulations by spinning about within the broader-scale general circulation. Unlike midlatitude wave cyclones, midlatitude anticyclones cannot be associated with trailing fronts, because the sinking air in an anticyclone does not provide adequate energy via the release of *latent heat* to develop the pressure gradients required to move fronts poleward or equatorward.

Occluded Fronts In general, a cold front progresses around a midlatitude cyclone faster than a warm front as the entire system moves eastward in the midlatitude circulation. The reason is that colder air can displace warmer air relatively easily because it is denser than the warmer air. Warm fronts migrate relatively slowly, because it is more difficult for the less dense, warmer air to displace the colder air. Also, the midlatitude cyclone migrates in the same general direction of warm front propagation, so the warm front does not radically change its position relative to the core of the system.

After a few days of traveling eastward in the midlatitude westerlies, the cold front actually catches up with the warm front near the cyclonic center, where the two fronts are closest to each other (**Figure 8.9a**). The result is an **occluded front**, which is depicted on a weather map by a purple line with alternating purple triangles and semicircles both pointing in the direction that the colder air is moving after having caught up with the warm front. Over time, the occlusion proceeds farther away from the cyclonic center (Figure 8.9b) as the **warm sector**—the area between the fronts—closes. At occluded fronts, both cumuliform clouds associated with the (former) cold front and stratiform clouds associated with the (former) warm front are present. By the time occlusion occurs, most of the storm's moisture has precipitated and most of its energy has been spent. The entire system will have also moved into an area of decreasing upper-air *absolute vorticity* as it approaches the upper-level *ridge*, altering the motion of air through the vertical. This causes the system to decay. Occluded fronts are usually associated with abundant clouds but neither intense precipitation nor severe weather activity.

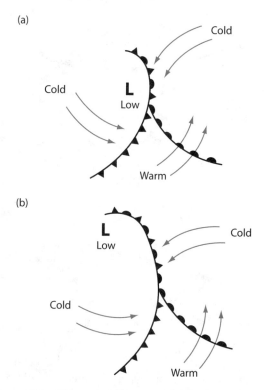

Figure 8.9 The occlusion process. (a) Occlusion begins nearest to the low pressure center and (b) then proceeds away from the center over time.

At an occluded front the warmer air mass is lifted over the surface by both the cold and the warm fronts. This results in a *stable stratification* of air, with colder, denser air lying beneath warmer, less dense air. This vertical lifting is also superimposed on the horizontal mixing of air induced by the cyclonic circulation of the entire system. Overall, the system worked to push colder air equatorward and warmer air poleward while lifting the less dense air over denser air. The system extracts energy from the differences between contrasting air masses. The occlusion represents the point when vertical and horizontal mixing has been accomplished. The temperature contrast between the air masses wanes and the fronts disappear, coincident with vorticity changes in the overlying atmosphere. Occlusion can be considered to be the end of the midlatitude wave cyclone's life.

■ Local and Regional Classifications

The **Lamb Weather Types** and **Muller Weather Types** are two climatic classification systems based on the midlatitude wave cyclone model. Although at their core they represent meteorological classification systems, these two systems can be used in climatic classification by tabulating daily frequencies of predefined weather patterns observed over the British Isles and the U.S. Gulf coast, respectively, in calendar form. These two systems are based on the position of a location relative to major secondary circulation features, including midlatitude cyclones and anticyclones. **Figure 8.10** provides an example of the Muller Types. Other similar manual climate classification systems have followed these by requiring manual analysis of weather maps to produce the classification.

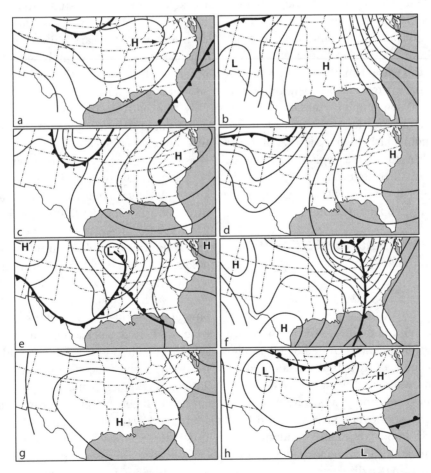

Figure 8.10   The Muller Weather Types for the central Gulf of Mexico coast. (a) Pacific high; (b) continental high; (c) coastal return; (d) Gulf return; (e) frontal Gulf return; (f) frontal overrunning; (g) Gulf high; and (h) Gulf tropical disturbance. *Data from*: The National Weather Service, NOAA.

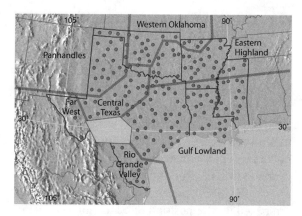

Figure 8.11 A climatic classification of temperature in the south-central United States. The light gray area in southwestern Texas indicates a region where lack of data precluded cluster assignment. Modified from Bohr, G.S., 2005. Trends in extreme daily temperature events in the south-central United States. Dissertation, Louisiana State University.

The Lamb and Muller Weather Types are genetic attempts to classify climates at local and regional scales. Of the many climatic classification schemes, the majority address very specific climatic topics for particular locations or regions. For example, Figure 8.11 shows a regionalization of temperature across the south-central United States, devised for analyzing changes in extreme temperature events across regions. When temperature is anomalously high (or low) at one location in Figure 8.11, it also tends to be anomalously high (or low) at all other locations within the same region. Meanwhile, temperatures in other delineated regions have as little in common as possible with concurrent temperatures in the group in question.

■ Quantitative Analysis to Derive Climatic Types

Eigenvector Analysis

The technique used to delineate the temperature regions in the south-central United States in Figure 8.11 is one of a family of multivariate mathematical techniques called **eigenvector analysis**. Although eigenvector analysis is fully discussed in Chapter 13, a brief overview is provided here because it is an important tool in contemporary climatic classification. Such analysis allows for relationships between two different sets of variables to be analyzed simultaneously. Climatologists are usually concerned with three types of variables:

atmospheric data (which may include variables such as surface temperature, 500-mb *geopotential height*, wind speed, and humidity), spatial variables (represented by weather stations), and temporal variables (such as monthly data from 1895 through 2009).

To date, no mathematician has identified a way to analyze all three sets of variables simultaneously while considering the variables as part of each respective set. But eigenvector analysis allows us to find relationships between any two sets of variables simultaneously while holding the third set constant. For example, a common type of eigenvector analysis (including the temperature study for the south-central United States discussed above) examines how one atmospheric variable (not a set of variables) varies across space and time. Such an analysis allows the climatologist to map the spatial distribution of that variable (such as surface air temperature) while simultaneously considering the temporal variation in that variable. Where in the study area is the greatest temporal variability in (say) surface temperature, and how does this variability change over time? This is a very powerful analytic tool, because no other family of quantitative techniques allows for simultaneous analysis across space and time. The only alternatives are visualization techniques, which could show a "movie" of how (say) temperature changes across space and time, but such changes would be difficult to quantify.

Alternatively, we could examine how a set of atmospheric variables (such as surface temperature, 500-mb geopotential height, humidity, and wind speed) interacts holistically at one station over time. This type of eigenvector analysis would hold space constant while examining the variability of atmospheric variables through time. Of course, such analysis would not allow for the mapping of the results per se, but if separate analyses were conducted, one station at a time, then atmospheric and temporal variables could be analyzed simultaneously at each site. Locations where similar characteristics of the atmosphere tend to occur simultaneously would be categorized as being dominated by the same air mass type at that time. This is the premise behind the **Spatial Synoptic Classification** designed by climatologists Laurence Kalkstein, Scott Sheridan, and others.

Time is seldom the entity held constant in climatology, because climatology is inherently concerned with conditions across time. However, multiple analyses could be conducted, perhaps one analysis per day (where time is held constant on each iteration). The distribution of synoptic types (represented by the various atmospheric variables) across space (represented by the weather stations) then can be mapped for each day, with temporal changes in the pattern noted.

In climatology the specific eigenvector technique that is most typically used is **Principal components analysis (PCA)**. For reasons beyond the scope of this discussion, this analytical tool is preferred over the two other types of eigenvector analysis: Common Factor Analysis and Empirical Orthogonal Functions. In each of the three types of examples described above, PCA can identify **components**—the main modes of variability in the data set.

Let's go back to the example shown in Figure 8.11. In analyses such as this one that examine spatial and temporal differences in only one atmospheric variable, the PCA technique provides coefficients for each spatial variable (called **loadings**) and each temporal variable (called **scores**) on the component. The loadings are proportional to the degree to which the eigenvector model represents the variability in each of one type of variable (say, temperatures at various stations)

collectively for all temporal observations in the data set. The scores are proportional to the degree to which the model represents variability in each of the other type of variable (in this case, each temporal observation) collectively for the first variable (temperature collectively, undistinguished by station). The PCA model (and all other eigenvector models) is set up so as to maximize the relationship between the component and as many locations and times as possible. Then, the eigenvector technique generates a second component solution that attempts to provide the strongest relationship possible between the atmospheric variable of interest (temperature) across the other two entities (space and time, in this example) that remains unaccounted for by the first component.

By analogy, imagine a series of data points in two-dimensional space (**Figure 8.12**). A **regression line** can be drawn to pass as closely as possible to as many points as possible, so as to minimize the squared sum of the vertical distances between each point and the line. If the line passes very close to many points, then the model represented by the line effectively "explains" the variance in the data points. By contrast, if the data points are scattered all over the graph, as they are in Figure 8.12, the regression line would not "explain" as much variance. If a second line is drawn that

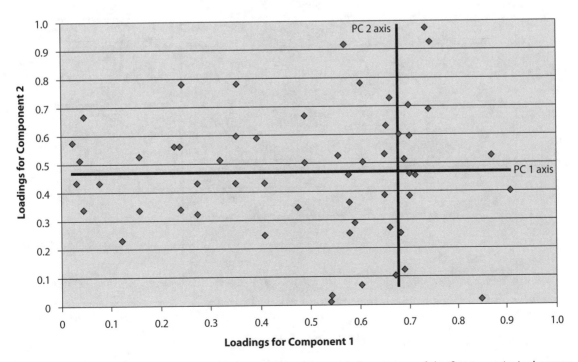

Figure 8.12 Schematic of a hypothetical data set with loadings and eigenvectors of the first two principal components plotted.

is perpendicular to the first line, it may explain the **residual**—"new" variance that remains unexplained by the first component. PCA uses eigenvectors in this manner to "draw lines" as near as possible to as many points as possible, represented in this case by spatial and temporal data. Each of these lines corresponds to a certain amount of variance present in the data set. There will be as many lines drawn through the data set as there are data points in the set so that 100 percent of the data set variance is explained.

Loadings and scores are also associated with the second component (line), which is perpendicular to the first component and accounts for a particular amount of the remaining data set variance (but less than the first component, which passes through the greatest concentration of data points). If a third component is also calculated, it is represented by a line that is perpendicular to both the first and the second lines. The process can again be repeated, but we cannot even imagine this line, because it would occur in four dimensions. Mathematically, and in a multidimensional space of a computer, drawing this line is completely possible, and it is "drawn" in such a way that it explains as much of the remaining (or residual) variance in the data set as possible. The process is again repeated with each successive component explaining less variability than the one before, as the unloaded (unrelated) data points are farther from the interior cluster. But it is, of course, impossible to visualize these perpendicular lines because they too exist in more dimensions than our brains can process.

Cluster Analysis

The actual classification process in any eigenvector analysis (including PCA) comes when the matrix of loadings or scores (which usually represents the spatial and temporal variability, respectively, in the data set) is subjected to a **cluster analysis**. In the example of temperatures across the south-central United States, cluster analysis could be used to categorize the loadings across space. Each station in the analysis would have a loading for each of the n components retained for analysis. We could plot the loadings of each station (at whatever the spatial scale of analysis was) in n-dimensional space. Obviously, because it is impossible to visualize more than three dimensions, this is a highly abstract concept. So let's just imagine a two-dimensional solution (**Figure 8.13**). If the loadings for the first two

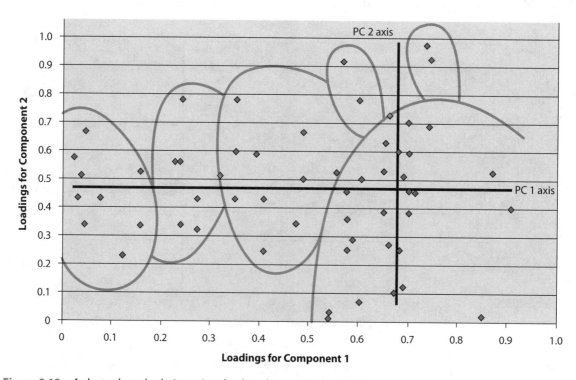

Figure 8.13   A cluster-based solution using the data shown in Figure 8.12.

components are plotted as shown, a clustering procedure can quantitatively group the resulting points into a manageable number of clusters, with the points (stations) whose loadings are closer to each other in the same group. Then boundaries can be drawn to separate the stations that fall into different categories. Just as the loadings can be clustered to delineate spatial regions, scores can be clustered to identify months (or days, or whatever the temporal scale of analysis) that have similar temporal properties of the atmospheric variable being analyzed (in this case, temperature).

Several types of cluster analysis can be used, and each gives a slightly different classification result. However, all have the common goal of selecting the groups so that the distance in n-dimensional space between the points within the cluster (within-group variability) is minimized. Likewise, the distance between points that lie in different clusters (between-group variability) should be maximized. The researcher may select the number of clusters to be created, either by using a preselected value or by looking for natural "break points" in the data. This could be determined by plotting the **eigenvalues**—the total variance accounted for by a component—for each component on a chart called a **scree plot**. At some point a break in the eigenvalue curve appears. The researcher should examine all components up to the break point while discarding all components that appear after that point. So, for example, if the breaking up of a data set into seven clusters does not minimize the distances between points within each cluster much better than a six-cluster solution, then the researcher may simply opt for the six-cluster solution because it is simpler to use.

Hybrid Techniques

In recent years some researchers have made use of **hybrid climatic classification techniques**—methods of climatic categorization that have the advantages of manual techniques such as the Lamb or Muller Weather Types and those of automated eigenvector analysis. Hybrid methods typically require the investigator to identify "prototype" atmospheric circulation patterns manually—a "textbook" case of each particular genetic category, perhaps. Then a computerized cluster analysis is used to categorize all other days (or months, if monthly mean data are

used) automatically, quickly, and efficiently into the group with the prototype day (or month) that the weather map for the day (or month) in question most resembles. This is done by comparing the pressure value at each data point on the map to the pressure value at each analogous point on each of the prototype maps. Such a map provides the smallest sum of differences for all pairs of data points with the day to be classified and is the circulation type for the day in question.

■ Summary

Climatologists classify climates for the same reason as other scientists—that is, to summarize and simplify our conception of a myriad of unique climates. The ancient Greeks divided Earth into three zones, based on their perception of habitability in those zones. In modern times Vladimir Köppen's classification system has become the most widely used system, largely because of its strong correspondence to vegetation and soil realms. Thornthwaite developed a classification system based on the surface water balance, thereby implicitly incorporating energy as a driver of evaporation. Thornthwaite's system is more detailed and precise than Köppen's classification, but the complexity of deriving the moisture indices has limited its utility, particularly among nonscientists. Other notable classifications have been derived by Holdridge and Budyko.

Genetic classifications are based on the causes of the climate rather than on the results as shown by climatic data. Such genetic classifications in the extratropical world often rely on the position of a location relative to air masses, fronts, and midlatitude wave cyclones. Air masses can be characterized as arctic, continental polar, maritime polar, maritime tropical, continental tropical, and equatorial.

When dissimilar air masses meet, fronts are formed, and these fronts represent parts of a midlatitude wave cyclone. At a cold front the colder air mass displaces a warmer air mass, and it is characterized by cumuliform cloudiness and possibly intense precipitation and severe weather. The warm front is the site of warmer air that is displacing colder air, and it is characterized by stratiform cloud cover and precipitation of light to moderate intensity across a widespread area. Eventually, the occlusion process ends the midlatitude wave cyclone's life span. Notable among the cyclone-based genetic classification systems are weather-typing schemes such as those devised by Lamb and Muller that can be adapted for use in regional climatological analysis.

Eigenvector-based techniques in combination with cluster analysis can be used to automate the classification process for the analysis of data across space and time. The Spatial Synoptic Classification is a notable example of such a classification. Eigenvector analysis can also be used to identify the major climatic types represented in the data set. This is done by manipulating a data matrix containing two types of entities (some combination of locational variables, temporal variables, and atmospheric variables) so that the major modes of variability in the data set are extracted. The resulting loadings and scores matrix can then be clustered to categorize the major climatic types in the data set. In recent years human observers can select the "prototype" of the major types, and then automated clustering techniques can be used to find the prototype to which a given day or month corresponds most closely.

▶ Key Terms

Absolute vorticity
Advection
Air mass
Arctic (A) air mass
Bergeron Climatic Classification System
Biome
Biotemperature
Budyko Climatic Classification System
Budyko ratio
Cluster analysis
Cold front
Component
Continental polar (cP) air mass
Continental tropical (cT) air mass
Cumuliform cloud
Cumulonimbus cloud
Cyclone
Deficit
Dryness index (DI)
Ecotone
Eigenvalue
Eigenvector analysis
Empirical climatic classification system
Equatorial (E) air mass
Evaporation
Evapotranspiration
Extreme

Frigid Zone
Front
Genetic classification system
Geopotential height
Holdridge Life Zones Climatic Classification System
Humidity index (HI)
Hybrid climatic classification technique
Köppen Climatic Classification System
Lamb Weather Types
Latent heat
Latitude
Leeward
Loading
Maritime polar (mP) air mass
Maritime tropical (mT) air mass
Mesothermal
Microthermal
Midlatitude wave cyclone
Moisture index (MI)
Muller Weather Types
Net radiation
Normal
Occluded front
Potential evapotranspiration
Principal components analysis (PCA)
Regression line

Residual
Ridge
Rossby wave
Score
Scree plot
Secondary circulation
Source region
Spatial Synoptic Classification
Stable stratification
Stationary front
Steppe climate
Stratiform cloud
Subtropical anticyclone
Summer thermal efficiency concentration
Surplus
Temperate Zone
Thermal efficiency index
Thornthwaite Climatic Classification System
Torrid Zone
Transpiration
Treeline
Tundra climate
Warm front
Warm sector
Terms in italics have appeared in at least one previous chapter.

▶ Review Questions

1. What purpose does climatic classification serve?

2. What is the goal of any climatic classification system?

3. How did the ancient Greeks classify the climates of Earth?

4. What is the goal of the Köppen Climatic Classification System?

5. What is the basis of the Thornthwaite Climatic Classification System?

6. In the Köppen system, what does the third letter usually represent?

7. What is the treeline and how is it used in the Köppen system?

8. How did Köppen define "dry" when he devised the arid or "B" climatic type?

9. What is the moisture index in the Thornthwaite system and how is it derived?

10. In what part of the world is the Holdridge Life Zones Classification most applicable? Why?

11. On what is the Budyko Climatic Classification System based, and how is it best used?

12. How does a genetic classification system differ from an empirical classification system?

13. Discuss how and why air masses form over source regions.

14. What are the four types of fronts? How does each form and what are the general characteristics of each?

15. Briefly describe the midlatitude wave cyclone model.

16. How would weather conditions (winds, temperature, moisture, clouds, and precipitation) change for a location if a midlatitude cyclone tracked over it with the core of the system passing to the north of the location?

17. What are the three entities that can be evaluated only two at a time in climatological classification using eigenvector analysis?

18. Why is it necessary to use cluster analysis after doing eigenvector analysis for climatic classification?

▶ Questions for Thought

1. In what sense could the argument be made that the Budyko Climatic Classification System is both empirical and genetic?

2. Draw a midlatitude cyclone model, complete with warm and cold fronts, for the southern hemisphere.

3. Do you believe that manual methods are better than the computer-generated eigenvector techniques for classifying climates? Why or why not?

4. Design a climatic classification system for tourists interested in outdoor urban sculpture.

---

http://physicalscience.jbpub.com/climatology

Connect to this book's Web site: http://physical science.jbpub.com/climatology. The site provides chapter outlines, further readings, and other tools to help you study for your class. You can also follow useful links for additional information on the topics covered in this chapter.

Extratropical Northern Hemisphere Climates

Chapter at a Glance

Climatic Setting of North America
   General Characteristics
   Severe Weather
   Role of the Gulf of Mexico and the Low-Level Jet
   Effect of Mountain Ranges
   Effect of the Great Lakes
   Ocean Currents and Land-Water Contrast
Climatic Setting of Europe
   General Characteristics
   Effect of Ocean Currents
   Effect of Mountain Ranges
   Blocking Anticyclones
Climatic Setting of Asia
   General Characteristics
   Monsoonal Effects
   Effect of Mountain Ranges
   Effect of Coastal Zones on Climate
Regional Climatology
   B—Arid Climates
   C—Mesothermal Climates
   D—Microthermal Midlatitude Climates
   E—Polar Climates
   H—Highland Climates
Summary
Review

Have you ever wondered whether some other place on Earth, perhaps thousands of miles away, has the exact same type of climate as your hometown? Because the fundamental forcing properties, such as day length, Sun angle, storm passages, elevation, *continentality*, and cloud cover, exert an influence everywhere on Earth, it is possible that distant locations may have the same fundamental climatic properties. This chapter describes and explains the major aspects of the climatic setting of the extratropical northern hemisphere, with a focus on the Köppen-derived climates of North America, Europe, and Asia. Tropical climates that occur in North America and Asia are discussed in Chapter 10.

■ Climatic Setting of North America

General Characteristics

The climates of North America may be described as diverse. Conditions range from extremely dry to exceedingly wet and from bitterly cold to oppressively hot. The extreme range of *latitudes* in North America, from the tropics to well within the *Arctic Circle*, causes a wide array of temperatures. In addition, the continent spans longitudinally from approximately 170°E to about 30°W, or from the western edge of the Aleutian Islands to the eastern shores of Greenland. Although the *international date line* is situated at approximately 180° *longitude*, it zig-zags around the Aleutians so that all of Alaska is on the same calendar day even though the westernmost part of the state is actually in the eastern hemisphere.

With such an expansive latitudinal and longitudinal span, one would expect a considerable continental influence. Although continentality is an important factor in the climatology of North America, the presence of the Gulf of Mexico and the North Atlantic Ocean gives the eastern half of North America a significant *maritime tropical (mT) air mass* influence as well. Furthermore, the north–south mountain chain alignment in North America allows mT air masses to infiltrate deep into the continental interior east of the Rockies, thus reducing the effect of continentality in comparison with Asia.

The most densely populated parts of North America lie within the northern hemisphere's midlatitudes and subtropics. The northern hemisphere's midlatitudes can be defined as the region between 30°N and approximately 60°N. The climate of this area is dominated by the *midlatitude westerlies*. The midlatitudes represent a zone of transition between the warm, moist tropics and the cold, dry polar region. Pronounced seasonality and a wide range of weather phenomena occur across this part of the continent.

Severe Weather

The midlatitude part of North America is the site of some of the most impressive severe weather on the planet. No other area of the world experiences such frequent juxtaposition of radically different *air masses* as midlatitude North America. This variety of air masses occurs primarily because of the orientation of the major north–south mountain ranges (**Figure 9.1**). This arrangement allows for *continental polar (cP)* and *Arctic (A) air masses* to penetrate to fairly low latitudes, where they mix freely with mT air masses that migrate deep into the continental interior. The resulting *baroclinicity*—the intermixing of different air masses—leads to the development of very strong *midlatitude wave cyclones*. Frequently, along the

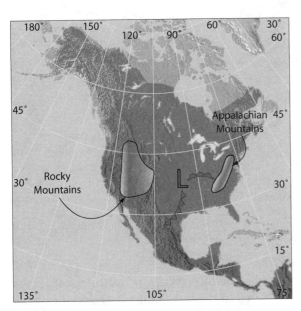

Figure 9.1 Physiographic features of North America, with a mature midlatitude cyclone over the Great Plains.

warm and (especially) the *cold front* region of these storm systems, tornadoes, hail, strong winds, heavy precipitation (rain, snow, hail, sleet, and/or freezing rain), and/or lightning occur.

Because of the high frequency and magnitude of severe frontal *cyclones* and *tropical cyclones* ranging from weak disturbances to major *hurricanes*, the United States leads the world in tornado frequency, recording an average of nearly 1000 per year. Tornadoes typically develop in the large *cumulonimbus clouds* that accompany a cold front, in association with a tropical cyclone, or (most frequently) in advance of a cold front along a **squall line**.

Squall line thunderstorms are set up as mT air meets a *continental tropical (cT) air mass* within the warm sector of a midlatitude wave cyclone, at a location termed the **dryline**. When these two warm air masses meet, the mT air is forced above the drier cT air because moist air is less dense and more buoyant than dry air. If this fact is surprising, recall that most of the dry atmosphere is composed of N_2 and O_2, while H_2O consists of two atoms of the lightest element and only one of oxygen (which is only slightly heavier than nitrogen). Thus, a molecule of water has a lower atomic mass than a molecule in dry air, even if the air masses are of the same temperature. As the moist air rises, it is often very unstable because of the greater amount of energy associated with air in the warm sector. Cold front thunderstorms create a cold downdraft of air—the **outflow boundary** or **gust front**—that wedges beneath warmer air in advance of the cold front line. The combination of lifting caused by the dryline and the outflow boundary leads to rapid uplift of the warm air in the warm sector. This can result in squall line thunderstorms that may be severe enough to spawn tornadoes.

Most severe frontal cyclones occur in the continental interior from south-central Canada southward through central Texas. This region is termed **Tornado Alley**, as the greatest frequency of tornadoes on Earth occurs within the area (**Figure 9.2**). In actuality, the high-frequency region extends eastward from the northern and southern sectors of the "alley," giving it a "C" shape. These eastern extensions include the Midwest and the Gulf South of the United States. The highest rate of tornado-related casualties in the United States is found in **Dixie Alley**—a region in the Gulf South

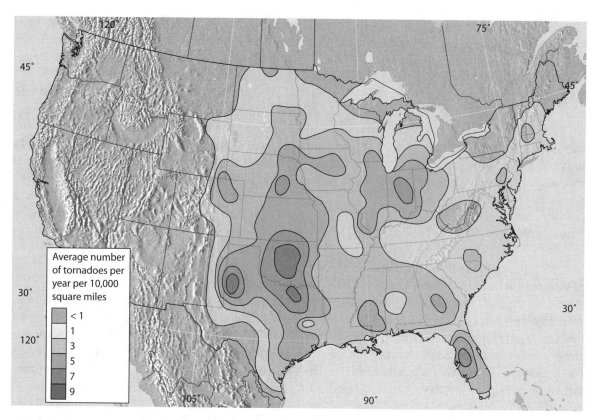

Figure 9.2 Average number of tornadoes per year per 10,000 mi² in the United States. Modified with permission. © 1997, Oklahoma Climatological Survey.

ranging from Texas to Florida. This pattern is due to a combination of factors, including complex forested terrain, low visibility from severe thunderstorm rainfall, and a relatively high percentage of residents in dwellings that lack stability and the safety of basements.

Role of the Gulf of Mexico and the Low-Level Jet

The Gulf of Mexico is of particular importance to North American climates. The Gulf represents the *source region* for the mT air masses that are so important east of the Rocky Mountains. The effect of Gulf moisture is augmented by the **low-level jet (LLJ) stream**, a river of air at approximately the 800- to 900-mb level, moving at perhaps 40–110 km hr⁻¹ (25–70 mi hr⁻¹). The LLJ advects copious amounts of moisture into the southern Great Plains.

The LLJ occurs either nocturnally from diurnal pressure changes associated with daily heating of the continent or through interactions with a cyclone. In the former case, the difference in eleva-

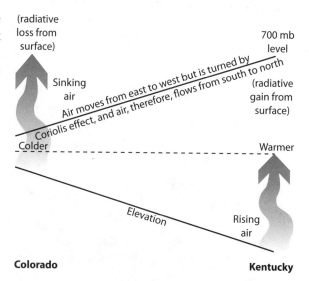

Figure 9.3 The formation of the low-level jet (LLJ) stream in the United States.

tion from west to east across North America plays a role in its formation (**Figure 9.3**). Specifically, as radiative cooling occurs nocturnally in the lower-elevation eastern part of North America, the *long-wave radiation* lost from the surface is gained by the atmosphere at approximately the same height as the surface of the western United States. At the same time, the loss of surface radiation in the western United States is occurring. This situation induces sinking of cold air over the western United States, while the relatively warm air at the same height over the eastern United States rises. This causes a *geopotential height* gradient from east to west (Figure 9.3). But as the air begins to move from east to west (against the midlatitude westerlies), the *Coriolis effect* turns the air stream to the right. The result is a southerly flow of air into the southern Great Plains. Atmospheric *stability* and an associated midlevel *temperature inversion* increase wind speeds through the *surface boundary layer* by restricting vertical fluid motion.

The effect happens at night because daytime *shortwave radiation* receipt from the Sun reduces the tendency for the surface cooling and sinking air in the western United States. However, the LLJ can occur during the daytime or nighttime in association with a midlatitude wave cyclone to the northwest of the Gulf of Mexico. In such cases the counterclockwise rotation around the cyclone draws moist air northward ahead of the cold front. The presence of the LLJ enhances the development of severe frontal cyclones and associated tornadoes.

Effect of Mountain Ranges

The high western cordillera of North America creates direct modifications to the climate. First, it effectively prevents warm, moist Gulf air from being pushed by the large semipermanent *Bermuda-Azores high* west of the Rocky Mountains. It also prevents cP and A air masses from spilling westward against the flow of the midlatitude westerlies. As a result, air mass contrast is weak west of the Rocky Mountains, which means that powerful fronts and associated severe weather are relatively rare in extreme western North America. Certainly, midlatitude wave cyclones occur west of the Rockies, but these are generally *occluded fronts* that move in the westerlies from the western Pacific and Asia. They are characterized by weak frontal uplift, cloudiness, and light precipitation. During *El Niño* events the mean storm track is displaced equatorward and is termed the **Pineapple Express** because of its origins closer to Hawaii. The Pineapple Express provides abundant moisture for

snowpack on the *windward* side of the high Sierra Nevada Mountains, which subsequently provides the water supply for densely populated southern California.

The Rocky Mountains also provide the mechanism for *ridges* to be locked in place over them, as was discussed in Chapter 7. Furthermore, because of conservation of *potential vorticity*, air that descends the *leeward* sides of high mountains such as the Rockies is likely to acquire *positive vorticity* on its descent. The resulting storms are often vigorous and may spawn tornadoes, particularly if associated with the dryline.

Cyclogenesis is frequent on the leeward sides of the Rockies, particularly when abundant moisture is present. This occurs most noticeably during the transition seasons of autumn and spring. Midlatitude wave cyclones that form in this region are called **Colorado wave cyclones** or, farther north, **Alberta clippers**. As discussed in Chapter 7, the arrangement of a trough-to-ridge side of the *Rossby wave* east of the mountain cordillera also generally provides these surface storms with adequate upper-level divergence needed to sustain them as they traverse eastern North America. Such a situation, aided by the topography, promotes the intermingling of contrasting air masses. These interactions help create the most violent weather on Earth as the atmosphere works to equalize energy and moisture inequalities.

Air moving downslope from high mountains warms at the *unsaturated adiabatic lapse rate* along its descent. As a result, mountains create a situation where air can be very dry and warmer on their leeward sides. Such winds are capable of evaporating snow or water quickly. In western North America these downslope winds are known as the **Chinook winds** (the word "Chinook" is derived from a Native American term for *snoweater*). *Evaporation* and melting caused by Chinooks can destabilize slopes, causing avalanches. They can also trigger extremely rapid changes in weather, as air may shift from northerly (cold) to westerly (warming Chinook winds) in only a matter of minutes to hours. The greatest 2-minute temperature change ever recorded on Earth, an astonishing 27 C° (49 F°), occurred in Spearfish, South Dakota, on January 23, 1943 in association with a Chinook wind.

Downslope winds also occur often on the western side of the Pacific coastal mountain ranges when a surface *anticyclone* is situated over Nevada or Utah. In such cases the clockwise flow of air around the anticyclone lifts air up the *eastern* sides of the slopes and down the *western* sides in southern California, where it dries because of *adiabatic motion*. On descent this air may be so dry, particularly in the late summer/autumn dry season, that forest fires and water shortages are a typical occurrence. These **Santa Ana winds** are a major hazard to life and property.

In some synoptic situations the alignment of mountains in North America also causes **cold air damming**—a condition whereby low-level cold air becomes trapped on the upwind side of highlands. This occurs from either a midlatitude surface anticyclone or cyclone forcing surface air toward a mountain range, most often along the eastern sides of both the Rocky and the Appalachian mountains when synoptic-scale systems migrate eastward. This synoptic situation pushes air from these secondary circulations against the windward side of the mountains, sometimes for days. Warm, moist air overruns the colder, denser air, enhancing cloudiness and precipitation. The damming event may cause a variety of precipitation types as the moist air is pushed upslope, with rain in lower elevations and freezing rain, sleet, and snow at successively higher elevations. Furthermore, in a damming event the affected region sees persistent cold for longer than would occur if the mountains were not present. Adiabatic cooling of the air mass moving upslope further decreases associated temperatures.

Effect of the Great Lakes

The Great Lakes are a significant source of moisture and energy in the continental interior. The presence of the lakes offsets some of the continental influence, as the lakes store energy throughout the summer and release it to the atmosphere slowly during winter. In the cold season the region surrounding the lakes, especially on the leeward areas to the south and east, remains milder than comparable regions upwind of the lakes.

During summer the opposite situation occurs. The downwind areas remain cooler and more humid than their continental position would indicate, as the lakes store energy and allow for a higher percentage of *latent heat* at the expense of

sensible heat. Similar to any other large body of water, the lakes also delay the onset of the seasons. Fruit trees and grape vines flourish adjacent to the lakes because the delayed onset of spring protects the fruits from early budding, which would leave them more susceptible to damage from late spring frosts.

The Great Lakes also cause significant precipitation modifications. *Lake effect snows* occur leeward of the lakes in autumn through spring, as the still-warm lake waters evaporate into A and cP air masses being advected over the region from the northwest. This induces rapid cloud and precipitation formation processes that cause significant snowfall along the leeward shores. Many leeward regions receive nearly continuous snow and cloud coverage for long periods during the cold season. These phenomena frequently occur with the trailing cold fronts from Alberta clippers, but they also occur when migratory anticyclones cause northwesterly airflow across the lake surfaces. Lake effect snows alter the local *surface water balance* of near-shore locations as affected regions record some of the highest snowfall totals on the continent. Mean annual snowfall totals typically decrease in the leeward direction away from the lakes by over $1.6 \, \text{cm} \, \text{km}^{-1}$ ($1 \, \text{in} \, \text{mi}^{-1}$). In other instances (especially in association with Lakes Superior and Michigan) snowfall is maximized in snowbelts farther inland because of topographic enhancement. The "snow machine" weakens if the lakes begin to freeze. Because of its shallowness, Lake Erie is more likely than the other lakes to freeze in winter, despite its more southerly location.

Ocean Currents and Land–Water Contrast

The eastern side of North America is affected by the warm *Gulf Stream* that results from the clockwise circulation around the Bermuda-Azores high. In winter, cP or A air masses, particularly those affected by cold air damming, are sometimes positioned adjacent to this warm, moist air near the Atlantic coast. This meeting of cold and warm air with accompanying moisture provides the ingredients necessary for strong northeast-moving storms known as **nor'easters**. These winter storms exert a significant impact on the extreme event climatology of the northeastern United States because they form and intensify very quickly, and they may drop copious amounts of snowfall on a densely populated region. Slight differences in a nor'easter's track can cause major differences in both the amount and the type of precipitation. A track to the east-northeast provides little or no snow on the United States, but a nor'easter with a track that is slightly more south-to-north in orientation may produce more than 90 cm (3 ft) of snow in New England and New York. Widespread areas may undergo changes in precipitation from rain to freezing rain to sleet and finally snowfall, depending on the exact location and trajectory of the storm and its associated fronts. Such a sequence of precipitation types may effectively halt transportation and cause massive power outages through much of the most densely populated region of the continent.

Farther north, the Gulf Stream curves eastward away from the continent, and the cold **Labrador current** replaces the departing Gulf Stream, at times reaching as far south as Maryland. This cold current often chills the warmer air overlying it, resulting in frequent coastal fogs in the Atlantic Provinces of New Brunswick, Nova Scotia, Prince Edward Island, and Newfoundland and in northern New England. This cold water may either stabilize the local atmosphere, trapping the fog near the surface, or produce storminess along the margins between the air over the cold water and the warmer air adjacent to it.

Ocean currents also exert significant climatic effects elsewhere in North America. For example, the **Loop current** branches off from the warm Gulf Stream west of the Florida coast, where it "loops" across a section of the Gulf of Mexico. It then moves into the Atlantic between Florida and Cuba, contributing its energy to the Gulf Stream. One effect of this warm water is the energy that it provides to developing tropical cyclones. The position of the warm water also influences tropical cyclone trajectories, with increased likelihood of intensification and occurrence over the Loop current.

The 2005 Atlantic **hurricane** season provides an excellent example of the influence of the Loop current. Although most Gulf–Atlantic tropical cyclones do not strike the U.S. mainland, with only about 25 percent striking the U.S. Gulf coast and slightly less striking the Atlantic coast, multiple strikes occur in the same season, particularly when the Loop current is advecting warm water into the Gulf of Mexico. Hurricanes Katrina, Rita, Wilma, and others of 2005 show the influence of anomalously high Gulf sea surface temperatures associated with the Loop current during the height of the hurricane season. All three of these storms contained 280 km hr^{-1} (175 mi hr^{-1}) winds at some point in their life cycle.

In winter the usually warm Gulf of Mexico waters sometimes meet with frigid cP air masses at the coastal zone, which can trigger nor'easter-like storms that originate in the northwestern Gulf of Mexico continental shelf. Like nor'easters, these **Gulf Coast cyclones** can be very dangerous, because they are nearly always accompanied by abundant moisture from which latent heat can be released to power the quickly intensifying storm. An example of an intense Gulf low was the March 1993 "Blizzard of the Century" that impacted most of the Gulf and Atlantic coasts. An extremely rapid drop in core pressure occurred over a very warm Gulf of Mexico eddy that spun from the warm Loop current in the eastern part of the basin.

The cold *California current* produced by the flow around the *Hawaiian high* near the U.S. Pacific coast exerts a more benign influence on the North American climate. This current stabilizes the surface atmosphere by reducing the *environmental lapse rate*, often even creating a surface temperature inversion. The result is a low frequency of severe weather on the Pacific coast, increased frequencies of fog as moisture is trapped near the surface, and increased air quality problems in heavily populated areas such as Los Angeles.

■ Climatic Setting of Europe

General Characteristics

Europe's climate may be described as milder than it should be, given the relatively high latitude of the continent. Latitudinally, the bulk of Europe exists near or poleward of the U.S.–Canadian border (**Figure 9.4**). Yet, by comparison, most of Canada experiences a harsher climate than most of Europe. For example, the southernmost region of Europe (southern Spain, Italy, and Greece) lies at a latitude similar to New York City. Images of these southern European locales usually include warm, sunny days. This is hardly an image that

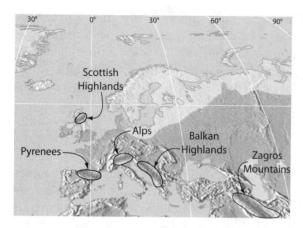

Figure 9.4 Physiographic features of Europe.

one would have of New York's climate, especially during winter.

Europe is generally not an ideal source region for very cold air masses, except for the inland areas of Russia and other eastern European countries in winter. Instead, *maritime polar (mP) air masses* of Atlantic origin are advected eastward in the midlatitude westerlies, south of the *Icelandic low*. These air masses dominate most of Europe, but occasionally mT and cT air masses can penetrate the southernmost sections of the continent. Two primary factors explain the relative mildness of European climates considering its latitude.

Effect of Ocean Currents

The first reason that Europe's climate is relatively mild involves the warm *North Atlantic Drift*—a warm surface ocean current that flows across the North Atlantic basin to the Arctic Ocean north of Scandinavia and into the Barents Sea as an extension of the Gulf Stream. As the Gulf Stream travels northward along the U.S. Atlantic coast, it slowly curves away from North America toward the mid-ocean basin. In the mid-ocean region the Gulf Stream is renamed the North Atlantic Drift. The bulk of the water in the North Atlantic Drift originates in the low-latitude *North Equatorial current*. Most of the remainder originates from the warm Caribbean Sea and the Gulf of Mexico via the Loop current.

The North Atlantic Drift is the northern extension of the circulation *gyre* that occupies the North Atlantic basin. This gyre owes its existence to the Bermuda-Azores high that occupies the central North Atlantic basin year-round. As we have seen,

the anticyclone waxes and wanes seasonally, reaching a maximum intensity in the summer months during the maximum *Hadley cell* intensity and a minimum during the late winter, when the Hadley cell weakens. The position of the Bermuda-Azores high fluctuates seasonally as well, reaching its maximum poleward position during the summer months and its most equatorial location during winter, when the Hadley cell weakens. The clockwise rotation of the ever-present Bermuda-Azores anticyclone imparts considerable shearing force on the waters of the North Atlantic. Warm waters originating in low latitudes flow poleward along the western ocean basin edge near North America, while cooler waters flow equatorward along the eastern basin edge along the coast of Africa, creating the *Canary current*. Because Europe lies poleward of the mean central gyre position, the coldest waters of the Canary current have minimal effect on Europe.

The North Atlantic Drift splits into two distinct currents near the southern coast of England. One part of the current flows southward along the coasts of France, Spain, and Portugal. This water is eventually incorporated into the cooler Canary current. The remaining portion of the North Atlantic Drift flows northward along the west coast of England and Norway, eventually flowing eastward north of the Scandinavian countries and into the Arctic Ocean. The waters are so warm that the area remains ice free even during the coldest winters. Murmansk, which lies poleward of the Arctic Circle, is Russia's only all-season port.

Effect of Mountain Ranges

The second major reason for Europe's mild climate is the position of its mountain ranges (Figure 9.4). The major mountain system of Europe, the Alps, extends along a west-to-east transect. This is a different alignment from the north–south oriented mountain axes of North America. In contrast to North America, European warm air masses originating over the Mediterranean basin are largely confined to the southern regions of the continent. Colder air, originating in higher-latitude polar locations, is primarily confined to the northern regions of the continent, but this air is generally not as frigid as air at comparable latitudes in North America, because of the oceanic influence in Europe. This segregation of dissimilar air masses keeps Europe relatively warm, especially in the southern sections of the continent. The mountains also have a direct effect on daily weather, especially storms. In contrast to the situation over central North America, European storms are rarely severe, because the mixing of radically dissimilar air masses is almost nonexistent, largely because of the mountain barrier. This is not to say that severe weather does not occur in Europe; rather, the frequency and magnitude of severe weather events are significantly reduced as compared with the volatile North American continent.

The Alps create a regional circulation known as **Föehn winds**, which are very similar to the Chinooks of North America. The Föehns are initiated when air traveling up the windward side of the Alps undergoes adiabatic cooling, usually to the *dewpoint temperature*, causing clouds and precipitation. As the air stream traverses the mountain region and begins to move down the leeward side, adiabatic warming occurs. The warming rate usually exceeds the adiabatic cooling rate on the opposite side of the mountain range, causing a dry, warm blast of air on the leeward side. During winter the Föehn winds may lead to remarkable short-term temperature increases. Although the Föehns are important regionally, other European mountain ranges are important for the local winds they generate. These are discussed later in the chapter.

Blocking Anticyclones

Both oceanic and topographic influences play a part in a relatively frequent European phenomenon known as the **blocking anticyclone**. Such anticyclones at the surface (**Figure 9.5**) or blocking ridges aloft originate primarily as a result of the persistent strengthening of the air–sea gyre. Atmospheric "blocks" form from the stalling of Rossby wave ridges over the North Atlantic Ocean and/or Europe. During winter and spring such a situation may occur with the migration of polar outbreaks from the east over northern Europe, in the zone of polar easterly circulation north of the midlatitudes. During warmer times the situation may develop as the Bermuda-Azores high-pressure cell expands over the North Atlantic and the continent. In either scenario the east-to-west aligned mountains help to maintain the position of the anticyclone.

During blocking events the continent is affected by largely cloudless skies and little chance of pre-

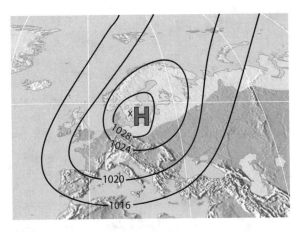

Figure 9.5   A blocking anticyclone at the surface over Europe.

cipitation. Approaching midlatitude wave cyclones either stall on the westward flanks of the blocking anticyclone or are forced to skirt the periphery of the blocking system. As a result, storm tracks are displaced either northward or southward of central Europe. Drought, heat waves, cold waves, and severe pollution events are common surface features beneath the blocking anticyclone or ridge, depending on the season and location of occurrence. Subsiding air associated with the blocking high decreases cloud formation by restricting vertical motions in the atmosphere, leading to drought. Extreme cold may result in winter as Arctic air remains stationary over a location. During warmer times the same situation can lead to a heat wave as the clear skies allow extreme amounts of *insolation* reaching the surface while clouds and precipitation are suppressed. Subsidence may also trap pollutants near the surface. These situations periodically but profoundly affect the European climate.

■ Climatic Setting of Asia

General Characteristics

The primary climatic attribute of Asia is continentality. Asia is the largest continent on Earth, spanning 160 degrees of longitude (**Figure 9.6**). The high-latitude interior of the continent is as far removed from major sources of water as can occur on the planet. This continentality promotes climates that vary extensively from season to season. The lowest temperatures recorded in the northern hemisphere have occurred in Asia. One would

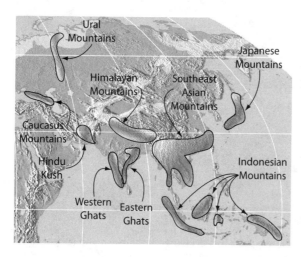

Figure 9.6 Physiographic features of Asia.

probably assume the polar area to be the coldest, but the cold extremes in the oceanic north polar region are not as cold as the most extreme temperature in inland Siberia. Even though the polar region is frozen throughout much of the year, the underlying ocean still imparts a moderating effect on the climate by acting as an energy source. The Asian land mass holds no such energy reservoir, resulting in extreme cold during the winter months. The cold air supports the *Siberian high*—a strong, thermally driven anticyclone. Development of the Siberian high in winter led to the highest surface air pressure ever recorded when on December 31, 1968, a sea-level equivalent pressure of 1083.8 mb was recorded at Agata, Siberia.

Summer months are marked by a strongly opposite circulation as the Asian interior heats appreciably under long days and abundant solar energy receipt. Large water bodies are too distant to store and tie up energy in latent heat, leaving the interior with an abundant sensible heat supply, which is manifested through very high temperatures. The average annual temperature range in a given year across Asia may reach 60 C° (108 F°), the most dramatic continental influence on Earth.

The size of the Asian continent also subjects it to a wide array of climatic types, including every category in the Köppen climate classification system. But overall, the continent is dominated by dry climates, due to the extensive continentality of the interior. *Microthermal* ("D") climates are also widespread at these high latitudes. Tropical climates are found in the southern parts of the continent. Of those, most are associated with the

east and south **Asian monsoons**—the seasonal reversal of continental-scale circulation.

Monsoonal Effects

A **monsoon** is defined as a seasonal reversal of wind that occurs because of a see-saw of atmospheric pressure between the Asian interior and the Indian and North Pacific Oceans. During winter, continentality causes the Asian interior to cool significantly, supporting the Siberian high by chilling the overlying air. Winter is also characterized by relatively low pressure over the Indian Ocean and associated water bodies such as the Arabian Sea and the Bay of Bengal, along with the western tropical Pacific Ocean. This occurs as water bodies retain heat in appreciable quantities. The warmer ocean surfaces heat the overlying atmosphere, thereby supporting lower atmospheric pressures. Because atmospheric mass is transferred from areas of high pressure to areas of lower pressure, surface air in winter circulates from the continental interior toward the tropical oceans even after Coriolis deflection (**Figure 9.7a**). Not surprisingly, this air is typically cold and dry. As long as this subsiding airflow persists, little precipitation occurs over most of Asia.

During summer the pressures see-saw in an opposite manner, as the Asian land mass heats in the absence of major water bodies. This surface heating causes the development of a surface-based **thermal low** called the *Tibetan low*. Conversely, the Indian and North Pacific Oceans are relatively cool in summer compared with the heated continent. The cooler air at the ocean surface supports generally higher atmospheric pressures. The resultant airflow over the continent is from the ocean to the continent even after Coriolis deflection (Figure 9.7b). This flow is manifested as the **southwesterly monsoon** of South Asia. Because this air is warm and moist, it promotes the development of clouds and precipitation over Asia.

Precipitation is further enhanced by orographic uplift over the high plateau that comprises the bulk of the Indian subcontinent and over the Himalayan mountain range. Areas on the windward (or southern, in this case) side of the Himalayas have some of the highest precipitation totals on Earth. Cherrapunji, India, holds the world record for the most precipitation recorded in a single month (929.99 cm, or 366.14 in) and year, when 2646 cm

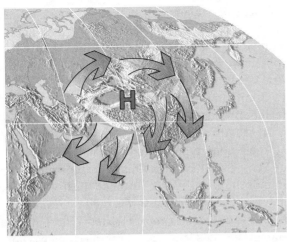

(a)

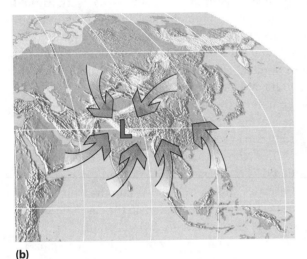

(b)

Figure 9.7　(a) Asian winter monsoon. (b) Asian summer monsoon.

(1042 in) of rain was recorded from 1860 to 1861. This excessive annual precipitation total, which equals 26.5 m (87 ft) of water, is even more impressive in that the vast majority of the precipitation occurred over the course of only about 4 months.

Because the summer phase of the Asian monsoon can produce remarkable precipitation totals, some people assume that the word "monsoon" indicates a type of storm system. Few remember that the monsoon also induces a distinct dry period that spans roughly half the year.

Effect of Mountain Ranges

Another physical characteristic of Asia that imparts a significant climatic influence are the Himalayas and the Hindu Kush (Figure 9.6). We just saw that these mountain chains (particularly the

Himalayas) promote very impressive precipitation totals on the windward side, especially during summer. The mountains also help to create and maintain the very dry interior of Asia through a **rain shadow effect**—the dryness experienced on the leeward side of a mountain, after the moisture and precipitation is left on the windward side, leaving sinking air that warms adiabatically on the leeward side. As air is warmed its **relative humidity** drops, discouraging cloud formation on the leeward side of the mountain barrier. Because winds are persistent throughout the course of the year, the Tibetan Plateau on the northern side of the Himalayas has a **True Desert (BW) climate**. This situation also causes the dry western interior of North America and several other deserts throughout the world. The rain shadow effect is enhanced in Asia by the extreme height of the Himalayas and the size of the land mass, because most of the moisture in air moving over the continent from a surrounding water body will surely be wrung out long before it reaches the continental interior. The Gobi Desert, which dominates the interior of Asia, is one of the driest locations on Earth.

The east-west orientation of the Himalayas and adjacent Tibetan Plateau also causes the *polar front jet stream* to split into two branches in winter—one branch north of the mountains and the other to their south. The northern jet's position fluctuates wildly, but the southern jet generally flows over northern India and then curves northward over southern China and Japan. In summer the southern branch of the jet generally disappears and the onshore southwesterly monsoon tends to dominate regions south of northern China.

Other mountain ranges exert important impacts in other parts of Asia. The mountain spines of the Japanese and Philippine Islands cause significant orographic impacts. The complicated topography of southeast Asia also causes a vast array of microclimates. Finally, the eastern and western Ghats of southern India create important impacts on local climates.

Effect of Coastal Zones on Climate

Some 30 percent of the world's tropical cyclones occur in the western Pacific Basin of the northern hemisphere, making this by far the most active tropical cyclone region in the world. Many of these storms strike the mainland, from southern Japan southward along the Chinese and Vietnamese coast. Others traverse the Philippines and Taiwan.

The two branches of the winter polar front jet merge over central China, where an upper-level trough is favored due to persistent baroclinicity. This trough tends to be anchored in place because of the sharp temperature gradients that exist between the continent and the ocean. The associated trough-to-ridge side of the Rossby wave occurs over coastal China and Japan, making it a favored zone for extratropical cyclogenesis. This is true particularly because of moisture enhancement from the adjacent East China Sea and other bodies of water in addition to the steep thermal gradient between the continent and ocean surfaces. These storms move across the Pacific Basin and occlude near the western coast of North America.

■ Regional Climatology

For the sake of simplicity, our investigation of specific climatic types concentrates on the Köppen system because it is the most widely recognized climatic classification system in use today. Because tropical (A) climates within North America and Asia are discussed in Chapter 10, our discussion of the regional climates of the extratropical northern hemisphere begins with dry (B) climates.

B—Arid Climates

Geographic Extent As discussed, dry climates can be defined as those in which *potential evapotranspiration* routinely exceeds precipitation. Dry climates occupy much of the western United States and a small section of adjacent interior western Canada, along with all of interior Asia. They are far less prevalent in Europe, with the Asian B climates spilling over into extreme southern Russia, the southern Ukraine, and very local isolated pockets in interior Spain (see Figure 8.1). On a global basis, B climates occupy about 26 percent of Earth's land area, the most of any of the five major Köppen categories.

One subcategory of B climates is the drier, or True Desert (BW) climate. In North America the BW climate is centered on the southwestern United States and adjacent northern Mexico. The BW regions of Asia are far more extensive and exist in three major areas: (1) the Gobi Desert of north-central China and south-central Mongolia;

(2) the Aral Sea area, encompassing most of Kazakhstan and parts of Uzbekistan and Turkmenistan; and (3) a large swath of southwestern Asia (popularly known as the Middle East) from the Arabian Peninsula across much of Iran, most of Pakistan, and extreme western India. This latter region extends westward to include nearly all of northern Africa.

The second type of B climate is the semiarid, or **Steppe (BS) climate**. This type generally surrounds the True Desert core for all the BW regions noted above. Steppes are regarded as transition regions—*deficits* are too persistent and long to support forested landscapes, yet the climate is not sufficiently arid to be considered a desert. Annual variability in precipitation and potential evapotranspiration can cause a given location in a BS climate to be sufficiently humid to be classified as a nonarid climate in 1 year but sufficiently arid to be classified as a True Desert in the next year. Long-term averages suggest that these regions are semiarid, but they are by no means consistently semiarid every year.

The North American Steppe is generally bordered by the western mountain ranges such as the Cascades and Sierra Nevadas on the west and the eastern Great Plains region on the east. The Steppe extends from southern Alberta southward to include most of Mexico south of the BW region. In Asia the Steppe is extensive through the southwestern and central areas of the continent. Most of Turkey, Iran, Afghanistan, Mongolia, Turkmenistan, and other central Asian republics fall into this category, as well as a sizable section of northeastern China. These climates occur despite the **summer monsoon** circulation regime, as the mountain chains induce a rain shadow effect across the region.

Forcing Mechanisms The aridity in B climates is caused by only a few climatic forcing mechanisms that act singularly or in some combination. Many dry regions are roughly centered along the 30° line of latitude, a region characterized by subsiding air associated with the subtropical anticyclones on the poleward edge of the Hadley cell. Other desert regions are located on the western portion of continents, near the cold eastern ocean basin currents. Many of the tropical and southern hemisphere deserts to be described in the next chapter fall into this category. This oceanic thermal influence reinforces the subtropical anticyclone through persistent atmospheric stability and associated high-pressure characteristics. Subsidence induced by the anticyclone suppresses cloud formation. Another factor is the rain shadow influence that was described earlier. Finally, the distance to a significant moisture source in continental regions such as central Asia can create B climates.

BW—True Desert The True Desert (BW) portions of extratropical North America and Asia are subdivided into **Subtropical Hot Desert (BWh)** and **Midlatitude Cold Desert (BWk)** in the Köppen classification system. In general, the lower-latitude deserts carry the BWh designation, while the higher-latitude areas are classified as BWk, but this pattern can be disrupted by high elevation or by cooling in subtropical areas from cold ocean currents.

Thermal Characteristics Temperatures in True Desert areas fluctuate considerably on both daily and seasonal time scales. Daytime summer temperatures may be exceedingly high, while nocturnal temperatures may be cool. This wide range is caused by the characteristic low humidity levels, because the primary *greenhouse gas* in the atmosphere—water vapor—is deficient. When few clouds and water vapor molecules are present to absorb or reflect insolation during the daytime hours, most insolation is transmitted to the surface as *direct radiation* and is converted to sensible heat. Because little surface water exists to be evaporated (which ties up energy in latent form), very high temperatures are possible. Sinking, or subsidence, in the subtropical latitudes at the poleward (sinking) side of the Hadley cell reinforces the tendency for hot, dry conditions, because air aloft that sinks usually contains little water vapor and warms adiabatically on its descent. The result is an enhancement of the dry, predominantly clear conditions and an even greater likelihood that incoming radiant energy will reach the surface.

The shortwave energy absorbed by the surface is stored and then reradiated as longwave energy from the surface to the atmosphere. At night, little atmospheric water vapor is available to absorb and reradiate longwave energy back down to the surface. The clear nights allow longwave radiation to escape to space readily, resulting in significant radiational cooling of the surface and the near-surface atmosphere. Extreme diurnal temperature ranges are normal because of this weakened local *greenhouse effect.*

BWh Climates Average summer monthly temperatures may exceed 38°C (100°F) for many Subtropical Hot Desert locations as daily maximum temperatures frequently climb to 44°C (110°F) or more. Winters are cool, with average monthly temperatures in January and February perhaps near 10°C (50°F). Vast differences between summer and winter lead to high annual temperature ranges, but in many locations the diurnal temperature range exceeds the annual range in mean monthly temperatures throughout the year.

Representative examples of BWh climates in the northern hemisphere's extratropics are at Phoenix, Arizona, and Las Vegas, Nevada. **Figures 9.8a** and **9.9a** show climographs for these cities, revealing several features of the temperature climatology. First, insolation is intense in summer. Second, the climograph shows that the diurnal and annual temperature ranges are also high. Figures 9.8b and 9.9b show mean monthly water balance diagrams for Phoenix and Las Vegas. The high radiation receipt causes temperatures and atmospheric demand for water to be so high that large deficits prevail throughout most of the year.

BWk Climates Midlatitude Cold Desert (BWk) climates are more extensive than BWh in the extratropical regions of the northern hemisphere, because high-latitude and/or nocturnal radiational cooling permits annual mean temperatures in many desert regions to dip below the 18°C (64°F) threshold—the boundary between BWh and BWk. Nevertheless, both BW climates share common features of high surface radiation receipt (especially in summer), high potential evapotranspiration, and little cloudiness and precipitation. Although summer temperatures are very similar, winter temperatures are lower in a midlatitude desert than in a subtropical desert. Locations with BWk climates typically experience greater annual temperature ranges than subtropical deserts. The decreased energy (and, therefore, potential evapotranspiration) totals in BWk climates may partially offset the effect of small precipitation totals in local water balances, thereby producing smaller deficits than in comparable BWh climates.

Moisture Characteristics Although precipitation totals are very small, both BWh and BWk climates

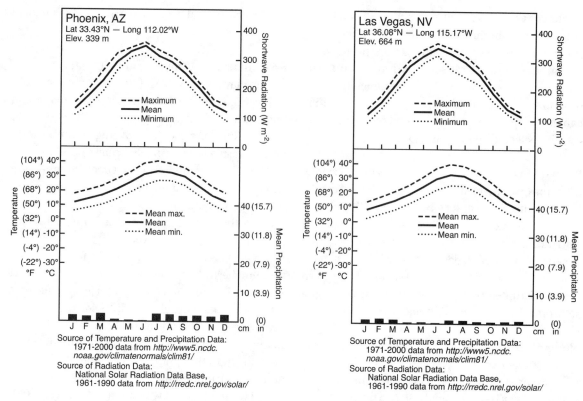

Figure 9.8a Climograph for Phoenix, Arizona (BWh climate). Average annual temperature is 22.7°C (72.8°F) and average annual precipitation is 21.1 cm (8.3 in).

Figure 9.9a Climograph for Las Vegas, Nevada (BWh climate). Average annual temperature is 20.1°C (68.1°F) and average annual precipitation is 11.4 cm (4.5 in).

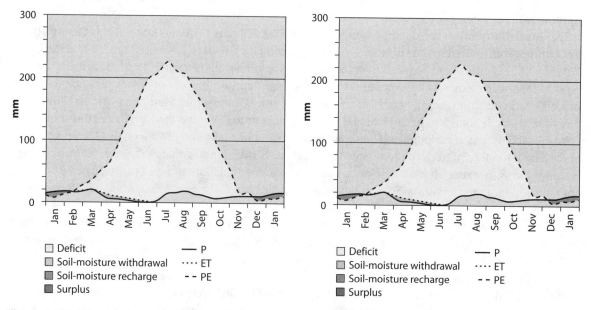

Figure 9.8b Water balance diagram for Phoenix, Arizona.

Figure 9.9b Water balance diagram for Las Vegas, Nevada.

All water balance diagrams in this chapter (part b of Figures 9.8–26, 28–56) are modified with permission using Web/WIMP version 1.01 implemented by K. Matsuura, C. Willmott, and D. Legates at the University of Delaware in 2003.

tend to experience more frequent, abundant, and intense precipitation during the warmest months. At that time of year precipitation is triggered by the extreme heat that increases the environmental lapse rate, thereby destabilizing the atmosphere and promoting vertical motion of surface air. This causes the development of a thermal low, as discussed previously.

In Asia the summer thermal low triggers the impressive monsoonal rains of that continent, but, as we have seen, little moisture is advected as far inland as the Gobi Desert area. Furthermore, the region receives little moisture from migratory midlatitude wave cyclones passing through the region because of the vast distances from upwind moisture sources in the midlatitude westerlies. The climograph for Hami, China (**Figure 9.10a**), reveals low average precipitation totals. Figure 9.10b shows that deficits in Hami are high throughout the year in this dry, desolate location with frigid winters. Similar onshore airflow occurs in association with the other major BWk region in Asia—the Aral Sea region—where precipitation increases slightly in late spring and early summer because of the development of the thermal low.

The North American summertime thermal low is far less extensive and influential than that in Asia. Nevertheless, low-level heating induces rising motion centered over Arizona. This causes the *advection* of surrounding surface air from the Gulf of California and the Gulf of Mexico, carrying some moisture with it. This air sweeps into the region, converges in the area of the thermal low, and rises. The rising motions cause the development of cumulonimbus clouds and some rainfall, giving Phoenix a slight late summer precipitation peak (Figure 9.8a), but the peak is not as distinct as in most BWk regions. The thermal low typically reaches its maximum intensity around August. This precipitation characteristic has caused some to refer to this phenomenon as the **North American summer monsoon.**

The BWk climate in North America is represented by Albuquerque, New Mexico. The climograph (**Figure 9.11a**) again suggests abundant summer radiation, wide diurnal and annual temperature ranges, and low precipitation totals with a peak in late summer. Figure 9.11b shows that the precipitation peak is too small to eliminate moisture deficits. Nevertheless, in contrast to the BWk region in Asia, represented by Hami, the summer

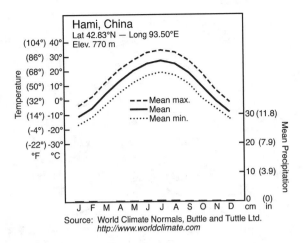

Figure 9.10a Climograph for Hami, China (BWk climate). Average annual temperature is 9.9°C (49.8°F) and average annual precipitation is 3.6 cm (1.4 in).

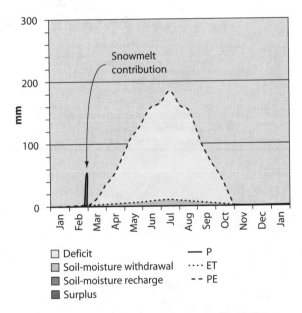

Figure 9.10b Water balance diagram for Hami, China.

precipitation in southwestern North America and cooling effect of high elevation do reduce the deficit despite the copious energy availability for evaporating water in late summer.

Although the North American monsoon region experiences pronounced precipitation seasonality, it differs from a true monsoon, which is characterized by a distinct seasonal reversal of prevailing surface winds. No such situation occurs in the southwestern United States and adjacent northern Mexico. During the cool season, rain and snow are generated by occasional migratory frontal cyclones that sweep through the region. Precipitation totals associated with these storm systems are generally

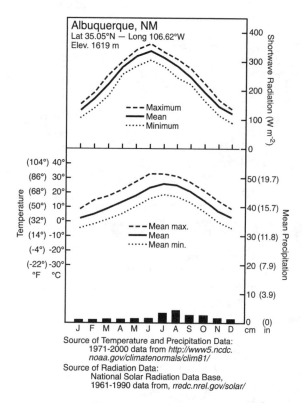

Figure 9.11a Climograph for Albuquerque, New Mexico (BWk climate). Average annual temperature is 13.8°C (56.8°F) and average annual precipitation is 24.1 cm (9.5 in).

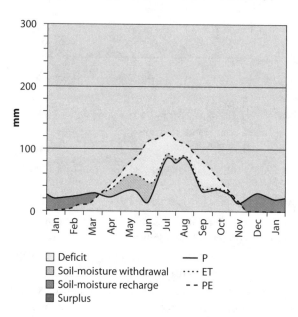

Figure 9.11b Water balance diagram for Albuquerque, New Mexico.

low. Interannual variability of precipitation is high, because many years may pass with virtually no precipitation, but a few other years experience much-above-normal precipitation. Much of this variability is tied to upper-atmospheric flow changes induced by El Niño events. Monthly and

annual precipitation averages may be somewhat misleading and unrepresentative of most years.

BS—Steppe The North American Steppes occur because of the increasing distance from the clockwise advection of moist air by the Bermuda-Azores high and because of a rain shadow produced by Pacific coastal mountain ranges. Steppe regions of the interior of Asia exist primarily because of continentality, because the regions are located far from major water sources such as the North Atlantic Ocean, the Mediterranean Sea, and the Indian Ocean. Mountains further deprive the regions of moisture through rain shadow effects. These eastern Steppe areas are true transition zones adjacent to the harsher and drier Gobi Desert core of central Asia and the deserts of southwest Asia. Iberian Peninsula (Spain) Steppe regions exist predominantly from rain shadow influences induced by the surrounding Pyrenees and Cantabrian Mountains to the north and the Betic Cordillera to the south.

In all these regions, annual precipitation totals—between 25 and 76 cm (10 and 30 in) on average—exceed those of a True Desert. Higher precipitation totals result in the proliferation of Steppe grasses that become more widespread, thicker, and taller with distance from the True Desert core. The grasslands give way to forests at the *ecotone* with the nonarid climatic zones.

Thermal Characteristics Steppe climates are subdivided into **Hot Steppe (BSh) climates** and **Cold Steppe (BSk) climates**, with the boundary again being the 18°C (64°F) annual mean temperature *isotherm*. In the extratropical northern hemisphere, BSh climates are rare. Two examples are the U.S.-Mexican border region (represented by Tucson, Arizona) and much of Iraq (represented by Baghdad, Iraq) (**Figures 9.12a** and **9.13a**, respectively). In addition, some BSh climates are found in sections of southern Iran, Afghanistan, Pakistan, and India, particularly in low elevations. Elsewhere, including the Spanish BS climates, winter temperatures are low enough to qualify extratropical northern hemisphere BS climates as BSk. The BSk climate is prevalent across North America and Asia.

Temperatures in Cold Steppe climates are similar to those in Midlatitude Cold Deserts except that they exhibit less pronounced seasonality. Both annual and diurnal temperature ranges are smaller than in Midlatitude Cold Deserts, as can be seen in the climographs for Lubbock, Texas (**Figure 9.14a**),

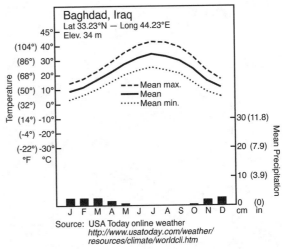

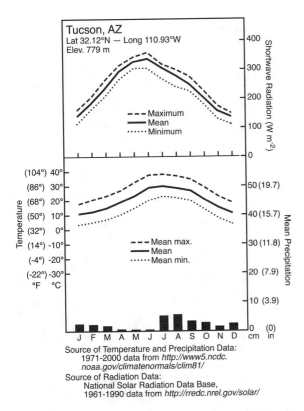

Source: USA Today online weather
http://www.usatoday.com/weather/
resources/climate/worldcli.htm

Figure 9.13a Climograph for Baghdad, Iraq (BSh climate). Average annual temperature is 22.3°C (72.2°F) and average annual precipitation is 15.7 cm (6.2 in).

Source of Temperature and Precipitation Data:
1971-2000 data from http://www5.ncdc.
noaa.gov/climatenormals/clim81/
Source of Radiation Data:
National Solar Radiation Data Base,
1961-1990 data from http://rredc.nrel.gov/solar/

Figure 9.12a Climograph for Tucson, Arizona (BSh climate). Average annual temperature is 20.4°C (68.7°F) and average annual precipitation is 30.9 cm (12.2 in).

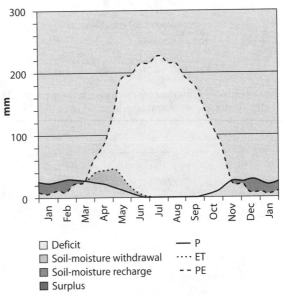

Figure 9.13b Water balance diagram for Baghdad, Iraq.

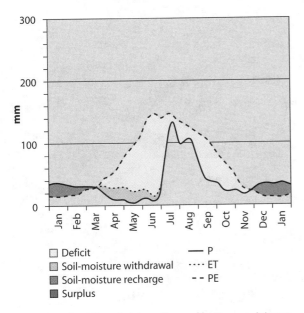

Figure 9.12b Water balance diagram for Tucson, Arizona.

Cedar City, Utah (**Figure 9.15a**), Boise, Idaho (**Figure 9.16a**), Calgary, Alberta (**Figure 9.17a**), and Rapid City, South Dakota (at the ecotone, **Figure 9.18a**). Similar temperature characteristics are evident by comparing the Asian sites of Baghdad (BSh climate, with average annual temperature of

22.3°C [72.2°F], Figure 9.13a), and Tashkent, Uzbekistan (BSk climate, with average annual temperature of 14.7°C [58.4 °C], **Figure 9.19a**).

Moisture Characteristics Precipitation totals typically reach a maximum during winter for Steppe locations at higher latitudes, because of the more frequent passage of migratory midlatitude wave cyclones in winter—Boise and Tashkent provide excellent examples. As the climographs show, the diurnal temperature range increases in summer and decreases in winter, as the clear summer skies allow for much radiation receipt by day and rapid losses at night. On the equatorward side of the True Desert,

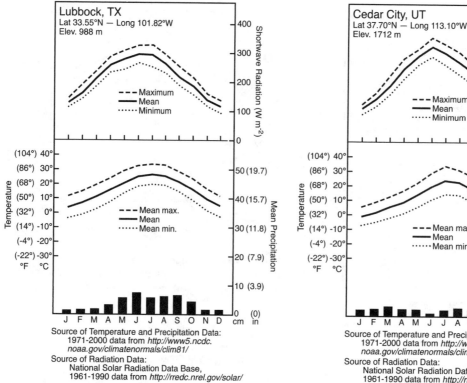

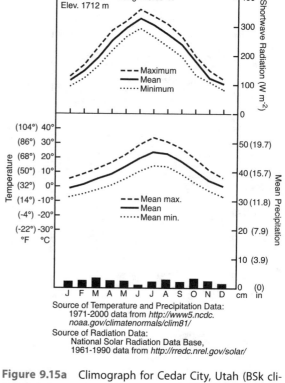

Figure 9.14a Climograph for Lubbock, Texas (BSk climate). Average annual temperature is 15.4°C (59.7°F) and average annual precipitation is 47.5 cm (18.7 in).

Figure 9.15a Climograph for Cedar City, Utah (BSk climate). Average annual temperature is 10.3°C (50.5°F) and average annual precipitation is 29.0 cm (11.4 in).

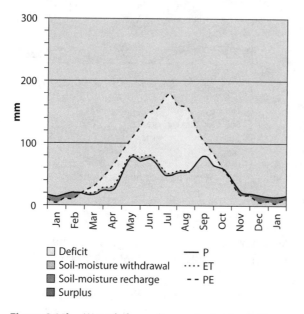

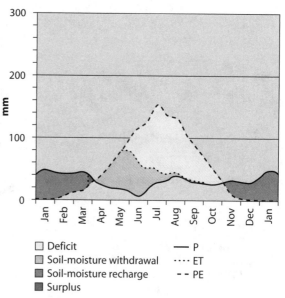

Figure 9.14b Water balance diagram for Lubbock, Texas.

Figure 9.15b Water balance diagram for Cedar City, Utah.

precipitation totals tend to increase during the summer season when convective thunderstorm activity is greater and when the Intertropical Convergence Zone makes its nearest approach to the region. These locations are discussed in more detail in Chapter 10.

The slightly reduced energy levels associated with the lower temperatures in winter generally eliminate the deficits in soil moisture during those months. A comparison of the water balance figures (Figures 9.12b through 9.19b) reveals that lower temperatures in higher-latitude Steppe climates

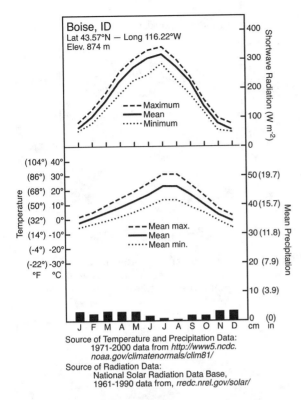

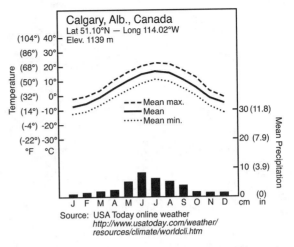

Figure 9.16a Climograph for Boise, Idaho (BSk climate). Average annual temperature is 9.7°C (49.4°F) and average annual precipitation is 31.0 cm (12.2 in).

Figure 9.17a Climograph for Calgary, Alberta (BSk climate). Average annual temperature is 9.7°C (49.4°F) and average annual precipitation is 31.0 cm (12.2 in).

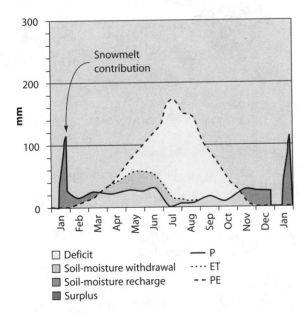

Figure 9.16b Water balance diagram for Boise, Idaho.

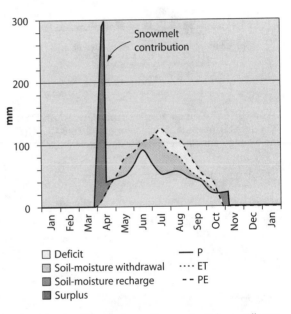

Figure 9.17b Water balance diagram for Calgary, Alberta.

C—Mesothermal Climates

Primarily because they are found in the most heavily populated parts of China and India (see Figure 8.1), the "C" (or *mesothermal*) family of climates is home to more of the world's people than any other of the Köppen categories. These climates are supportive of intense human activity because of their long growing season, mild winters, and moderate to abundant moisture. Four major types of C climates exist: Humid Subtropical, Subtropical Dry Winter, Mediterranean, and Marine West Coast.

Cfa—Humid Subtropical The **Humid Subtropical (Cfa) climates** have long, hot summers, relatively

produce more winter *surplus* than in locations farther equatorward. Although Baghdad experiences only 15.7 cm (6.2 in) of precipitation annually and Tashkent receives only 37.5 cm (14.8 in), the more equatorward site, Baghdad, shows far more deficit and less surplus (Figures 9.13b and 9.19b, respectively).

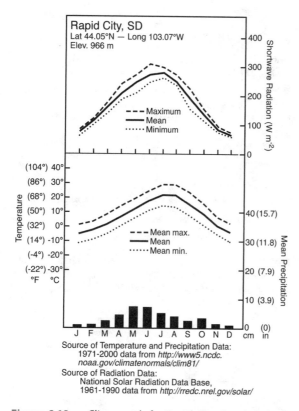

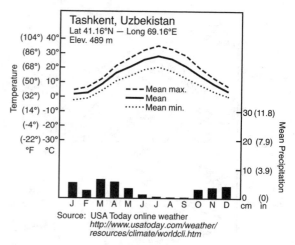

Source: USA Today online weather
*http://www.usatoday.com/weather/
resources/climate/worldcli.htm*

Figure 9.19a Climograph for Tashkent, Uzbekistan (BSk climate). Average annual temperature is 14.7°C (58.4°F) and average annual precipitation is 37.5 cm (14.8 in).

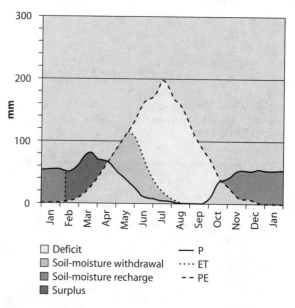

Figure 9.19b Water balance diagram for Tashkent, Uzbekistan.

Figure 9.18a Climograph for Rapid City, South Dakota (BSk climate). Average annual temperature is 8.1°C (46.6°F) and average annual precipitation is 42.3 cm (16.6 in).

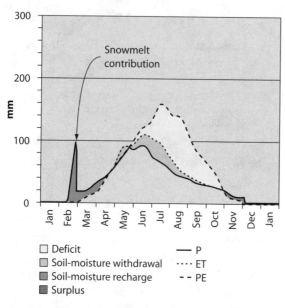

Figure 9.18b Water balance diagram for Rapid City, South Dakota.

mild winters, and abundant precipitation with a relatively even distribution of precipitation across all months. In North America this climatic type occurs through nearly all of the U.S. South, from central Texas to the middle Atlantic coast. Cfa climates

also dominate southeastern Asia, extending from areas of India to the central Chinese coast, including Taiwan and the southern half of Japan. The region is sandwiched between the tropical climates of extreme southeastern Asia and the Highland and more severe midlatitude climates to the north.

Thermal Characteristics In both North America and Asia the Cfa climate is associated with a *maritime effect* from warm ocean (or Gulf of Mexico) waters. This is especially true of North America, where the clockwise flow of the Bermuda-Azores high pumps mT air far inland. Summers are normally hot, long,

Color Plate 2.1   The Orion Nebula.

Color Plate 2.7   The aurora borealis.

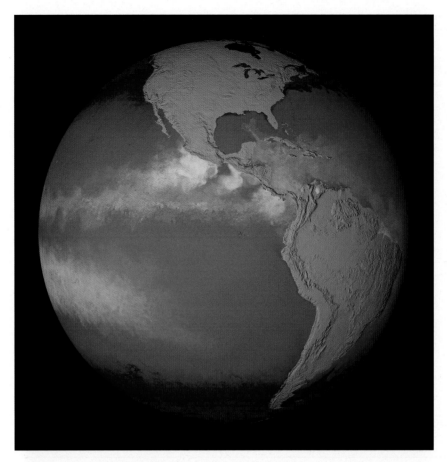

Color Plate 4.3 Cold water upwelling along the South American coast.

Color Plate 4.9 Satellite image of water temperatures in the Pacific Ocean during an El Niño event. White represents unusually warm water.

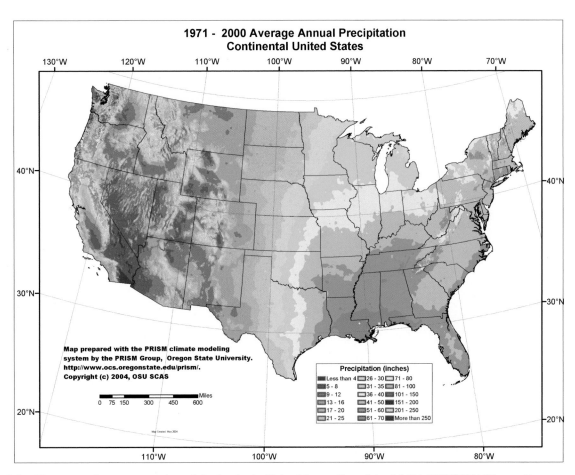

Color Plate 6.6 Average annual precipitation in the United States. Copyright © 2004, OSU PRISM Climate Group.

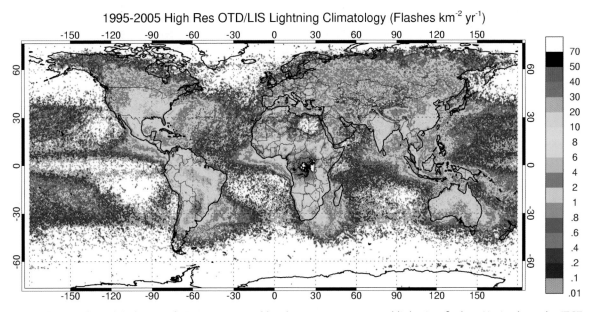

Color Plate 7.5b Global view of ITCZ represented by the most concentrated lightning flashes. Notice how the ITCZ is most prominent at the longitudes experiencing afternoon at a given time. Courtesy of Global Hydrology Resource Center/NASA.

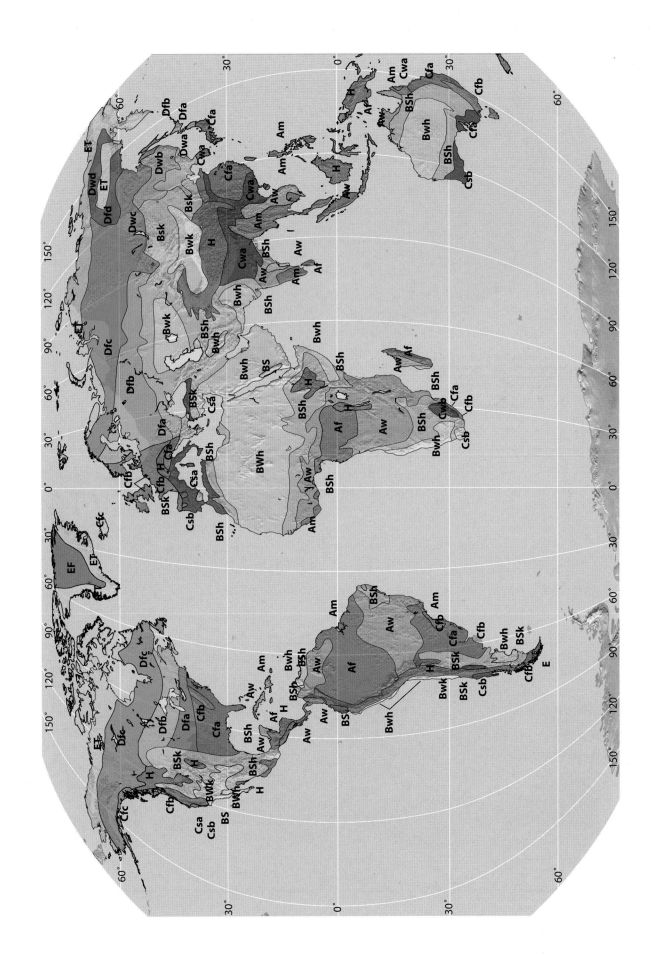

A TROPICAL CLIMATES

Af Tropical rain forest climate

Am Tropical monsoon climate

Aw Tropical savannah climate

B DRY ARID AND SEMIARID CLIMATES

BW Desert climate

BS Steppe climate

h Low-latitude hot

k Midlatitude cold

C MESOTHERMAL CLIMATES

Cfa Humid subtropical, without dry season, hot summers

Cwa Humid subtropical, winter-dry

Cwb

Cfb Marine west coast, without dry season,

Cfc warm to cool summers

Csa Mediterranean summer-dry

Csb

D MICROTHERMAL CLIMATES

Dfa Humid continental, hot summers

Dwa

Dfb Humid continental, warm summers

Dwb

Dfc Subarctic, cool summers

Dwc

Dfd Subarctic, cold winter

Dwd

w Winter dry

f Without a dry season

E POLAR CLIMATES

H HIGHLANDS

ET Tundra climate

EF Ice cap and sheets

H Denotes cold climate due to elevation

Color Plate 8.1 Köppen climatic regions of the world. *Adapted from Goode's World Atlas, 19th ed. Edited by E. B. Espenshade. Skokie, IL: Rand McNally and Co., 1995.*

Moisture Regions in the United States

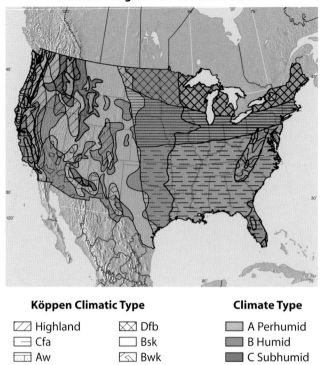

Köppen Climatic Type

		Climate Type
⊿ Highland	⊠ Dfb	▢ A Perhumid
▭ Cfa	▢ Bsk	▢ B Humid
▭ Aw	◹ Bwk	▢ C Subhumid
▨ Cfb	◿ Csa	▢ D Semiarid
▤ Dfa		▢ E Arid

Color Plate 8.3 Overlay of the Köppen and Thornthwaite climatic classication for the United States. Courtesy of Timothy Melancon.

Color Plate 11.4 The effect of a 6-m sea level rise in Florida. Courtesy of Jeremy Weiss and Jonathan Overpeck, University of Arizona.

Color Plate 11.6 A view of the Grand Canyon, showing colorful walls that are the result of differences in the mineral composition of the rock. Many of these contrasts are attributable to the differences in marine versus continental deposition, as oceans covered this region at various times in geologic history.

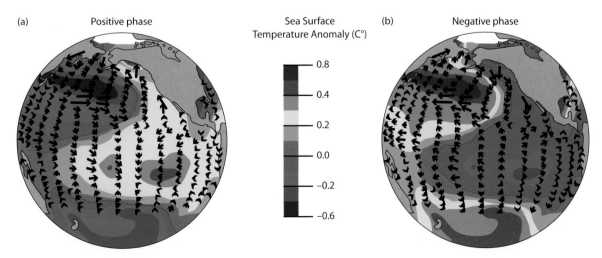

Color Plate 13.5 Comparison of sea surface temperature anomalies during the (a) positive and (b) negative phases of the Pacific Decadal Oscillation. *Modified from:* Nathan Mantua at the University of Washington's Joint Institute for the Study of the Atmosphere and Oceans.

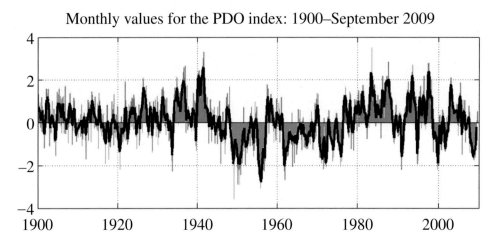

Monthly values for the PDO index: 1900–September 2009

Color Plate 13.6 Time series of the monthly Pacific Decadal Oscillation index between 1900 and 2004. *Modified from:* Nathan Mantua at the University of Washington's Joint Institute for the Study of the Atmosphere and Oceans.

and continuous, featuring average temperatures of perhaps 26–28°C (79–82°F) with high relative humidity values. High humidity in conjunction with high temperatures produces sultry, oppressive weather with a high *heat index.* Daily maximum temperatures typically approach 35°C (95°F) during the summer months, with extreme temperatures near 39°C (102°F) for many locations. Such conditions are discernible in climographs for several locations in North America, including Baton Rouge, Louisiana (**Figure 9.20a**), Little Rock, Arkansas (**Figure 9.21a**), Raleigh-Durham, North Carolina (**Figure 9.22a**), and Richmond, Virginia (**Figure 9.23a**). A representative example in Asia is Tokyo, Japan (**Figure 9.24a**). Notice the moderating effect on coastal sites such as Baton Rouge, Raleigh, Richmond, and Tokyo.

Freezing temperatures are generally not uncommon. The coldest month on average is January, with an average temperature of perhaps 10°C (50°F) in the mildest Cfa regions and only slightly above 0°C (32°F) in the harshest. January is also generally the month of the most frequent freezes, with freezes occurring later in the season nearest to the coasts. Winter in the true sense of the word does exist, but cold weather is interrupted by frequent warmer periods. Extreme cold occurs occasionally.

Moisture Characteristics Precipitation is generally evenly distributed throughout the year, averaging 102 to 152 cm (40 to 60 in) annually. High-magnitude precipitation events are relatively common, with most associated with convective thunderstorms that can occur on most summer afternoons as a result of diurnal heating of the land surface. The water balance diagrams for the North American sites (Figures 9.20b through 9.23b) show relatively small summer deficits, with higher deficits inland, such as at Little Rock. Convective precipitation is sometimes too intense to allow sufficient *infiltration* into the soil. Some Cfa climates are subject to fairly lengthy intervals between summer precipitation events. Thus, wilting vegetation is fairly common even though the monthly precipitation totals may suggest a humid environment. For Tokyo (Figure 9.24b), the summer monsoon prevents climatological deficits from occurring.

Frontal cyclones are the most common precipitation-forcing mechanism during winter and spring as the polar front jet stream reaches its farthest equatorial position during those seasons.

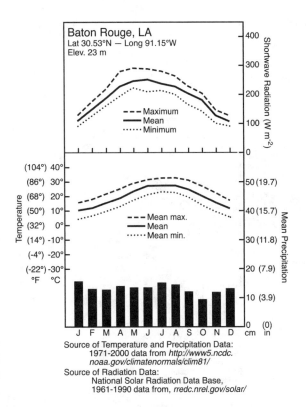

Source of Temperature and Precipitation Data: 1971-2000 data from *http://www5.ncdc. noaa.gov/climatenormals/clim81/*
Source of Radiation Data: National Solar Radiation Data Base, 1961-1990 data from, *rredc.nrel.gov/solar/*

Figure 9.20a Climograph for Baton Rouge, Louisiana (Cfa climate). Average annual temperature is 19.4°C (67.0°F) and average annual precipitation is 160.2 cm (63.1 in).

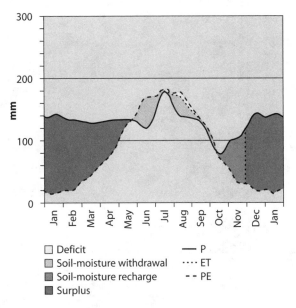

Figure 9.20b Water balance diagram for Baton Rouge, Louisiana.

The frequency of frontal cyclones also increases during those times as the latitudinal thermal gradient is maximized. The steeper thermal gradient triggers greater Rossby wave amplitude and increased storm frequencies and magnitudes to

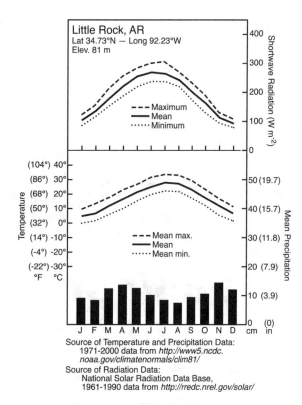

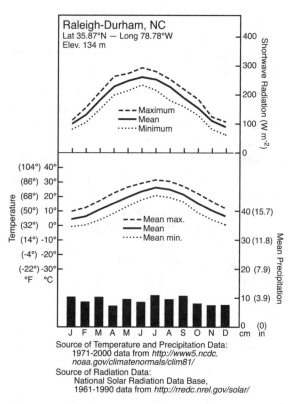

Figure 9.21a Climograph for Little Rock, Arkansas (Cfa climate). Average annual temperature is 16.7°C (62.1°F) and average annual precipitation is 129.4 cm (50.9 in).

Figure 9.22a Climograph for Raleigh-Durham, North Carolina (Cfa climate). Average annual temperature is 15.3°C (59.6°F) and average annual precipitation is 109.3 cm (43.1 in).

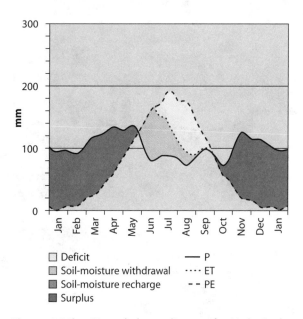

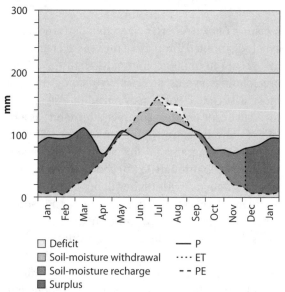

Figure 9.21b Water balance diagram for Little Rock, Arkansas.

Figure 9.22b Water balance diagram for Raleigh, North Carolina.

equalize energy inequalities across the latitudes. Moderate reductions in precipitation usually occur in October and November because those months represent a transition between the two main precipitation-forcing mechanisms: afternoon convection in the warm season and midlatitude

wave cyclones in the cold season. As with tropical climates, the tropical cyclone remains an important component of the summer and autumn season precipitation regime, particularly in east Asia. Although destructive, tropical cyclones have some

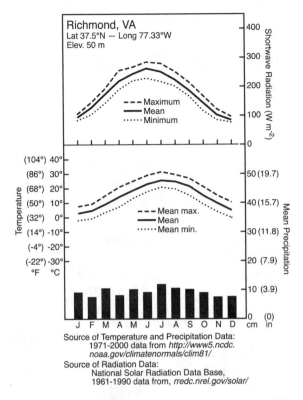

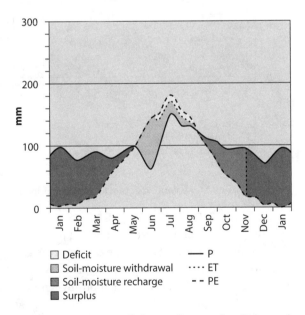

Figure 9.23a Climograph for Richmond, Virginia (Cfa climate). Average annual temperature is 14.2°C (57.6°F) and average annual precipitation is 111.5 cm (43.9 in).

Figure 9.23b Water balance diagram for Richmond, Virginia.

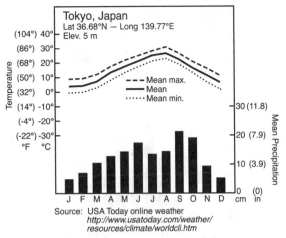

Figure 9.24a Climograph for Tokyo, Japan (Cfa climate). Average annual temperature is 14.5°C (58.1°F) and average annual precipitation is 151.9 cm (59.8 in).

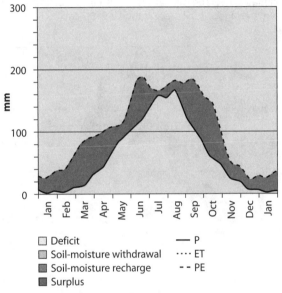

Figure 9.24b Water balance diagram for Tokyo, Japan.

benefit to the regional water balance inland, especially during the somewhat drier autumn season.

Cwa, Cwb—Subtropical Dry Winter In Asia a distinct summer monsoon influence is present. If the monsoon season provides an even more distinct increase in summer precipitation than is seen at locations such as Tokyo, it is classified as a **Humid Subtropical Winter Dry (Cwa) climate.** By definition, in this climatic type the month with the highest mean precipitation has at least 10 times the average precipitation as the driest winter month. Cwa climates are among the wettest on Earth in the summer season, especially if the monsoonal rains are augmented by orographic uplift.

Although Cwa and Cwb climates share a dry winter, the primary difference lies in average summer temperatures. The Cwa climate experiences higher temperatures and is far more common than the Cwb counterpart, which is more typically

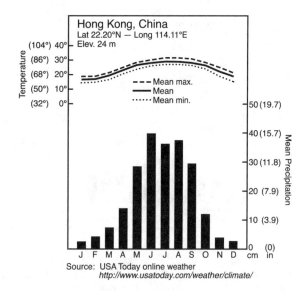

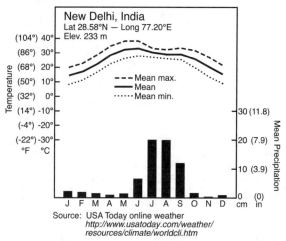

Figure 9.25a Climograph for Hong Kong, China (Cwa climate). Average annual temperature is 23.7°C (74.6°F) and average annual precipitation is 218.7 cm (86.1 in).

Figure 9.26a Climograph for New Delhi, India (Cwa climate). Average annual temperature is 25.0°C (77.1°F) and average annual precipitation is 70.6 cm (27.8 in).

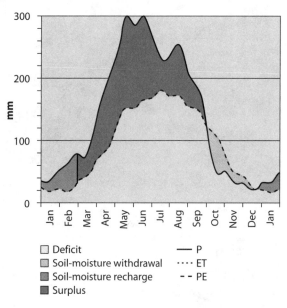

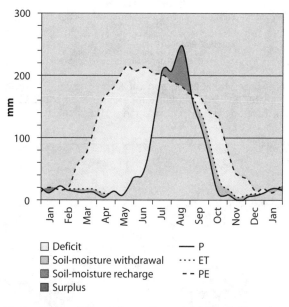

Figure 9.25b Water balance diagram for Hong Kong, China.

Figure 9.26b Water balance diagram for New Delhi, India.

found at higher elevations and where temperature modification by the maritime effect occurs. The Cwa climate is found along an extensive portion of northern India, extending into southeast Asia. Two examples of locations with a Cwa climate are Hong Kong, China (**Figures 9.25a** and 9.25b), and New Delhi, India (**Figures 9.26a** and 9.26b).

Thermal Characteristics   Daily maximum temperatures during winter may average near 21°C (70°F) for Cwa locations, while summer temperatures generally peak between 35°C and 43°C (95°F and

110°F). These high temperatures are usually much higher than those found in the Cwb zones. The climograph of New Delhi (Figure 9.26a) illustrates this concept. The highest average monthly temperatures occur in May and June, just before the onset of the summer monsoonal rains. Cloud cover associated with the monsoon drives down average temperatures during the middle of summer.

Moisture Characteristics   Very impressive seasonal precipitation totals are recorded within some Cwa climate zones. The seasonality of monthly precipitation totals of Cherrapunji, in northeast India, exemplifies the monsoonal moisture pulse within this region, but a strong *orographic effect* on the normal monsoon-influenced precipitation regime makes these totals atypical compared with most monsoonal locations. The climograph for New Delhi is much more representative of a Cwa climate, with annual precipitation totals below 100 cm (40 in). The associated water balance diagram indicates strong soil moisture deficits in the Cwa location during the dry spring.

Csa, Csb—Mediterranean Possibly the most identifiable climate in Europe and likely the best-known climatic type in the world is the **Mediterranean climate**. Like Humid Subtropical climates, temperatures are warm to hot in summer and mild in winter. Unlike Humid Subtropical climates, however, Mediterranean climates are characterized by distinct dryness in summer and a relatively wet winter. The agricultural landscape of locations with Csa and Csb climates is distinct, because only certain types of crops can thrive in such an environment of extreme aridity in the warm season. Any other crops grown must bring a sufficiently high price to make irrigation feasible. Perhaps the crop that best indicates the presence of a Mediterranean climate is the olive, but others such as artichokes, figs, and certain varieties of grape are also well suited to this distinctive climate. As a result the customary food and drink in cultures of the regions having Mediterranean climates include olive oil, artichokes, figs, and grapes for wine production.

This climate is found in southern Spain, southern France, Italy, and Greece, and the remainder of the crescent around the Mediterranean Sea (see Figure 8.1). It is also found along the western edges of continents in the lower midlatitudes. Mediterranean climates in North America are limited to the southwestern coast of the United States, exclusively within California. Asia has no regions that meet the climate classification criteria for Mediterranean type climates except for the coastal zone of Turkey, Syria, Lebanon, and Israel.

Two associated features encourage sinking atmospheric motions on the eastern sides of the subtropical anticyclones (and, therefore, the western sides of continents) where Mediterranean climates are found. First, the adjacent cold surface ocean currents stabilize the atmosphere, and perhaps even generate a temperature inversion. By contrast, the warm currents on the western sides of the subtropical anticyclones (east sides of conti-

nents) destabilize and energize storm systems associated with the Humid Subtropical climates.

Second, sinking motion at subtropical latitudes is more prevalent on the west coasts of continents than east coasts because the eastern sides of the subtropical anticyclones support surface divergence, which draws in subsiding air from aloft. By contrast, the western sides of the subtropical anticyclones support surface convergence, which augments rising motion. In other words, the effect of the subtropical high tilts with height, such that sinking motion occurs down to the surface on the eastern sides (western sides of continents), but only occurs aloft (in fulfillment of the tendency for sinking motion by the general circulation) on the western sides of the anticyclones.

To see why this is the case, we can invoke the **gradient wind equation**, which specifies motion in curved flow, such as around the subtropical anticyclones. The wind speed in gradient flow (V_G) relates to the *geostrophic wind* (V_g) (i.e., flow with no curved *isohypses*) as

$$V_G = \frac{2V_g}{1 + \sqrt{1 + \dfrac{4V_g}{fr}}}$$

where f is the *Coriolis parameter* ($2\Omega\sin\varphi$), and r is the radius of curvature of the flow. Smaller r values represent "tighter-turning" flow. Notice from the gradient wind equation that if r is infinite (i.e., if it has no curvature at all), $V_G = V_g$. Positive values of r are used for counterclockwise flow and negative r values represent clockwise flow.

The gradient wind equation suggests that for a negative r (because air moves clockwise around the subtropical anticyclone in the northern hemisphere), the term inside the radical must be less than one, so the overall denominator in the equation will be less than two. Because 2 divided by a number less than 2 is greater than 1, V_G is bigger than V_g for a given pressure gradient, for clockwise (anticyclonic in the northern hemisphere) flow. As we saw in Chapter 7, air flows relatively fast around ridges (which have clockwise flow in the northern hemisphere), compared with the flow around troughs, assuming a constant pressure gradient.

Furthermore, for a given pressure gradient at the poleward side of the subtropical anticyclone, f is relatively large. This causes the $4V_g/fr$ term to be nearer to zero than at the equatorward side (but negative), which means that the term inside the radical will be 1 minus a very small number, and the entire denominator will be almost 2. Thus, at the poleward side of the high, V_G is only slightly larger than V_g. But at the equatorward side, where f is a negative number near zero, the $4V_g/fr$ term is more negative. This causes the term inside the radical to be 1 minus a bigger number, and the entire denominator will be much lower than 2. Thus, at the equatorward side of the high, V_G is much larger than V_g. So the gradient wind is larger as air flows around the equatorward side of the high and smaller as it moves clockwise around the polar side of the high. This difference in gradient wind at the surface causes divergence on the eastern sides of the subtropical anticyclones (western sides of continents) and a stable atmosphere and surface convergence on the western side (**Figure 9.27**).

Thermal Characteristics   The warmer subcategory of Cs climates is the **Mediterranean Hot Summer (Csa) climate**, which has the warmest month with an average temperature exceeding 22°C (72°F). The **Mediterranean Warm Summer (Csb) climate** is characterized by a warmest month averaging below 22°C (72°F). In general, these milder summer temperatures occur near the coast and at higher elevations, with Csa climates becoming more dominant farther inland. Summer coastal temperatures are typically in the range of 20°C (70–80°F), while inland locations may have daily highs near 38°C (100°F).

Radiation and temperature seasonality in Mediterranean climates in North America are repre-sented by three California sites: Long Beach (**Figure 9.28a**), where summer temperatures are just high enough to be classified as Csa; Sacramento (**Figure 9.29a**), a warmer inland location; and the Csb coastal site of Santa Maria (**Figure 9.30a**).

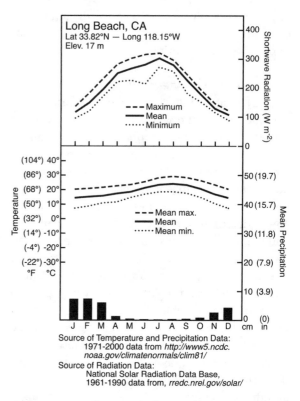

Source of Temperature and Precipitation Data:
    1971-2000 data from *http://www5.ncdc.*
    noaa.gov/climatenormals/clim81/
Source of Radiation Data:
    National Solar Radiation Data Base,
    1961-1990 data from, *rredc.nrel.gov/solar/*

Figure 9.28a   Climograph for Long Beach, California (Csa climate). Average annual temperature is 18.5°C (65.3°F) and average annual precipitation is 32.9 cm (12.9 in).

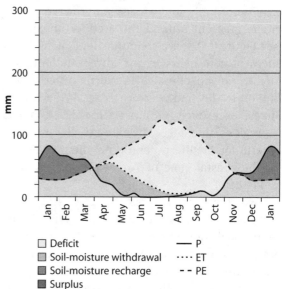

Figure 9.28b   Water balance diagram for Long Beach, California.

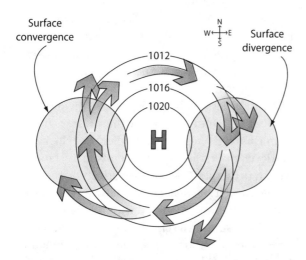

Figure 9.27   Surface convergence and divergence associated with flow around a subtropical anticyclone.

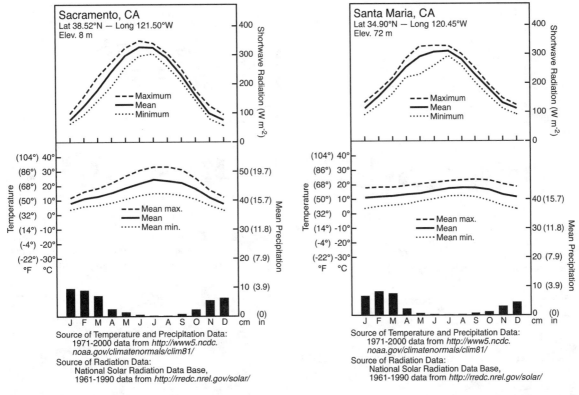

Figure 9.29a Climograph for Sacramento, California (Csa climate). Average annual temperature is 16.2°C (61.1°F) and average annual precipitation is 45.5 cm (17.9 in).

Figure 9.30a Climograph for Santa Maria, California (Csb climate). Average annual temperature is 14.3°C (57.7°F) and average annual precipitation is 35.6 cm (14.0 in).

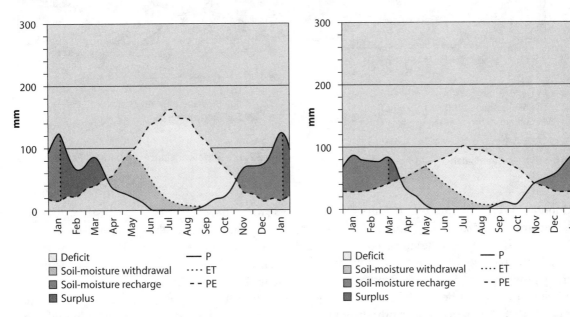

Figure 9.29b Water balance diagram for Sacramento, California.

Figure 9.30b Water balance diagram for Santa Maria, California.

Summer temperatures in Mediterranean Europe are moderated by the maritime effect as inlets of the Mediterranean Sea give a large surface area of maritime-influenced lands. Representative examples of Mediterranean climates in Europe are shown by Rome, Italy, with a Csa climate (**Figure 9.31a**), and Faro, Portugal, with a Csb climate (**Figure 9.32a**).

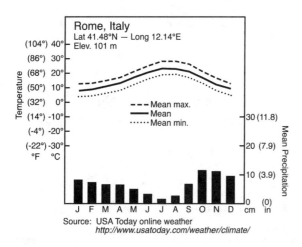

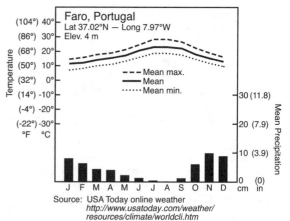

Figure 9.31a   Climograph for Rome, Italy (Csa climate). Average annual temperature is 15.5°C (60.0°F) and average annual precipitation is 80.3 cm (31.6 in).

Figure 9.32a   Climograph for Faro, Portugal (Csb climate). Average annual temperature is 17.4°C (63.3°F) and average annual precipitation is 52.8 cm (20.8 in).

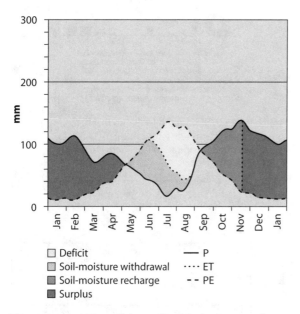

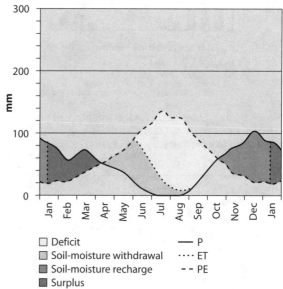

Figure 9.31b   Water balance diagram for Rome, Italy.

Figure 9.32b   Water balance diagram for Faro, Portugal.

Monthly average winter temperatures in most Mediterranean climates may approximate 10°C (50°F), while summer averages may approach 27°C (80°F) with daytime highs significantly higher. During the winter months inland temperatures may dip below freezing, while coastal locations, moderated by the relative ocean warmth, have mean temperatures that remain in the 10–16°C (50–61°F) range. Fogs are common, especially in winter, as air moving onshore is chilled to its dewpoint temperature by the cold ocean currents. These **advection fogs** generally increase the minimum temperatures in the early morning hours but delay the warming of the surface as morning progresses.

Moisture Characteristics   The onset of the summer drought occurs as the subtropical anticyclones expand and move poleward, along with the Hadley cells. These subtropical anticyclones then approach the midlatitude west coasts of continents. Along with the cold ocean currents, the subsidence side of the anticyclone suppresses rising motion that might otherwise support precipitation-producing mechanisms.

The summer dryness in North America is induced by the presence of the semipermanent subtropical anticyclone that dominates the central North Pacific Ocean. During the warm season this Hawaiian high becomes particularly well established, especially along the eastern edge of the ocean basin where the cold California current helps to stabilize the overlying atmosphere. During winter the anticyclone weakens and migrates equatorward, allowing for the passage of migratory frontal cyclones embedded in the polar front jet stream flow. The frequency and magnitude of these storm systems are variable. Some winters may be relatively dry, while others produce copious precipitation totals. In general, precipitation increases with latitude in both the Csa and the Csb types because poleward areas are more affected by midlatitude wave cyclones and their trailing cold fronts that traverse the region.

In the case of Europe, high pressure builds over the Mediterranean Sea as summer approaches. The strength of the eastern edge of this anticyclone is enhanced by thermal differences between the warmer land regions and the cooler Mediterranean waters. In this area the high may be viewed as being an extension of the Bermuda-Azores high that dominates the North Atlantic basin during summer. The polar front jet stream, which follows the upper atmospheric boundary between cold air and much warmer air, typically induces much of the precipitation falling on Europe and other midlatitude locations. It retreats north of the Alps in summer as the *polar cell* and *circumpolar vortex* contract and the subtropical anticyclone expands. As a result, summer precipitation caused by midlatitude wave cyclones embedded within the jet stream is confined to northern and central Europe. Occasional rainfall does occur in Mediterranean Europe as a result of surface heat-induced convective thunderstorms, and these regions are generally wetter in summer than their North American counterparts. Likewise, a comparison of the water balance diagrams for the five cities (Figures 9.28b through 9.32b) suggests that deficits are somewhat smaller and less prominent in Europe's Mediterranean climates than in those of North America.

During winter, cold polar air builds in higher latitudes and expands toward lower latitudes. In Europe the polar front jet expands southward, ultimately positioning itself south of the Alps and over the Mediterranean basin. In these instances cyclogenesis is common over the warm waters of the Mediterranean Sea, as the warm surface destabilizes the overlying atmosphere. Midlatitude wave cyclones provide the area with rainfall throughout the winter. Even with these precipitating mechanisms in place, however, annual rainfall amounts are relatively low, ranging from 25 to 89 cm (10 to 35 in).

Cfb, Cfc—Marine West Coast   The **Marine West Coast climate** occupies the coastal strip of North America north of Mediterranean climates along the Pacific coast, from northern California through Alaska (see Figure 8.1). The maritime effect does not extend to the leeward sides of the coastal mountains, however, so the **Marine West Coast Warm Summer (Cfb) climate** and the **Marine West Coast Cool Summer (Cfc) climate** occupy a relatively narrow zone in North America. The climatic type is found prominently throughout western Europe, from northern Spain through France, England, Norway, and across the northern tier of Europe. The European Cfb climatic type is extensive because of the prevalence of the warm North Atlantic Drift and the east–west alignment of the Alps, which allows for the penetration of mP air deep into the continental interior via the global midlatitude westerlies, pushed eastward on the southern side of the Icelandic low. Winds and storms that traverse this region from west to east carry the ocean-moderated winds far inland.

Thermal Characteristics   As mentioned above, Marine West Coast temperatures in both North America and Europe are mild considering the latitude of the regions, as the moderating effects of the sea are transported inland. In Europe, in addition to the North Atlantic Drift, the North Sea and Baltic Sea also reinforce the moderating influence of this climatic type through much of Europe. As "C" climates, they are characterized by average monthly temperatures above freezing for all months. Daily and annual temperature ranges are also suppressed by persistent cloud cover virtually year-round.

Mean monthly temperatures in the warmest month rarely exceed 24°C (75°F), while annual averages are typically between 8 and 13°C (46 and 55°F). Monthly temperature averages during winter remain above 0°C (32°F), and daily winter minima rarely dip below −7°C (20°F). Extreme temperatures do occur, but they are rare. North America's Marine West Coast Warm Summer region is represented by Seattle (**Figure 9.33a**), while Manchester, England (**Figure 9.34a**), and Berlin, Germany (**Figure 9.35a**), represent the European Cfb climate.

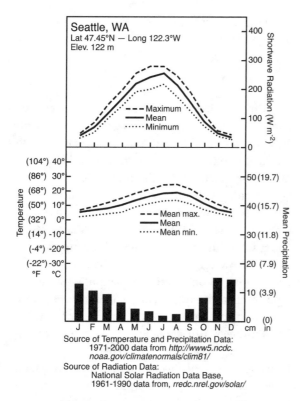

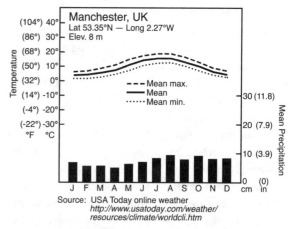

Source: USA Today online weather
http://www.usatoday.com/weather/
resources/climate/worldcli.htm

Figure 9.34a Climograph for Manchester, England (Cfb climate). Average annual temperature is 9.5°C (49.1°F) and average annual precipitation is 89.9 cm (35.4 in).

Figure 9.33a Climograph for Seattle, Washington (Cfb climate). Average annual temperature is 11.3°C (52.3°F) and average annual precipitation is 94.2 cm (37.1 in).

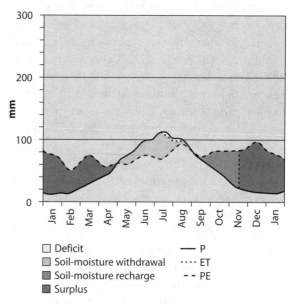

Figure 9.34b Water balance diagram for Manchester, England.

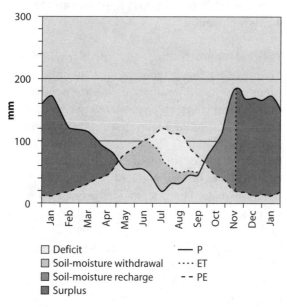

Figure 9.33b Water balance diagram for Seattle, Washington.

Increasing continentality can be seen in the Berlin climograph, as compared with that of Manchester, as the two cities are at nearly the same latitude. Berlin falls just inside the transition zone between the "C" and "D" climatic boundaries. Manchester is especially influenced by the effects of the ocean both in the annual cycle and in the daily cycle, as the temperature range remains small.

The Cfc climate is distinguished from Cfb by its cooler summer, with only 1 to 3 months having average temperatures above 10°C (50°F). This climatic type is very rare globally, because locations with so few warm months are likely to have at least 1 month with a temperature below freezing and, therefore, fall into the "D" climate regime. Southern Iceland is the largest continuous land mass

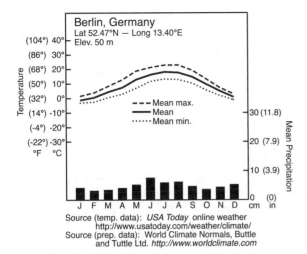

Source (temp. data): *USA Today* online weather
http://www.usatoday.com/weather/climate/
Source (prep. data): World Climate Normals, Buttle
and Tuttle Ltd. *http://www.worldclimate.com*

Figure 9.35a Climograph for Berlin, Germany (Cfb climate). Average annual temperature is 8.8°C (47.9°F) and average annual precipitation is 58.4 cm (23.0 in).

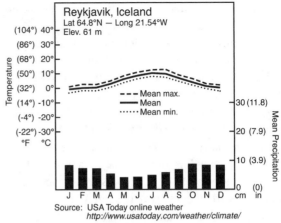

Source: *USA Today* online weather
http://www.usatoday.com/weather/climate/

Figure 9.36a Climograph for Reykjavik, Iceland (Cfc climate). Average annual temperature is 4.4°C (39.8°F) and average annual precipitation is 82.0 cm (32.3 in).

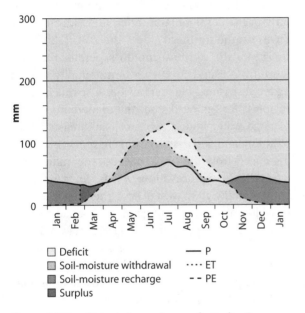

Figure 9.35b Water balance diagram for Berlin, Germany.

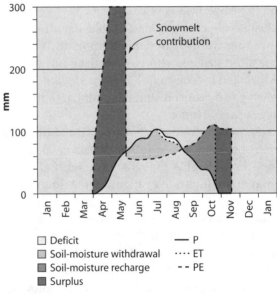

Figure 9.36b Water balance diagram for Reykjavik, Iceland.

containing Marine West Coast Cool Summer climates on Earth, and Reykjavik's climograph is representative of the climatic type (**Figure 9.36a**).

Moisture Characteristics As is shown in Figures 9.33a through 9.36a, precipitation is evenly distributed through all months of the year in Cfb and Cfc climates. Midlatitude wave cyclones embedded within the polar front jet stream traverse the region nearly year-round and produce precipitation at least every few days. Annual precipitation typically ranges between 75 and 125 cm (30 and 50 in) for

most locations, but much higher totals may be common in highland areas.

Many assume that these climates receive higher precipitation totals, but the typically light intensity of precipitation maintains surprisingly low annual totals. Vertical motions are usually forced by a frontal boundary and/or orographic effect without being accompanied by strong instability. Furthermore, the contrast between air masses is generally too weak to support vigorous uplift and intense precipitation. The result is abundant, horizontally oriented *stratiform clouds* and light precipitation, and less vertically oriented *cumuliform clouds* and severe weather. Marine West Coast climates have a high percentage of days with precipitation. Maritime polar air masses originating over the North Pacific and North Atlantic Oceans are generally cool and wet, and they are associated with frontal cyclones that undergo **west-coast occlusion** over western North America and Europe. As we saw in Chapter 8, **occlusions** represent the decaying phase of a midlatitude wave cyclone's life cycle. A west-coast occlusion occurs with general warm front characteristics (gentle uplift of moist cool air, stratiform clouds, and drizzle) induced by the movement of maritime air intruding upon continental air as the system moves inland. This type of occlusion is different from an **east-coast occlusion**, which occurs on the eastern sides of continents. East-coast occlusions are characterized by cold, dry air masses invading warmer, wetter maritime air. In east-coast occlusions, stronger vertical lift occurs, which can induce cumulonimbus cloud development and heavier precipitation, similar to the passage of a cold front.

Snow is uncommon in many Cfb and Cfc locations but common in interior areas or locations with considerable topographical relief. Farther inland, in higher elevations, more impressive annual precipitation totals may occur with local orographic enrichment, with some locations exceeding 254 cm (100 in), similar to tropical rain forest totals. These precipitation totals support the coniferous forests of the Pacific Northwest region. In North America a slight autumn maximum occurs, while a minimum occurs during the summer months. In general, though, there is a relatively even distribution of monthly precipitation. The associated water balance diagrams for the Cfb and Cfc sites also reflect the effects of abundant precipitation and low input energy amounts, because only a minimal summer water balance deficit, if any, occurs (see Figure 9.33b through 9.36b).

D—Microthermal Midlatitude Climates

Dfa, Dfb, Dwa, Dwb—Humid Continental The **Humid Continental (Dfa, Dfb) climates** extend through the central to east-central portion of the North American continent between the Great Plains and the Atlantic Ocean south of Lake Superior (see Figure 8.1). They occur north of the Cfa climatic type and east of the BSk region. In Europe the Dfa region is small, but the Dfb region is vast, extending in a wide swath from western Europe (Poland, Slovakia, and the northern Balkan nations) to south-central Asia (central Russia) and also occupying much of the eastern areas of Scandinavia. Small areas of Dfa and Dfb occur in the northern half of Japan and adjacent islands. The climatic regime is similar to the Humid Subtropical climate except that temperatures and precipitation totals are generally lower. Asia also experiences the **Humid Continental Winter Dry (Dwa, Dwb)** forms of these climates along its eastern seaboard.

Thermal Characteristics Summer temperatures are considered "hot" (Dfa, Dwa) in the southern parts of the Humid Continental climates, particularly in the continental interior, and "warm" (Dfb, Dwb) on the northern sides. The 22°C (72°F) isotherm for the warmest month separates the Dfa/Dwa from the Dfb/Dwb. Continentality causes extreme seasonal fluctuations because much of the region is far removed from the moderating effects of large water bodies. Even coastal areas such as New England and eastern China have few moderating effects from the ocean because the midlatitude westerlies carry prevailing winds, influenced by continentality, from west to east. The Great Lakes do provide a bit of moderating influence and a delayed onset of the seasons over the adjacent leeward lands.

In summer when high solar angles prevail, the land heats quickly as sensible heating increases at the expense of latent heating. Low solar angles and the reduction in daylight hours in winter cause little surface heating. The distance from large water bodies upwind augments warming and/or chilling of the land masses and the overlying air, depending on the season. Many locations in this climatic zone have summertime maximum temperatures of 27–29°C (81–84°F), while monthly winter averages dip below −10 to −1°C (14 to 30°F).

Because the Dfa climatic type is so widespread in eastern North America, several examples are shown via climographs. Des Moines, Iowa (**Figure 9.37a**), Columbus, Ohio (**Figure 9.38a**), Erie, Pennsylvania (**Figure 9.39a**), and Hartford, Connecticut (**Figure 9.40a**), all illustrate features of the Dfa climate. The Dfb type is represented by Duluth, Minnesota (**Figure 9.41a**), Caribou, Maine (**Figure 9.42a**), and Ottawa, Ontario (**Figure 9.43a**).

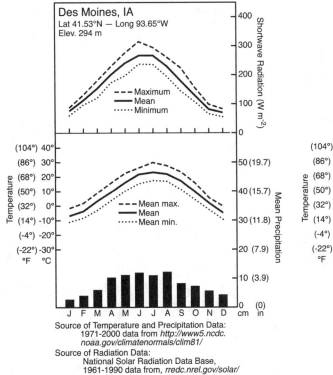

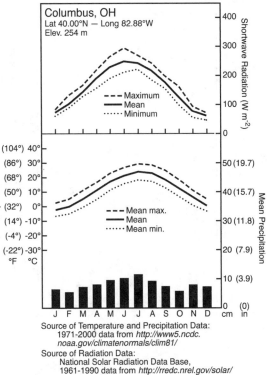

Figure 9.37a Climograph for Des Moines, Iowa (Dfa climate). Average annual temperature is 10.0°C (50.0°F) and average annual precipitation is 88.2 cm (34.7 in).

Figure 9.38a Climograph for Columbus, Ohio (Dfa climate). Average annual temperature is 11.6°C (52.9°F) and average annual precipitation is 97.8 cm (38.5 in).

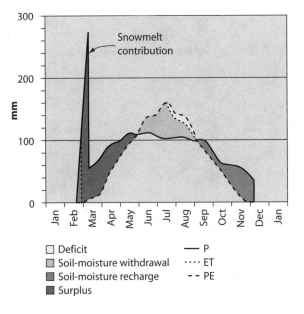

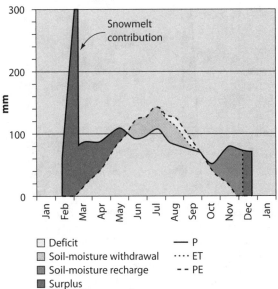

Figure 9.37b Water balance diagram for Des Moines, Iowa.

Figure 9.38b Water balance diagram for Columbus, Ohio.

A special case of Dfb also occurs in the area surrounding Elkins, West Virginia (**Figure 9.44a**), because the high elevation causes winter temperatures to fall below the freezing level and prevents

summer temperatures from reaching an average of 22°C (72°F), despite its relatively low latitude. Notice that except in the case of Elkins, insolation is slightly higher for the Dfa climates than for the Dfb climates. This is particularly true toward the

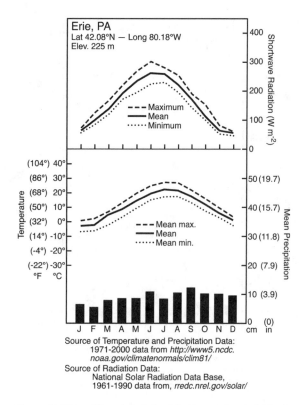

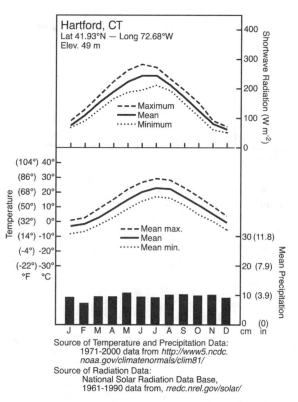

Figure 9.39a Climograph for Erie, Pennsylvania (Dfa climate). Average annual temperature is 10.0°C (50.0°F) and average annual precipitation is 108.6 cm (42.8 in).

Figure 9.40a Climograph for Hartford, CT (Dfa climate). Average annual temperature is 10.1°C (50.2°F) and average annual precipitation is 117.2 cm (46.2 in).

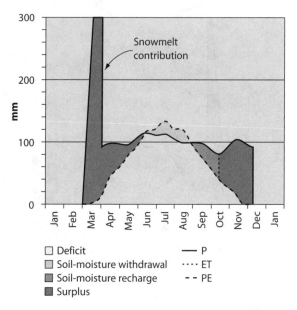

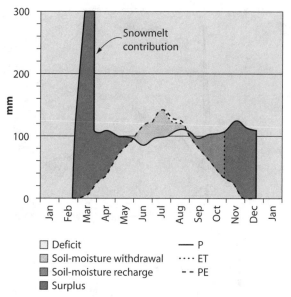

Figure 9.39b Water balance diagram for Erie, Pennsylvania.

Figure 9.40b Water balance diagram for Hartford, Connecticut.

west, where cloud cover is generally less than in the east.

The Eurasian Humid Continental climate is represented by a Dfb climate at Minsk, Belarus

(**Figure 9.45a**), and Moscow, Russia (**Figure 9.46a**). Winter dry versions are represented by Beijing, China's Dwa climate (**Figure 9.47a**), and by a Dwb type at Vladivostok, Russia (**Figure 9.48a**).

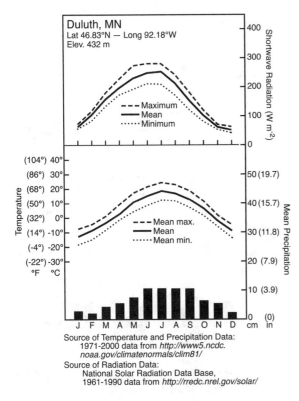

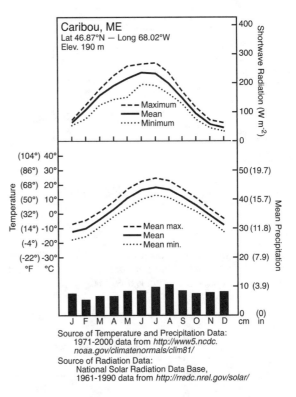

Figure 9.41a Climograph for Duluth, Minnesota (Dfb climate). Average annual temperature is 3.9°C (39.1°F) and average annual precipitation is 78.7 cm (31.0 in).

Figure 9.42a Climograph for Caribou, Maine (Dfb climate). Average annual temperature is 4.0°C (39.2°F) and average annual precipitation is 95.1 cm (37.4 in).

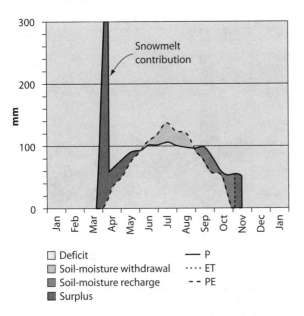

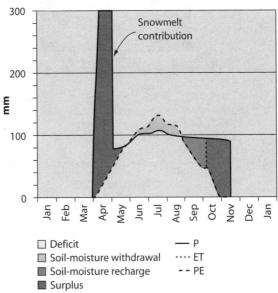

Figure 9.41b Water balance diagram for Duluth, Minnesota.

Figure 9.42b Water balance diagram for Caribou, Maine.

Moisture Characteristics Precipitation in all Dfa and Dfb climates is well distributed throughout the year, but a slight maximum generally occurs during the summer months for locations that are inland or upwind from major water bodies. For example, a summer precipitation peak is obvious in the climograph for Des Moines, Iowa (Figure 9.37a). Adjacent to the Great Lakes the relative warmth of the lakes in winter destabilizes the atmosphere as cold

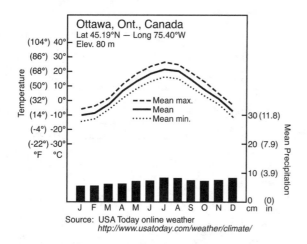

Figure 9.43a Climograph for Ottawa, Canada (Dfb climate). Average annual temperature is 6.1°C (43.0°F) and average annual precipitation is 87.9 cm (34.6 in).

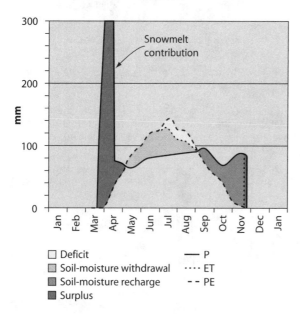

Figure 9.43b Water balance diagram for Ottawa, Canada.

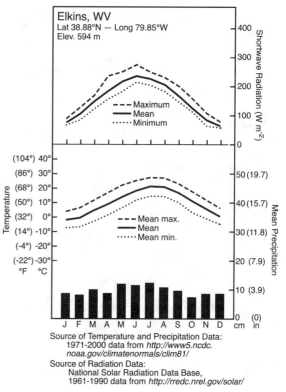

Figure 9.44a Climograph for Elkins, West Virginia (Dfb climate). Average annual temperature is 9.9°C (49.8°F) and average annual precipitation is 117.1 cm (46.1 in).

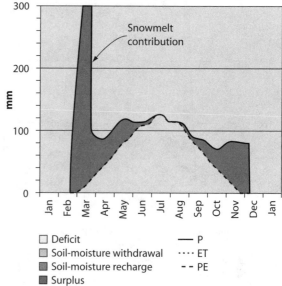

Figure 9.44b Water balance diagram for Elkins, West Virginia.

cP and A air masses cross over them. As a result, winter precipitation is as abundant, or even more abundant, than summer precipitation in such locations. The climograph for Erie (Figure 9.39a) shows this lake effect moisture. Much of this winter precipitation occurs in the form of lake effect snow.

Precipitation in both North America and Eurasia is largely generated by migratory frontal cyclone passage during all seasons except summer, when convective thunderstorm activity becomes more important. Like Humid Subtropical areas, the summer convective storms are controlled by intense surface heating that initiates high *evapotranspiration* rates. Advection of moisture from

the Gulf of Mexico, the North Atlantic, and the North Pacific Ocean, largely provided by strengthening of the Bermuda-Azores and Hawaiian high pressure systems, is of great importance to sum-

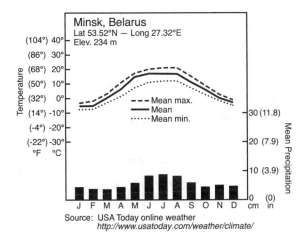

Figure 9.45a Climograph for Minsk, Belarus (Dfb climate). Average annual temperature is 6.0°C (42.8°F) and average annual precipitation is 67.8 cm (26.7 in).

Figure 9.46a Climograph for Moscow, Russia (Dfb climate). Average annual temperature is 4.3°C (39.8°F) and average annual precipitation is 59.9 cm (23.6 in).

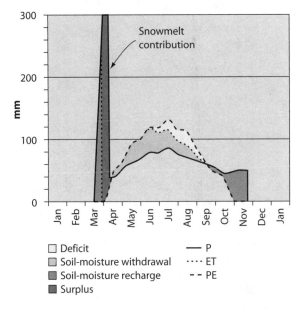

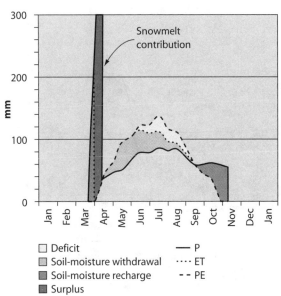

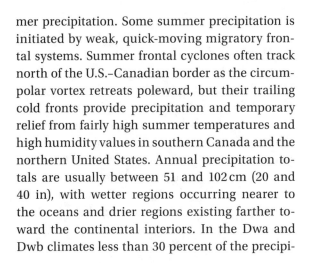

Figure 9.45b Water balance diagram for Minsk, Belarus.

Figure 9.46b Water balance diagram for Moscow, Russia.

mer precipitation. Some summer precipitation is initiated by weak, quick-moving migratory frontal systems. Summer frontal cyclones often track north of the U.S.–Canadian border as the circumpolar vortex retreats poleward, but their trailing cold fronts provide precipitation and temporary relief from fairly high summer temperatures and high humidity values in southern Canada and the northern United States. Annual precipitation totals are usually between 51 and 102 cm (20 and 40 in), with wetter regions occurring nearer to the oceans and drier regions existing farther toward the continental interiors. In the Dwa and Dwb climates less than 30 percent of the precipi-

tation occurs in winter. In Asia the dry monsoon season and associated Siberian high suppress precipitation.

Water balance diagrams for all the Humid Continental climates (Figures 9.37b through 9.48b), as well as all climatic types poleward of Dfa/Dwa, have missing data in winter, which results from the storage of water in the form of snow. As spring approaches, snowmelt produces abundant water at the surface, much of which fell as snow months before. This phenomenon is depicted by the very prominent spike in surplus, with the spikes occurring later in the spring as temperature decreases. The surface water balances in general suggest

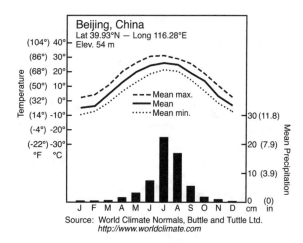

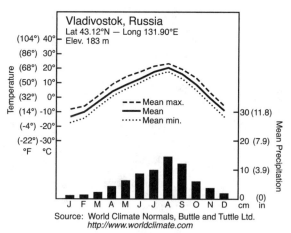

Figure 9.47a Climograph for Beijing, China (Dwa climate). Average annual temperature is 11.7°C (53.1°F) and average annual precipitation is 63.0 cm (24.8 in).

Figure 9.48a Climograph for Vladivostok, Russia (Dwb climate). Average annual temperature is 4.4°C (39.9°F) and average annual precipitation is 74.4 cm (29.3 in).

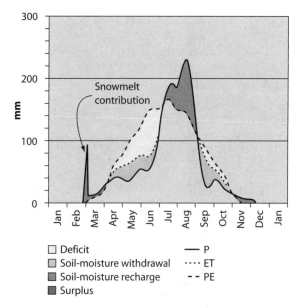

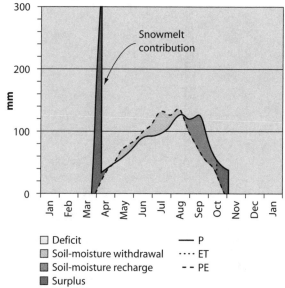

Figure 9.47b Water balance diagram for Beijing, China.

Figure 9.48b Water balance diagram for Vladivostok, Russia.

that mean deficits are small or nonexistent in Humid Continental climates. The largest deficit occurs at Beijing (Figure 9.47b), at the end of the dry monsoon season and also after the wet monsoon season ends (but while the Sun is still high in the sky and days are still relatively long). Notably, at Elkins, West Virginia (Figure 9.44b), no soil-moisture deficit is apparent in the mean monthly data.

Dfc—Subarctic The **Subarctic (Dfc, Dwc, Dwd) climates** exist poleward of Humid Continental climatic regions (see Figure 8.1). All Subarctic areas in

North America receive similar precipitation totals year-round and, therefore, Dwc and Dwd types are found. There, the Dfc region covers more than half of Alaska and Canada. The vegetation in this North American region is typically referred to as the **boreal forest**—an area dominated by extensive, slow-growing coniferous forest. In Europe the Subarctic climate is found in the sparsely inhabited northern reaches of Scandinavia, while in Asia the four Subarctic types occupy the northern interior tier of the continent (see Figure 8.1). Vegetation in the Asian Subarctic climatic zone is termed the **taiga**, the equivalent Russian term for the North American boreal forest. The Dfc climatic type is the second-most extensive on Earth, falling behind only the BWh climate.

Thermal Characteristics Summer temperatures are lower than those of the Humid Continental climatic region. Winter temperatures are generally far below those of the adjacent climate, with subfreezing average monthly temperatures a common occurrence. Extreme seasonality of temperatures occurs with the annual temperature range may be as high as 60 C° (108 F°). The highest annual temperature range on Earth occurs at Oymyakon, Siberia, where the range is 64 C° (115 F°).

The lowest temperature ever recorded in the northern hemisphere occurred within this climatic zone at two Siberian locations: Verkhoyansk and Oimekon. On February 6, 1922 the mercury plummeted at these locations to –68°C (–90°F). In some Subarctic locations the average temperature may remain below freezing for up to 7 months. The mean temperature of the coldest month may approach –38°C (–36°F) for some locations. The transition seasons of autumn and spring are relatively short in comparison with the length of summer and winter, with winter being extremely long. Average summer temperatures show a dramatic reversal with monthly averages perhaps approaching 20°C (68°F).

Climographs representing Dfc climates are shown for the coastal locations of Cold Bay, Alaska (**Figure 9.49a**), Kotzebue, Alaska (**Figure 9.50a**), and Murmansk, Russia (**Figure 9.51a**). Inland Dfc sites shown include Fairbanks, Alaska (**Figure 9.52a**), Yellowknife, Northwest Territories (**Figure 9.53a**), and Moosonee, Ontario (**Figure 9.54a**). Dwc, the dry winter equivalent, is represented by Tura, Russia (**Figure 9.55a**).

Moisture Characteristics Annual precipitation amounts are typically 12 to 50 cm (5 to 20 in) per year. These sparse totals are not surprising, because little moisture is present in such cold air, even at saturation. Summer precipitation maxima are pronounced in most locations, while winter remains dry and cold. Warm-season precipitation is primarily frontally induced, because convection is not strong enough to produce much precipitation, even in summer, and summer is the only time when reasonable air mass contrast is present.

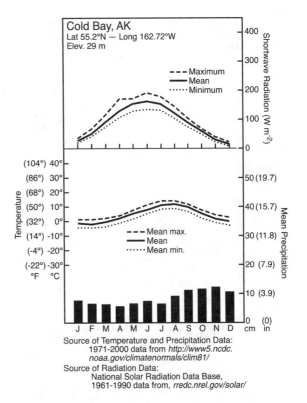

Figure 9.49a Climograph for Cold Bay, Alaska (Dfc climate). Average annual temperature is 3.6°C (38.4°F) and average annual precipitation is 102.3 cm (40.3 in).

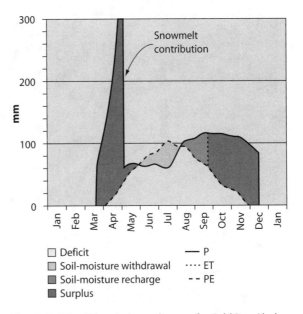

Figure 9.49b Water balance diagram for Cold Bay, Alaska.

Despite the low precipitation totals, the low amounts of available energy for evapotranspiration generally allow for sufficient soil moisture to support the boreal forests. Moisture also remains frozen in the ground as **permafrost** for extensive

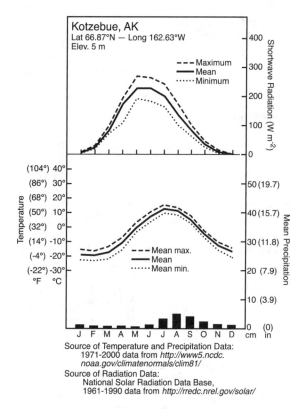

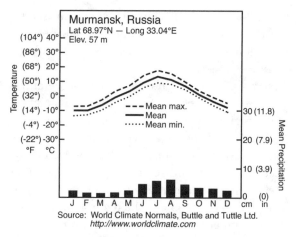

Figure 9.51a Climograph for Murmansk, Russia (Dfc climate). Average annual temperature is 0.3°C (32.5°F) and average annual precipitation is 41.4 cm (16.2 in).

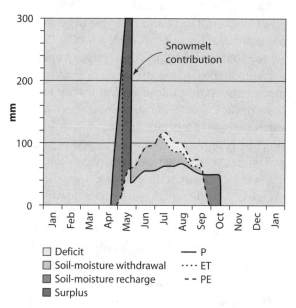

Figure 9.50a Climograph for Kotzebue, Alaska (Dfc climate). Average annual temperature is −5.7°C (21.8°F) and average annual precipitation is 25.5 cm (10.1 in).

Figure 9.51b Water balance diagram for Murmansk, Russia.

ture is partially replaced through precipitation, but typically this water input is insufficient to off-set soil moisture deficits (Figures 9.49b through 9.55b).

E—Polar climates

ET—Tundra The first of the **Polar climates** is the **Tundra (ET) climate**, which exists poleward of the Subarctic climate. The boreal and taiga forests of North America, Europe, and Asia terminate at the 10°C isotherm for the warmest summer month. This isotherm represents the *treeline*—the border between the Subarctic and Tundra (ET) climates. The

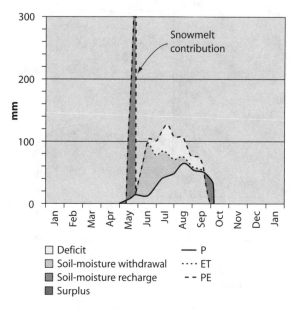

Figure 9.50b Water balance diagram for Kotzebue, Alaska.

periods. Fairly impressive snow depths are common in some areas because an interannual accumulation may result from incomplete melting during the short summers. When temperatures rise suitably for evaporation to occur, the lost mois-

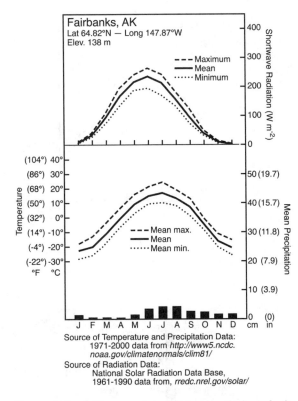

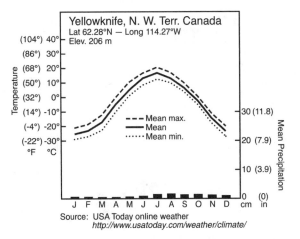

Figure 9.53a Climograph for Yellowknife, Northwest Territories, Canada (Dfc climate). Average annual temperature is −4.7°C (23.5°F) and average annual precipitation is 25.7 cm (10.1 in).

Figure 9.52a Climograph for Fairbanks, Alaska (Dfc climate). Average annual temperature is −2.9°C (26.7°F) and average annual precipitation is 26.3 cm (10.3 in).

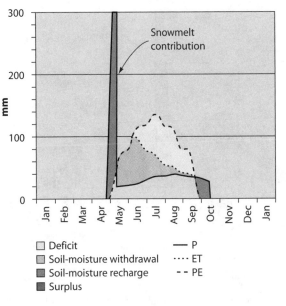

Figure 9.53b Water balance diagram for Yellowknife, Northwest Territories, Canada.

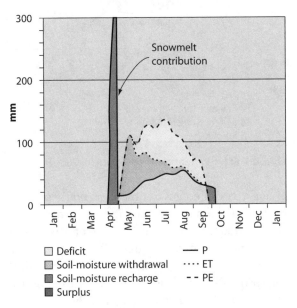

Figure 9.52b Water balance diagram for Fairbanks, Alaska.

ET climate in North America is found in the highest latitudinal locations of Alaska and the Canadian territories. European Tundra is limited to the extreme northern strip of the continent—namely, northern Iceland, Scandinavia, and Russia. The Asian Tundra climate extends through northern Russia to the Kamchatka Peninsula of eastern Russia (see Figure 8.1).

The Tundra climatic type is named for tundra vegetation, which consists of stunted low-growing flowering plants, bushes, shrubs, lichens, and mosses. Low temperatures combine with low Sun angles and a short period of extensive day lengths to produce specially adapted vegetation types. The low-growing lichens, mosses, and shrubs undergo rapid growth cycles once temperatures climb above freezing. The plants flower almost immediately in the presence of above freezing temperatures, ensuring the propagation of the species.

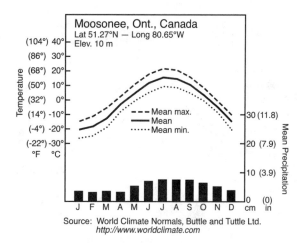

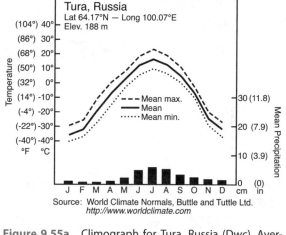

Figure 9.54a Climograph for Moosonee, Ontario (Dfc climate). Average annual temperature is −1.1°C (30.1°F) and average annual precipitation is 66.3 cm (26.1 in).

Figure 9.55a Climograph for Tura, Russia (Dwc). Average annual temperature is −9.2°C (15.4°F) and average annual precipitation is 33.3 cm (13.1 in).

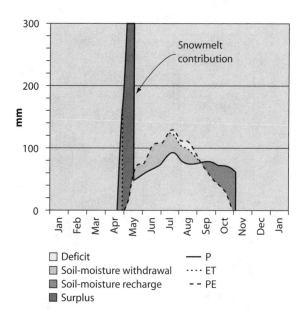

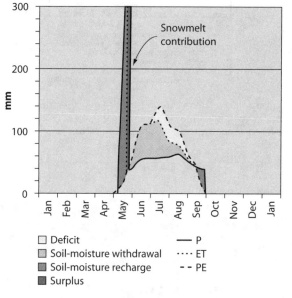

Figure 9.54b Water balance diagram for Moosonee, Ontario.

Figure 9.55b Water balance diagram for Tura, Russia.

Thermal Characteristics Because of the high latitude, winter temperatures are extremely low. The Sun shines for only a few hours each day, if at all. Even in summer significant beam depletion, low solar angles, and high *albedo* combine to keep temperatures relatively low even in the presence of a nearly constant influx of solar radiation. Resulting summer temperatures remain only slightly above freezing on average. Because freezes may occur in any month of the year, the growing season is effectively nonexistent, but in some locations summer temperatures may become relatively warm over short periods.

During winter all soil is frozen, but in summer the layers above the permafrost thaw. This may create a saturated environment, as liquid water is prohibited from penetrating into the frozen subsurface. Floods can occur because of this and/or the combination of ice dams blocking stream paths into the Arctic Ocean. The latter occurs during spring thaw, with most rivers draining northward toward the Arctic Ocean where the river mouths remain frozen long after the main river body (which lies southward) thaws. The surrounding region thaws as well, leading to a high volume of water moving northward toward an ice-blocked river. Permafrost also leads to complications in construction. Local heating of concrete, for example, may melt permafrost below it, leading to foundation failure of buildings and roads.

Moisture Characteristics Because of the extreme cold, precipitation is sparse, with annual totals below 25 cm (10 in) and most locations receiving less than 13 cm (5 in). Although some precipitation typically falls in every month, a distinct warm-season maximum is discernible. **Figure 9.56a** shows this feature in the climograph for Thule Air Force Base, Greenland. Snow remains at the surface for long periods, with some areas receiving little melting from year to year. Most locations usually see spring water surpluses in conjunction with low evapotranspiration rates, increasing precipitation, and melting snow (Figure 9.56b).

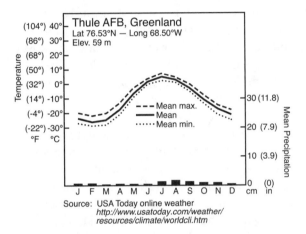

Figure 9.56a Climograph for Thule Air Force Base, Greenland (ET climate). Average annual temperature is −11.6°C (11.1°F) and average annual precipitation is 10.2 cm (4.0 in).

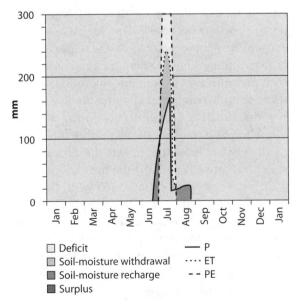

Figure 9.56b Water balance diagram for Thule Air Force Base, Greenland.

EF—Ice Cap The final Köppen category is the **Ice Cap (EF) climate**. This climate is found in the northern hemisphere extratropics only in interior Greenland. The moderating presence of the Arctic Ocean prevents these climates from existing in Eurasia. Significant sea ice does accumulate throughout the Arctic Ocean region, except along the Scandinavian coast, which is warmed by the North Atlantic Drift.

In the EF climate, ice covers the ground throughout the year as mean monthly temperatures remain below freezing in every month. Cold, dense air forms above the Greenland ice cap and spills toward lower coastal elevations, ensuring frigid temperatures even near the ocean. These **katabatic winds** are especially prominent at night when the air adjacent to the surface increases in density as it undergoes radiative cooling. The region receives very little precipitation, because the extreme cold induces sinking motions in the atmosphere and the saturation vapor pressure is so low in such cold conditions. The little precipitation that does occur is mainly limited to late summer and autumn during the times of highest temperatures. Annual average precipitation is typically less than 12 cm (5 in). The notion of this region being one of raging blizzards is extremely exaggerated. Instead, ice cap growth largely occurs through the *deposition* of atmospheric water vapor directly onto the surface in the presence of very low temperatures. This growth is approximately balanced by the very slow downslope motion of the ice and its subsequent calving off into the ocean. This climatic type is treated more extensively in Chapter 10.

H—Highland Climates

Areas of high elevations have climatic characteristics that do not allow them to be classified neatly into one of the climatic types discussed previously. Local-scale elevation changes—**vertical zonation**—create such vast differences in climates that it is impossible to categorize them on a planetary scale of analysis. Such zones can be classified as **Highland (H) climates**, a catch-all category. The progression of climatic zones with height largely mimics the progression of latitudinal climatic zones. For instance, a high mountain near the equator likely has a tropical climate regime and associated vegetation at the base, then progresses through C, D, and finally E climatic zones and associated vegetation, ascending toward an ice-capped (EF) pinnacle.

In North America, Highland climates exist mainly along the spine of the Rocky Mountains from Alaska through Canada and into the western United States and Mexico (see Figure 8.1). Another area of Highland climates exists in the coastal mountains of Alaska and western Canada, extends through the Cascade Mountains of the Pacific Northwest, and finally terminates at the southern portion of the Sierra Nevadas of California. In Europe, Highland climates prevail through the central Alps region, centered in Switzerland. The Scottish Highlands and dissected areas of Scandinavia have Highland climates as well. Asia contains the greatest expanse of Highland climates on Earth, centered on the massive Himalayan range and the adjacent Tibetan Plateau. Extending from this zone are the western mountain extensions of the Pamirs, the Hindu Kush, and the Tien Shan mountain ranges. Other significant mountain ranges such as the Caucasus, the Zagros, and the highlands of Japan also have regions with radical climatic contrasts across small spatial zones.

Local differences in slope, orientation, and elevation make specific categorization of these regions problematic. Vegetation in extratropical H climates, particularly at higher elevations, is often largely limited to equator-facing slopes, because of the prevalence of afternoon sunlight in the southern sky (in the northern hemisphere). Low-latitude mountain regions experience different insolation characteristics than high-latitude mountain areas because the Sun shines on north-facing slopes at one time of year and south-facing slopes at the opposite time of year. However, both extratropical and tropical highland regions support complicated local climate features.

Of particular note in Highland climates is the development of **wind channeling**—the constriction of mountain-induced winds that triggers a velocity increase. In addition, high elevations combine with abundant radiational cooling to create very cold, dense air masses over the mountains. The dense air is pulled to lower elevations by gravity through the process of **cold air drainage**. This triggers the development of a katabatic wind, or **mountain wind** similar to the katabatic winds that form over the ice sheets in EF climates. During winter the downslope-moving katabatic winds are common, especially during the early morning

when diurnal cooling is maximized. Mountains may channel these winds through narrow canyons. Katabatic winds are most prevalent under regional high pressure conditions that allow the cold air pools to form under relatively clear skies over the high elevations. The air pools may be as thick as 600 m (2000 ft) and may affect very large regions of surrounding plains and valleys. Although adiabatic warming occurs as the air pool descends, the warming is generally not significant enough to offset the chilling effects of the air pool. This movement leads to rapid reductions in air temperature over the affected areas.

A significant katabatic wind in the Adriatic region of Europe is the **bora wind**, which forms in the Balkan highlands and spills toward lower elevations through gaps in the Dinaric Alps toward the Adriatic Sea. The word "katabatic" is derived from the Greek word *katabasis*, which refers to an army marching toward a coast, much as the wind spills out of the highlands and progresses toward the sea. The word "bora" is derived from "Boreas," the god of the north wind in Greek mythology. The bora also may be induced when a cyclone occurs over the Ionian Sea region, resulting in a counterclockwise circulation that leads to the funneling of air through the Strait of Otranto between the "heel" of Italy and Albania. Such a situation may lead to treacherous water conditions in the Ionian Sea.

Another significant European katabatic wind is the **meltemi**, which is also associated with the Balkan region. This wind develops when high pressure occurs over the Balkan region near Hungary and a cyclonic center lies to the southeast near or over Turkey. The circulation characteristics of each pressure system combine to produce an offshore wind blowing over the Aegean Sea. Such a synoptic situation usually occurs during the summer when a thermal low develops over Turkey.

The **mistral wind** that affects the Rhone River valley of France is perhaps the best-known katabatic wind system in Europe. The wind flows directly from the Alps and spills through the Rhone valley toward Marseilles, where it spreads along the Mediterranean coast of France. Note that the katabatic winds detailed above differ from Föehn winds—warm, dry winds that affect the regions north of the Alps, mainly Germany.

The winter monsoon in Asia is technically a katabatic wind, as much of the offshore airflow that affects the southern portions of Asia stems from cold air flowing down from the high Tibetan plateau. Other than the monsoonal system itself, the most prominent Asian katabatic wind is the **oroshi** of Japan, an equivalent of the European Bora.

■ Summary

Climates in the extratropical sections of North America and Eurasia are influenced by semipermanent pressure features, ocean currents, topography, continentality, and water bodies. The geography of Central North America is more supportive of severe weather than any other location. Other important features of the North American climate include the LLJ and cyclogenetic regions on the leeward sides of the Rocky Mountains and along the Gulf of Mexico and Atlantic coasts. The Alps of Europe prevent the interaction of the coldest and warmest air masses. The size of the Asian land mass allows for a seasonal shift in circulation known as the Asian Monsoon, and the Himalayas of Asia alter the position of the polar front jet stream and create a strong orographic effect on that continent.

The factors noted above, along with others, produce a wide variety of climates in the extratropical northern hemisphere. True Desert climates are generally located in the southwestern United States and adjacent northern Mexico, along with several zones within a wide swath of Asia from the Red Sea to central China. Semiarid Steppe climates surround the True Deserts and occupy a much more sizable area in western North America than the True Deserts. Most of these True Deserts and Semiarid Steppes are classified as "cold" in the extratropical part of the world. Sinking air from the poleward edge of the Hadley cell, along with continentality and rain shadow effects, are largely responsible for the aridity.

Mesothermal climates in the extratropical northern hemisphere are also widespread. These areas include the Humid Subtropical climates of the southeastern United States, eastern China, southern Japan, and the winter-dry variety in parts of south Asia. Return flow of warm, humid air around subtropical highs, abundant annual precipitation, and occasional winter cold weather outbreaks characterize this climate. Mediterranean climates persist in coastal southern California and the area surrounding the Mediterranean Sea. These are characterized by very dry summers and wet winters, due to the seasonal influence of the subsidence side of the subtropical highs. Finally, the Marine West Coast climates are found in Pacific coastal North America and northwestern Europe. These well-watered, mild summer and mild winter climates are influenced strongly by the frequent passage of midlatitude wave cyclones.

Microthermal climates occupy vast expanses of northern North America and Eurasia. The Humid Continental type occurs from the Ohio River northward into southern Canada, east of the semiarid climates, in a wide area from eastern Europe toward central Russia, and in northeastern China and the adjacent Russian Far East. Summers are warm to hot, while winters are long, cold, and uninterrupted. An even harsher climate is the Subarctic type, which sees shorter summers and less precipitation than the Humid Continental climates.

Polar climates are found along the northern fringes of the northern hemisphere. Tundra climates may have a freeze in any month of the year, while the Ice Cap type of polar climate sees an average temperature below freezing in every month of the year. Finally, Highland climates exhibit tremendous climatic variety in short lateral distances. The topography in these climates gives rise to several well-known wind systems of local and regional importance.

▶ Key Terms

Adiabatic motion	*Arctic (A) air mass*	*California current*
Advection	*Arctic Circle*	*Canary current*
Advection fog	Asian monsoon	Chinook wind
Air mass	*Baroclinicity*	*Circumpolar vortex*
Albedo	*Bermuda-Azores high*	Cold air damming
Alberta clipper	Blocking anticyclone	Cold air drainage
Aleutian low	Bora wind	*Cold front*
Anticyclone	Boreal forest	Cold Steppe (BSk) climate

Colorado wave cyclone
Continental polar (cP) air mass
Continental tropical (cT) air mass
Continentality
Convection
Coriolis effect
Coriolis parameter
Cumuliform cloud
Cumulonimbus cloud
Cyclogenesis
Cyclone
Deficit
Deposition
Dewpoint temperature
Direct radiation
Dixie Alley
Dryline
East-coast occlusion
Ecotone
El Niño
Environmental lapse rate
Evaporation
Evapotranspiration
Föehn wind
Geopotential height
Geostrophic wind
Gradient wind equation
Greenhouse effect
Greenhouse gas
Gulf Coast cyclone
Gulf Stream
Gust front
Gyre
Hadley cell
Hawaiian high
Heat index
Highland (H) climate
Hot Steppe (BSh) climate
Humid Continental (Dfa, Dfb)
    climate
Humid Continental Winter Dry
    (Dwa, Dwb) climate
Humid Subtropical (Cfa) climate
Humid Subtropical Winter Dry
    (Cwa) climate
Hurricane
Ice Cap (EF) climate

Icelandic low
Infiltration
Insolation
International date line
Isohypse
Isotherm
Katabatic wind
Labrador current
Lake effect snow
Latent heat
Latitude
Leeward
Longitude
Longwave radiation
Loop current
Low-level jet (LLJ) stream
Marine West Coast climate
Marine West Coast Cool Summer
    (Cfc) climate
Marine West Coast Warm Summer
    (Cfb) climate
Maritime effect
Maritime polar (mP) air mass
Maritime tropical (mT) air mass
Mediterranean climate
Mediterranean Hot Summer (Csa)
    climate
Mediterranean Warm Summer
    (Csb) climate
Meltemi
Mesothermal
Microthermal
Midlatitude Cold Desert (BWk)
Midlatitude wave cyclone
Midlatitude westerlies
Mistral wind
Monsoon
Mountain wind
Nor'easter
North American summer
    monsoon
North Atlantic Drift
North Equatorial current
Occluded front
Occlusion
Orographic effect
Oroshi wind

Outflow boundary
Permafrost
Pineapple Express
Polar cell
Polar climate
Polar front jet stream
Positive vorticity
Potential evapotranspiration
Potential vorticity
Rain shadow effect
Relative humidity
Ridge
Rossby wave
Santa Ana wind
Sensible heat
Shortwave radiation
Siberian high
Source region
Southwesterly monsoon
Squall line
Stability
Steppe (BS) climate
Stratiform cloud
Subarctic (Dfc, Dwc, Dwd)
    climate
Subtropical Hot Desert (BWh)
Summer monsoon
Surface boundary layer
Surface water balance
Surplus
Taiga
Temperature inversion
Thermal low
Tibetan low
Tornado Alley
Treeline
Tropical cyclone
Trough
True Desert (BW) climate
Tundra (ET) climate
Unsaturated adiabatic lapse rate
Vertical zonation
West-coast occlusion
Wind channeling
Windward
*Terms in italics have appeared in at
    least one previous chapter.*

▶ Review Questions

1. How does the alignment of mountains affect the climates of North America and Europe?
2. What role does the low-level jet play in the climate of North America?
3. What is Tornado Alley and why does it exist?
4. Discuss the oceanic circulation regime responsible for the creation of the North Atlantic Drift.
5. Describe the importance of the North Atlantic Drift for the climate of Europe.
6. Compare/contrast the role of continentality in North America and Asia.
7. Why doesn't Europe have a significant continental effect?
8. What is a monsoon?
9. Discuss the influence of the Himalayan mountain range on precipitation in Asia.
10. Discuss the similarities and differences between Desert and Steppe climates.
11. Discuss the various climatic causes of arid climates.
12. Compare and contrast Mediterranean and Marine West Coast climate regions.
13. Why do Mediterranean climates exist?
14. Why are Marine West Coast climates so warm for their latitudinal locations?
15. Compare and contrast microthermal and mesothermal climatic regimes.
16. Discuss the characteristics of the transition from the Subarctic climate to the Tundra and then Ice Cap climates.
17. Why is the growth of tundra vegetation stunted?
18. Explain why Highland climates constitute a specific climate regime.
19. Detail the katabatic and regional mountain winds of each of the continents discussed in this chapter.

▶ Questions for Thought

1. Discuss katabatic wind formation and how these winds differ from mountain-induced winds such as the Chinooks.
2. Identify the climatic type in your hometown and find a town on another continent that has the most similar climatic regime as your town. Why are the climatic regimes so similar?

http://physicalscience.jbpub.com/climatology

Connect to this book's Web site: http://physicalscience.jbpub.com/climatology. The site provides chapter outlines, further readings, and other tools to help you study for your class. You can also follow useful links for additional information on the topics covered in this chapter.

10 | Tropical and Southern Hemisphere Climates

Chapter at a Glance

Contrasts Between Extratropical and Tropical
   Atmospheric Behavior
Contrasts Between Northern and Southern
   Hemisphere Atmospheric Behavior
Climatic Setting of Africa
   General Characteristics
   Intertropical Convergence Zone
   Air Mass and General Circulation Influences
Climatic Setting of Australia and Oceania
   General Characteristics
   El Niño-Southern Oscillation Influences
   South Pacific Convergence Zone
   Madden-Julian Oscillation
   Quasi-Biennial Oscillation
Climatic Setting of Latin America
   ENSO Contributions
Climatic Setting of Antarctica
Regional Climates
   A—Tropical Climates
   B—Arid Climates
   C—Mesothermal Climates
   E—Polar Climates
   EF—Ice Cap
   H—Highland Climates
Summary
Review

■ Contrasts Between Extratropical and Tropical Atmospheric Behavior

In the previous chapter we focused on climates of the northern hemisphere. This chapter shifts to the southern hemisphere and also includes the tropics. Atmospheric features in the tropics differ markedly from those in the extratropics. Annual temperature ranges are small in the tropics because day length and Sun angle vary little throughout the year. Also, the *Coriolis effect* is minimal because, as we have seen, its magnitude is proportional to *latitude*. The weakness of the Coriolis effect causes air to flow according to different rules than in the extratropics. Outside the tropics *geostrophic balance* resulting from an equilibrium between the *pressure gradient force* and the Coriolis effect occurs far above the surface, where *friction* is negligible. Flow in the tropics cannot be assumed to be geostrophic and winds cannot be inferred as being parallel to the *isohypses* (lines of constant *geopotential height*). Instead, **streamline analysis** must be used to deduce winds in the tropics.

Furthermore, little or no *air mass* contrast is present in the tropics because cold air masses cannot penetrate to such low latitudes. Tropical storm systems cannot originate from the juxtaposition of cold and warm air masses, and *fronts* are nonexistent, at least in the form that they appear in the midlatitudes. In general, large-scale tropical weather and climate are much less predictable than in the midlatitudes, where climates are largely controlled by large-scale, upper-atmospheric interactions and *Rossby waves* mix warm air of tropical origin with cold air of polar origin. The resulting flow patterns govern where and when midlatitude storms or clear skies occur. The dynamics involved have been worked out mathematically, and fairly accurate short-term predictions are possible in the midlatitudes. By contrast, the tropics have little similar overriding upper-air forcing. Instead, climate is largely influenced by local and/or regional factors related to daily heating of the surface. This introduces a level of daily variability that is experienced in the midlatitudes only during the summer season, if at all. Still, some large-scale factors control and contribute to elements of the tropical climatic landscape.

There are only three ways for a tropical air parcel to rise so that it can cool to the *dewpoint temperature* and produce a cloud that might generate precipitation. Local instability can cause **convective precipitation**, uplift associated with topography can cause an *orographic effect*, and two streams of air might collide to produce **horizontal convergence**, which results in uplift after the collision. In the latter case horizontal convergence may also result from two near-surface air streams moving laterally in the same direction but at different speeds. The faster-moving parcel could "catch up with" the slower-moving parcel, thereby causing horizontal convergence. Regardless of whether precipitation results from *convection*, the orographic effect, or convergence, the energy that

is derived to fuel storms in the tropics must be enormous because the lack of air mass contrast gives no mechanism for producing the dynamic lows that are produced in the middle and high latitudes.

Tropical climates also differ from extratropical climates in the orientation of *ridges* and *troughs*. In the northern hemisphere's midlatitudes, ridges—elongated zones of high pressure—in Rossby waves are areas where isohypses bulge toward the north, because higher pressure associated with *subtropical anticyclones* is to the south. By contrast, ridges in the northern hemisphere tropics appear as southward "dips" in **easterly waves** in the *northeast trade winds* (**Figure 10.1a**), because the elongation of the effect of the subtropical anticyclones extends from the north.

Likewise, any elongated area of low pressure is a trough. In the northern hemisphere's midlatitudes, the lower pressure aloft is always associated with the *subpolar lows* to the north, rather

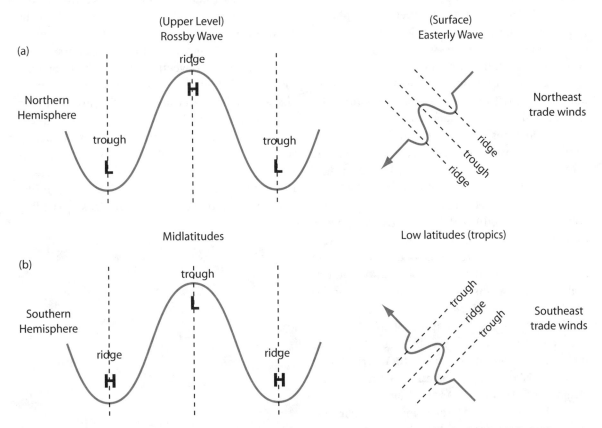

Figure 10.1 Rossby waves and easterly waves in the (a) northern and (b) southern hemispheres. Midlatitude waves are shown on the left, and low latitudes (tropics) are shown on the right.

than to the south as in the tropics, where the *intertropical convergence zone* (ITCZ) provides the low pressure. This causes troughs to appear as "dips" southward in the northern hemisphere's midlatitude Rossby waves but "bulges" northward in the northern hemisphere's tropical easterly waves. The latter, often called **inverted troughs** in the tropics, represent *tropical cyclone* genesis areas.

■ Contrasts Between Northern and Southern Hemisphere Atmospheric Behavior

Southern hemisphere climates are distinct from northern hemisphere climates in several important ways. First, the overwhelming dominance of water surface in the southern hemisphere limits the effectiveness of continental air masses and allows ocean currents to redistribute energy very effectively between the tropics and the South Polar latitudes. Second, the seasons are inverted from the time of the year when they occur in the northern hemisphere. Because Earth is nearest to the Sun (*perihelion*) during the southern hemisphere's summer and farthest from the Sun (*aphelion*) during southern hemisphere's winter, summer is a bit warmer and winter a bit colder in the southern hemisphere than would otherwise be expected. In addition, the leftward (rather than rightward) deflection of the Coriolis effect causes secondary circulation features (i.e., *cyclones* and *anticyclones*) to spin in opposite directions than expected in the northern hemisphere. Finally, Rossby wave ridges in the midlatitude southern hemisphere actually appear the same as troughs in the northern hemisphere (Figure 10.1b). In the southern hemisphere's midlatitudes the lower pressure associated with the subpolar lows is to the south (rather than to the north as in the northern hemisphere), so southern hemispheric troughs are similar in appearance to northern hemispheric ridges.

This chapter describes and explains the tropical and southern hemisphere climates of the world, including climates of Africa, South America, Australia, the islands of the western equatorial Pacific Ocean, and Antarctica. As in Chapter 9, the primary physical causes for the climates of each continent are discussed before investigating the specific Köppen climatic regions for these areas.

■ Climatic Setting of Africa

General Characteristics

The African continent straddles the tropics, extending poleward to approximately 35° of latitude in both hemispheres (**Figure 10.2**). The northern edge of the continent lies at the latitudinal equivalent of Virginia and southern Japan, whereas the southernmost extent lies at similar latitudes to southern Australia and central Argentina. Because of the shape of the continent, there is more than twice as much land area in the northern hemisphere than in the southern hemisphere. The popular notion of Africa as an exclusively equatorial or southern hemisphere continent is inaccurate.

Intertropical Convergence Zone

Tropical influences, particularly the ITCZ, are important to the climatic setting of much of the continent. As was discussed in Chapter 7, the ITCZ represents a region of convergence between the northeast trades and the *southeast trade winds*. The ITCZ largely follows the vertical ray of the Sun as it migrates between the *Tropic of Cancer* and *Tropic of Capricorn* through the course of the year. Afternoon convection from the high solar angles and the resultant surface heating destabilize the atmosphere and augment the lifting due to convergence.

The ITCZ does not migrate uniformly across African latitudes through the course of the year as the direct rays of the Sun shift seasonally. Instead, there is a high degree of variability in the timing and motions through the year. This variability

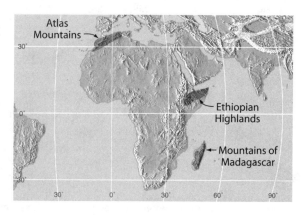

Figure 10.2 Physiographic features of Africa.

affects regional climates in that "summer" rains may be delayed or completely diminished for some areas for any given year, leading to extended drought conditions and its associated environmental and social stresses.

The ITCZ affects more than just the amount of rainfall in a given region; it also contributes directly to the associated thermal characteristics. During the "high-sun" period (summer), there is more energy in the environment as compared with other times. The nearly unbroken area of afternoon convection and convective thunderstorms along the ITCZ not only produce much-needed rainfall, but they also shade the surface from the otherwise high input of *insolation*. If the ITCZ moves abnormally slowly and fails to reach a particular area at the climatological average time of year, greater amounts of surface energy are received, leading to increased *evapotranspiration* and stress on the local or regional ecosystem.

In such times, affected areas may have local vegetation cover denuded by human and animal activities. Such a situation changes the local *surface water balance*, usually by decreasing the percentage of precipitation that goes into *soil moisture recharge* and evapotranspiration back to the atmosphere and increasing the percentage that becomes *runoff*. If vegetation removal is extensive enough, the regional water balance may also change in a similar fashion. These hydroclimatological changes may lead to expansion of the drier climates adjacent to those that experience rainfall from the ITCZ. Such *desertification* of a region may lead to further water balance changes on the regional level. A *positive feedback system* may occur, where interactions between the surface and the overlying atmosphere then feed back on circulation, potentially leading to further changes in the domain of the ITCZ. If this occurs, then the desert "permanently" expands over a previously wetter region. This notion was expressed succinctly by climatologist Michael Glantz in the title of his book, *Drought Follows the Plow*.

Desertification caused by natural regional drought in combination with *anthropogenic* land cover changes occurred in recent decades in the *Sahel* region on the southern edge of the Sahara desert (see Figure 4.16). Once vegetation was eliminated in the Sahel, there was less evapotranspiration into the regional atmosphere, leading to a net decrease in cloud formation. This led to expansion of the nearby desert (desertification) into the normally wetter Steppe region and an increase in dependency on humanitarian relief aid among the indigenous population. Despite worldwide aid, the region continues to be one plagued by famine, disease, and water shortages.

Air Mass and General Circulation Influences

Africa is affected largely by tropical and equatorial air masses (see Figure 8.7). *Continental tropical (cT) air masses* are dominant in the northern third of the continent and in other smaller patches, particularly in southwestern Africa. *Maritime tropical (mT)* and *Equatorial (E) air masses* tend to be most influential in coastal Africa, especially south of the equator, and also in equatorial Africa. *Midlatitude westerlies* exert a role in the extreme northern and southern latitudes in winter.

The influence of the subtropical anticyclones is strong. Because of the latitudinal extent of Africa, significant portions of the continent fall under the influence of the subtropical anticyclones in both hemispheres. Subtropical anticyclones sit over ocean basins to the northwest, southwest, and southeast of the continent. Of these, the *South Atlantic high* is most persistent. The counterclockwise rotation around this southern hemisphere semipermanent anticyclone produces the **Benguela current**—a cold ocean current that runs along the southwestern coast of Africa. This current persists nearly year-round and causes the convection associated with the ITCZ off Africa's west coast to remain displaced to the north, in the northern hemisphere, year-round. It also stabilizes the atmosphere, creating a *temperature inversion* and dry conditions in adjacent southwestern Africa. The counterclockwise circulation around the *Indian Ocean high* brings the **Aghulas current**—warm equatorial water—to the southeastern African coast. The clockwise flow around the *Bermuda-Azores high* in the northern hemisphere advects cold water southward in the form of the *Canary current* to northwestern coastal Africa, stabilizing the atmosphere and supporting dry conditions in Morocco and adjacent lands.

Much of northern and southern Africa falls under the influence of these high pressure zones. However, the northeastern quadrant of Africa is not influenced by a subtropical anticyclone. Rather, the hot, dry conditions of the Arabian Peninsula are advected toward Africa by the northeast trade winds. These conditions are reinforced by strong atmospheric stability around the "horn" of Africa (Somalia) by the effect of the *Ekman spiral* as surface waters are deflected offshore and replaced by *upwelling* of colder waters from beneath the surface. In northwestern Africa the northeast trade winds are known as the **Harmattan winds**. These winds can sometimes be especially persistent in blowing dry Saharan air southwestward, resulting in very hazy conditions over western coastal Africa, particularly along the Gulf of Guinea region, and they can even occasionally deliver dust and sand to the Caribbean and northern South America.

The Sahara Desert—the largest desert on Earth—dominates northern Africa, spanning approximately 65° of *longitude* and 20° of latitude. The deserts of southwestern Asia, such as the Arabian Desert and the Rub-al-Khali, are extensions of the Sahara. The lack of Saharan precipitation and cloud cover makes the Sahara among the most inhospitable places on Earth. The region is also arguably the hottest on Earth. The highest temperature reading on record occurred within this area on September 13, 1922, when a temperature of 58°C (136°F) was recorded at El Azizia, Libya. The deserts of the southern hemisphere are only slightly less brutal. Harsh conditions exist in the Kalahari Desert, which occupies much of Botswana, Namibia, Zambia, and the Republic of South Africa. Still, permanent, although largely nomadic, populations have found ways to adapt to these harsh conditions.

When the *solar declination* is in the northern hemisphere (as it is in June and July), the trade winds generally converge at the ITCZ north of the equator. When the southeast trade winds cross the equator on their trek to the ITCZ, they are influenced by a rightward (rather than leftward) Coriolis deflection. They, therefore, become southwesterly rather than southeasterly winds, and they are termed **counter-trade winds (Figure 10.3)**. These southwesterly counter-trades offer some respite from the heat in the Sahel and southern Sahara region, because they can push the oppressively hot cT air masses back toward the northeast. Likewise, when the solar declination is

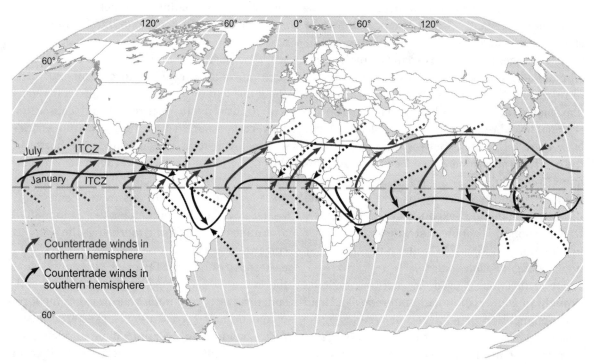

Figure 10.3 Counter-trade winds.

in the southern hemisphere, the northeast and southeast trade winds generally converge in the southern hemisphere at the ITCZ's position at that point in the year. As this happens the northeast trades change direction after crossing into the southern hemisphere to become northwesterly counter-trades.

Another important circulation feature is the **Turkana jet stream**—a low-level (generally from about the 950- to 650-mb level) jet stream that exists between the Ethiopian and East African highlands. Taking its name from the nearby Lake Turkana basin, this easterly or southeasterly jet peaks in intensity between February and March. Although most of Africa is devoid of high mountains, the orography of this section of Africa is most important in the development of the jet. The highlands constrict flow through the region, forcing *wind channeling* as air speed increases when flow is confined. Jet speeds vary diurnally, with the strongest winds occurring during the morning and the weakest winds in the afternoon. The jet is partly responsible for dry conditions over the Lake Turkana basin, as the jet inhibits the development of mesoscale cyclonic circulations over the region.

The Turkana jet differs from the **East African low-level jet**, which is also influenced by regional orography. The East African jet develops during the daytime hours across the East African coastline. Airflow is directed inland from east to west toward an active convection zone governed by surface heating. The jet is responsible for water vapor *fluxes* and resultant precipitation across East Africa.

The influences of the larger-scale monsoonal flow regime and thermal and frictional forcing factors are also important across the region. The monsoon circulation triggers the **Somali jet**, also known as the **Findlater jet**, another important feature of the tropical circulation in Asia. This low-level jet provides southwesterly flow paralleling the Somali coast, in association with the summer *Asian monsoon* system, advecting moisture into south Asia. Later in the summer monsoon season, the Somali jet and the entire *southwesterly monsoon* give way to the southeasterly flow over much of south and southeast Asia caused by the clockwise flow around the *Hawaiian high*.

■ Climatic Setting of Australia and Oceania

The Australian continent is the largest island in a region collectively known as **Oceania (Figure 10.4)**. The western portion of this region is sometimes referred to as the **Maritime Continent**—the southeastern continuation of the Asian land mass, including Australia and the thousands of smaller islands occupying the southwestern Pacific Ocean basin. These islands include those that make up Indonesia, the Philippines, and Papua New Guinea. Oceania also includes the many island chains elsewhere in the tropical western Pacific basin, including those comprising Melanesia, Micronesia, and Polynesia (the latter of which includes the Hawaiian Islands). For simplicity, the term "Oceania" is used to refer to all islands extending from the southeast Asian mainland except Australia and New Zealand. Much of the region was once connected to southeast Asia during the last *glacial advance*, approximately 18,000 years ago. During that time sea levels were low enough to expose land bridges between many of the larger islands.

General Characteristics

Most of the climatic landscape in this part of the world falls into only two major climatic types: tropical and dry. Virtually all of Oceania is classified

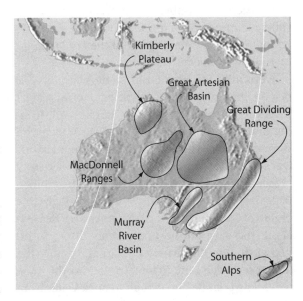

Figure 10.4 Physiographic features of Australia and New Zealand.

under the heading of tropical climates, whereas most of Australia is defined as being dry, and New Zealand is largely *mesothermal*. As in Africa, seasonal migration of the ITCZ plays an integral role in the tropical climates. In this region of the world the position of the ITCZ is influenced by the development of the Asian monsoon and determines moisture **anomalies**—departures from the expected atmospheric conditions—both intra- and interannually.

Some regions of Oceania, however, receive only minimal seasonal temperature and precipitation variation. This is especially true of regions lying along, or very near, the equator. Precipitation in these locations is forced by thermal characteristics and the development of afternoon thunderstorms. Many areas in this region receive precipitation on most days in the year.

El Niño-Southern Oscillation Influences

The greatest single cause of climatic variability in this region is the *El Niño-Southern Oscillation (ENSO) event*. As discussed in Chapter 4, an *El Niño*, also known as a warm-ENSO event, refers to a shifting of warm equatorial waters from the western equatorial Pacific Ocean to the eastern equatorial Pacific. A similar but less intense shift is caused annually during the transition between northern hemisphere summer–southern hemisphere winter to northern hemisphere winter–southern hemisphere summer. Such a transition causes a weakening of the Hawaiian high and the *South Pacific high*, which weakens the trade winds.

Normally, the equatorial trade wind flow is responsible for pushing warm equatorial waters from the eastern Pacific basin to the west, leading to the normal pile-up of very warm water off the eastern coasts of Australia and Oceania. This warm-water accumulation induces low sea-level pressure and low-level atmospheric instability, triggering diurnal thunderstorms through this region of Oceania and northern Australia. Sea levels in the west are normally about 20 cm (8 in) higher than in the eastern equatorial Pacific. The higher sea levels remain intact as long as the trade wind flow is sufficient enough to push water from east to west. When the trades weaken during the normal seasonal transition, some of this warm water migrates back toward the eastern Pacific ba-

sin. This slight weakening is typical in northern hemisphere autumn.

Approximately every 3 to 7 years a warm-ENSO event occurs in association with a massive reduction in the strength of the trade wind flow. During these times warm water migrates back eastward in autumn. This flow induces significant changes in the overlying atmosphere in both the western and the eastern equatorial regions. The normally warm western basin waters with low atmospheric pressures and abundant convective thunderstorms give way to greater atmospheric *stability* and higher pressure in the western tropical Pacific Ocean. The low atmospheric pressure and convective activity follow the warm surface waters eastward. This brings precipitation to the central and perhaps even the eastern tropical Pacific basin, where higher atmospheric pressures forced by cold oceanic upwelling usually occur. In short, a warm-ENSO event reverses the normal atmospheric pressures and winds in the tropical Pacific basin. The atmospheric "see-saw" in sea-level pressures between the tropical western and eastern Pacific Ocean is the *Southern Oscillation*.

Australia and Oceania are influenced directly by the frequency and severity of warm-ENSO events. The normally wet regions of the western equatorial Pacific (as well as much of monsoonal Asia) become drought-ridden, whereas the normally dry eastern Pacific basin experiences abundant precipitation. These conditions normally last between 8 and 12 months, but longer periods have occurred. Significant environmental stress occurs through eastern Australia and most of Oceania during warm-ENSO events, as the tropical ecosystems depend on a high amount of precipitation during each month of the year.

Figure 10.5a shows the winter/spring mean rainfall anomalies for Australia during the 12 strongest warm-ENSO events on record. Most of eastern Australia is significantly drier than normal. The northern reaches of Australia are normally dry during the winter/spring time period, partly mitigating the impact of a warm-ENSO event. The ENSO influence in northern Australia is felt more during the summer months, as Figure 10.5b indicates. Water shortages in the entire region can become a problem during these events, particularly on small islands where recent local precipitation represents the source of available fresh water.

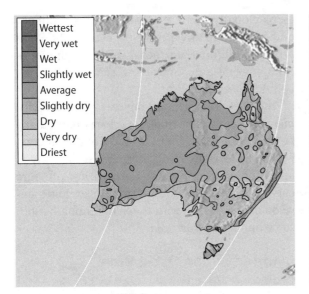

Figure 10.5a Winter/spring mean precipitation anomalies during warm-ENSO years. *Data from*: The Australian National Climatic Centre, Bureau of Meteorology, Melbourne, Victoria, Australia.

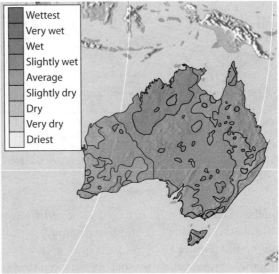

Figure 10.6 Winter/spring mean precipitation anomalies during cold-ENSO years. *Data from*: The Australian National Climatic Centre, Bureau of Meteorology, Melbourne, Victoria, Australia.

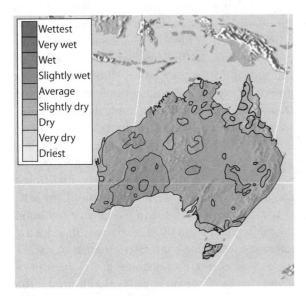

Figure 10.5b Summer mean precipitation anomalies during warm-ENSO years. *Data from*: The Australian National Climatic Centre, Bureau of Meteorology, Melbourne, Victoria, Australia.

Australia/Oceania is also directly affected by *La Niña* or cold-ENSO conditions. La Niña is a strengthening of the "normal" atmospheric and oceanic circulation conditions. The normally wet areas of northern Australia and virtually all of Oceania become wetter as regions in the eastern Pacific become even drier. **Figure 10.6** shows precipitation anomalies for Australia during the cold-ENSO winter/spring months. Figure 10.6 indicates

that virtually all of eastern Australia and especially the northern region of the Northern Territory become exceedingly wet during La Niña. The region of positive precipitation anomalies extends northward into the bulk of Oceania.

The environmental stress of El Niño and La Niña is significant across the region. This area may be viewed as one of "feast or famine" regarding precipitation. Both extreme phases stress the physical environment and human activities. The ENSO influence extends to drought and fire, which occur across broad areas during the drier warm-ENSO periods, and severe thunderstorms, increased tropical cyclone activity, and floods, which occur largely in association with cold-ENSO events.

The "flip side" of events indicates the "good" manifestations of the extreme climatic reversals. The Australians often view every situation with an eye toward both "bad" and "good" (or "flip side") inspection. Drought may be viewed as being detrimental for farmers, but it may also be seen as advantageous for owners of beach resorts. El Niño and La Niña events offer many opportunities for flip-side viewpoints.

South Pacific Convergence Zone

The **South Pacific Convergence Zone (SPCZ)** is another important contributor to the climatic setting of Australia and Oceania. The SPCZ is an offshoot

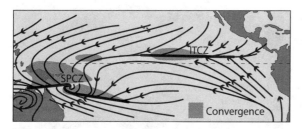

Figure 10.7 The South Pacific Convergence Zone. *Adapted from*: C.S. Ramage, in Barry, R.G. and R.J. Chorley (Eds.), *Atmosphere, Weather, and Climate* (8th ed.). New York: Routledge.

of the ITCZ, generally oriented in a northwest-southeast alignment or perhaps a north-south alignment (**Figure 10.7**), and it is most active during the summer months when surface heating is maximized. Its band of convection is approximately 200 to 400 km (125 to 250 mi) wide, stretching from the ITCZ near the Solomon Islands to the region of Fiji, Samoa, Tonga, and islands southeastward. Like the ITCZ, diurnal cloud cover and afternoon and early evening precipitation are abundant.

The SPCZ forms as a convergence zone between the South Pacific high and the frequent anticyclones that tend to remain in place over or near Australia. Because both of these anticyclones are in the southern hemisphere, they support counterclockwise-rotating air around them. The SPCZ is formed where the weak southwesterly or southerly flow from the eastern side of an Australian anticyclone meets with the very strong northeasterly or northerly flow around the western side of the South Pacific high. Because of variability in both strength and position of the two anticyclones that form the convergence zone, the SPCZ itself may fluctuate in orientation from northwest-southeast to north-south, with the northern part acting almost as a "hinge" with more "swinging" on the southern side. It also varies in its longitudinal location. The position where the SPCZ meets the ITCZ is generally a zone of maximum convection, and it usually occurs in the western tropical Pacific near Indonesia.

Migrations of the SPCZ affect the moisture regimes across the eastern regions of Oceania. A significant shift in the position of the SPCZ occurred during the late 1970s, coincident with a major climatic adjustment throughout North America and the Pacific. The large-scale circulation shift caused higher surface temperatures in most re-

gions globally and an increase in the frequency and magnitude of warm-ENSO events. The SPCZ typically migrates northeastward during warm-ENSO events and toward the southwest during cold-ENSO events.

The SPCZ is a primary region for the development of South Pacific tropical cyclones, particularly beyond about 5° of latitude south of the equator, where the Coriolis effect increases sufficiently to allow for rotating winds around a low-pressure core. Typically four to five tropical cyclones develop in association with the SPCZ annually, with some impacting Australia.

Madden-Julian Oscillation

Another important contributor to the climates of Oceania and coastal Australia is the **Madden-Julian Oscillation (MJO)**, a phenomenon that explains much of the intra-annual fluctuations in tropical climate. The MJO is often referred to as the **30–60 Day Oscillation** or the **40–50 Day Oscillation** because of the time scales on which it generally operates, with some variability present in the timing of the oscillation. The MJO involves a band of cloudiness and precipitation traveling from west to east across the tropical oceans, sandwiched between bands of suppressed cloudiness and precipitation. Upper- and lower-level winds and sea surface temperatures undergo fluctuations with the presence or absence of the associated convections. Because clouds are involved, the MJO is usually investigated using *longwave radiation* data detected by satellites. Variations in outgoing longwave radiation—terrestrial radiation escaping to space—are proportional to the amount of cloud cover in a region because clouds hinder the loss of terrestrial radiation to space.

The eastward-propagating outgoing longwave radiation anomalies across the tropical Indian and Pacific Ocean recur at a 30- to 60-day interval at a given tropical location. Whenever a band of convective cloudiness rolls through a region, precipitation peaks occur. Between convection bands there are periods of low precipitation. Winds are directed toward the MJO-related convective area from the east and west. Once inside the convective area, convergence increases the uplift of air, triggering thunderstorm development. A region of increasing convection occurs in advance of the main convection zone (between the thunderstorms at the convection zone and the clear skies at the suppressed convection zone) as greater amounts of *evaporation* enhance low-level instability and low-level moisture storage, both resulting from higher amounts of insolation in the suppressed convection region. This typically occurs east of the convection zone. The wave then propagates eastward over the enhanced convection zone that then becomes the primary area of convection. The progression then develops anew and the line continues its eastward propagation.

The MJO may be thought of as being a smaller-scale version of the air–sea interactions that drive El Niño events. In warm-ENSO events, eastward-propagating convection areas are driven by the movement of warm-water *equatorial Kelvin waves*. The warm surfaces trigger overlying convective thunderstorms that move eastward. MJO propagation is analogous to this because the clear skies of the suppressed convection zone allow sea surface temperatures to increase in advance of the line of thunderstorms (the convection zone). This creates the enhanced convection zone (which is still marked by clear skies). The line of thunderstorms then migrates toward this area of enhanced convection just as the convection areas migrate along with an equatorial Kelvin wave associated with a warm-ENSO event. The MJO completes its cycle of wave propagation over a 30- to 60-day period, at which point the cycle begins anew.

Because the MJO involves the continuous migration of a convective zone, it introduces a large-scale overriding climatic forcing mechanism to the affected tropics. Knowledge of the MJO leads to forecasts of increased or decreased convective thunderstorm development for particular regions. Thus, it is an important climatic component for much of Oceania. Such short-term precipitation oscillations are superimposed on larger-scale phenomena such as monsoon rains and ENSO-induced precipitation anomalies. The MJO may either dampen or intensify the precipitation anomaly produced by the larger-scale phenomena. For instance, the MJO convective zone mitigates La Niña–influenced droughts in western Oceania.

Quasi-Biennial Oscillation

Another circulation feature important to tropical climates is the **Quasi-Biennial Oscillation (QBO)**.

Before the use of radiosonde balloons, satellites, and other modern data-gathering tools, little was known about upper-atmospheric motions, particularly in the tropics. During the late 1800s it was assumed that winds in the equatorial *stratosphere* blew from east to west. This assumption was supported by observation of the east-to-west drift of the dust cloud ejected into the stratosphere on August 27, 1883 by the eruption of the Krakatau volcano in Indonesia—the largest in modern times. The dust cloud encircled the planet over the low latitudes in only 13 days. These stratospheric winds were then named the **Krakatau easterlies**.

In 1908 data balloons launched above Lake Victoria in Africa recorded westerly winds in the equatorial stratosphere, a finding that contradicted the previous assumption that the Krakatau easterlies prevail unobstructed. The winds are known as the **Berson westerlies**. The nature of these Berson westerlies in the equatorial stratosphere was not resolved until the early 1960s, when it was discovered that stratospheric winds above the equator oscillate in direction with a period of about 26 months. This oscillation of the equatorial zonal stratospheric wind flow came to be known as the QBO.

Further investigations revealed that winds associated with the easterly or westerly phase of the QBO propagate downward from the upper levels of the stratosphere at a rate of about 1 km per month until they reach the tropical *tropopause*. At that point they weaken and the opposite phase is initiated, with westerly (or easterly) winds in the upper stratosphere propagating downward slowly over a period of several months. The winds in the easterly phase of the QBO are approximately twice as strong as those in the westerly phase. A complete explanation of the cause of this oscillation has not yet been found.

The effects of the QBO are numerous. In addition to mixing gases, including *ozone*, throughout the otherwise-stagnant stratosphere, the QBO also combines with ENSO to influence east Asian monsoon precipitation. Its variation has also been linked to Sahelian precipitation totals. The strongest influence of the QBO is related to tropical cyclone activity in the western Pacific, the western Indian, and the North Atlantic basins. Pacific tropical cyclone activity increases in the westerly phase of the QBO, whereas activity in the western Indian Ocean basin increases with the easterly phase. Increases in *hurricane* activity in the North Atlantic are linked to the westerly QBO phase, with strong decreases in the easterly phase. One reason for the unprecedented 2005 North Atlantic hurricane season was that the QBO was in a strong westerly phase. This combined with very high sea surface temperatures across the tropical North Atlantic, the Caribbean, and the Gulf of Mexico to produce a record-setting 28 named tropical cyclones, including 15 hurricanes.

■ Climatic Setting of Latin America

The northernmost point in Latin America is in the northern hemisphere subtropics, with the poleward tip of South America extending to approximately 55°S latitude (**Figure 10.8**). Because of the shape of the continent, the bulk lies within the tropics, with only a tiny strip approaching the high latitudes.

Perhaps the most striking feature of the physiography of Latin America is the western cordillera, from the Sierra Madre Occidental in Mexico along the entire western edge of Central America and continuing as the Andes Mountains of South America. The Andes run the length of western South America from the Llano Highlands of Venezuela and Colombia to southern Chile, covering

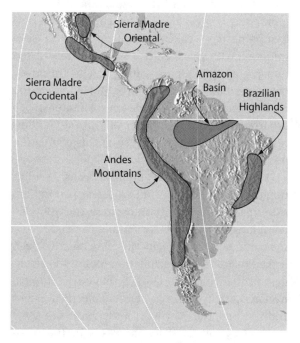

Figure 10.8   Physiographic features of Latin America.

most of Ecuador and a significant portion of both Peru and Bolivia.

East of the Andes are lowlands across most of Latin America. These include the Central Valley of Mexico, the vast Amazon basin that dominates the continental interior of South America, and the grasslands known as the Pampas and the Patagonian Highlands of Argentina. The eastern side of Mexico and parts of Brazil, particularly the southeast, are elevated, though the Sierra Madre Oriental and the Brazilian Highlands are not as high, steep, or continuous as the western mountains. In parts of Central America the Occidental and Oriental essentially continue as a central highland region.

Of the mountain features, the Andes exert by far the most influence on climate. They act as a barrier to moderating air masses of Pacific origin and also prevent the trade winds from advecting moisture to the west coast of South America. The mountains combine with the very cold *Humboldt (Peru) current* off the western coast to keep the atmosphere very stable. Such conditions promote clear skies and dry conditions through the narrow coastal band.

The Atacama Desert of southern Peru and northern Chile is arguably the driest desert on Earth. Entire ecosystems there have evolved around the prevalence of fog, the only moisture source through much of the coastal zone. *Advection fogs* are frequent and persistent along this region, forming as warm, moist air blows across the waters of the South Pacific Ocean. Just before reaching the coast, the air blows over the cold Humboldt (Peru) current. Combining waters originating from relatively high latitudes near Antarctica with coastal upwelling resulting from Ekman transport, this current creates an ocean surface that is cold enough to chill air passing over it to the dewpoint temperature. The resultant fog blows inland where trees and animals are adept at catching the fog droplets. For instance, spiders string webs between vegetation such that the webbing catches and collects water droplets that then converge into small holes created by the repeated dripping onto the soil. Plants then grow around the hole, taking advantage of the water source that nourishes the immediate area.

A cold surface ocean current also exists off the west coast of North America—the California current. This feature does not affect the west coast of Mexico and Central America significantly, because the coastline curves away from the *gyre* in the northern tropical Pacific Ocean. This contrasts with the shape of the South American western coastline, which conforms very well to the South Pacific gyre.

Despite its interior location, the Amazon basin acts as a primary moisture source for a large segment of South America. Abundant heat energy combines with efficient soil and biomass moisture storage to create a very humid atmosphere. Water evapotranspires throughout the day, increasing local water vapor content in the atmosphere. By the afternoon cumulus cloud development becomes extensive enough to create copious precipitation. This system is sometimes termed the *rain machine* because such a high percentage of the local rainfall is recycled continuously from local moisture sources. Additional water is advected into the continent by the trade winds and some arrives as meltwater from snowpack in the high Andes Mountains. Local water also runs out to sea via the extensive fluvial network.

Significant disruption of the rain machine occurs through agricultural and other developmental practices. *Deforestation* of the Amazon basin is a primary concern for several reasons, including its influence on climate. This largest region of rain forest on Earth acts as the "lungs" of the atmosphere by supplying the global atmosphere with oxygen from *photosynthesis*. It is also feared that decreases in the forest extent will cause a fundamental change in the regional atmospheric moisture balance. Decreasing moisture in the region will subject the area to higher temperatures, which will lead to increased drying. Some believe that deforestation may one day lead to a South American landscape that resembles the extensive deserts of northern Africa.

ENSO Contributions

Like most tropical regions the greatest source of climatic variability is caused by El Niño and La Niña events. The migration of a warm water pool from the western to the eastern equatorial Pacific during warm-ENSO events displaces the very cold surface waters along the western edge of South America. The warm water pool overrides the colder waters because of its decreased density. In addition, the warm water pool causes a deepening near the coast of the *thermocline*—the boundary between cold water in the deep ocean and much warmer water near the ocean surface. Although coastal upwelling still occurs during warm-ENSO events due to the Ekman transport, this water is much warmer than normal and the upwelling does not penetrate all the way up to the surface. The result is that the horizontal and vertical sea surface temperature regimes are fundamentally changed along much of the western South American coastline.

The impact of such a change is profound. First, ecosystem balance is altered, as the indigenous ecosystem largely depends on the existence of upwelling of nutrient-rich cold water. Warming of the sea surface and a net decrease in nutrients fundamentally change the regional oceanic food chain. Second, the increased sea surface temperatures destabilize the overlying atmosphere, leading to low-pressure anomalies that induce clouds and precipitation in a normally very stable, clear, and dry environment. The positive precipitation anomalies wreak havoc on local ecosystems. Flooding and mudslides are commonplace throughout western South America during warm-ENSO events.

The El Niño impact carries to other regions as well. Displacement of the usual high pressure over western South America causes a displacement of atmospheric flow across the tropics, including the area over the Amazon Basin. The typically low atmospheric pressure of the Amazon Basin gives way to much higher pressures, clearer skies, and reduced precipitation during warm-ENSO events. Because of the high temperatures, the basin depends on frequent and heavy daily rainfall totals to avoid *deficit*. Without diurnal thunderstorms the region may dry in a very short period of time. The resulting ENSO-induced drought causes considerable stress on the ecosystem.

La Niña events also cause environmental stresses throughout South America. The anomalously cold waters off the western coast of the continent cause unusually dry conditions (even in environments that are normally dry) through the western regions. Over Amazonia the opposite occurs, with very heavy afternoon rains.

■ Climatic Setting of Antarctica

Antarctica is the fifth-largest continent on Earth. Because it sits over the South Pole (**Figure 10.9**), it is characterized by very low temperatures and high winds. Rapid changes in weather occur near the coastal fringes, but the interior is marked by persistent cold. The region is the coldest on Earth, being far colder than its Arctic counterpart. The reason is that the Arctic is composed of ocean waters that retain some heat energy for transfer to the overlying atmosphere, even in winter when the Arctic Ocean is frozen.

Not only is Antarctica affected by high latitudes, but it is also strongly influenced by *continentality* and very high elevations. Despite its relatively modest high point (the Vinson Massif) of 4897 m (16,066 ft), the mean elevation of 2200 m (7200 ft) far exceeds that of any other continent. Such high elevations lead to low temperatures and low atmospheric pressures. At an elevation of 2800 m (9200 ft), the South Pole has an average surface pressure of only 680 mb. Ironically, though, the continent also contains the lowest point on Earth that is not under seawater. The Bentley Subglacial Trench extends 2555 m (8382 ft) below sea level and is currently locked beneath an extensive ice sheet.

Not surprisingly, the very high latitudes, continental influence, and high elevations combine to produce extremely cold conditions. The lowest temperature ever recorded on Earth, −89°C (−129°F), occurred at Vostok, Antarctica, on July 21, 1983. The highest average summer temperatures are around −7°C (20°F), whereas winter temperatures for much of the continent may hover around or below −34°C (−30°F).

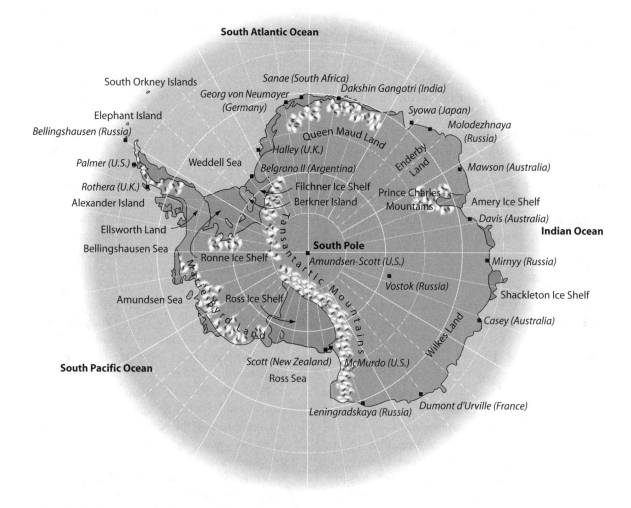

Figure 10.9   Physiographic features of Antarctica.

A noticeable precipitation gradient exists between coastal locations and the interior. At coastlines, not only is more moisture present in the warmer coastal air, but the air mass contrast is likely to be greater, as relatively warmer maritime air masses interact with the bitter cold continental air masses. The extreme low temperatures in the continental interior dictate that *vapor pressures* are very low, even at saturation. Blizzards and **whiteouts**—atmospheric conditions that cause the sky to blend with the snow-covered surface, eliminating the horizon—are common, but they mainly occur from blowing ice, not newly precipitated snow. All clouds are *stratiform* in nature because vertically oriented clouds are impossible given the low energy amounts for uplift.

There are two major reasons that Antarctica is extensively studied for its potential role in anthropogenic *global warming*. First, the high latitudes of the continent are intriguing because modeling studies of the impacts of human-influenced global warming show that most of the global warming signal is concentrated at relatively high latitudes. Slight warming produces melting, exposing bare soil, thereby decreasing the *albedo* and allowing further warming to occur in a positive feedback system. Second, the continent is overlain with a huge continental ice sheet that has increased *ablation* during warmer climatic conditions, thereby contributing to global—or **eustatic**—sea level increases.

Less than 5 percent of the continent is ice free, with the ice-free portions being small coastal locations, islands, and some mountain areas. The present ice sheet is similar in size and thickness to that which covered North America at the peak of the *Wisconsin Glacial Phase* 18,000 years ago. The region of East Antarctica is covered by a layer of ice that is 4000 m (13,000 ft) thick. This ice sheet is so large that liquid lakes have formed beneath it in some locations. These lakes are composed of basal meltwater formed from the friction between the moving ice sheet and the continental bedrock. The largest is Lake Vostok, which lies approximately 4 km (2.5 mi) beneath the Vostok Research Station Ice in East Antarctica. Lake Vostok is approximately 280 by 60 km (174 by 36 mi) and is 500 m (1600 ft) deep.

Two major coastal indentations occur in the Ross Sea, which opens to the Pacific Ocean, and the Weddell Sea, which opens to the Atlantic Ocean. Both seas are covered by extensive ice shelves, with the Ross Sea covered by the **Ross Ice Shelf** and the Weddell Sea covered by the **Ronne Ice Shelf** and the **Filchner Ice Shelf**. The shelves are between 180 and 1220 m (600 and 4000 ft) thick. Ablation rates in this region are studied extensively as possible indicators of global warming. Sea level increases resulting from melting of these shelves would inundate vast coastal zones of the present-day continents. It must be noted that the ice shelves are distinct from the seasonal sea ice that forms around the continent during the cold season. However, the health of the ice shelves is related to the number of icebergs that typically form as ice chunks calve off of the edge of the shelves.

Considerable variation occurs in Antarctic temperatures, but a notable warming trend has been recorded in recent years. Most of the warming has occurred near the coast, whereas interior temperatures have remained stable. No appreciable temperature change has occurred at the South Pole since the 1970s, but some coastal locations have warmed at a rate of about three times higher than the global average.

There is some correlation between temperatures at the South Pole and the Southern Oscillation, but the correlations are evident only by lagging the temperatures by 1 year. After lagging, the correlation is statistically significant. It, therefore, takes about a year for the anomalies produced by the Southern Oscillation to have an influence on temperatures at the South Pole. This suggests at least some degree of large-scale forcing in continental interior temperature variation. More recent research has linked temperature variation in nonpolar areas to the Southern Oscillation.

■ Regional Climates

A—Tropical Climates

Tropical (A) climates occur throughout Southeast Asia, Indonesia, northern Australia, Oceania, Central Africa, Central America, the Caribbean Sea, and most of South America (see Figure 8.1). The southern tip of Florida is also classified as tropical. The tropical climates in Asia dominate an arc from Indonesia through India. Central America is almost exclusively tropical in classification, except for the highland spine.

Tropical Rain Forest (Af) and **Tropical Monsoon (Am) climates** dominate throughout most of the equatorial tropics. The Tropical Monsoon (Am) climate is found mainly along coastal margins and in larger sections of South and Southeast Asia. Inland areas, particularly those located several degrees of latitude from the equator and affected by seasonal wind shifts, are classified as **Tropical Savanna (Aw) climates**. The "winter" dry season that characterizes Aw climates marks a transition between the very wet tropical climates and the drier climates of interior poleward locations. In Asia these climates are caused by the monsoonal wind shift associated with the development of the *Siberian high* during winter and the *thermal low* during summer.

Af—Tropical Rain Forest Tropical Rain Forest climates are characterized by hot, humid, and rainy conditions that persist year-round. The climatic type is mainly associated with low-elevation areas bounded by higher topographical regions. The main attribute of the climate is a total lack of seasonality. In the tropics, neither day lengths nor the amount of insolation receipt at a given time of day over the course of the year changes significantly. Other tropical climates have pronounced seasonal pulses of higher rainfall, but this does not occur in most Af climates. Slight seasonal variations in precipitation do occur in some locations, as shown in the climographs of Af locations discussed on page 235.

Thermal Characteristics By definition, in tropical climates the coldest month must have an average temperature over 18°C (64°F). In Af climates the diurnal range in temperature, typically about 8 to 10 C° (14 to 18 F°), is much greater than the annual range in mean monthly temperature, which may be as small as 2 to 3 C° (3 to 5 F°).

Despite the generally sultry conditions, daily maximum temperatures are often much lower than may be expected, with temperatures rarely exceeding 32°C (90°F) at most Af locations. Extensive cloud cover and abundant late afternoon precipitation explain these temperature maxima. Remember that insolation peaks at solar noon, but this energy receipt does not produce a temperature peak at that time. Instead, a lag of 3 to 4 hours occurs between maximum solar reception and maximum air temperatures, because abundant insolation continues to be absorbed at Earth's sur-

face for several hours after solar noon. Longwave energy is continuously emitted to the atmosphere, which absorbs some wavelengths of energy, resulting in surface air temperature increases for as long as the shortwave and longwave energy gained by the surface exceeds the longwave energy emitted to the atmosphere.

The increase in surface temperatures occurs more effectively in Af climates than in most, but the abundant heat energy combines with high humidity values to generate convective thunderstorms that peak during the time of maximum air temperature. The *cumuliform clouds* then cool the area by shading the surface from additional solar energy receipt and by precipitation generated from the thunderstorms. This precipitation cools the surface as the water evaporates, thus transferring energy from the surface to the atmosphere as *latent heat*. Less energy is left as *sensible heat* at the surface, reducing air temperatures.

Nocturnal temperatures in Af climate regions are also consistently high because so much water exists in the atmosphere. This leads to an enhanced local *greenhouse effect*, which allows for emission of longwave radiation back toward the surface by water vapor and/or cloud cover. Nighttime temperatures are usually in the twenties Celsius (seventies Fahrenheit), and they again show remarkable consistency through the year.

Temperature and precipitation consistency throughout the year promotes the widest **diversity** of life forms of any climate because the limitations to growth are minimized. However, competition, predation, and disease limit the number of individuals of any species present. If you were to list every species found in an area of 1 km^2 of tropical rain forest, the list would be very long, but only a few individuals of each species would be present.

Moisture Characteristics The "classic" Af precipitation regime is shown by the climographs of Hilo, Hawaii (**Figure 10.10a**), and Singapore, Singapore (**Figure 10.11a**). Seasonal variations in precipitation are more notable in the climographs of Colombo, Sri Lanka (**Figure 10.12a**), and Belize City, Belize (**Figure 10.13a**). Even stronger seasonality is present in the precipitation regime of Manaus, Brazil (**Figure 10.14a**), which is not as wet from June through September when the ITCZ migrates to the northern hemisphere.

Rainfall amounts are very high, leading to annual totals over 200 cm (80 in) for most locations.

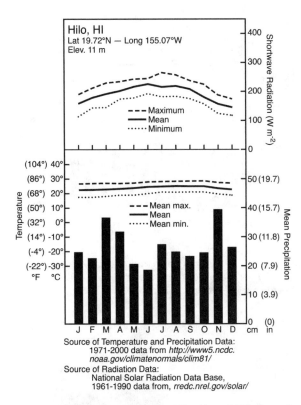

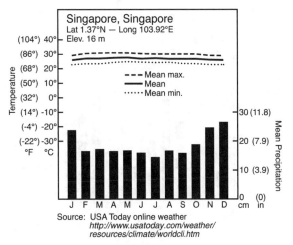

Figure 10.10a Climograph for Hilo, Hawaii (Af climate). Average annual temperature is 23.3°C (73.9°F) and average annual precipitation is 320.7 cm (126.3 in).

Figure 10.11a Climograph for Singapore, Singapore (Af climate). Average annual temperature is 27.1°C (80.8°F) and average annual precipitation is 226.8 cm (89.3 in).

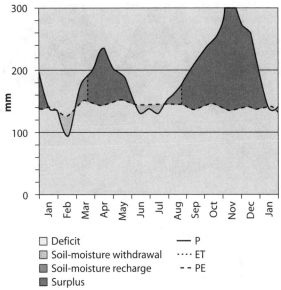

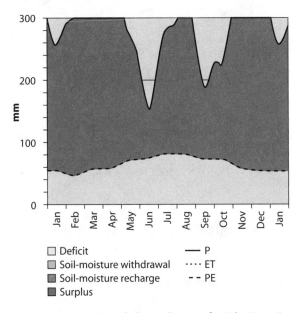

Figure 10.10b Water balance diagram for Hilo, Hawaii.

Figure 10.11b Water balance diagram for Singapore, Singapore.

However, precipitation amounts vary, extending from annual mean lows of about 150 cm (60 in) to over 1000 cm (400 in) across given locations. Precipitation is tied to the development of afternoon convective thunderstorms along the ITCZ. During summer the superheated continent of Asia, and the associated Tibetan low, draws the ITCZ up to its farthest poleward location—to nearly 30°N latitude. Convergence of the trade winds (or counter-trades) associated with the ITCZ typically enhances thunderstorm activity during that time, leading to pronounced summertime peaks in precipitation.

Topography can also exert a significant impact on precipitation in Af regions. High-elevation regions enhance precipitation totals. The cooling in high elevations also results in greater cloud and fog cover. This is especially true for most Caribbean

All water balance diagrams in this chapter (part b of Figures 10.10–17, 19–40, 43–45) are modified with permission using Web/WIMP version 1.01 implemented by K. Matsuura, C. Willmott, and D. Legates at the University of Delaware in 2003.

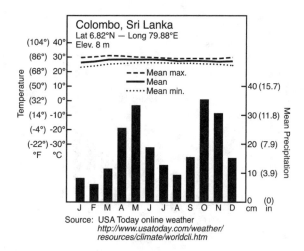

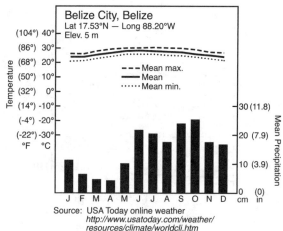

Figure 10.12a Climograph for Colombo, Sri Lanka (Af climate). Average annual temperature is 27.8°C (82.0°F) and average annual precipitation is 223.8 cm (88.1 in).

Figure 10.13a Climograph for Belize City, Belize (Af climate). Average annual temperature is 26.4°C (79.5°F) and averge annual precipitation is 181.9 cm (71.6 in).

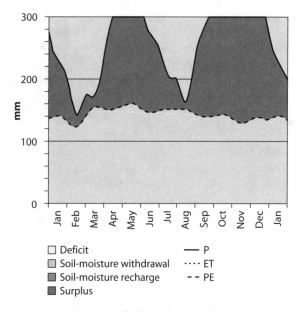

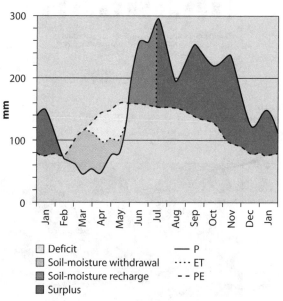

Figure 10.12b Water balance diagram for Colombo, Sri Lanka.

Figure 10.13b Water balance diagram for Belize City, Belize.

islands, which typically have volcanic mountain cores. The resulting orographic effect creates islands that are dominated by Af climates on the *windward* side (as in Hilo) and much drier climatic types on the *leeward* side.

Another important component of the precipitation regime of Af climates is the tropical cyclone. These storms—called **typhoons** in the western tropical North Pacific Ocean and simply cyclones in the Indian Ocean and tropical South Pacific Ocean—are normally stronger and more frequent than their North Atlantic Ocean hurricane counterparts. The lowest sea level pressure on Earth

was recorded over the North Pacific Ocean during Typhoon Tip when on October 12, 1979 the internal pressure dipped to 870 mb. Most people correctly view tropical cyclones as being harbingers of destruction, but these systems are also integral to the water balance maintenance for many locations. Given the high evapotranspiration rates associated with these lush, low-latitude locations, vegetation stress may occur with only a few dry days. Tropical cyclones help replenish moisture shortfalls.

As expected, the water balance at Af climates shows few, if any, deficits. Hilo (Figure 10.10b) and

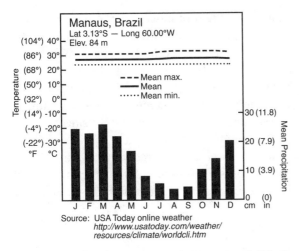

Figure 10.14a Climograph for Manaus, Brazil (Af climate). Average annual temperature is 27.9°C (82.2°F) and average annual precipitation is 181.1 cm (71.3 in).

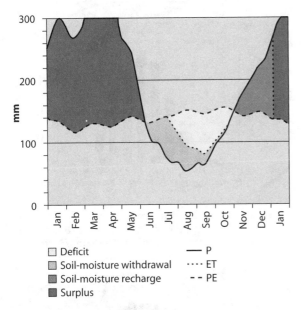

Figure 10.14b Water balance diagram for Manaus, Brazil.

Colombo (Figure 10.12b) have no signs of significant soil moisture shortage throughout an average year. Hilo is on the windward side of the northeast trade winds and receives orographic enhancement of precipitation year-round. In the case of Colombo, the Asian monsoon suppresses precipitation in winter, but not to the extent of reducing precipitation below the rate of *potential evapotranspiration*. Precipitation also decreases in late summer when the ITCZ is displaced northward of Colombo, but precipitation nevertheless manages to exceed potential evapotranspiration. Some soil moisture decrease (but no deficit) is characteristic

of Singapore (Figure 10.11b), and brief deficit occurs in both Belize City (Figure 10.13b) and Manaus (Figure 10.14b) at the end of the dry season. The deficits are not sufficiently long or intense enough to classify these cities outside of the Tropical Rain Forest climate, even though they are near the edge of the climatic zone, but some years may show climatic conditions that are more characteristic of climates other than Af climates.

Am—Tropical Monsoon Tropical Monsoon (Am) climates differ from Tropical Rain Forest areas only in that they exhibit a short dry season on a climatological average. Because of the consistency of temperatures, this dry season constitutes the only real seasonal variation in climate. This climatic type may be thought of as being a transition between the Tropical Rain Forest and the Tropical Savanna (Aw), because general characteristics fall between those two climatic types. The Am climate is present in South and Southeast Asia and parts of Oceania, the Caribbean, and South America.

The climatic region is usually found along coastal areas that receive persistent onshore winds for most, but not all, of the year. For a short period each year the trade winds may not advect moist air into these regions. In Am climates of Asia a winter dry season is typical because of the effects of the Asian monsoon, but in many locations, such as the Caribbean, this climate may be found in locations without a true monsoonal wind shift. In such cases, onshore flow regimes wax and wane seasonally along with the intensity of the subtropical anticyclones. "Weak" periods in the subtropical anticyclones cause a short transitional dry season as the trade winds flowing out of the anticyclones subside. This period is then followed by a **trade surge** in moisture-laden air, inducing a precipitation peak.

Thermal Characteristics Temperatures are very similar to those of the Af climate, with mean monthly averages exceeding 18°C (64°F) in each month. The primary temperature difference between the two climatic types involves the annual variation in temperature. Monsoon climates tend to be more variable in this regard, because of the influence of the dry season. During dry times the lack of afternoon cloud cover allows for higher daily temperature maxima, whereas enhancement of nocturnal cooling occurs due to the dampened local greenhouse effect. Because the ITCZ typically dominates the summer season, the annual temperature peak (and heightened diurnal variability) usually occurs just before that time, during late spring.

Moisture Characteristics Precipitation amounts vary from location to location, but the general rule is that a distinct dry season occurs just before a notable wet-season peak in precipitation. This pattern is evident in the climographs for this climatic type (**Figures 10.15a, 10.16a,** and **10.17a**). The Asian monsoon system produces a brief winter dry period in Manila, Philippines, and Dhaka, Bangladesh, whereas the southerly displacement of the trade winds and weakening of the Bermuda-Azores high leaves Tegucigalpa, Honduras, relatively dry in winter. Tropical cyclones provide moisture in late summer and autumn at all three sites.

Precipitation totals are very similar to those of the Af climatic type, resulting in similar rain forest ecosystems. Although the short dry season does occur, local water supplies are seldom fully

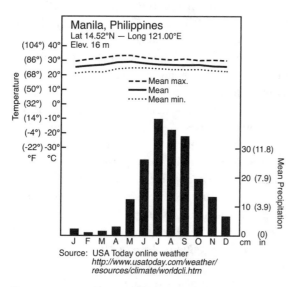

Figure 10.15a Climograph for Manila, Philippines (Am climate). Average annual temperature is 27.5°C (81.5°F) and average annual precipitation is 197.4 cm (77.7 in).

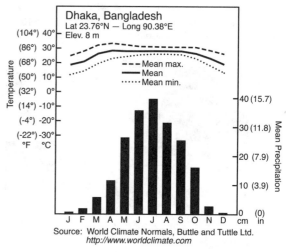

Figure 10.16a Climograph for Dhaka, Bangladesh (Am climate). Average annual temperature is 25.8°C (78.5°F) and average annual precipitation is 199.9 cm (78.7 in).

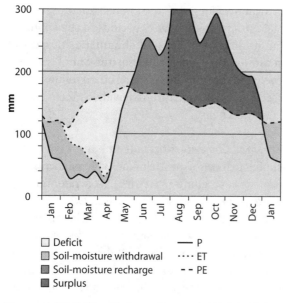

Figure 10.15b Water balance diagram for Manila, Philippines.

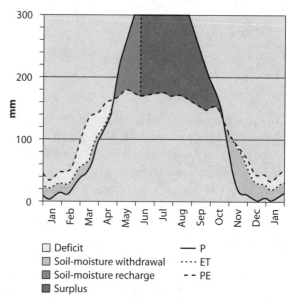

Figure 10.16b Water balance diagram for Dhaka, Bangladesh.

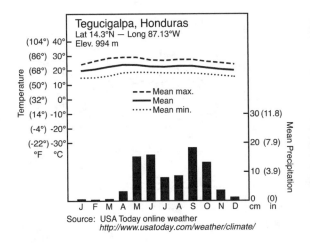

Figure 10.18   An acacia tree in a tropical savanna.

Figure 10.17a   Climograph for Tegucigalpa, Honduras (Am climate). Average annual temperature is 22.3°C (72.2°F) and average annual precipitation is 90.7 cm (35.7 in).

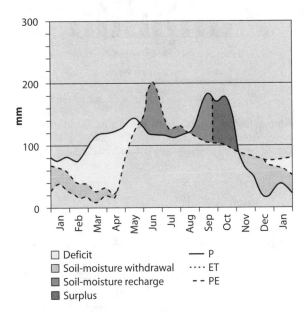

Figure 10.17b   Water balance diagram for Tegucigalpa, Honduras.

depleted, and the rain forests receive adequate water through soil moisture storage. The ITCZ typically produces the distinct summer precipitation peak for most locations. Other times are marked by lighter rainfall amounts as compared with the Af climate zone. In some cases, the drier months produce water balance deficits, particularly toward the end of the dry season, as indicated in the associated water balance diagrams (Figures 10.15b through 10.17b).

Aw—Tropical Savanna   Tropical Savanna (Aw) climates occur in regions poleward of Af and Am

climatic types. The most extensive of the tropical climates, the Aw climate is characterized by distinct wet and dry seasons of approximately equal length. The Aw dry season is typically longer and more severe than that associated with the Am climate. During the low-precipitation season insolation remains high, and therefore potential evapotranspiration rates are high. The result is a desiccated environment for several months.

The distinct dry season creates a landscape with far fewer trees than in the other tropical climates. Instead of rain forest, the regional vegetation is dominated by tall savanna grasses that grow through the wet season and become dormant during the dry season. Trees do exist, particularly along areas of local moisture accumulation, but the canopy is not closed, as in a forested environment. Instead, a "park-like" environment exists—the type of landscape that would be seen on a safari. A typical tree of this climatic type is the **acacia** (**Figure 10.18**). These trees tend to grow "outward" rather than upward because there is no competition for sunlight from adjacent trees. This form allows giraffes to eat leaves from a large part of the tree.

Many anthropologists believe that the Aw climate provided the optimal environment for the survival and proliferation of early hominids. Disease would be less problematic, and predators would be seen more easily in this environment than in a rain forest environment; early humans could have fled into one of the sporadic trees to avoid predators. Standing on two legs, rather than four, and walking upright was necessary to see over the tall grasses, leading to the development of permanent bipedal locomotion. Furthermore, some anthropologists suggest that the hair on our heads and eyebrows evolved as protection from the high solar angles in Aw climates for early humans, when they began to walk on two legs through the savanna grasses.

Thermal Characteristics Despite the fact that tropical climates are characterized by an absence of winter, temperatures in Aw climates show more annual variation than the other tropical climates, largely the result of strong differences in cloud cover and humidity between the wet and dry seasons. This climatic type also supports higher diurnal temperature variations with higher daily maximum air temperatures and lower minima than the other tropical climates. The diurnal range peaks during the dry season. This may be seen in the climographs for selected locations having Aw climates (**Figures 10.19a** through **10.26a**). In general, temperature maxima may push into the thirties Celsius (upper nineties Fahrenheit), or even near 38°C (100°F), whereas minima may dip into the teens Celsius (fifties Fahrenheit). Monthly means are typically between 19 and 28°C (66 and 82°F).

Moisture Characteristics Precipitation in Aw climates is largely driven by the ITCZ and the influence of the large-scale monsoonal wind shift for these regions. During the high-sun "summer" period the ITCZ approaches, increasing the chance of convective precipitation. The dry season occurs during the low-sun "winter" as the ITCZ migrates into the opposite hemisphere. Particularly in Asia, the low-sun "winter" period is characterized by relatively dry offshore airflow from the monsoon, which decreases the chances of cloud and precipitation development. Some African and northern Australian locations experience less-pronounced monsoon-like shifts.

Annual precipitation totals typically fall between 75 and 175 cm (30 and 70 in). Many locations have two distinct precipitation maxima, once when the ITCZ traverses the region on its poleward migration and again when it migrates equatorward. Poleward edges of the Aw climatic zone typically display a single precipitation maximum because the ITCZ generates abundant precipitation over the region at only one time during the year.

Tropical savanna locations are drier than the more equatorial regions, with some months receiv-

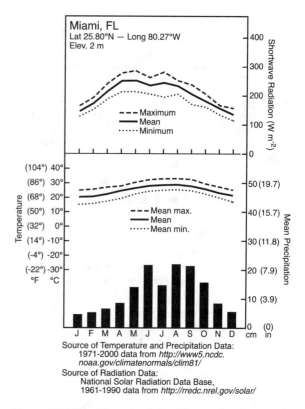

Source of Temperature and Precipitation Data:
  1971-2000 data from *http://www5.ncdc. noaa.gov/climatenormals/clim81/*
Source of Radiation Data:
  National Solar Radiation Data Base,
  1961-1990 data from *http://rredc.nrel.gov/solar/*

Figure 10.19a Climograph for Miami, Florida (Aw climate). Average annual temperature is 24.8°C (76.7°F) and average annual precipitation is 148.7 cm (58.5 in).

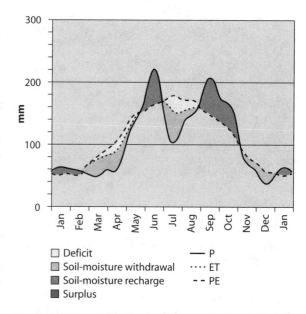

Figure 10.19b Water balance diagram for Miami, Florida.

ing virtually no precipitation. These poleward areas may be thought of as transition zones between the low-latitude rain forests and the much drier higher-latitude continental interiors. Because ITCZ

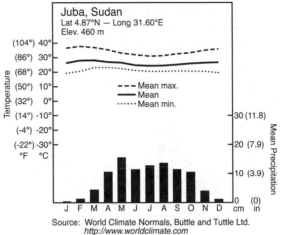

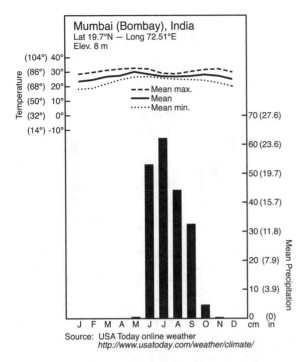

Source: USA Today online weather
http://www.usatoday.com/weather/climate/

Figure 10.20a Climograph for Mumbai (Bombay), India (Aw climate). Average annual temperature is 27.2°C (81.0°F) and average annual precipitation is 216.7 cm (85.3 in).

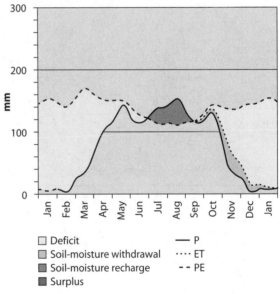

Source: World Climate Normals, Buttle and Tuttle Ltd.
http://www.worldclimate.com

Figure 10.21a Climograph for Juba, Sudan (Aw climate). Average annual temperature is 26.4°C (79.5°F) and average annual precipitation is 97.0 cm (38.2 in).

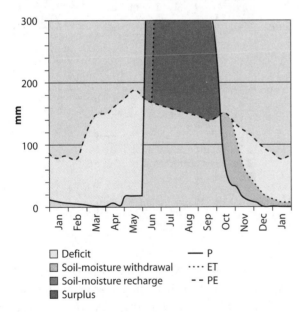

□ Deficit   —— P
▨ Soil-moisture withdrawal   ···· ET
▨ Soil-moisture recharge   – – PE
■ Surplus

Figure 10.21b Water balance diagram for Juba, Sudan.

□ Deficit   —— P
▨ Soil-moisture withdrawal   ···· ET
▨ Soil-moisture recharge   – – PE
■ Surplus

Figure 10.20b Water balance diagram for Mumbai (Bombay), India.

migration is variable, annual precipitation totals show a high degree of interannual variability. Extreme variability is also evident in the intra-annual precipitation totals, which may fluctuate between intense flooding and prolonged drought during the same year.

The distinct dry season combines with high energy and temperatures to produce significant soil moisture deficits during that time, as noted in the associated water balance diagrams (Figures 10.19b through 10.26b). The Asian monsoon influence is particularly evident in the water balance diagram for Mumbai, India (Figure 10.20b). During summer, the local water balance indicates strong *surpluses* despite higher energy values. This stems from the overabundance of ITCZ- and monsoon-influenced precipitation. Soil moisture decreases rapidly after the onset of the dry season.

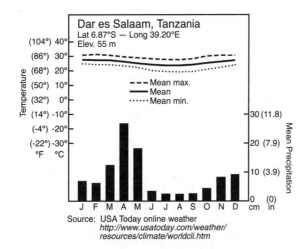

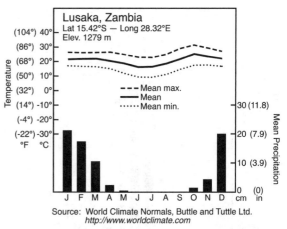

Figure 10.22a Climograph for Dar es Salaam, Tanzania (Aw climate). Average annual temperature is 25.8°C (78.5°F) and average annual precipitation is 105.4 cm (41.5 in).

Figure 10.23a Climograph for Lusaka, Zambia (Aw climate). Average annual temperature is 20.6°C (69.0°F) and average annual precipitation is 81.8 cm (32.2 in).

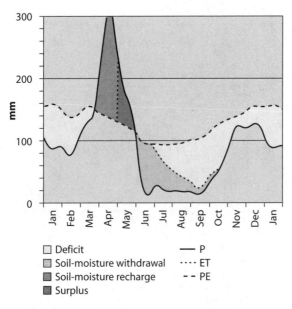

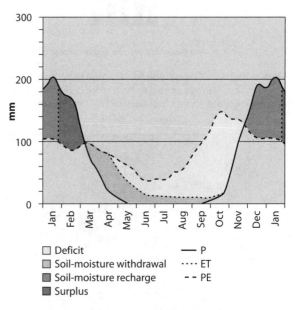

Figure 10.22b Water balance diagram for Dar es Salaam, Tanzania.

Figure 10.23b Water balance diagram for Lusaka, Zambia.

B—Arid Climates

The dry regions that fall within the scope of this chapter are the most extensive on Earth. The largest single dry region on the planet is centered in the Sahara Desert, which extends latitudinally across northern Africa. An extension of the Sahara spills into southwestern Asia, from the Arabian Peninsula into Syria, Iraq, Iran, Afghanistan, Pakistan, and western India (see Figure 8.1). Other dry regions of note include most of southern Africa, the bulk of Australia, and the western fringe of South America.

BW—True Desert Most of the *True Desert (BW) climates* covered in this chapter are classified as the *Subtropical Desert (BWh) climate*, because of the relatively low latitudes of the regions. The only region of the *Midlatitude Cold Desert (BWk) climate* exists in southern Chile as part of the Atacama Desert. The Sahara Desert dominates the BWh zone through northern Africa, encompassing over 8.6 million km² (3.3 million mi²). The desert stretches 4000 km (2500 mi) from east to west and between 1200 and 2000 km (750 and 1250 mi) north to south, or about the size of the conterminous United States.

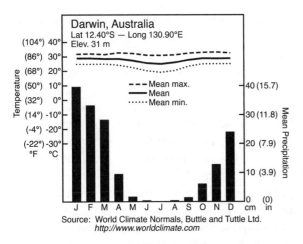

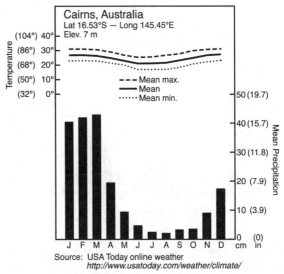

Figure 10.24a Climograph for Darwin, Australia (Aw climate). Average annual temperature is 27.8°C (82.0°F) and average annual precipitation is 157.2 cm (61.9 in).

Figure 10.25a Climograph for Cairns, Australia (Aw climate). Average annual temperature is 24.7°C (76.5°F) and average annual precipitation is 198.4 cm (78.1 in).

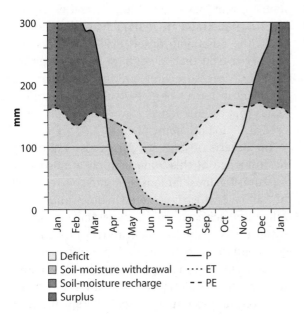

Figure 10.24b Water balance diagram for Darwin, Australia.

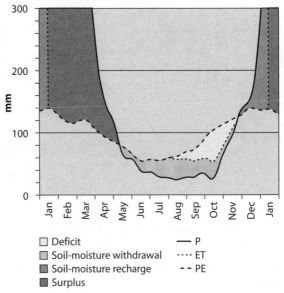

Figure 10.25b Water balance diagram for Cairns, Australia.

Although the Sahara region is the hottest large region on the planet, and obviously one of the driest, this was not always the case. The desert began to form only about 3 million years ago. Before that time, the region was lush, supporting a wide variety of plants and animals.

Contrary to popular belief, the Sahara is not a vast expanse of sand dunes. Only about 25 percent of the region is classified as sand dunes. Instead, most of the Sahara consists of gravelly plains, rocky plateaus, and mountains. Still, single dunes may become enormous; the Libyan Erg dune is approximately the size of France. The desert landscape is spotted with many small oases that support freshwater springs and vegetation. Collectively, about 200,000 km² (77,000 mi²) of the region are oasis. Over the centuries, nomadic people have learned the shortest routes between these areas. Of the 2.5 million people of many different groups who inhabit the region, the Tuareg are perhaps the most famous for their adaptation to the dry central desert core.

Most of the region receives an average of less than 7 cm (3 in) of precipitation per year. Average

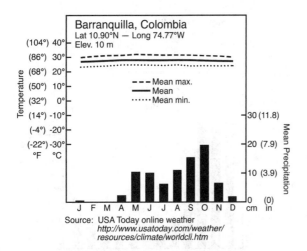

Figure 10.26a   Climograph for Barranquilla, Colombia (Aw climate). Average annual temperature is 27.9°C (82.3°F) and average annual precipitation is 85.9 cm (33.8 in).

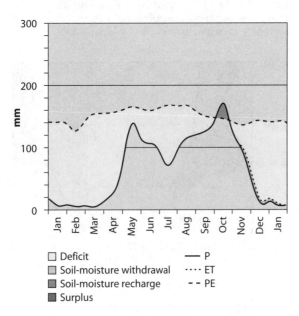

Figure 10.26b   Water balance diagram for Barranquilla, Colombia.

precipitation totals may be misleading, because a substantial amount of rain may occur at one time, with little or none returning for months or even years. Several years of below-average precipitation may be needed to offset only a single year of above-average precipitation. Paradoxically, more people are believed to die from drowning in the Sahara than from thirst because of the nature of the individual rainfall events, which are typically convective thunderstorms that produce copious downpours.

Another broad desert region exists across the bulk of Australia. Perhaps the most famous Australian desert is the Great Sandy Desert, which occupies the northwestern interior. The continental core may be subdivided into the Great Artesian Basin, which occupies the northeastern interior; the Great Victoria Desert, which spans the southern interior; the Gibson Desert in the southwest; the Simpson Desert in the central interior; and the Great Sandy Desert in the northwest. All regions are actually part of one large, dry landscape. Other names for the dry interior include the popular Outback, the Dead Heart, and simply The Desert. All these names conjure images of a barren, life-devoid landscape even though the region, like most deserts, abounds with specially-adapted plants and animals. Like the Sahara Desert, the deserts of Australia are marked by high temperatures. The Australian desert exists for the same primary reason that the Sahara exists—high atmospheric pressure and sinking air from the persistent influence of global subtropical anticyclones that extend across a latitudinal zone centered at 30° in both hemispheres.

The deserts of southern Africa also fall under the influence of this zone of high atmospheric pressure. The most famous desert in this region is the Kalahari, which is centered in Botswana. Other influences such as the cold Benguela current off the western coast of the continent and the *rain shadow effect* combine with the normally high atmospheric pressures to keep the regions dry. The Namib and Atacama deserts of Africa and South America, respectively, are examples of cold current–induced deserts.

The Atacama Desert extends through approximately 32° of latitude southward from near the equator. The desert begins in Peru and extends through much of the length of Chile. As noted previously, the Atacama is the driest desert on Earth because of a combination of the very cold Humboldt current and the nearby Andes Mountains that limits the *advection* of warm and moist air masses originating in the Amazon basin through the western coastal plain of South America. A rain shadow effect, therefore, helps to create the desert through blockage of the easterly trade wind flow across the low latitudes, whereas the mountains trap the cooler oceanic air on their western side, ensuring stable atmospheric conditions. On the eastern side of the mountains, moisture is effectively wrung out of the air by the orographic influence, marking the western edge of the Amazon rain forest. A further influence is provided by a temperature inversion again triggered by the nearby mountains. Cold air drainage from the higher altitudes effectively caps the lower atmosphere, stabilizing the region and reducing the possibility of cloud formation. The higher-latitude regions of the Atacama are classified as BWk because of lower annual average temperatures than the lower-latitude regions.

Thermal Characteristics Temperatures in the BWh regions are the highest on Earth, due to clear skies and high solar angles. Most regions display a fairly wide annual range in the monthly mean temperature. The diurnal range of temperatures is considerable, because of the low atmospheric moisture content. This decreases the local greenhouse effect, allowing for efficient radiational cooling of the surface at night.

As the climographs for representative locations indicate (**Figures 10.27a** through **10.30a**), the lower-latitude locations show more monthly consistency in average temperatures than poleward locations. The climographs also show that with the exception of coastal Mogadishu, Somalia (**Figure 10.29a**), the BWh climates experience a greater daily range in temperatures than locations in A climates.

Moisture Characteristics Despite the very low precipitation, some locations may be humid. For example, a region that records high average annual humidity is located around the Red Sea. The dampness results from a combination of high energy amounts for evaporation and a nearby water source (the sea). The dryness results from the lack of a precipitation-generating mechanism rather than a lack of atmospheric moisture.

Rainfall does occur in even the driest locations, but much is required to overcome the physical and atmospheric barriers that inhibit the precipitation-generating mechanisms in True Desert environments. BW areas display a wide range in the amount and timing of precipitation. The one unifying factor is that the totals are low.

Average monthly precipitation totals may be misleading, because a high amount of precipitation

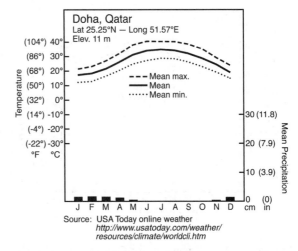

Source: USA Today online weather
*http://www.usatoday.com/weather/
resources/climate/worldcli.htm*

Figure 10.27a Climograph for Doha, Qatar (BWh climate). Average annual temperature is 27.1°C (80.8°F) and average annual precipitation is 8.1 cm (3.2 in).

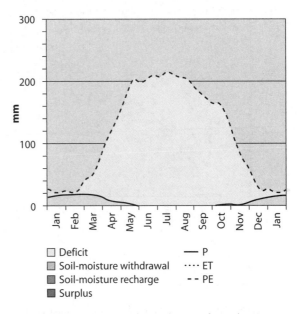

Figure 10.27b Water balance diagram for Doha, Qatar.

may occur in a singular precipitation event. Precipitation may then be absent for many years during that month, although the long-term average may indicate that the month in question receives rain. This may skew average precipitation totals.

Virtually all these locations receive very sporadic precipitation events. As the climographs show, total annual precipitation tends to be below 10 cm (4 in) for most locations, although some locations may receive as much as 20 cm (8 in) of annual rainfall. The low precipitation amounts combine with high energy inputs to produce large

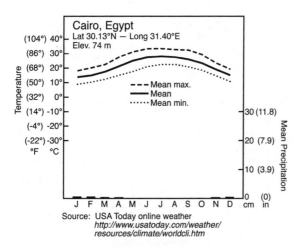

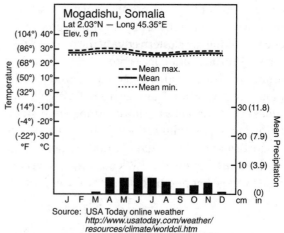

Figure 10.28a Climograph for Cairo, Egypt (BWh climate). Average annual temperature is 21.7°C (71.0°F) and average annual precipitation is 2.3 cm (0.9 in).

Figure 10.29a Climograph for Mogadishu, Somalia (BWh climate). Average annual temperature is 27.7°C (81.9°F) and average annual precipitation is 41.1 cm (16.2 in).

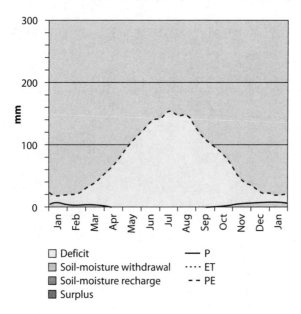

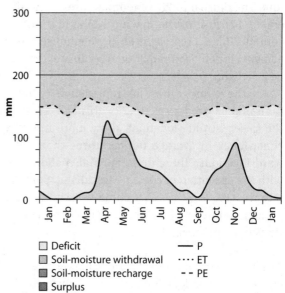

Figure 10.28b Water balance diagram for Cairo, Egypt.

Figure 10.29b Water balance diagram for Mogadishu, Somalia.

water balance deficits, as depicted in the associated water balance diagrams (Figures 10.27b through 10.30b).

BS—Steppe   As with the True Desert locations, the *Steppe (BS) climates* of the tropics and southern hemisphere are vast. Steppe regions exist poleward of the True Desert cores. This zone represents a transition region between the deserts and the wetter climates. Steppe areas occur in a narrow coastal band on the northwest side of the Sahara Desert, through southwestern Asia, along a narrow band of the interior plateau of India, in a circular ring around the core of Australia, around the Kalahari

Desert in southern Africa, the interior of southern South America, and in the Sahel region that lies south of the Sahara.

Thermal Characteristics   Accra, Ghana (**Figure 10.31a**), is classified as a *Hot Steppe (BSh) climate.* The temperature profile indicates a very small annual average range because of the moderating influence of the ocean in this coastal city affected by the regional **west African monsoon system**. A more representative temperature profile is indicated by the climograph of Broome, Australia (**Figure 10.32a**), which shows a wide annual temperature

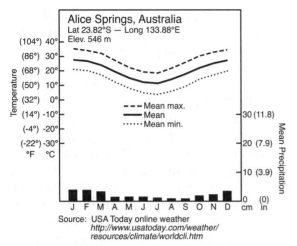

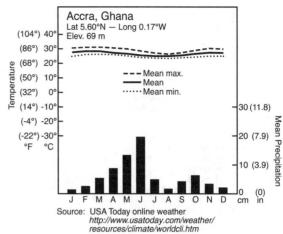

Figure 10.30a Climograph for Alice Springs, Australia (BWh climate). Average annual temperature is 20.8°C (69.3°F) and average annual precipitation is 28.7 cm (11.3 in).

Figure 10.31a Climograph for Accra, Ghana (BSh climate). Average annual temperature is 27.2°C (80.9°F) and average annual precipitation is 74.9 cm (29.5 in).

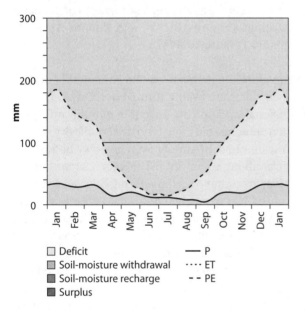

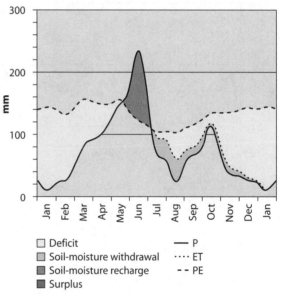

Figure 10.30b Water balance diagram for Alice Springs, Australia.

Figure 10.31b Water balance diagram for Accra, Ghana.

range and distinct seasonality in monthly average temperatures. The *Cold Steppe (BSk) climate* in the southern hemisphere is largely restricted to mid-latitude Argentina.

Moisture Characteristics Precipitation in the Steppe, although relatively low, is higher than True Desert locations. Furthermore, a distinct seasonality of precipitation is normally noticeable. Such is the case with both of the locations depicted in the climographs. Precipitation in the low-latitude Steppe is largely driven by seasonal passage of the ITCZ, which triggers daily afternoon and evening thun-

derstorms. Like the True Desert, there is a wide range in monthly precipitation totals across the years. The water balance diagrams (Figures 10.31b and 10.32b) indicate severe soil moisture deficits throughout the year, but some surpluses may exist if a distinct wet season occurs.

C—Mesothermal Climates

The climates classified as mesothermal in the part of the world covered by this chapter are limited (see Figure 8.1). Only a sparse amount of land exists at the appropriate southern hemisphere

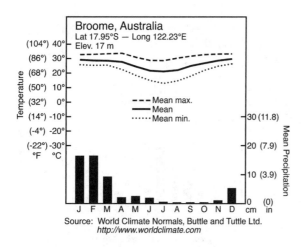

Figure 10.32a Climograph for Broome, Australia (BSh climate). Average annual temperature is 26.5°C (79.8°F) and average annual precipitation is 57.9 cm (22.8 in).

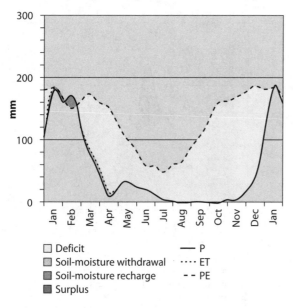

Figure 10.32b Water balance diagram for Broome, Australia.

latitudes. The eastern and southern coasts of Australia and all of New Zealand carry mesothermal climate designations. In addition, only parts of the extreme coastal zones of southern and northern Africa and southeastern and southwestern sections of South America have mesothermal climates.

Cfa—Humid Subtropical The *Humid Subtropical (Cfa) climates* comprise a large percentage of the area of mesothermal climate regions of the southern hemisphere. These climates are found along an extensive region of subtropical eastern Brazil, Uru-

guay, and Argentina. In addition, small sections of the central eastern coast of Australia and the southeastern tip of South Africa have Cfa climates.

Thermal Characteristics Like their extratropical northern hemisphere counterparts, Cfa climates have hot summers and moderate winter temperatures. Average monthly means typically peak in the latter portion of summer (January or February), with averages hovering around 25 to 29°C (77 to 84°F). Daily maxima over 35°C (95°F) are common in summer. Cfa areas are also humid, although generally not as humid as in the southeastern United States. Nevertheless, summertime conditions are generally uncomfortable. Temperatures during winter may be comfortable, with averages near 10 to 15°C (50 to 59°F). The annual temperature range is reflected in the associated climographs of Brisbane, Australia (**Figure 10.33a**), and Buenos Aires, Argentina (**Figure 10.34a**).

Moisture Characteristics Precipitation is distributed relatively evenly throughout the year. Some regional variation does occur in monthly precipitation amounts, but there is no discernible dry season. A wide range of annual averages may be present, from about 50 cm (20 in) to over 180 cm (70 in). Most precipitation falling during the cool season is caused by the passage of frontal cyclones embedded within the global westerly wind flow, whereas summer precipitation is normally associated with sporadic convective thunderstorm activity.

About 12 percent of the world's tropical cyclones occur in the South Indian Ocean, approximately the same number as in the South Pacific. Interestingly, virtually no tropical cyclones occur in the South Atlantic Ocean, leaving eastern South America immune to their devastation. One exception occurred in April 2005, leading some to speculate on whether global warming was the culprit in this unusual occurrence.

Large water balance deficits in the southern hemisphere's Cfa climates are uncommon, although they do occur. Periodically substantial surpluses of surface water are more common, occurring during the cooler winter season (Figures 10.33b and 10.34b).

Csa, Csb—Mediterranean As discussed in Chapter 9, *Mediterranean climates* are generally found

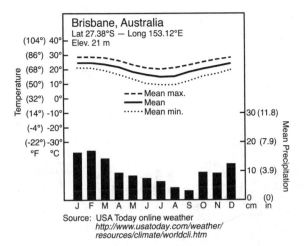

Figure 10.33a Climograph for Brisbane, Australia (Cfa climate). Average annual temperature is 20.6°C (69.1°F) and average annual precipitation is 119.6 cm (47.1 in).

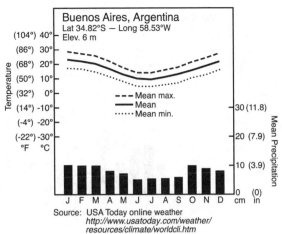

Figure 10.34a Climograph for Buenos Aires, Argentina (Cfa climate). Average annual temperature is 16.5°C (61.7°F) and average annual precipitation is 98.0 cm (38.6 in).

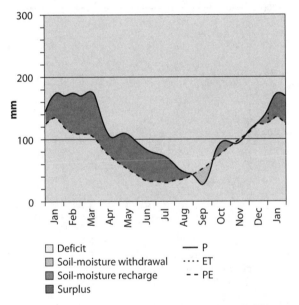

Figure 10.33b Water balance diagram for Brisbane, Australia.

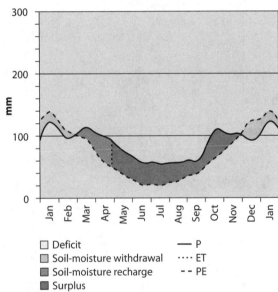

Figure 10.34b Water balance diagram for Buenos Aires, Argentina.

along the western coasts of continents in the mid-latitudes. Coastal northwestern Africa falls into this category. The Mediterranean climatic type is also found along the southern coast of Australia, the southwestern tip of southern Africa, and central Chile west of the Andes.

Thermal Characteristics Like similar areas in North America and Europe, Mediterranean climates are marked by very comfortable and pleasant temperature regimes. Temperatures seldom remain very hot or very cold for long periods, although un-

comfortable extremes may occur for short periods of time. In the *Mediterranean Hot Summer (Csa) climate*, daily maximum temperatures may spike above 38°C (100°F). The cold ocean currents associated with the *Mediterranean Warm Summer (Csb) climate* keep summer average temperatures below 22°C (72°F). For Mediterranean climates in general, annual temperature ranges usually fall between a moderate 7 and 21C° (13 and 38F°). Winter mean monthly temperatures rarely fall below about 4°C (40°F). The Csa regime is represented by Perth, Australia (**Figure 10.35a**), whereas Capetown, South

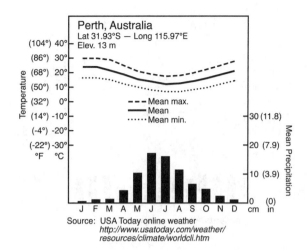

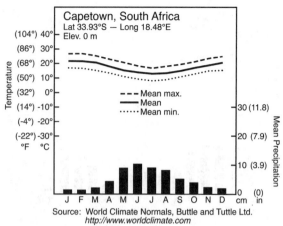

Figure 10.35a Climograph for Perth, Australia (Csa climate). Average annual temperature is 18.0°C (64.4°F) and average annual precipitation is 79.8 cm (31.4 in).

Figure 10.36a Climograph for Capetown, South Africa (Csb climate). Average annual temperature is 17.6°C (63.7°F) and average annual precipitation is 61.2 cm (24.1 in).

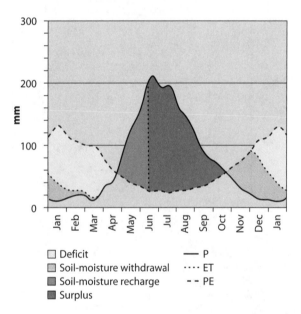

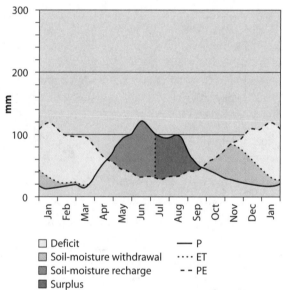

Figure 10.35b Water balance diagram for Perth, Australia.

Figure 10.36b Water balance diagram for Capetown, South Africa.

Africa (**Figure 10.36a**), and Santiago, Chile (**Figure 10.37a**), are typical examples of Csb climates.

Moisture Characteristics Precipitation in Mediterranean climates is largely a winter phenomenon, when the regions are influenced by the passage of midlatitude frontal systems. Summer convection typically forces whatever small precipitation totals occur during warmer times. Precipitation totals are normally between 50 and 70 cm (20 and 30 in) for most locations. Given the fact that the period of highest temperatures (summer) coincides with the

driest time of the year, it is no surprise that soil moisture deficits are common in December through February (Figures 10.35b through 10.37b). The water balance profile leads to stunted natural vegetation types, as most plant growth occurs during the time of year with the lowest input of solar radiation.

Cfb—Marine West Coast The *Marine West Coast Warm Summer (Cfb) climates* are found poleward of Humid Subtropical (Cfa) climates. Paradoxically, the Marine West Coast climate is not restricted to the west coasts of continents in the southern hemi-

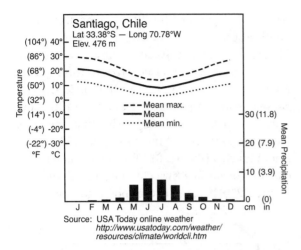

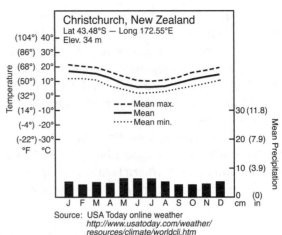

Figure 10.37a Climograph for Santiago, Chile (Csb climate). Average annual temperature is 14.6°C (58.3°F) and average annual precipitation is 33.8 cm (13.3 in).

Figure 10.38a Climograph for Christchurch, New Zealand (Cfb climate). Average annual temperature is 11.6°C (52.8°F) and average annual precipitation is 65.3 cm (25.7 in).

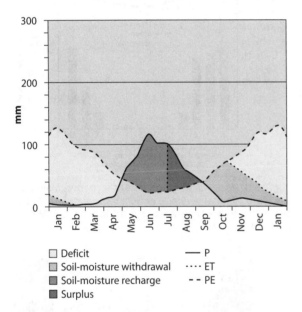

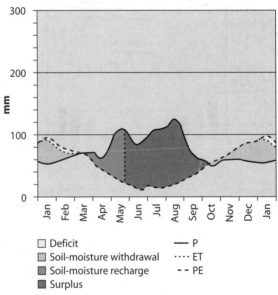

Figure 10.37b Water balance diagram for Santiago, Chile.

Figure 10.38b Water balance diagram for Christchurch, New Zealand.

sphere. In addition to its location on the western coasts of midlatitude Chile and Argentina, the Cfb climate occurs along the southeastern coast of Australia, in virtually all of Tasmania and New Zealand, and along a small sliver of southern Africa (see Figure 8.1). In addition, the climate occurs along scattered coastal (or near coastal) locations of southern Brazil and Argentina.

Thermal Characteristics The Cfb climate is distinguished from Cfa climates by the magnitude of summer heating. Cfb regions are somewhat cooler, with

the coolest month averaging below 22°C (72°F) and average daily maxima usually below 27°C (80°F). Monthly averages during summer are typically around 15 to 20°C (59 to 68°F), whereas winter months dip to approximately 0 to 10°C (32 to 50°F). Christchurch, New Zealand (**Figure 10.38a**), provides an example of the annual cycle of temperature in these climates.

Moisture Characteristics As in northern hemisphere Cfb climates, precipitation is ample year-round, but summer tends to be drier than winter because the subtropical highs are sufficiently near to suppress large precipitation totals. Cfb climates in New Zealand are affected strongly by orographic enhancement and suppression of precipitation on the windward and leeward sides and (respectively) of the Southern Alps range that runs perpendicular to the westerlies.

A variant of the Marine West Coast climate is the **Marine Dry Winter (Cwb) climate**. This climate is not found in the northern hemisphere; instead, it is located in southeastern Africa and a small part of Madagascar. The climographs of Pretoria, South Africa (**Figure 10.39a**), and Antananarivo, Madagascar (**Figure 10.40a**), show distinct winter dry periods, in stark contrast to the adjacent Mediterranean climates. Precipitation in Cwb locations is forced primarily by summertime convection associated with surface heating. The water balance diagrams for Pretoria (Figure 10.39b) and Antananarivo (Figure 10.40b) show smaller deficits than occur in a Cwa climate (such as those in Figure 9.27b) because of the cooler summer.

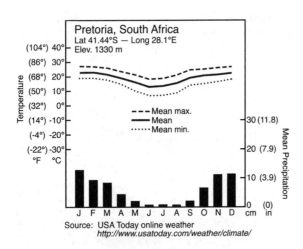

Figure 10.39a Climograph for Pretoria, South Africa (Cwb climate). Average annual temperature is 19.0°C (66.2°F) and average annual precipitation is 70.9 cm (27.9 in).

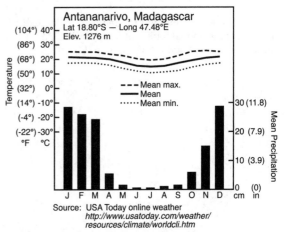

Figure 10.40a Climograph for Antananarivo, Madagascar (Cwb climate). Average annual temperature is 19.0°C (66.2°F) and average annual precipitation is 136.1 cm (53.6 in).

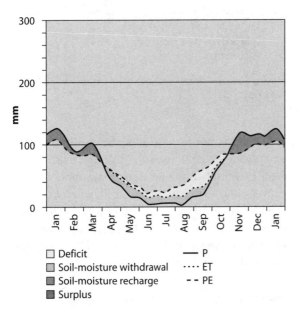

Figure 10.39b Water balance diagram for Pretoria, South Africa.

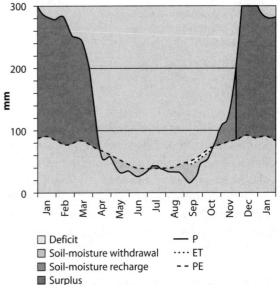

Figure 10.40b Water balance diagram for Antananarivo, Madagascar.

E—Polar Climates

Because of the lack of high-latitude land surfaces in the southern hemisphere, D climates do not exist. Instead, *polar climates* encompass all of Antarctica and the southernmost tip of South America, known as Tierra del Fuego (see Figure 8.1). The two types of E climates are discussed below.

ET—Tundra The *Tundra (ET) climates* of the southern hemisphere are limited in extent because of the general absence of land masses at the latitudes in which tundra vegetation thrives. Tundra exists only on the Antarctic Peninsula extending from Ellsworth Land, Antarctica, and on some of the ice-free islands surrounding Antarctica. These locations have the most moderate climates of the Antarctic continent, because the Antarctic Peninsula represents the farthest equatorward position of Antarctica, pushing to nearly 60°S latitude, just south of the South American coast.

The small ice-free region abounds with marine-adapted life, as do the many ice-free islands of the region. Life persists because of the abundant nutrient supplies in the cold ocean waters. Many scientists regard this region as being the basis of the entire oceanic food chain because it includes a source of phytoplankton and zooplankton for much of the world ocean.

The stability of this food chain is of particular concern to many scientists, because both phytoplankton and zooplankton are susceptible to excessive amounts of *ultraviolet radiation.* The high latitudes of the southern hemisphere are impacted by the development of a seasonal ozone hole. The hole forms during winter in part as a result of the development of the *circumpolar vortex* over the cold Antarctic continent. The circumpolar vortex limits advection across the latitudes, effectively trapping gases over the continental *troposphere* and stratosphere through the cold season. Ozone amounts decrease naturally under these conditions, because the sunlight required for ozone production is absent during the very long high-latitude winter.

At the same time, however, as we saw in Chapter 2, humans have accelerated the growth of the hole through the production of *chlorofluorocarbons* that react with the oxygen atoms that would otherwise produce ozone. These gases also become trapped over Antarctica during winter. This has led to a net increase in the size of the ozone hole in addition to the extension of its existence into the spring months. This is problematic considering that ultraviolet radiation is incoming during a time when ozone molecules in the stratosphere are absent. Increases in surface ultraviolet radiation have led to decreases in the production of phytoplankton and zooplankton in the oceans. The result has been deleterious for the ecosystem. For example, some changes in the DNA of icefish (an Antarctic fish lacking hemoglobin), and the animals that feed on those plants and animals, have already been linked to changes in ultraviolet radiation receipt. Most Earth and environmental scientists agree that if this trend continues, the entire oceanic food chain will be adversely affected.

Thermal Characteristics As with the northern hemisphere's Tundra climates, temperatures are very low throughout the year, but they do reach above the freezing point for a short time during the summer. An explosion of life occurs during this "warm" period. To be classified as Tundra, the average temperature of the warmest month must be between 0°C (32°F) and 10°C (50°F), which in the southern hemisphere occurs only on the sliver of land that comprises the Antarctic Peninsula. The highest temperature ever recorded on Antarctica, approximately 15°C (58°F), occurred at Esperanza, at 63°S latitude.

Moisture Characteristics Precipitation is very low, with totals typically being 25 cm (10 in) or less. Precipitation is almost exclusively a summer phenomenon in this region given the extensive cold. Summer is the only season when appreciable amounts of water vapor are present in the air.

EF—Ice Cap

Approximately 98 percent of Antarctica has an *Ice Cap (EF) climate.* The continental ice sheet is the largest on the planet, averaging approximately 5 km (3 mi) in thickness over an area equivalent in size to the conterminous United States. The floating Ross, Ronne, and Filchner Ice Shelves increase the area of the continent by 11 percent.

Katabatic winds are a prominent feature of the Antarctic climate, given the high altitude of the land mass. Like Greenland, the katabatic winds originate over the continental ice sheet where air chills and spreads toward the low-elevation coastal locations. The extreme winds, blowing across the featureless ice sheet in a low-friction environment, reduce the **wind-chill factor**—a contrived "temperature" that represents human comfort by taking both temperature and wind speed into account. The low wind-chill factor makes for even harsher conditions for living creatures than would exist through low temperatures alone. Not surprisingly, the average wind speed is the highest for any continent. The high winds cause frequent blizzards of blowing ice and snow. Because of the physical geography of the continent, the area of unimpeded ocean surrounding the land mass produces the most violent seas on Earth. These can be especially treacherous during storms.

Thermal Characteristics　The extreme cold results from the very high latitude of the continent. The latitude effect combines with the large land mass to produce a significant continentality effect. However, even mean summer temperatures remain below freezing, due to the latitude and resultant low solar angles. January temperatures for most inland areas average below −18°C (0°F).

Moisture Characteristics　Because of the size of the land mass and higher latitude, Antarctica is much colder and drier than Greenland. The extremely low saturation vapor pressure of such cold air causes the continent to have the lowest *specific humidity* on Earth. The extreme dryness is noticeable in the climographs of McMurdo Research Station (**Figure 10.41**) and the South Pole (**Figure 10.42**), the latter of which indicates only a trace of summer (February) average monthly precipitation. Annual precipitation totals throughout the continental areas are less than 4 cm (2 in). An occasional cyclonic storm may affect the ice cap region of the continent after formation over the southern oceans. In general, **riming**—the *deposition* of water vapor directly onto the ice surface—leads to subsequent ice accumulation, which is balanced by the slow, downward migration and subsequent calving off of ice at the coasts. In recent years the rate of accumulation is believed to be less than the rate of calving. Water balance plots are not included for these sites, because they are meaningless in such environments.

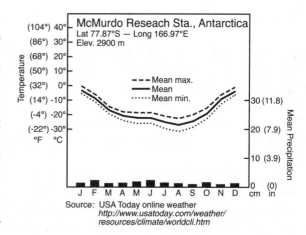

Figure 10.41　Climograph for McMurdo Research Station, Antarctica. Average annual temperature is −17.2°C (1.1°F) and average annual precipitation is 19.1 cm (7.5 in).

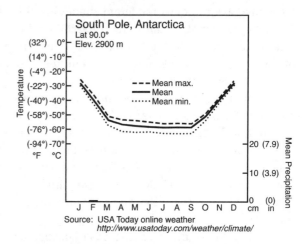

Figure 10.42　Climograph for South Pole, Antarctica. Average annual temperature is −49.4°C (−56.9°F) and average annual precipitation is 0.3 cm (0.1 in).

H—Highland Climates

Highland (H) climates are scarcer in the regions covered in this chapter than they are in the midlatitude northern hemisphere. Still, highland regions occur in Ethiopia in east-central Africa, the core of the Indonesian island of Borneo, and the spine of the western cordillera of North and Central America, continuing through the Andes mountain chain in western South America. The latter represents, by far, the most extensive highland influence. As mentioned in Chapter 9, highland climates are marked by vertical zonation in the climatic regimes, along with radically variable climatic conditions across short spaces. The climographs for Mexico City, Mexico (**Figure 10.43a**),

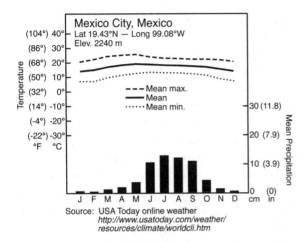

Source: USA Today online weather
http://www.usatoday.com/weather/
resources/climate/worldcli.htm

Figure 10.43a Climograph for Mexico City, Mexico (H climate). Average annual temperature is 17.1°C (62.7°F) and average annual precipitation is 62.7 cm (24.7 in).

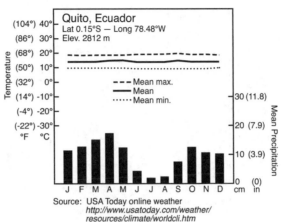

Source: USA Today online weather
http://www.usatoday.com/weather/
resources/climate/worldcli.htm

Figure 10.44a Climograph for Quito, Ecuador (H climate). Average annual temperature is 14.5°C (58.1°F) and average annual precipitation is 120.9 cm (47.6 in).

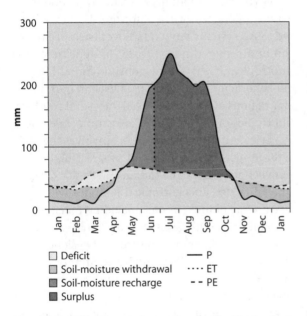

Figure 10.43b Water balance diagram for Mexico City, Mexico.

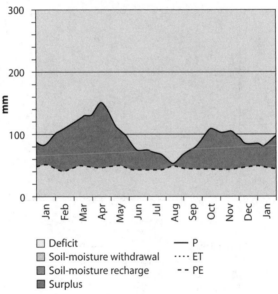

Figure 10.44b Water balance diagram for Quito, Ecuador.

Quito, Ecuador (**Figure 10.44a**), and Arequipa, Peru (**Figure 10.45a**) show the temperature and precipitation variability normally associated with highland influences. Although these regions show very little variation in temperature, due to relatively low latitudes and nearly constant energy receipt through the year, the locations vary widely in precipitation. Comparison of resultant water balances (Figures 10.43b through 10.45b) reflects such differences, with Mexico City and Quito having large surpluses and Arequipa experiencing massive deficits.

■ Summary

This chapter detailed the causes of regional climate variations for Africa, Oceania, South America, and Antarctica. The daunting array of regions is linked by similar climate forcings, because most of the regions exist in tropical and subtropical latitudes. Specifically, the ITCZ influences many of the climatic variations present throughout all the regions except Antarctica. Migration of the ITCZ triggers high-sun precipitation through the affected regions, whereas associated afternoon cloud cover

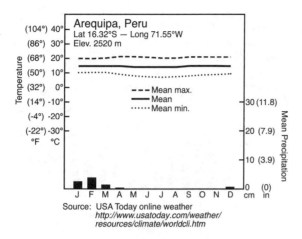

Figure 10.45a Climograph for Arequipa, Peru (H climate). Average annual temperature is 15.0°C (58.9°F) and average annual precipitation is 9.7 cm (3.8 in).

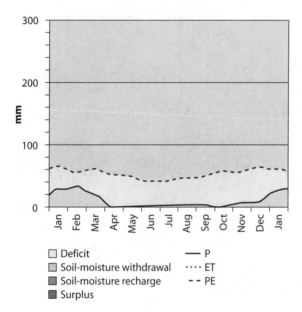

Figure 10.45b Water balance diagram for Arequipa, Peru.

influences regional temperatures. The annual migration of the ITCZ causes variations in both temperature and precipitation, mainly across savanna grasslands that separate drier Steppe and True Desert regions from wetter Rain Forest or Tropical Monsoon regions. These transition zones are of particular concern, given their fragility. Long-term water shortages in these regions contribute, along with human activities, to desertification of the region. In Oceania the position of the SPCZ, an extension of the ITCZ in the South Pa-

cific Ocean, provides another source of moisture variability.

Another common link for most of the areas discussed in this chapter is the influence of warm- and cold-ENSO events. Because ENSO forces large-scale atmospheric and oceanic changes through the tropics, most tropical regions are affected directly. Most locations see an "opposite" climate regime during warm-ENSO events from what is considered "normal." For instance, an El Niño event usually triggers drought through northern and eastern Australia, southeastern Asia, the Amazon, equatorial Africa, and Oceania, while at the same time causing copious precipitation in western South America, eastern Africa, and western Australia.

Further commonalities for many locations include influences of the subtropical anticyclonic pressure systems embedded over each ocean basin and centered near the 30° latitude area. The subtropical highs are responsible for sinking atmospheric motions leading, at least in part, to the dry regions of Africa (both northern and southern subtropics of the continent), Australia, and South America. Another commonality involves the MJO, a 30- to 60-day oscillation of progressing lines of convection through the Indian and Pacific Oceans. The QBO also affects tropical latitudes with a diverse array of interactions, including precipitation variability in Southeast Asia, monsoonal variations, Sahelian precipitation, and tropical cyclone activity in the Pacific, Indian, and North Atlantic basins. A final link involves cold ocean currents, which stabilize the atmospheres of western continental coasts and contribute to the formation of deserts in western Africa, Australia, and South America.

The only region that does not share these causes of climate variation is Antarctica. The geographical isolation of this high-latitude continent, along with the presence of an extensive ice sheet and high elevation, causes different forcing features to dominate there. Very little precipitation occurs in Antarctica because of the very low temperatures and the related low saturation vapor pressure of that cold air. Instead, the region is marked by extreme katabatic winds that blow existing ice and snow downslope. New ice accumulation comes primarily through riming of water vapor directly upon the cold surface.

▶ Key Terms

30–60 Day Oscillation
40–50 Day Oscillation
Ablation
Acacia
Advection
Advection fog
Aghulas current
Air mass
Albedo
Anomaly
Anthropogenic
Anticyclone
Aphelion
Asian monsoon
Benguela current
Bermuda-Azores high
Berson westerlies
Canary current
Chlorofluorocarbon
Circumpolar vortex
Cold Steppe (BSk) climate
Continental tropical (cT)
   air mass
Continentality
Convection
Convective precipitation
Coriolis effect
Counter-trade winds
Cumuliform cloud
Cyclone
Deficit
Deforestation
Deposition
Desertification
Dewpoint temperature
Diversity
East African low-level jet
Easterly wave
Ekman spiral
El Niño
El Niño-Southern Oscillation (ENSO)
   event
Equatorial (E) air mass
Equatorial Kelvin wave
Eustatic
Evaporation
Evapotranspiration
Filchner Ice Shelf
Findlater jet

Flux
Friction
Front
Geopotential height
Geostrophic balance
Glacial advance
Global warming
Greenhouse effect
Gyre
Harmattan winds
Hawaiian high
Highland (H) climate
Horizontal convergence
Hot Steppe (BSh) climate
Humboldt (Peru) current
Humid Subtropical (Cfa) climate
Hurricane
Ice Cap (EF) climate
Indian Ocean high
Insolation
Intertropical convergence zone
Inverted trough
Isohypse
Katabatic winds
Krakatau easterlies
La Niña
Latent heat
Latitude
Leeward
Longitude
Longwave radiation
Madden-Julian Oscillation (MJO)
Marine Dry Winter (Cwb) climate
Marine West Coast Warm Summer
   (Cfb) climate
Maritime Continent
Maritime tropical (mT) air mass
Mediterranean climate
Mediterranean Hot Summer (Csa)
   climate
Mediterranean Warm Summer (Csb)
   climate
Mesothermal
Midlatitude Cold Desert (BWk)
   climate
Midlatitude westerlies
Northeast trade winds
Oceania
Orographic effect

Ozone
Perihelion
Photosynthesis
Polar climate
Positive feedback system
Potential evapotranspiration
Pressure gradient force
Quasi-biennial oscillation (QBO)
Rain machine
Rain shadow effect
Ridge
Riming
Ronne Ice Shelf
Ross Ice Shelf
Rossby wave
Runoff
Sahel
Sensible heat
Siberian high
Soil moisture recharge
Solar declination
Somali jet
South Atlantic high
South Pacific Convergence Zone
   (SPCZ)
South Pacific high
Southeast trade winds
Southern Oscillation
Southwesterly monsoon
Specific humidity
Stability
Steppe (BS) climate
Stratiform cloud
Stratosphere
Streamline analysis
Subpolar low
Subtropical anticyclone
Subtropical Desert (BWh)
   climate
Surface water balance
Surplus
Temperature inversion
Thermal low
Thermocline
Trade surge
Tropic of Cancer
Tropic of Capricorn
Tropical (A) climate
Tropical cyclone

Tropical Monsoon (Am) climate
Tropical Rain Forest (Af) climate
Tropical Savanna (Aw) climate
Tropopause
Troposphere
Trough
True Desert (BW) climate

Tundra (ET) climate
Turkana jet stream
Typhoon
Ultraviolet radiation
Upwelling
Vapor pressure
West African monsoon system

Whiteout
Wind channeling
Wind-chill factor
Windward
Wisconsin Glacial Phase
Terms in italics have appeared in at least one previous chapter.

▶ Review Questions

1. Discuss the importance of the ITCZ as a climatic control throughout the tropical regions discussed in this chapter.
2. How does the SPCZ influence the climate of Oceania?
3. Discuss the importance and impacts of ENSO on each of the regions discussed in this chapter.
4. How does the Madden-Julian Oscillation form and propagate over time? What are the climatic contributions of the MJO on affected regions?
5. Discuss the importance and ramifications of the QBO on tropical climates.
6. Detail the impacts of the subtropical anticyclones on the regions covered in this chapter.
7. What is the contribution of cold ocean currents to the climatic landscape of the associated regions?
8. Discuss the recycling of environmental water in the rain forest regions.
9. Discuss recent changes in Alpine glaciers to changes in the Antarctic ice sheet. Are the changes consistent?
10. Why is it important to monitor the ice shelves extending from the Antarctic land mass?
11. Detail important aspects of the Tropical Rain Forest climatic type.
12. Discuss Tropical Monsoon climates.
13. Why are Tropical Savanna regions considered climatic transition zones?
14. Why are deserts so extensive throughout the regions covered in this chapter?
15. What factors contribute to desertification in Steppe climate areas?
16. Discuss the causes of mesothermal climates in the associated regions.
17. Why are there no "D" climates in the regions associated with this chapter?
18. Why is stratospheric ozone so important to the high-latitude regions of the southern hemisphere?

▶ Questions for Thought

1. After studying Chapters 9 and 10, discuss the climatic accuracies and inaccuracies portrayed in three popular movies. Would the plot of the movie have changed if the climates had been portrayed more accurately?
2. Assume that a 7000-m (23,000-ft) mountain suddenly appears at your current location. Using the normal environmental lapse rate, calculate the vertical average temperature from your current elevation to the mountain peak. Based on that temperature, calculate which climatic zones would appear on the mountain, and where they would be located both vertically and horizontally (assuming shading influences, etc.).

http://physicalscience.jbpub.com/climatology

Connect to this book's Web site: http://physical science.jbpub.com/climatology. The site provides chapter outlines, further readings, and other tools to help you study for your class. You can also follow useful links for additional information on the topics covered in this chapter.

4

Climates Through Time

Chapter 11—Climatic Change and Variability
Climatic Changes in Geological History
How Do We Know What We Know About Past Climatic Changes?
Natural Causes of Climatic Change and Variability
Summary

Chapter 12—Anthropogenic Climatic Changes
Global Warming
Atmospheric Pollution
Classifying Air Pollutants
Reactions and Attitudes to Climatic Change
Summary

Chapter 13—Linking Spatial and Temporal Aspects of Climate Through Quantitative Methods
Computerized Climate Models
Seven Basic Equations
Statistical Techniques
Atmospheric Teleconnections
Summary

11 Climatic Change and Variability

Chapter at a Glance

Climatic Changes in Geological History
Temperature
Ice Ages and Sea Level
Recent Trends
How Do We Know What We Know About Past
Climatic Changes?
Basic Principles
Radiometric Dating
Lithospheric and Cryospheric Evidence
Biological Evidence
Historical Data
Converging Proxy Evidence
Natural Causes of Climatic Change and Variability
Continental Drift and Landforms
Milankovitch Cycles
Volcanic Activity
Variations in Solar Output
El Niño-Southern Oscillation Events
Summary
Review

Climatology is about much more than just the average weather conditions over a long period of time. Climatology considers both average conditions and variability as well as *extremes*, long-term trends, and impacts of atmospheric conditions. In general, the greatest impacts occur during periods of extremes, and these extremes may provide evidence of climatic change or climatic variability. Chapters 9 and 10 discussed spatial variations in climate; Chapters 11 and 12 focus on temporal changes and variability in climate.

Temporal climatic changes and variability may take many forms (**Figure 11.1**). **Random variability** (sometimes referred to as **noise**) occurs when an atmospheric variable undergoes fluctuations that do not appear to be caused systematically. Fluctuation, therefore, does not repeat at a distinct interval and is considered unimportant. **Periodic variability** takes the form of an abnormality that recurs with some relatively constant regularity. **Variability changes** may or may not involve a change in mean but include a temporal drift to larger or smaller variability about the mean.

More permanent changes include **temporal trends**, which are represented by a slow, temporal drift toward an increasing or decreasing mean in some atmospheric variable. **Step changes** involve a sudden shift to a different mean, with no slow drift toward that mean. Of course, random or periodic variability or variability change may be superimposed on any change. It is also important to note that the variability and changes need not be toward increases, as shown in Figure 11.1. Decreasing values of some atmospheric variable are just as likely to occur.

Regardless of the form of climatic variability or change, more than just the atmosphere is involved. Instead, the entire climate system is impacted by atmospheric variability and changes because the system undergoes exchanges of energy, moisture, or *momentum* (which may change in abundance) among the atmosphere, *hydrosphere*, *lithosphere*, *biosphere*, and *cryosphere*. Climatic changes do not occur in isolation.

The form of temporal change and variability being experienced can be complicated and confusing, not only because *anomalies*—observations that do not match the general prevailing trend—may obscure the pattern but also because several different forms of change and variability may exist simultaneously. Trends and cycles can be super-

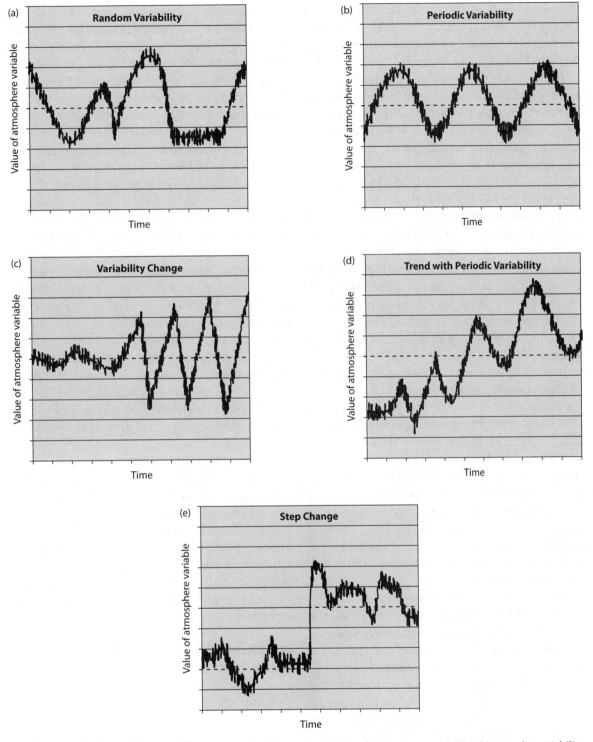

Figure 11.1   Many forms of temporal climatic changes and variability: (a) random variability, (b) periodic variability, (c) variability change, (d) trend with random variability embedded, and (e) step change with random variability embedded.

imposed on other trends and cycles operating at longer or shorter time scales.

Another complicating effect of temporal climatic change is that the prevailing definition of "change" has itself changed over the last few decades. In the 1960s climatic change was largely used to refer to what we would now call "climatic variability"—that is, climatic inconsistencies or anomalies. In the 1970s climatic change generally connoted very long-term changes, perhaps on the order of thousands of years. The prevailing idea was that climatic change existed on time scales that were too long to be noticed during the course of a human lifetime.

In the 1980s the meaning of climatic change again shifted, this time to refer to differences in 30-year averaging periods. Traditionally, climatologists have used data sets of 30 years, terminating at the end of the last full decade. For example, for discussing trends in annual average temperatures from 1991 through 2000, the 30-year averaging period in use was 1961–1990, and from 2001 through the present, the 1971–2000 interval was used for the construction of "annual climatic normals." Thus, climatic change came to refer to differences in atmospheric variables from one 30-year period to another, such as the 1951–1980 period versus the 1961–1990 period, even though this comparison of "normals" may indicate short-term variability trends rather than long-term climatic change.

By the 1990s climatic change came to imply that humans were having an impact, whether through the combustion of *fossil fuels*, *deforestation*, land use change, or deliberate attempts to alter the climate directly or through some combination of one or more anthropogenic causes. In the 2000s climatic change continues to imply a human influence, but with the realization that this human influence can occur on much shorter time scales than previously believed. In this light, climatic change is sometimes even seen as the difference *within* averaging periods rather than the differences *between* them.

Despite all the connotations about climatic change that have prevailed since the 1960s, the notion that climate varies is a relatively new one. In general, before the 1960s little thought was given to the idea that climates change. Even today, climate is erroneously assumed by many in the general public to be static. Even among people who recognize that climate varies, many do not realize that climatic variation and change can affect them in their lifetimes, despite the current prevailing theme in the scientific community that human impact on climate, across short time scales, is real. As a result, change and variability are all too often overlooked in environmental planning and hazard mitigation.

In many ways the distinction between climatic variability and climatic change has been blurred in recent years, and it is often impossible to detect whether we are experiencing variability or change as we live through the phenomena. This reality complicates the formulation of policy to deal with the phenomena. For example, should we enact long-term policy that would be necessary if the recent trend in active Atlantic *hurricane* seasons continues unless we believe that increased Atlantic hurricane activity is a long-term reality? Or should we just "ride it out" until Atlantic hurricanes become less frequent and intense once again?

Both formal and informal policy is often dictated by the persistence of an anomaly. For example, people are usually more willing to help victims of climatic variability, such as a flood, than they would be to help victims who have repeatedly been hit by the same type of disaster. This persistence of the anomaly also affects how we experience variability and change. Variability is experienced as uncharacteristically intense floods, storm frequency, heat waves, and other individual extreme events. But change is experienced more insidiously, as the barely noticeable shifting of agricultural and bioclimatic zones, increasingly more extreme events than usual occurring over time, and so forth.

We are probably more vulnerable to the effects of extreme events related to climatic variability and change than we once were. More people than ever before are living in "fringe" areas such as fluvial and coastal floodplains and in arid environments. However, we may not be as vulnerable to minor fluctuations and changes, because of the introduction of hybrid crops and improved science and technology for warning the public about potential dangers.

■ Climatic Changes in Geological History

Temperature

Converging lines of scientific evidence show that Earth is approximately 4.6 billion years old and the oldest terrestrial rocks identified are approximately 4.0 billion years old. Although it is difficult to identify changes in climate across such an incomprehensibly long period of time, little evidence has been found to suggest that the average global temperature was ever below freezing or above a few tens of degrees above today's average temperatures. It is more likely that Earth's average temperature has remained within an even smaller range (about 15C° [27F°]), although some evidence to the contrary is present.

As was discussed in Chapter 2, scientific evidence suggests that in Earth's early history, solar output was about 25 to 30 percent less than present because of the predictable pattern of stellar evolution that the Sun has undergone. Yet no conclusive evidence suggests that Earth was much colder when the Sun was emitting less energy. This *faint young Sun paradox* remains a topic of wide-ranging geoscientific research today. It is currently believed that *methanogens*—a family of anaerobic bacteria that produces *methane*—may have allowed for an accumulation of atmospheric methane. This methane would have absorbed terrestrial *longwave radiation* and reradiated it back downward effectively as *counter-radiation*, enhancing the *greenhouse effect* and keeping Earth relatively warm even while the Sun was emitting less radiation. This hypothesis is supported by processes observed on Titan—Saturn's largest moon—which displays characteristics similar to early Earth and has an abundance of methane in its atmosphere. For this reason scientists are eager to probe Titan for signs of life, specifically methanogens.

The methanogens are believed to have been intolerant to atmospheric oxygen, which would have become more abundant after the onset of *photosynthesis* some 3 billion years ago when the first green plants evolved. As a result, just as the Sun began emitting more energy to Earth, the increasing oxygen concentration would have killed the methanogens, making methane less abundant as a *greenhouse gas* and causing Earth's atmosphere to become less effective as a greenhouse system. At times, the weakening of the greenhouse effect may have overshadowed the consequences of increasing solar intensity. It is suggested that the global demise of the methanogens may have caused Earth's first global *ice age*—any period in geological history in which ice exists continuously for a sustained period of time over some section of Earth. Furthermore, "pulses" of methanogen demise may have triggered other ice ages.

Regardless of whether the methanogen hypothesis is scientifically valid, most of geological history is believed to have been dominated by somewhat higher temperatures than exist today. Furthermore, there is no convincing scientific evidence for widespread glaciation for at least the first half of Earth's existence. Instead, a sequence of relatively low temperatures and intermittent glaciations has been identified in the geological record, beginning at about 2.5 billion years ago—about halfway through Earth's existence—coinciding with the arrival of abundant atmospheric oxygen. This first known major ice age occurred at the beginning of the **Proterozoic Eon**, the last broad division of geological time before the current one—the **Phanerozoic Eon (Figure 11.2)**. **Eons** are subdivided into **eras**, which are subdivided into **periods**.

Evidence indicates that the most impressive ice age may have occurred in the **Cryogenian Period** of 800 to 600 million years ago at the end of the Proterozoic Eon. Some geological evidence indicates that the presence of large land masses near the poles during this period may have caused the removal of carbon dioxide (CO_2) from the atmosphere in the rock-forming (sedimentary) processes, thereby weakening the greenhouse effect and triggering a massive ice age. Proponents of this **snowball earth hypothesis** assert that conditions were so cold during the Cryogenian Period that ice even existed toward the equatorial land masses and oceans. Although evidence of glacial conditions during this period has been found globally, little convincing evidence of the subsequent warming from such a massive freeze has been identified. As such, the snowball earth hypothesis remains a topic of intense geological research. The end of the Cryogenian glaciation marked the beginning of the Phanerozoic Eon—characterized by the renewed warmth that coincided with the evolution of fish and shellfish.

A third glaciation occurred about 460 to 430 million years ago, mostly during the **Ordovician Period** of the **Paleozoic Era** of the Phanerozoic Eon. Although the Ordovician glaciation was relatively short-lived and minor, it resulted in massive extinctions on the planet. The end of this glaciation was followed by warmer, ice-free conditions through the next hundred million years or so (shown in Figure 11.2 as the Silurian and Devonian Periods of the Paleozoic Era within the current Phanerozoic Eon). During that time terrestrial plants, insects, trees, and then amphibians appeared and evolved to more complex forms.

Another period of largely permanent ice occurred in the latter part of the Paleozoic Era, specifically in the **Mississippian, Pennsylvanian,**

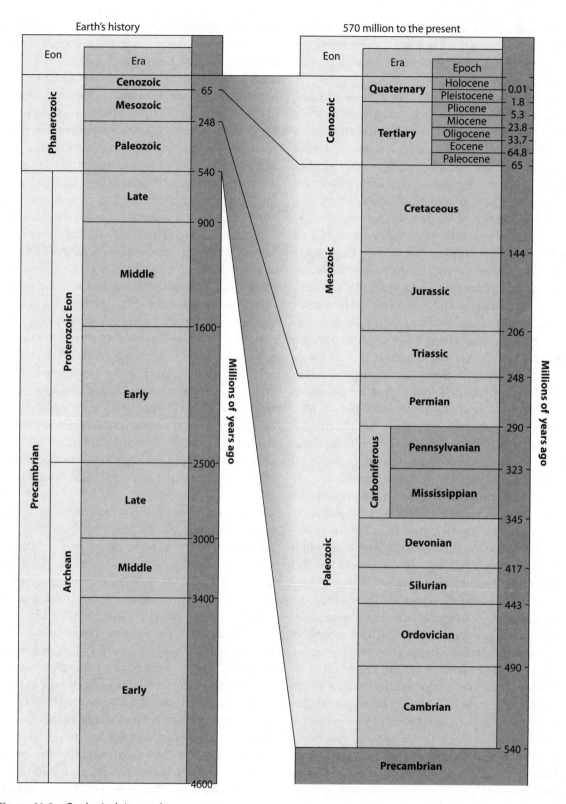

Figure 11.2 Geological time scale.

and **Permian Periods**, from about 320 to 250 million years ago. This **Gondwanan Ice Age**—so-named for the large southern hemisphere continent that resulted from the breakup of the supercontinent **Pangaea**—also culminated in major extinc-

tions around Earth. The extinctions separate the Paleozoic Era from the subsequent, warmer **Mesozoic Era**.

The warmth of the Mesozoic coincided with the development of large reptiles, particularly the dinosaurs, as well as birds, small mammals, and the **angiosperms** (flowering plants). But the Mesozoic Era ended suddenly around 65 million years ago with the **K-T extinction**, which derives its name from the German abbreviations for the **Cretaceous** (abbreviated "K") **Period** of the Mesozoic Era and the succeeding **Tertiary Period**, which began the **Cenozoic Era**—the present **era** of geological time. Prevailing theory suggests that a comet or asteroid impact, or **bolide**, which struck Earth near the tip of the Yucatan Peninsula of Mexico produced massive heat, dust, chemical changes, and volcanic activity that impacted the atmosphere and caused the extinction of perhaps 85 percent of the species on Earth, including all large mammals and dinosaurs.

Because "permanent" ice exists on Earth today (in the ice sheets of Greenland and Antarctica, along with numerous mountain glaciers), we are currently in another ice age in geological history. This ice age, which may have begun as far back as 40 to 50 million years ago in the Tertiary Period of the Cenozoic Era, began to intensify at around 1.8 million years ago—coinciding with the beginning of the **Quaternary Period** of the Cenozoic Era. The large mammals had largely evolved before the onset of this ice age, but some, such as the giant wooly mammoth and sabertooth tiger, did not survive.

Ice ages are subdivided into **glacial phases**—intervals when the "permanent" ice is displaced equatorward and downslope—and *interglacial phases*—intervals when the ice retreats poleward and upslope. In general, as we saw in Chapter 4, glacial phases appear to develop slowly and irregularly, and they may persist for perhaps 50,000 to 150,000 years. By contrast, interglacial phases arrive relatively suddenly and tend to exist for shorter time intervals of only 8000 to 12,000 years (**Table 11.1**). Because the global annual average temperature may vary by perhaps only about 5 C° (9 F°) between glacial and interglacial phases, concern exists that Earth could easily slip into a step change generating a new set of mean conditions, particularly if *positive feedback systems* and human interference are involved.

Currently, we are in an interglacial phase, because the "permanent" ice is currently confined and displaced to include only the polar and highest-altitude areas. The present *Holocene Interglacial Phase* (named for the geological **epoch** that coincides with this interglacial phase) is believed to have begun around 9000 years ago, with abrupt warming that coincided with the first attempts at human agriculture. The knowledge that interglacial phases last only about 8000 to 12,000 years caused some climatologists in the late 1970s to speculate that Earth was slipping into a major, long-term cooling phase. This assumption was largely driven by the very cold northern hemisphere winters of 1978 and 1979. Such fears are in stark contrast to our present concerns with global warming.

Before the onset of the Holocene Interglacial Phase, the *Wisconsin Glacial Phase* in the *Pleistocene Epoch* peaked at approximately 18,000 years ago, when "permanent" ice covered most of North America north of the Ohio River and extended across the pole to cover much of Europe and Asia. The Wisconsin Glacial Phase and subsequent Holocene Interglacial Phase left numerous signs of their presence on the landscape in the form of glacial landforms. This extensive glacial phase obliterated many of the signs of previous glacial–interglacial

Table 11.1   Major Glacial and Interglacial Phases Within the Pleistocene Ice Age

Phase*	Approximate Number of Years Before Present (in thousands)	Deposits that Provide the Name of the Phase
Wisconsin Glacial	9–70	
Sangamon Interglacial	70–130	Sangamon County, Illinois
Illinoian Glacial	130–180	
Yarmouth Interglacial	180–230	Yarmouth, Nova Scotia
Kansan Glacial	230–300	
Aftonian Interglacial	300–330	Afton, Iowa
Nebraskan Glacial	330–480	

*Different names are used in northern Europe, Alpine Europe, Britain, eastern Europe, and other parts of the world to correspond with major lines of evidence for the phases found in other locations.

sequences (Table 11.1). Some of the more prominent resulting landforms include the final formation phase of the Great Lakes, Martha's Vineyard, Long Island, the Finger Lakes Region of New York, and Yosemite Valley in California.

Ice Ages and Sea Level

During glacial phases and ice ages in general, worldwide sea level falls as more of the global water is stored in the cryosphere in the form of continental ice sheets. During the Wisconsin Glacial Phase, *eustatic* sea level was as much as 170 m (550 ft) below today's level, exposing large sections of the continental shelves (**Figure 11.3a**). At that time the Bering Strait was a land bridge between Asia and North America. Mongoloid traits of Native Americans suggest that the ancestors of the Native American population migrated to the Americas via this land bridge.

As Pleistocene ice melted during the *Holocene Epoch*, eustatic sea level rose, drowning low-lying

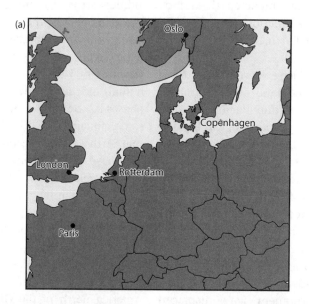

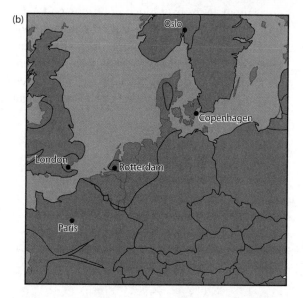

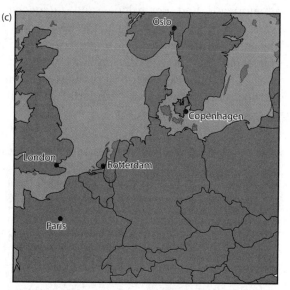

Figure 11.3   Impacts of sea level changes on coastlines in northwestern Europe. (a) Coastline approximately 20,000 years ago, during Wisconsin Glacial Advance; (b) coastline if the East Antarctic Ice Sheet melted, causing a 50-m (170-ft) sea level increase; and (c) coastline if West Antarctic Ice Sheet melted, causing a 7-m (22-ft) sea level increase. *Adapted from*: "Water World," William Haxby, Lamont-Doherty Earth Observatory, for PBS, NOVA Web site, http://www.pbs.org/wgbh/warming/waterworld/, New Content Copyright ©2000 PBS Online and WGBH/NOVA/FRONTLINE.

coastal plains and reducing the slopes of major rivers as they flowed to sea. This caused the deposition of sediments from these rivers, forming vast deltaic land masses and coastal plains through the Holocene. For example, the southernmost part of the Mississippi River valley was formed over the last 10,000 years as conditions warmed, ice melted, and the Mississippi River flooded over into its floodplain and the Gulf of Mexico. Other ice-melt remnants include the drowned river valleys that comprise Chesapeake Bay and the St. Lawrence Seaway and Great Salt Lake, which is a **pluvial lake**—a lake that existed in much greater size in the past, when conditions were colder and/or wetter than today. A common misconception concerning glaciation sea level adjustment is that sea ice displaces the same amount of water as its liquid equivalent (according to **Archimedes' Principle**), so its formation and melting do not contribute to sea level changes. In reality, because fresh water is less dense than saltwater, freshwater sea ice occupies a greater volume than an equal mass of seawater in which it floats. The melting of all current freshwater sea ice alone, therefore, would add about 4 cm (1.5 in) to global sea level, according to recent research. This total does not include the much greater amounts of sea level increase from meltwater of continental ice sheets and glaciers. However, increased precipitation under global warming scenarios could actually increase sea ice extent, according to other recent research by scientists at the University of Maryland.

If all the world's current cryosphere were to melt, then sea level would rise by about 67 m (220 ft), and substantial areas of the coastal plains would be flooded worldwide. Even if only the East Antarctic Ice Sheet melted, large areas of the coastal zone would be flooded (Figure 11.3b). Smaller sea level increases, such as the 5-m (17-ft) increase expected if the West Antarctic Ice Sheet melted, would be problematic as well (Figure 11.3c). **Figure 11.4** depicts the impacts of such sea level increase on coastal Florida.

Some places in Scandinavia and Alaska that were covered by substantial depths of ice during the Wisconsin Glacial Phase are now rising, or

Figure 11.4 Effect of a 6-m sea level rise in Florida. (See color plate 11.4) Courtesy of Jeremy Weiss and Jonathan Overpeck, University of Arizona.

rebounding, because of the release of the tremendous pressure/weight of the ice that has been removed very recently in geological time. In such places sea level is not currently rising relative to the land as elsewhere in the world. Instead, the **isostatic rebound** (rising continental crust) occurs mainly along such coastal locations in Scandinavia and Alaska. Due to the preponderance of major coastal cities and recent coastal development, sea level changes threaten the well-being of a sizable share of Earth's population. In addition, marine ecological systems are affected directly by sea level changes, and changes in the extent of inland seas affect other species.

Recent Trends

Following the peak in the Wisconsin Glacial Phase 18,000 years ago, a period of warming began, signifying the onset of the Holocene Interglacial Phase. However, a brief but very sudden and notable "lapse" back to cold conditions followed from about 13,000 to 11,500 years ago. This cold phase is termed the **Younger Dryas**, except in Europe where it is known as the **Loch Lomond Stadial**. The Younger Dryas is named after the dryas, a herbaceous tundra plant whose pollen grains in sediment and ice layers identified its chronological occurrence. Its name also distinguishes it from the Older Dryas, a less-impressive geological phase that occurred previously. Much research is being invested in determining whether the Younger Dryas was a true global event. Regardless of its geographical extent, the Younger Dryas is important because it provides concrete evidence that very rapid changes in climate have occurred in the past. Presumably, they could also occur in the future.

After the cold conditions associated with the Younger Dryas, a warm period ensued. This warming phase culminated between 8000 and 5000 years ago, a time known as the **Hypsithermal**, in which global average temperatures were approximately 2 C° (4 F°) above today's levels. Gradual cooling followed the Hypsithermal, with a desiccation of the areas now covered by desert in southwestern Asia and northern Africa. A warmer interval followed shortly after the beginning of the first millennium AD (though not as warm as the Hypsithermal), coinciding with the time of Roman expansion.

By AD 500 a cooler and stormier period began in Europe—the Dark Ages. This period came to an end in approximately AD 900, when a warm interval known as the **Little Climatic Optimum (LCO)**, also known as the **Medieval Warm Period**, began. It was during that time that the Vikings colonized (and named) Greenland—a name that would hardly seem appropriate today. During the LCO, crops in Europe grew farther north and higher in elevation than the same crops are known to grow now. **Paleoclimatologists**—scientists who study climates of the preinstrumental record—disagree about whether global temperatures in the LCO were higher than today's values.

From approximately 1450 to 1850 the **Little Ice Age (LIA)** occurred. The LIA is an unfortunately named phenomenon because it does not follow the connotation that "ice ages" last for millions of years. During the LIA streams in Europe that do not freeze today were observed to freeze regularly. Likewise, the notorious winter at Valley Forge for General Washington and his troops seems to have been more severe than most winters experienced today. Another example of the impact of the LIA was the "Year without a Summer" of 1816, in which frosts occurred in every month of the year in New England—an unrealistic phenomenon today. It must be noted that the year without a summer followed the massive eruption of *Tambora*, an Indonesian volcano that exacerbated the resulting temperature decrease by spreading a huge dust plume through the *stratosphere*. Considerable debate exists about whether all these occurrences were parts of the same truly global phenomenon.

The end of the LIA largely coincides with the onset of formal instrumental records from at least some parts of Earth. Three major global climatic periods can be identified from these modern records (**Figure 11.5**). Warming occurred from the late 1800s to about 1945, and then a short but distinct period of cooling was apparent from 1945 to about 1980. Abrupt and dramatic warming has occurred since 1980. Most notably, the 10 warmest years of the instrumental record have all occurred since 1990! An estimated global sea level rise of approximately 15 cm (6 in) has accompanied the warming of the past century. This amount may seem small, but on small islands and coastal zones with little topography, small increases in sea level can produce large increases in beach

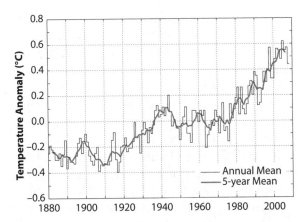

Figure 11.5 Global mean annual average temperature between 1880 and 2008. Anomalies are calculated from mean temperatures between 1951 and 1980. *Source:* J.E. Hansen, R. Ruedy, M. Sato, and K. Lo/NASA/Goddard Institute for Space Studies.

erosion, saltwater intrusion into fresh or brackish streams, ecosystem disruption, and coastline loss. In recent times Earth's climate is generally believed to be more rigorous, with greater extremes of heat and cold during most times in the last 500 million years of geological history.

How Do We Know What We Know About Past Climatic Changes?

As noted above, formal meteorological records go back only to the mid- to late-1800s. Data from upper-level, balloon-based *radiosondes* extend back only to the mid-1900s, and reliable satellite data exist only as far into the past as the 1970s. Nevertheless, paleoclimatologists can make inferences about climatic conditions into the distant geological past. These estimates are made using *proxy evidence* to ascertain climatic conditions of the past and assess the timing and magnitude of climatic change or variability. These clues are hidden in the geological, geophysical, biological, chemical, and historical record. This section reviews some of the major lines of proxy evidence and how they are used.

Basic Principles

All scientific techniques for paleoclimatic reconstruction rely on some basic principles. First, a preponderance of evidence is used to reach conclusions about the climate of a particular place at a particular time in the past. No single study can provide results with enough certainty to lead to any firm conclusions. Our knowledge of paleoclimates is the result of the majority of conclusions from dozens to thousands of independent scientific research studies spanning numerous scientific disciplines. A corollary to this idea is that no single research manuscript could be used to refute the preponderance of evidence.

All scientific techniques for paleoclimatic reconstruction rely on the **Principle of Uniformitarianism**. This axiom states that the processes that occurred in the past are the same ones that occur in the present, at similar rates. Any evidence (such as vegetation types or soil types) that would accompany a certain climatic type today would be assumed to suggest the same kind of climate at the time in the past when similar evidence existed. The Principle of Uniformitarianism also assumes that the same environmental conditions required by a given species today were also favorable for similar species in the past. The Principle of Uniformitarianism, therefore, states that the present is the key to the past.

The opposite of uniformitarianism is **catastrophism**—the belief that changes do not occur uniformly but will instead reshape Earth's environment in a step change in only a short time geologically. Catastrophic changes certainly do occur in nature, as they did during Hurricane Katrina in 2005, and they can be helpful in providing clues to the timing and forcing mechanisms of paleoclimatic change. For example, **paleotempestology**—the science of identifying signatures of ancient storms primarily from sediment records—has been used to infer changes in the atmospheric circulation patterns.

Catastrophic events also can complicate our understanding of the paleoclimate record, because evolutionary processes would not necessarily have time to provide a signature of the increasingly stressful or favorable changes that were undergone by organisms in such a short period of geological time. As a result, catastrophism can leave an unresolved discontinuity in the puzzle of climatic change. A good example of catastrophism in the paleoclimate record is the sudden disappearance of any evidence of living dinosaurs after the K-T extinction. We have seen no convincing evidence that the dinosaurs disappeared slowly in geological time. Instead, the bolide that obliterated the dinosaurs may have also destroyed signs of whether climatic changes occurring up to the time of the bolide were favorable to their existence. It must be noted that most uniformitarian views include, and account for, catastrophic elements.

Radiometric Dating

For most scientific forms of paleoclimatological evidence, **radiometric dating** is used to determine the approximate time at which the evidence was preserved. Radiometric dating is based on the principle of **radioactive decay**—the rate at which spontaneous nuclear disintegration of certain (usually rare) **isotopes** of certain elements occurs. Isotopes are different varieties of the same element, having the same number of protons but different numbers of neutrons from each other. Protons and/or neutrons spontaneously leave the nuclei of radioactive elements at very precise and known rates. When protons leave, the **parent isotope** changes into a different element, resulting in a **daughter isotope**. The inevitable release of energy in the decay process (in fulfillment of the *first law of thermodynamics*) can be detected by a Geiger counter.

To determine the age of a sample, the amount of parent isotope remaining is compared with the amount of resulting daughter isotope. The amounts of each isotope are found by using **mass spectrometry**—instrumental laboratory techniques that can determine the chemical composition of a substance. The rate of decay of each radioactive element is known, and the amount of daughter isotope can be used to infer the amount of time since the rock containing the radioactive element formed. The **half-life** of a radioactive element is the amount of time required for one-half of the parent atoms to decay into daughter product. Because radioactive decay occurs independently of temperature, pressure, or any other known variable, it is a very useful tool in providing relatively precise dates of paleoclimatological evidence.

Dozens of radioactive elements are useful for radiometric dating. Some elements are used more commonly than others for a variety of reasons, such as relative abundance and convenience of half-life for measuring time on geological time scales. Half of a sample of uranium-238 (U_{238}, so-named because the sum of protons and neutrons—the **atomic mass**—is 238) decays radioactively to its daughter product, lead-206 (Pb_{206}), in 4.46 billion years. All the Pb_{206} in a given rock or mineral sample can be inferred to have been U_{238} at the time that the rock solidified. Thus, the ratio of the amount of parent isotope present to the original amount of parent isotope can be deduced. This ratio can be used along with the known half-life to determine the time elapsed since the sample formed. Specifically:

$$\text{Time elapsed} = -\frac{\ell n\left(\dfrac{\text{Number of atoms of parent isotope remaining}}{\text{Original number of atoms of parent isotope}}\right)}{\dfrac{\ell n\,2}{\text{Half-life}}}$$

So, for example, if a rock sample containing evidence of paleoclimatic activity has radioactive elements in that sample with 90 percent U_{238} and 10 percent Pb_{206}, then the equation above tells us that the climate regime identified by the sample must have occurred at about 678 million years ago because

$$-\frac{\ell n\left(\dfrac{90}{100}\right)}{\dfrac{0.693147}{4.46\times10^9 \text{ years}}} = 6.78\times10^8 \text{ years}$$

Another element commonly used in radiometric dating in the geosciences is a rare, radioactive form of potassium-40 (K_{40}), which has the same atomic mass as nonradioactive K_{40}. The radioactive K_{40} decays in a different manner—when one of its protons becomes a neutron, its nucleus changes from 19 protons and 21 neutrons to 18 protons and 22 neutrons. With 18 protons, the daughter isotope becomes argon instead of potassium, even though its atomic mass remains 40. The half-life of this radioactive decay is approximately 1.3 billion years.

Although U_{238} and K_{40} are both useful for many types of radioactive dating, their very long half-lives present a cumbersome problem for relatively recent evidence. Nearly all the parent material of these radioactive elements would remain present if the sample were formed only a few thousands of years ago, and the error in determining the tiny fraction of daughter material may be far too large to date such a recent sample accurately. Radioactive isotopes that have relatively short half-lives are more useful for relatively young samples.

A commonly used isotope with such a short half-life is C_{14}, with a half-life of approximately 5730 years. C_{14} is an extremely rare isotope, with six protons and eight neutrons, whereas most natural carbon (C_{12}) has six protons and six neutrons. Because carbon is used by all living things, it exists in the same concentration in living tissue as in the atmosphere at the time it was ingested. Although the rare C_{14} immediately begins to decay in living tissue, it is continually being replaced by more C_{14} that is being added to the biomass or body of the organism, so a living organism maintains the same ratio of C_{14} to C_{12}. Upon death, the organism ceases to assimilate carbon and the radioactive isotope decays. Radioactive dating can then be used to determine the amount of time since death. Because the amount of C_{14} becomes too small to detect after approximately 10 half-lives, radiocarbon dating is useful only for dating items of origin in the Holocene and late Pleistocene Epochs (the last 50,000 to 60,000 years). But this range provides very useful chronologies for many types of paleoclimatic evidence as well as applications in archaeology, hydrology, oceanography, biochemistry, sedimentology, and a host of other disciplines.

Radiometric dating techniques have strong advantages, because they are the only **absolute dating** techniques that go back farther in geological history than a few hundred years. As such, we rely heavily on them for what we know about the geological past. The technique is also scientifically solid if it can be assumed that the rock or mineral is in a "closed system," meaning that radioactive parent or daughter isotopes are not brought in from other sources during the time that the originally formed radioactive isotopes are decaying. Some argue that we do not know for certain whether the rate of decay is truly independent of all other factors, by drawing in arguments that some poorly understood processes have not been shown for certain to leave half-lives unaffected. But the over-

whelming preponderance of scientific evidence indicates that radiometric dating is a valid, reliable, and precise technique.

Lithospheric and Cryospheric Evidence

If geological samples are not dated radiometrically, the *relative* (rather than *absolute*) ages of various samples in sedimentary rock can be estimated based on the **principle of superposition**—an axiom stating that older evidence is found successively lower in sedimentary rock layers. This **relative dating** of evidence is far simpler and less expensive than absolute dating techniques, but it is less precise.

A related concept is **cross-cutting relationships**—the notion that igneous rock (rock formed from solidifying magma) formations cross-cut (exude into) preexisting sedimentary rock layers. Because the igneous rock cuts through the sedimentary rocks, we know that the sedimentary rock layers must be older than the igneous intrusions. Cross-cutting can be used as both a relative dating technique and an absolute dating technique if samples of the igneous rock layer are subjected to precise radiometric dating techniques.

Relic landforms can also suggest features of past climates. For example, the minerals comprising the breathtaking layers along the walls of the Grand Canyon in Arizona (**Figure 11.6**) provide evidence that the region experienced alternating periods of being exposed above the ocean and being submerged below it; some of these features are related to climatic changes. Another example is provided by pluvial lakes. The Great Salt Lake is merely a remnant of the much larger pluvial Lake Bonneville, which formed from melting glaciers associated with the Wisconsin Glacial Phase approximately 18,000 years ago. The *evaporation* of Lake Bonneville left behind the salt flats surrounding the Great Salt Lake that are used for automobile racing today, providing convincing evidence of a very different paleoclimate. Likewise, environmental features in now-arid locations that required humid conditions for their formation, such as Carlsbad Caverns in New Mexico and petrified forests such as those found in Arizona, provide further evidence of past climatic change.

Figure 11.6   View of the Grand Canyon, showing colorful walls that are the result of differences in the mineral composition of the rock. Many of these contrasts are attributable to the differences in marine versus continental deposition, as oceans covered this region at various times in geological history. (See color plate 11.6)

Glaciers provide many other forms of physical evidence. Their formation, motion, and *ablation* leave behind erosional and depositional landforms or features that provide geological hints of their presence in the past. These include glacial **moraines**, which are large piles of unsorted rock and soil debris deposited by a glacier; **rock striations**, which are grooves cut into bedrock by rocks embedded within the base of moving glaciers; and deposits of **loess**, a soil formed by the grinding of rocks by glaciers into a fine, powdery consistency.

More direct evidence of the past presence of glaciers can also be found. For example, **glacial erratics** provide clues that glaciations occurred in Earth's past. Erratics are rocks that differ in composition from the surrounding geology, suggesting that a glacier must have moved the rock to its present location from somewhere near other rocks similar to its composition (**Figure 11.7**). Erratics can also provide clues on the movement of gla-

ciers. Evidence of a glacier's advance equatorward and/or downslope suggests a period that was cooling and/or becoming wetter. Likewise, signs of glacial retreat poleward and/or upslope suggest that the climate was warming and/or drying. In some cases the discrepancy between local soils and vegetation characteristic of colder climates and the warmer regional climate present in their locations today can give hints of a past climate to which the local environment has not yet adjusted. These situations are especially common where the surface is shaded (particularly during summer afternoons) or in high elevations, and the microclimate has not yet had time to respond to the sudden warming since the end of the LIA, or even since the Wisconsin Glacial Phase.

Geological and glacial evidence can be advantageous to the paleoclimatologist because of the availability of clues extending back several hundred million years. In fact, geological records

Figure 11.7   Glacial erratic.

provide our only reliable glimpse into Earth's distant past. Although evidence directly produced by glaciers does not occur for such an extended portion of geological history, glaciers do provide clear clues about paleoclimates for recently glaciated areas.

Major limitations of geological and glacial evidence also exist. Convincing geological evidence of paleoclimates is difficult to find, as it is often buried deeply or obliterated over time, sometimes by more recent glaciers. Furthermore, it can be a major challenge to date such ancient occurrences radiometrically, because little of the parent isotope may remain or some other process may have introduced parent or daughter isotopes after the evidence in question was produced.

In the case of direct glacial evidence, major parts of Earth that have not been glaciated in the Pleistocene Epoch cannot be searched for clues. This complicates the identification of paleoclimates in tropical and many midlatitude locations, as well as at all oceanic locations. In addition, glacial evidence can lead to questions of why the glacier existed. Was it because of the low temperature, wetter conditions, or some combination of the two?

Finally, the remnants of glaciations before the Pleistocene are largely erased by more recent glacial episodes, restricting the evidence toward those latter events.

Evidence of climatic change can also come from existing ice. Clues about the paleochemistry of the atmosphere are preserved within air bubbles trapped in ice when the ice formed. To identify these clues, deep holes can be drilled into the ice sheets in Greenland and Antarctica, and the **ice cores** can be sampled subsequently in a laboratory. Variations in the ratio of a rare but natural and stable oxygen isotope—oxygen-18 ($\delta^{18}O$, pronounced "del-oh-18")—to "normal" oxygen ($\delta^{16}O$) in the ice layers can reveal clues about the temperature of Earth. Oxygen-18 contains 8 protons and 10 neutrons, rather than 8 protons and 8 neutrons as in "normal" Oxygen-16 ($\delta^{16}O$). Oxygen-16 is nearly a thousand times more common than $\delta^{18}O$, but *relatively* more $\delta^{18}O$ is present in air trapped in polar ice cores when the global temperature is high. The reason for this feature is that during warm periods the high amount of energy available for evaporation allows the heavier $\delta^{18}O$ a greater opportunity to be evaporated than normal, on a global average basis. This $\delta^{18}O$ is then advected toward the polar regions, where it is trapped in air bubbles as ice forms. The ratio of $\delta^{18}O$ to $\delta^{16}O$ is even lower than usual during cold periods in geological history, because the decrease in evaporation leaves more of the heavy $\delta^{18}O$ in the ocean.

Ice cores also provide our only reliable evidence for the concentration of atmospheric CO_2 in the preinstrumental record. The Vostok ice core from the Russian research station near the geographical center of Antarctica contained trapped samples of air from about 160,000 years ago. It was found that atmospheric CO_2 levels ranged from 180 to over 290 parts per million before the Industrial Revolution (compared with over 380 parts per million today!). The Vostok core also revealed that periods of higher atmospheric CO_2 concentrations tend to correspond to the warmer interglacial periods, and lower CO_2 concentrations are linked to glacial advances.

A major advantage of ice core data is the scientific soundness of this technique. Another advantage lies in its ability to provide a reliable record of climatic change dating back several hundred thousand years. A shortcoming is that it is difficult and expensive to obtain cores from the Greenland or Antarctic ice sheets.

A different type of cryospheric evidence is provided by *varves*, which are annual sets of silt and clay layers deposited on the bottoms of lakes and ponds that freeze in winter and thaw in summer. In a frozen lake only fine suspended clay can be deposited—the winter varve—because the surface ice prevents other materials from penetrating into the lake. When thawing begins in spring, fresh water and coarse sediments that had been deposited on the ice surface sink to the lake bottom—the summer varve. Thus, varve thickness can be related to climatic fluctuations. An abnormally thick summer varve may represent a longer freeze-free period. Lakes beneath glaciers do not have summer varves because of the necessity for annual freezing and thawing. A series of "missing" years in the dating of the sediments, therefore, identifies a time during which a glacier existed over the water body. Unlike any of the other techniques discussed to this point, varves offer the advantage of an annual record. The most extensive record of varves is a German data set that extends back about 23,000 years.

Biological Evidence

Fossils   Biological evidence for climatic changes in geological history takes several forms. First, fossilized rock allows paleoscientists to infer the climatic conditions that existed at the time (as determined by radiometric dating) and place that the fossil was formed. The Principle of Uniformitarianism allows us to assume that the same conditions that support organisms similar to the fossilized organism must have existed when the fossil formed.

Certain specific forms of life provide especially useful clues about Earth's climatic history. These include microfossils produced by **foraminifera**, known as marine protozoans. During times of warm oceans, some species of foraminifera tend to have more individuals whose shells developed with a coiling on the right side (**dextral**). On the other hand, during times of colder oceanic conditions left-coiling (**sinistral**) individuals are more common. By mapping the percentage of dextral and sinistral individuals over space and dating the samples, the sequence of past oceanic temperature distribution can be deduced. In addition, marine microfossils can provide clues about the salinity of the oceans, which is related to the energy and water budgets of the atmosphere. Furthermore, because foraminifera are so abundant, their shells are believed to represent a significant sink for atmospheric carbon and may play a role in the regulation of atmospheric CO_2, thereby providing evidence for the strength of the greenhouse effect at various times in geological history.

The composition of the exoskeletons of other fossilized microscopic marine organisms also provides clues about ancient atmospheric temperature. Specifically, the ratio of oxygen isotopes reveals the temperature distribution at the time when the shell was formed, as part of the same process that causes differences in isotope ratios in ice cores. During warm periods in geological history, tropical evaporation is especially intense. The ratio of $\delta^{18}O$ to $\delta^{16}O$ in microfossils in the *tropical* oceans is, therefore, lower than in cold periods, because the heavier $\delta^{18}O$ can be evaporated more easily than at cooler times. In contrast, as we saw previously, the $\delta^{18}O$ to $\delta^{16}O$ ratio is higher in polar ice cores during warm periods as winds in the general circulation carry the oxygen poleward. By contrast, colder periods allow for more "selectivity" in evaporation, causing the heavier $\delta^{18}O$ to be less likely to be evaporated in the same proportion. Cold periods are identified by a relatively high concentration of $\delta^{18}O$ in tropical marine microfossils and a simultaneous relatively low concentration of $\delta^{18}O$ in air bubbles trapped in polar ice cores.

The use of microfossils for a paleoclimatic signature has several benefits. These techniques are scientifically sound and can give reasonably accurate records of climatic change dating back at least 400,000 to 500,000 years—enough to identify previous glacial and interglacial sequences. A major disadvantage of using oceanic microfossil evidence is that the sampling area must be in remote areas to minimize the possibility of transport from other areas.

Palynology   **Palynology** is the scientific study of pollen and spores. Angiosperms and **gymnosperms** —cone-bearing vegetation that does not have seeds enclosed within an ovary—produce pollen grains, and lower plants produce spores. Because pollen and spores are especially resistant to decay and have different characteristics for different species, the presence of pollen and spores can reveal the natural vegetation types during the time that the pollen was produced. Lakes are especially useful catchment basins for pollen and spores because once they reach the water surface, they settle down to the lake bottom. Coring and radiocarbon dating can identify a temporal sequence of pollen/spore types. Some of the changes in the accumulation of pollen and spores are attributable to climatic changes, and they can give scientifically valid clues about Holocene paleoclimates.

Some complications are present in palynological analysis. Because plant species and even individuals of the same species differ markedly in the amount of pollen or spores produced, it is difficult to generalize about whether a certain amount found is indicative of the abundance of that type of vegetation. So, for example, if a small amount of grass pollen is found, it is difficult to know whether abundant grasslands existed far away and a small amount of its pollen drifted to the site or if a small amount of the grass was present near the site. In arid areas pollen and spores are dispersed far from the source, but in rainy areas they remain nearby. Again, it is difficult to ascertain whether a small amount of grass pollen present may indicate

that the environment was too humid to advect more of the grass pollen from the source to the location where it was found. Nevertheless, palynology is a useful tool that can provide important information about the paleoclimatic conditions, including circulation (unlike most of the other techniques), especially when used in combination with other types of paleoclimatic evidence.

Dendroclimatology   The characteristics of a tree's annual growth rings can give clues about the climate during the year when that ring was produced. Because tree rings from the early years of living trees can be matched with the rings from the late living years of dead trees, which can in turn be correlated with rings from trees that lived even earlier, **dendroclimatology** can provide climatic evidence dating back hundreds or even thousands of years. Like varves, dendroclimatology offers the advantage of an annual and even a subannual record. Dendroclimatological analysis has been especially useful in precipitation and streamflow reconstructions.

Dendroclimatology also has its complications. First, other factors besides climate influence the growth of tree rings. These factors include the age, health, and species of a tree as well as characteristics of the soil such as nutrient availability and the availability of groundwater. An additional complication is that the climatic features that the tree ring is indicating may be unclear. For example, if a particular tree shows a series of abnormally narrow rings and it can be assumed that climate produced the stress indicated by the rings, what climatic features limited the growth of the tree? Possibilities may include high summer temperatures, low winter temperatures, lack of precipitation, extreme severe weather events, or recurring periods of minor disturbing events.

A few research groups around the world, most notably at the University of Arizona and the University of Arkansas, have amassed libraries of tree samples for dendroclimatological analysis. The best samples are from trees under stress, because many forms of stress are climate related. Because the features of tree rings are species dependent, dendroclimatologists must understand the characteristics of each tree species sampled, just as palynologists must have great expertise in identifying the species that produced the pollen or spores in a sample.

Other Sources of Biological Evidence Several other forms of biological proxy evidence exist. For example, analysis of the proportion of calcium carbonate ($CaCO_3$) generated from exoskeletons in marine sediments can provide clues about ancient ocean temperatures. Dissolution of $CaCO_3$ is more rapid in warmer conditions than in colder conditions. The distribution of $CaCO_3$ remaining in a sediment sample can, therefore, suggest past thermal conditions where the sample was found.

Likewise, when water temperatures exceed 32°C (90°F) for extended periods of time, algae on coral reefs die (which eventually causes the death of the reef because the energy provided by the algae is lost). The coral then appears white or "bleached." Thus, coral reefs can provide evidence of changes in temperature, particularly if the reef is in a "fringe" area of the ocean where temperature could be a limiting factor to its health.

Finally, fluctuations in the *treeline* also provide proxy evidence of climatic change. When climate is cooling, the "line" separating the zone in which trees can survive from the zone in which trees cannot survive moves equatorward and downslope. When the climate is warming, the treeline moves poleward and upslope. Identification of past forested landscapes is relatively straightforward and evidence is abundant. However, the poleward (or upslope) movement of trees is quicker during warming periods than is the equatorward (or downslope) movement during cooling periods, so the timing of the climatic change may be tricky to discern. Furthermore, the location of a treeline does not necessarily index climatic conditions, as climate may not be the only factor involved in treeline fluctuations.

Historical Data

Historical data refer to written information about climate, but they do not include formal meteorological records. Diaries, ship logs, newspaper accounts, and other sources can provide at least some climatic information. Some of these writings directly pertain to atmospheric phenomena. For example, heat waves, hurricanes, and snowfall totals often may be described in historical records. Other types of written records may be directly or indirectly related to weather and climate, such as observations of animal migrations, agricultural and fisheries yields, and the dates of the flowering of various types of trees.

A major advantage of historical data is the abundance of evidence for recent times. A disadvantage is that the results are not usually quantified, so it can be difficult to ascertain meaning. To alleviate this problem, historians often quantify written observations through **content analysis**— an objective method of converting text into discrete categories to infer climatic conditions. For example, if someone says that it is "very cold this winter," what does that mean? "Very cold" may mean different things to different people. However, recurring patterns of text can be used to glean more meaningful information that can be used to estimate climatic conditions in the historical past.

Converging Proxy Evidence

Paleoclimatological reconstruction requires a wide variety of expertise. Geologists, glaciologists, geochemists, oceanographers, botanists, paleontologists, vulcanologists, foresters, sedimentologists, and historians have all contributed significantly to our understanding of paleoclimates. Because of the inherent difficulty and contradictions in the geological, biological, biochemical, and historical records, the converging evidence from several of the proxy data sources must be used before reaching a firm conclusion. Nevertheless, when the limitations of the various techniques are recognized and conclusions are drawn cautiously, the use of proxy evidence can provide a clear picture of the climates of the past, especially when many forms of proxy evidence converge toward the same conclusions.

■ Natural Causes of Climatic Change and Variability

Several geological, astronomical, solar, and oceanographic mechanisms are known to contribute to climatic change and variability, on time scales ranging from millions of years to a few months. This section details the most well-known natural forcing mechanisms, beginning with those that operate at the longest time scales and proceeding to those that operate at the shortest.

Continental Drift and Landforms

As we saw in Chapter 4, Earth's crustal plates are slowly but constantly moving. Over millions of years this slow drift has moved continents great distances. During times in geological history when continents straddle the polar areas (as they do now), the continents cool far more easily than when water is located over the poles. This has global climatic implications, as the bitter cold polar land masses provide a positive feedback system whereby small amounts of ice and snow cause more reflection (higher *albedo*) of *insolation*, which enhances global cooling. Eventually, ice sheets develop on terrestrial polar locations.

It has been suggested that continental position may explain the ice age/non–ice age sequences. The aggregation of land masses over the South Pole may have led to the Gondwanan Ice Age of the late Paleozoic. Likewise, the present ice age may be attributable to the presence of Antarctica at the South Pole and the large land masses surrounding the high-latitude Arctic Ocean.

Even without the direct influence of continental positions on ice sheets, the positions of continents can affect ocean circulation(and therefore temperature) to a significant extent. When ocean currents are allowed to mix tropical waters with the colder waters of the high latitudes (such as the Gulf Stream/North Atlantic Drift today), ice sheets are less likely to form. In contrast, when continents are positioned so as to hinder the low-latitude waters from mixing with the high-latitude waters (as occurs between Siberia and Alaska today), ice sheets are more likely to form (**Figure 11.8**).

Visit

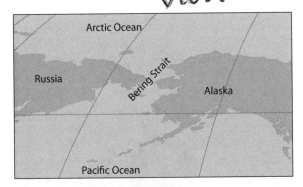

Figure 11.8 The Bering Strait, which nearly cuts off the flow of water between the Pacific Ocean and the Arctic Ocean.

Continental position can affect ocean temperatures in other ways too. Merely 3 million years ago the isthmus of Panama had not yet connected North America to South America and warm tropical waters streamed between the Caribbean Sea and the Pacific Ocean. This likely impeded the cold currents in the eastern Pacific and probably influenced atmospheric circulation, including paleohurricanes.

In addition to the positions of continents on Earth, the positions of mountainous regions on Earth can be important producers of climatic change. During periods after **orogeny**—the process by which mountains form—more of Earth's surface area has sufficiently high elevations to produce colder conditions. Once the cold conditions occur, the presence of snow and ice once again reinforces the cold by reflecting more insolation, setting up a positive feedback system. Mountain uplift can also affect the synoptic-scale atmospheric circulation, which can in turn alter climates. For example, we have seen in Chapter 10 how the presence of the Himalayan Mountains affects Asian precipitation and jet stream circulation. Because of the conservation of *potential vorticity*, discussed in Chapter 7, winds that approach north-south mountain ranges from the west are likely to encourage the formation of long-wave ridges upstream and over the mountain ranges and longwave troughs downstream from the mountains. The amplification of such a ridge–trough configuration caused by the mountains can spill cold air equatorward to a greater extent than would occur if the mountain range did not exist or was lower in elevation. In today's environment, for example, Arkansas is probably colder than it would be without the Rocky Mountains to induce an upper-level ridge that spills cold air on its ridge-to-trough side.

Milankovitch Cycles

In the early 1900s the Serbian astronomer Milutin Milankovitch mathematically calculated several cyclical changes in Earth's orbit around the Sun, axial tilt, and axial orientation. He postulated that these **Milankovitch cycles** that operate on the order of tens of thousands of years may be responsible for cycles in climate, including glacial and interglacial sequences. Milankovitch identified three specific cycles: (1) eccentricity, (2) tilt, and (3) precession of the equinoxes.

Eccentricity One cycle is in the **eccentricity**—circularity—of Earth's orbit around the Sun. Earth's orbit is not centered on the Sun, and the amount of eccentricity varies over time. At one extreme of the eccentricity cycle, the Sun is at almost the exact center of Earth's orbit. At the other extreme, Earth is approximately 11 percent closer to the Sun at *perihelion*—the day of the year when Earth reaches its closest distance to the Sun—than at its opposite position 6 months later—*aphelion*. Eccentricity gradually changes from a more circular to a more elliptical orbit, with a periodicity of approximately 95,000 years.

Presently, we are in a period of relatively low eccentricity. At perihelion, which falls on or about January 3, Earth is approximately 3 percent closer to the Sun than at aphelion, which falls on or about July 4. Because the **inverse square law** states that the intensity of radiation received by a body is inversely proportional to the square of the distance between the emitting and receiving body, this difference in distance results in Earth receiving about 7 percent less insolation in July than it does in January. The fact that July is a summer month in the northern hemisphere implies that eccentricity is not the major cause for the seasons. The cycle of eccentricity (acting alone) implies that northern hemisphere summers may be moderated more significantly than they would be if the orbital eccentricity were higher at this point in geological history.

Periods of low eccentricity are more likely to coincide with glacial periods within ice ages on Earth, whereas periods of high eccentricity are more likely to be linked to interglacial phases. To understand why, it is important to note that summer temperatures are perhaps a more important determinant than winter temperatures of whether ice is expanded equatorward or retreated poleward. Within an ice age, winters will certainly be cold enough for ice to exist over a relatively large portion of Earth's surface. But the degree to which summer is warm enough to melt the ice is the more important question in determining whether the ice age is characterized by a glacial or an interglacial phase. A more eccentric orbit changes the length of seasons in each hemisphere by changing the length of time between the vernal and autumnal equinoxes (**Figure 11.9**). This would provide

for longer summers (and shorter winters) in one hemisphere, thereby providing more opportunity for ice to melt in that hemisphere year-round. Furthermore, periods of higher eccentricity place one of the hemispheres much closer to the Sun during the summer. Such a situation makes it difficult for ice to persist throughout the year in that hemisphere. Apparently, these effects tend to be sufficient to offset the tendency in the opposite hemisphere for a longer winter and shorter summer, regardless of which hemisphere contains more land. The result, assuming that all other factors are equal, is a colder Earth under more circular orbital conditions.

Tilt In Chapter 3 we saw that Earth's axis is tilted from the vertical by 23.5° (**Figure 11.10**) and that this angle of tilt does not change through the course of the year. However, Milankovitch noted that in reality on geological time scales the tilt does slowly change in a cyclical fashion, from a minimum of about 21.8° from the vertical to a maximum of about 24.4° from the vertical. The periodicity of this Milankovitch cycle is approximately 44,000 years.

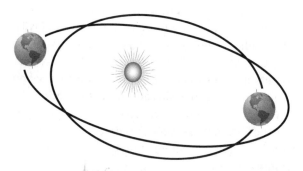

Figure 11.9 Eccentricity of Earth's orbit over time (exaggerated here).

Figure 11.10 Variation in the obliquity of Earth's axis.

This cycle implies that the *Tropic of Cancer* and *Tropic of Capricorn*, which are currently at 23.5°N and 23.5°S latitude, respectively, vary from about 21.8° latitude to about 24.4° latitude. The Tropics of Cancer and Capricorn are significant because they represent the most poleward extent that the *solar declination* can reach at some point throughout the year. Figure 3.5 shows that on or about June 21 Earth is positioned in its orbit such that the northern hemisphere is tilted toward the Sun to its maximal extent (even though the north polar axis itself remains pointed at Polaris, the North Star, year-round). At the same time the southern polar axis is maximally tilted away from the Sun. Likewise, on or around December 22 Earth's position in its orbit is such that the northern hemisphere is maximally tilted away from the Sun (and the southern hemisphere is maximally tilted toward the Sun).

The cyclical changes in the angle of tilt affect the amount of summer to winter contrast in heating, and, therefore, seasonality. The smaller the tilt, the closer to the equator the Sun's direct rays remain throughout the year. In periods of geological history when the tilt is relatively small, summers in middle and high latitudes are cooler (all other factors being equal) but winters milder because the solar declination is not so far away in the opposite hemisphere. A smaller tilt is associated with less seasonal temperature variation between summer and winter in middle and high latitudes. Such a situation encourages glaciation (assuming that all other factors are equal) because, as stated earlier, cooler summers tend to support glaciation in the high latitudes to a greater extent than relatively mild winters (which will be below freezing even when they are "mild").

By contrast, periods of high tilt allow the Sun's direct rays to penetrate farther poleward than average. This causes summer temperatures to be high, because the solar declination approaches the midlatitudes more closely. But it also causes very cold winters, when the solar declination is farther away in the opposite hemisphere.

Precession of the Equinoxes   The final Milankovitch cycle refers to **precession**—the changes in the direction that Earth's axis points (but not the angle—which represents tilt). Earth's axis wobbles like a spinning top because Earth is not a perfect sphere (**Figure 11.11**). Precession refers to this

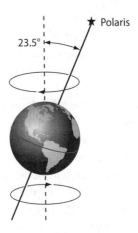

Figure 11.11   Precession, or wobble, of Earth on its axis.

gyrating type of motion. The axis points in a varying direction about the same angle of tilt (although, as we just saw, the tilt simultaneously undergoes cyclical changes) over time, with a periodicity of approximately 23,000 years. This implies that Earth will not always point to our present north star, Polaris. In fact, 11,500 years from now the north star will be Vega, but by that time the coauthor of this textbook will not be here to enjoy his newfound fame.

A progressive change in the dates of the solstices, equinoxes, perihelion, and aphelion (see Chapter 3) occurs because of this Milankovitch cycle and because Earth's elliptical orbit itself is rotating. In 23,000 years the northern hemisphere's summer solstice will again occur on or about June 21. Because 23,000 years is required to advance the cycle by 365 days, we can say that about 63 years are required to advance the cycle by 1 day. In other words, the dates of the solstices, equinoxes, perihelion, and aphelion all change by a little more than 1 day per human lifetime.

This cycle also implies that in about 11,500 years, Earth will experience perihelion in July and aphelion in January. Obviously, if Earth is closest to the Sun at the same time that a hemisphere is experiencing summer, that hemisphere could be expected to become hotter. This tendency is particularly true if the summer hemisphere has abundant land surfaces. At this point in geological history, the hemisphere that has most of the land happens to be farthest from the Sun when it is tilted toward the Sun. Summers in the northern hemisphere are milder than they would be if the axis were positioned so that December 22 was its summer solstice, because in that case the northern hemisphere would be pointing toward the Sun near the same time of year that it experiences perihelion. The same situation also applies to winter cooling for the northern hemisphere.

Combined Effects of the Milankovitch Cycles It has been proposed that when all three of the above factors are in phase, such that they all point to a warmer or colder Earth, the apexes of the glacial–interglacial sequence seem to be reached. The high latitudes may be the key in regulating the thermal conditions associated with these sequences. Most climatologists believe that tropical areas do not show the climatic fluctuations nearly as much as higher latitudes, because changes in the cycles do not affect solar radiation receipt much in those areas. The Milankovitch cycle hypothesis has been disputed in recent years, but it does provide a viable explanation for some types of long-term climatic changes and periodicities.

Volcanic Activity

Earth is believed to undergo intervals of relatively frequent volcanic activity and other periods of less frequent volcanic eruptions. The uneven frequency of volcanic activity through geological time may be explained by differences in the number and strength of **subduction zones**—locations where a denser crustal plate is displaced beneath an adjacent less-dense plate through tectonic forces driven by interior geological processes. As a plate subducts, partial melting of the subducted plate, the overlying plate, or both occurs to form **magma**—molten rock material beneath Earth's surface. The buoyant fluid rises through the overlying plate, and if the surface is reached a volcano is born. Volcanism can also occur where two crustal plates are spreading apart and magma rises from beneath the surface to fill in the "missing" land.

On long time scales we know that atmospheric CO_2 is produced in volcanic eruptions, and it is eventually removed from the atmosphere primarily by absorption in the oceans. Thus, periods of active volcanism may enhance the greenhouse effect by contributing CO_2 to the atmosphere, leading Earth into an ice-free period, and the slow removal of that CO_2 may allow for the onset of an ice age. Similarly, a drop in the production rate of CO_2 by a relative lack of volcanism might reduce

the atmospheric greenhouse effect and lead to lower temperatures. The Gondwanan Ice Age may have begun when the moving continental plates first assembled in Pangaea and then underwent a readjustment period before drifting apart again. Perhaps during the readjustment period the production of CO_2 dropped during a period of infrequent and weak volcanism along subduction zones, leading to cooling.

Paradoxically, on shorter time scales the high concentrations of dust ejected during periods of intense and frequent volcanism may diminish radiant energy reaching the surface, cooling the climate. In severe eruptions such as Tambora, dust can be ejected into the stratosphere and remain suspended for a year or more. Increased particulates can also alter global atmospheric circulation patterns, resulting in a higher frequency of unusual and extreme weather events. Similar short-term effects are believed to be caused by meteor impacts. One example is Barringer Crater near Winslow, Arizona (**Figure 11.12**), where the impact of a meteor of 40 to 50 m (130 to 160 ft) in diameter is likely to have provided global dust-related cooling some 50,000 years ago.

Several examples in addition to that of the "Year without a Summer" presented earlier suggest that volcanism is an agent of short-term "global cooling" and a culprit of severe and unusual weather. The eruption of El Chichon in Mexico in 1982 may have contributed to the unusual global weather of 1982 and 1983, capped by the strongest *El Niño* episode in the meteorological record. Atmospheric extremes included severe drought in South Asia, southern Africa, and Australia, with heavy rains and flooding in the United States Gulf Coast and western South America.

Figure 11.12 Barringer Crater near Winslow, Arizona.

Likewise, the eruption of Mount Pinatubo in the Philippines in 1991 may have been related to some of the unusual weather experienced in the United States in 1992 and 1993. Specifically, heavy rainfall preceded the catastrophic flood on the upper Mississippi River, a very heavy snow event occurred on the Atlantic coast of the United States, and other unusual weather events were recorded around the world. Those were the 2 coolest years globally of the last 20. The effect of the 2010 eruptions of Eyjafjallajökull on the global atmosphere remains uncertain as of this writing.

Variations in Solar Output

Solar output seems like a straightforward factor as a cause of climatic variability and change. It is generally assumed that increased solar output is associated with a warmer climate. Many assumptions about the effects of the Milankovitch cycles rely on this supposition. However, as we have seen with the faint young Sun paradox, reduced solar output does not always lead to colder conditions because other factors may compensate.

There is even some evidence to suggest that an *inverse* relationship between solar output and temperature could exist. With increased insolation, more latitudinal temperature differences would exist, because increases in heating in the tropics would be *relatively* greater than at the poles. A wider disparity in radiation receipt would occur than we see today. This increased energy gradient would cause stronger circulation with more vigorous contrast between tropical and polar *air masses*. This increased air mass contrast, along with the additional evaporation from a warmer tropical ocean, would create more precip-

itation. Some of this extra precipitation would be in the form of high-latitude snow and ice, which would increase the albedo of the surface. With higher albedo, a greater percentage of the Sun's energy would be reflected from the surface, thereby causing colder polar conditions even with greater input of solar energy. This scenario indicates that it is dangerous to oversimplify a problem; the climate system is very complicated.

Like volcanism, variations in solar output can cause both long-term changes over geological history and short-term variability. The Sun experiences small short-term fluctuations in output associated with **sunspots**—huge magnetic storms that appear as darker regions on the Sun's surface, varying in number and intensity over time. The sunspots themselves are "cooler" locations on the Sun's surface, but poorly understood reactions above the sunspots more than offset this tendency for lower solar output. Maximum sunspots imply increased solar energy and recur on average every 11 years (**Figure 11.13**). The Sun's equivalent of the *Coriolis effect*—the differential rate of rotation with latitude—is believed to cause the sunspot cycle by altering the solar magnetic field. Every 22 years or so, the Sun's magnetic field reverses during a period of minimum sunspots. The maximum part of the cycle was last reached in 2001, which was also the time of the last solar magnetic field reversal. Another sunspot maximum should occur again in 2012.

Solar activity indicated by sunspots correlates with some aspects of climate. Supporting evidence includes the fact that few sunspots were observed to exist from about 1645 to 1715 during the **Maunder Minimum**. This period corresponds to the peak of the LIA. Additionally, droughts in the Great

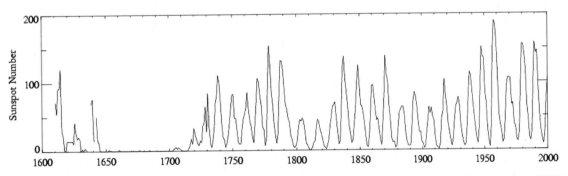

Figure 11.13 Number of sunspots observed since the late Maunder Minimum. Courtesy of D. Hathaway, NASA/ Marshall Space Flight Center, Solar Physics Group.

Plains seem to recur on an approximately 22-year cycle, including the Dust Bowl years of the mid-1930s, the mid-1950s (when drought was actually more severe than in the Dust Bowl years), the mid-1970s, and the severe drought of 1995 and 1996 on the southern plains. Nevertheless, correlations between sunspot cycles and climate are tenuous at best. A more formidable argument can be made that longer-term changes in solar output drive climatic changes.

El Niño-Southern Oscillation Events

As discussed previously, El Niño, *La Niña*, and the associated *teleconnection*—the *Southern Oscillation*—can have drastic consequences near both the tropical Pacific core of activity and around the globe. El Niño-Southern Oscillation (ENSO) affects the midlatitude atmospheric circulation systems because the shearing motion associated with circulation at the poleward edges of the *Had-*

ley cells alters midlatitude circulation. Oceanic circulation is also altered in the Pacific.

Impacts may be far-reaching during El Niño. For instance, *monsoon* circulations tend to weaken around the world, and Atlantic *tropical cyclone* activity tends to be suppressed because of the abnormally strong westerly winds, which shear off the tops of developing tropical cyclones. Other specific important impacts from ENSO have been discussed previously. In general, the effects of El Niño vary according to the strength of the episode. During an El Niño event an area may even experience flooding one time and drought the next.

Because El Niño and La Niña events usually persist for approximately 6 to 18 months and recur every 3 to 7 years on average (**Figure 11.14**), an individual event can hardly be said to bring about climatic change. Instead, ENSO is a producer of climatic variability. However, some evidence suggests that the frequency of ENSO events may have

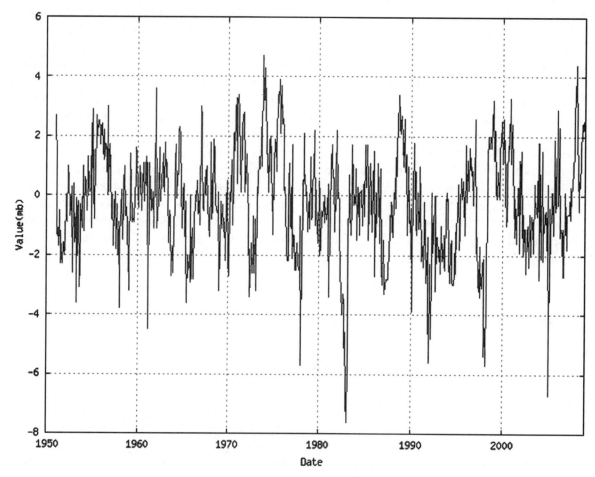

Figure 11.14   Southern Oscillation Index, 1950 through 2008. *Data generated from*: NOAA/Earth System Research Laboratory/Climate Diagnostics Center/Climate Indices Plotting Page [http://www.cdc.noaa.gov/ClimateIndices/].

changed over very long time scales. If true, then ENSO could be a producer of both short-term climatic variability and long-term climatic changes.

Other mechanisms producing climatic change and variability may also be important. One feature in particular is the *quasi-biennial oscillation*. As we saw in Chapter 10, the quasi-biennial oscillation is a periodic reversal of the equatorial stratospheric winds between westerly and easterly, with a cycle of approximately 26 months. Although the quasi-biennial oscillation is known to be related to Atlantic tropical cyclone frequency, the *Asian monsoon*, and winter temperature variability in the northern hemisphere, a precise cause–effect relationship has not been established with confidence. Similarly, its influence on other features of the climate system may exist but is poorly understood at this time.

■ Summary

Climates vary and change through time, as have people's perceptions of what constitutes climatic variability and change. This complicates the establishment of policy related to climate, because climatic variability and change mean different things to different people. Our best scientific evidence suggests that the climate has fluctuated significantly but within a reasonable range of the present temperature, despite the evidence that the solar output was significantly lower in the early part of geological history.

Ice ages are periods of approximately 50 million years when "permanent" ice exists somewhere on Earth's surface. No credible scientific evidence for ice ages in the first half of Earth's 4.6 billion years has been identified. However, since that time there have been at least five ice ages, with several hundred million years separating major ice ages. The present ice age may have begun as far back as 40 to 50 million years ago, but it escalated in intensity over the last 1.8 million years. Ice ages are subdivided into glacial phases—when ice extends equatorward—and shorter interglacial phases—when the ice retreats but is not absent. We are currently in an interglacial phase of an ice age—a rather unusual time in geological history. During periods when ice is abundant, sea level decreases as more of the global water budget is stored in the cryosphere. Rising sea levels are a concern if Earth continues to warm.

The Wisconsin Glacial Phase represents the most recent glacial advance, when at its peak 18,000 years ago large ice sheets covered North America approximately north of the Ohio River. Abrupt warming occurred after that point, except for a few significant cooling periods. Most notably, the Younger Dryas represented rapid and pronounced cooling, perhaps on a global scale, and showed that the onset of climatic change can occur relatively quickly. Since the Younger Dryas, warming has generally resumed, with the Hypsithermal having been even warmer than today's climate. In more recent times the cold conditions during the Little Ice Age have been followed by warming from the latter part of the 1800s to about 1945, cooling from 1945 to about 1980, and very abrupt and significant warming since 1980.

Various forms of proxy evidence are used to provide information about Earth's paleoclimates dating back hundreds of millions of years. The time at which such evidence was formed can be ascertained by using radiometric dating techniques that rely on the constant rate of decay of radioactive elements that make up a rock sample. Geological evidence allows for estimates to be made in Earth's most distant past, with glacial, ice core, and some forms of biological evidence such as oceanic microfossils allowing for estimates dating back hundreds of thousands of years. Other forms of biological evidence, including pollen, tree rings, treeline fluctuations, and coral, can provide information dating back thousands of years, but with more detail in the record. Sediments in the form of varves can also provide evidence from the past several thousand years with an annual record. Historical records round out the sources of data for paleoclimatic reconstruction, providing evidence over the past several centuries.

Several natural mechanisms have been identified as possible contributors to climatic change and variability. At the broadest time scales, continental drift and mountain-building episodes have been proposed as major causes. Cycles in the orbital and axial features of Earth, known as Milankovitch cycles, are believed to cause changes on time scales of tens of thousands of years because of their role in regulating the range in earth-Sun distance throughout the year, the range of latitudes that can experience the Sun's direct rays, and the time of year when the solstices occur.

Variations in volcanic activity and solar output can cause both short-term variability and long-term changes. Finally, very short-term events such as ENSO and perhaps the quasi-biennial oscillation may cause climatic variability on seasonal to biennial time scales.

▶ Key Terms

Ablation
Absolute dating
Air mass
Albedo
Angiosperm
Anomaly
Aphelion
Archimedes' principle
Asian monsoon
Atomic mass
Biosphere
Bolide
Catastrophism
Cenozoic Era
Content analysis
Coriolis effect
Counter-radiation
Cretaceous Period
Cross-cutting relationships
Cryogenian Period
Cryosphere
Daughter isotope
Deforestation
Dendroclimatology
Dextral
Eccentricity
El Niño
Eon
Epoch
Era
Eustatic
Evaporation
Extreme
Faint young Sun paradox
First law of thermodynamics
Foraminifera
Fossil fuel
Glacial erratic
Glacial phase
Gondwanan Ice Age
Greenhouse effect
Greenhouse gas
Gymnosperm
Hadley cell
Half-life

Holocene Epoch
Holocene Interglacial Phase
Hurricane
Hydrosphere
Hypsithermal
Ice age
Ice core
Insolation
Interglacial phase
Inverse square law
Isostatic rebound
Isotope
K-T extinction
La Niña
Lithosphere
Little Climatic Optimum
    (LCO)
Little Ice Age (LIA)
Loch Lomond Stadial
Loess
Longwave radiation
Magma
Mass spectrometry
Maunder Minimum
Medieval Warm Period
Mesozoic Era
Methane
Methanogen
Milankovitch cycle
Mississippian Period
Momentum
Monsoon
Moraine
Noise
Ordovician Period
Orogeny
Paleoclimatologists
Paleoclimatology
Paleotempestology
Paleozoic Era
Palynology
Pangaea
Parent isotope
Pennsylvanian Period
Perihelion

Period
Periodic variability
Permian Period
Phanerozoic Eon
Photosynthesis
Pleistocene Epoch
Pluvial lake
Positive feedback system
Potential vorticity
Precession
Principle of superposition
Principle of uniformitarianism
Proterozoic Eon
Proxy evidence
Quasi-biennial oscillation
Quaternary Period
Radioactive decay
Radiometric dating
Radiosonde
Random variability
Relative dating
Rock striations
Sinistral
Snowball earth hypothesis
Solar declination
Southern Oscillation
Step change
Stratosphere
Subduction zone
Sunspot
Tambora
Teleconnection
Temporal trend
Tertiary Period
Treeline
Tropic of Cancer
Tropic of Capricorn
Tropical cyclone
Variability change
Varve
Wisconsin Glacial Phase
Younger Dryas

*Terms in italics have appeared
    in at least one previous
    chapter.*

▶ Review Questions

1. Describe the different forms that climatic variability and climatic change can take in a data set.

2. How has the meaning of climatic change evolved since the 1960s?

3. What are the major ice ages that are known to have affected Earth? When did they occur?

4. What theory prevails regarding the cause of the K-T extinction?

5. How can melting sea ice contribute to global sea level rise?

6. What is the significance of the Younger Dryas?

7. Describe the sequence of global climatic conditions since the Younger Dryas.

8. How does radiometric dating work?

9. Name three radioactive elements that can be used in radiometric dating and describe their uses.

10. What are the advantages and disadvantages of using glacial evidence as a sign of climatic changes?

11. Contrast the ratio of $\delta^{18}O$ to $\delta^{16}O$ in tropical marine microfossils and polar ice cores. Why does this difference occur?

12. What are varves and how are they used to identify climatic changes? What pitfalls exist in the use of varve data?

13. What are the advantages and disadvantages of using dendroclimatological evidence for identifying climatic change?

14. What is the purpose of content analysis?

15. What explanations exist for the formation of ice ages in geological history?

16. How is eccentricity affected by the inverse square law?

17. How does the tilt of Earth's axis vary through time, and how is this assumed to cause climatic changes?

18. Why do the dates of the equinoxes and solstices change throughout geological time?

19. How is volcanism a cause of both climatic cooling and climatic warming?

20. How might increased solar output result in a decrease in global temperatures?

21. How does ENSO alter the extratropical atmospheric circulation?

▶ Questions for Thought

1. If you filled a glass with ice and then poured water into the glass until it was flush with the rim of the glass, would water spill over the edge of the glass when the ice melted? Would the water level decrease when the ice melted? Why?

2. What would be the resulting climatic conditions for both hemispheres given a hypothetical highly eccentric orbital variation of 20 percent, an increased tilt to 25° from vertical, and a reversal of the current perihelion/aphelion?

3. You are working for a research institution that is trying to ascertain climatic changes 250 to 500 million years ago. What data sets would you examine and what analytical methods would you use? Explain why you would use them.

4. You are now working to ascertain climatic changes that occurred over the past 25,000 years. Again, what data and methods would you use, and why?

5. You are now attempting to ascertain recent climatic changes over the past 250 years. What types of data would you examine (assume no access to instrumental records), what methods would you use, and why?

6. Compare the 2004 through 2009 Atlantic hurricane seasons using available information online. How are our perceptions of climate variability influenced by the locations where hurricanes strike?

---

http://physicalscience.jbpub.com/climatology

Connect to this book's Web site: http://physical science.jbpub.com/climatology. The site provides chapter outlines, further readings, and other tools to help you study for your class. You can also follow useful links for additional information on the topics covered in this chapter.

12 | Anthropogenic Climatic Changes

Chapter at a Glance
Global Warming
   Greenhouse Effect
   Greenhouse Gases
   Urban Heat Island
   Global Warming: The Great Debate
Atmospheric Pollution
   Global Dimming
   Atmospheric Factors Affecting Pollution
     Concentrations
   Air Quality Legislation in the United States
Classifying Air Pollutants
   By Response
   By Source
Reactions and Attitudes to Climatic Change
   Prevention
   Mitigation
   Adaptation
   Continued Research
Summary
Review

The previous chapter analyzed the major climatic changes throughout geological history, the natural mechanisms that cause climates to change, and the evidence for identifying climatic changes in the past. However, not all forcing mechanisms for climatic change are natural. As we saw in Chapter 4, humans exert a tremendous influence on their environment, including its climate, through land-use changes, and also through activities contributing to *deforestation* and *desertification.*

This chapter investigates some of the most significant *anthropogenic* mechanisms that contribute to climatic change on the planetary scale. Two

of these mechanisms (the anthropogenic *greenhouse effect* and the *urban heat island*) lead to warming and are considered to be irreversible—or more-or-less permanent (at least on human time scales)—in that they are considered to cause climatic change rather than climatic variability. The third mechanism (atmospheric pollution) may lead to cooling and may be reversible. Because the greatest human impact has occurred in the last century, evidence for these changes comes mostly from formal meteorological records rather than from *proxy evidence* such as that discussed in Chapter 11.

The nature of the material in this chapter is very controversial in that scientists tend to disagree on the degree of impact that humans have on their environment. Additional controversy arises because so much is at stake. Domestic and international policy hinges on the degree of seriousness associated with the problem of anthropogenic climatic change—a topic that has become the central issue not only in climatology but also in all earth and environmental sciences over the last 20 years.

■ Global Warming

Greenhouse Effect

As we have seen, the greenhouse effect is a natural process that can be intensified by human activity. The greenhouse effect occurs as certain atmospheric gases absorb *longwave* (terrestrial) *radiation* and hinder its transmission from Earth's surface to space. The portion of this energy emitted downward, back toward the surface, is known as *counter-radiation*, and it is responsible for keeping Earth warmer than it would be if there were no greenhouse effect. It is tempting to think of *greenhouse gases* as simply reflecting the long-wave radiation back to the surface, but in reality the gases absorb the energy and then reemit energy in all directions, as any object with a temperature above *absolute zero* does. This process was first described by the Swedish chemist Svante Arrhenius in 1896.

The term "greenhouse effect" was chosen because a greenhouse operates in a somewhat similar manner. *Shortwave radiation* from the Sun can enter a greenhouse easier than longwave energy can escape it, making the air inside the greenhouse much warmer than it would be if the greenhouse did not exist. A similar process occurs when the inside of a car is superheated while sitting in the Sun. The analogy between the "real" atmosphere and a greenhouse breaks down when *advection* and *convection* are considered. In the "real" atmosphere, the atmospheric circulation tends to mix warmer air with colder air, both horizontally and vertically. In a greenhouse the warmed air cannot mix with colder air outside the greenhouse. The greenhouse heats by suppressing advection and convection, holding energy in place inside the structure. Of course, no such suppression exists in the real atmosphere, but the analogy remains. In the real atmosphere greenhouse gases decrease longwave radiation losses from Earth's surface to space, which significantly raises the average surface temperature.

Greenhouse Gases

Water Vapor   Several gases play a major role in the greenhouse effect (**Table 12.1**). Water vapor is by far the most abundant and important greenhouse gas. It varies greatly in concentration across Earth, from nearly 0 percent of the mass of the lower atmosphere (near the poles in winter, where temperatures are too low to allow much water to exist as a vapor, even at saturation) to about 4 percent (in tropical oceanic areas) of the lower atmosphere. Water vapor is particularly effective at absorbing longwave radiation from Earth's surface at *wavelengths* between 5 and 8 *micrometers* (μm) and at wavelengths above 17 μm (Table 12.1). When cloud cover increases, water vapor and liquid water behave as effective greenhouse "traps" at all wavelengths at which any substantial terrestrial energy is radiated. By contrast, a lack of atmospheric moisture causes temperatures to decrease more at night than

Table 12.1 Wavelengths at Which Greenhouse Gases Absorb and Emit Radiation Under Cloudless Conditions

Wavelength	Primary Greenhouse Gas
4–5 μm	CO_2
5–8 μm	H_2O vapor (nearly complete at 5–6 μm)
8–13 μm	Atmospheric window—some H_2O vapor absorption and a narrow but strong O_3 absorption band between 9.4 and 9.8 μm
10 μm	Peak emission of energy from Earth and atmosphere*
13–17 μm	O_2—strong band
17–24 μm	O_2 and H_2O vapor but partially transparent
>24 μm	H_2O vapor (complete)

Under an overcast sky, atmospheric emission/absorption is augmented, primarily in the 8- to 13-μm "window."

*Note that Earth emits the greatest amount of energy at a wavelength of about 10 μm.

during times when water vapor is more abundant. This is because less atmospheric *absorption* can occur under such circumstances, particularly through the **atmospheric window** of 8 to 13 μm, where Earth emits its maximum amount of radiation. It just so happens that few gases absorb radiation effectively in the wavelengths of the atmospheric window.

Even though water is the most abundant greenhouse gas, it is mentioned little regarding the problem of anthropogenic greenhouse warming because the atmosphere is efficient at ridding itself of excess water vapor. This occurs through the precipitation process. Humans only play a small role in the global water cycle.

Carbon Dioxide   The second-most abundant greenhouse gas is carbon dioxide (CO_2), which accounts for only about 0.038 percent of the atmosphere. This means that only 380 molecules out of every million in the atmosphere are CO_2. Although this concentration hardly seems significant, CO_2 has a disproportionately large influence on temperature regulation of Earth. Chemical analysis of air bubbles trapped in *ice cores* tells us that this concentration represents an increase from about 0.028 percent in 1860—just the blink of an eye in geological time. The abrupt increase in concentration has occurred since the onset of large-scale combustion of *fossil fuels*—primarily coal, oil, and natural gas—since the beginning of the Industrial Revolution in the late eighteenth century.

Carbon dioxide represents by far the largest share of anthropogenic greenhouse gas emissions in the United States and in the world, with fossil fuels contributing the vast majority of CO_2 emissions (**Figure 12.1**). The concentration is projected to grow to 0.060 percent by 2050. According to computer models, this concentration would produce a globally averaged temperature increase of 1.5 to 3.5 Celsius degrees (2.7–6.3 Fahrenheit degrees) (**Figure 12.2**). A worst-case scenario puts the amount of warming equal to that experienced since the *Wisconsin Glacial Phase* but in a drastically shorter period of time.

Fossil fuel emission is the only significant source of atmospheric CO_2 that is primarily unnatural (see Figure 2.2). Although the combustion of wood and other organic matter for heating is not natural and emits CO_2 to the atmosphere, it is not considered a major source because the material being burned already removed CO_2 from the atmosphere in a relatively short time during its life cycle. So, little net long-term addition of atmospheric CO_2 from this combustion occurs, unless there is large-scale deforestation of old-growth forests. By comparison, the carbon stored in fossil fuels had been sequestered for millions of years beneath the surface until the CO_2 was released in the combustion process. Natural sources of atmospheric CO_2 include the decomposition of organisms, plant and animal respiration, and some oceanic processes.

Several important CO_2 *sinks*—mechanisms that remove more CO_2 from the atmosphere than they emit to it—are noteworthy. On geological time scales sedimentary rock formation represents by far the largest sink. But because this process of

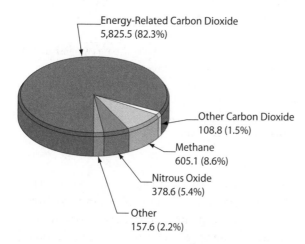

Energy-Related Carbon Dioxide
5,825.5 (82.3%)

Other Carbon Dioxide
108.8 (1.5%)

Methane
605.1 (8.6%)

Nitrous Oxide
378.6 (5.4%)

Other
157.6 (2.2%)

Figure 12.1 U.S. anthropogenic greenhouse gas emissions by gas in 2006 (million metric tons of carbon equivalent). *Data from*: Energy Information Administration. *Emissions of Greenhouse Gases in the United States 2006* (Washington, DC: 2007).

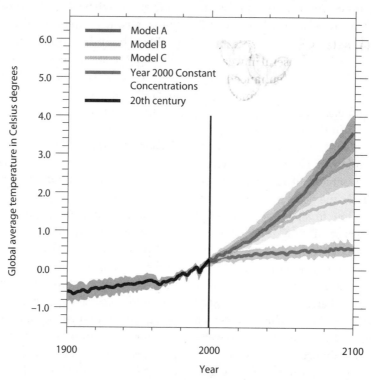

Figure 12.2 Global temperature curve: historical and projected. Courtesy of Windows to the Universe [www.windows 2universe.org].

lithification requires extraordinarily long time periods to remove CO_2 from the atmosphere effectively, we will not consider this to be an important process for mitigating the anthropogenic greenhouse effect.

On human time scales the oceans represent the largest sink. Figure 2.2 shows that the oceans represent a tremendous *reservoir* in the global carbon cycle, with 52 times as much carbon as the atmosphere and 19 times as much as in the combined *biosphere* and soils at any given time. Marine phytoplankton take in atmospheric CO_2 and subsequently release carbon in the form of calcium carbonate ($CaCO_3$) that eventually sinks to the ocean floor, where it is sequestered indefinitely through the lithification of sedimentary rocks. This process is estimated to reduce atmospheric CO_2 concentrations by one-third of what they would be otherwise.

Some researchers have suggested that fertilizing the ocean surface with iron sulfate will allow marine phytoplankton to reproduce rapidly and remove even more atmospheric CO_2, thereby curtailing *global warming*. Even the removal of all atmospheric CO_2 would result in an increase in the ocean storage of less than 2 percent. But the removal of all atmospheric CO_2 would be catastrophic. Notwithstanding the fact that without any atmospheric CO_2 Earth would cool greatly, CO_2 combines with water to produce **carbonic acid (H_2CO_3)**, a weak acid that is useful in weathering rock and forming soils. But if "enhanced" biological processes in the oceans could remove only the extraneous CO_2 that was released through human activities, a large part of the global warming problem might be solved. Of course, the negative impacts that may result from such a fertilization program have prevented the idea from being implemented. Regardless of whether such a plan to remove atmospheric CO_2 is ever carried out, researchers are currently working to achieve a better understanding of the ocean's capacity to store atmospheric CO_2.

Other atmospheric CO_2 sinks are minor compared with the oceans. Soil is believed to be the second-most important sink on human time scales. Soils sequester atmospheric carbon largely through their containment of partially decomposed organic matter known as **humus**. Although the total amount of potential storage of carbon in soil is unknown, with improved management practices,

soils have the capability to sequester more carbon than at present. As improved soil conservation practices gain higher acceptance for their own merits, soils are anticipated to, at least slightly, mitigate the problem of increasing atmospheric CO_2.

The third important sink for anthropogenic CO_2 is the nonsoil component of the terrestrial biosphere. Actively growing vegetation absorbs CO_2 during *photosynthesis*, and it stores carbon in its biomass as it continues to grow. But because only 30 percent of Earth's surface is land and only 30 percent of that total is covered by forests, the biosphere is a far smaller sink than the ocean and slightly smaller than the soils. Within the biosphere the most critical region for CO_2 sequestration is the tropical rain forest, and these areas are disappearing, largely because of human-related deforestation and desertification. Some evidence even suggests that replanting new forests initially acts as a local source of atmospheric CO_2 because carbon tied in the soil is released to the atmosphere. Other evidence suggests that the effect of deforestation may be somewhat offset by the more rapid growth of forests because of the nourishing effect of the extra atmospheric CO_2. Outside the rain forest region, seasonal changes in vegetative cover render CO_2 fixation by the biosphere less effective in winter, even if the environment becomes more favorable for forest growth under global warming scenarios.

Methane   Until recently, *methane* (CH_4) was assumed to play a relatively minor role in the greenhouse effect because its concentration in the atmosphere is only 1.85 parts per million (ppm). But recent research suggests that it is much more important than previously believed because each CH_4 molecule acts as a far more effective absorber of terrestrial radiation than an individual CO_2 molecule. In addition, as discussed in Chapter 2, CH_4-emitting bacteria, called *methanogens*, are increasingly recognized for their role in maintaining Earth's early "thermostat." Furthermore, CH_4 is seen as being an important player in the anthropogenic greenhouse effect because, like CO_2, its concentration has increased substantially since the Industrial Revolution. Ice core analysis suggests that atmospheric CH_4 concentrations changed very little over the last 160,000-year period, remaining near 0.8 ppm, until experiencing an abrupt rise in the last 200 years. This increase is far more rapid than the increase in atmospheric CO_2.

The details of the global CH_4 budget have been understood only for the last decade. Methane is released to the atmosphere through several processes (**Figure 12.3**), especially through **anaerobic** decomposition—the decomposition of organic matter in the absence of oxygen. Anaerobic decomposition occurs readily in waterlogged environments. On a global basis, about 27 percent of the atmospheric CH_4 comes naturally from wetlands, including bogs, *tundra*, and swamps, where methane-producing bacteria thrive in the rich organic matter and poor oxygen content. The next leading source, at about 22 percent of the total, is from the bacteria present in the anaerobic digestive process of animals, particularly ruminants such as cattle and certain species of termites. Livestock emit 18 percent of the atmospheric CH_4, with termites alone emitting another 4 percent. Although other animals also emit CH_4, the total is thought to be insignificant. A third major source (11 percent) of CH_4 is from the flooding of rice fields, where certain types of bacteria thrive. Biomass burning, landfills, fossil fuel combustion, and biological oceanic processes are believed to account for most of the rest of the atmospheric CH_4.

A final important source of methane is **methane hydrates**, which comprise only a small percentage of atmospheric CH_4 today but may be of great importance in the future. Methane hydrates are frozen lattices of water molecules surrounding CH_4 molecules. These solids are found extensively worldwide throughout ocean sediments and deep underground in polar regions. Methane hydrates are so abundant that some believe these provide far more energy potential than the world's known reserves of coal, oil, and natural gas combined. Methane is released from the hydrates naturally with changes in temperature, pressure, salt concentrations, and other factors, but to date no method of extracting the CH_4 safely, efficiently, and economically for energy consumption has been devised. If such methods are developed, far more rapid increases in atmospheric CH_4 would occur.

Other Greenhouse Gases   Other greenhouse gases are thought to play minor roles in the anthropogenic greenhouse effect. **Nitrous oxide (N_2O)** is a non-natural greenhouse gas that contributes a tiny percentage to the anthropogenic effect. It is released to the atmosphere primarily through the combustion of fossil fuels and industrial and agricultural activities. Also, various types of **fluorocarbons** are emitted to the atmosphere by industrial activities. Although very minute in concentration, they are extremely efficient absorbers of terrestrial energy. Other gases such as **carbon monoxide (CO)** and various types of **nitrogen oxides (NO_x)** other than N_2O can influence chemical cycles in the atmosphere that affect the concentrations of greenhouse gases.

Trends in Greenhouse Gas Emissions   Greenhouse gas emissions are increasing for several reasons. First, sources of CO_2, CH_4, N_2O, and fluorocarbons continue to increase worldwide, despite the ratification of the **Kyoto Treaty** in 1997, which sought to limit emissions of these gases by seven of the eight leading industrial nations of the world (the United States being the exception). Second, large-scale deforestation is decreasing an important CO_2 sink. Finally, the conversion of mature forests to younger forests may result in a reduced overall reservoir for carbon in the biosphere, even though the rate of carbon fixation by younger trees is faster than for mature trees.

The United States currently produces 25 percent of the world's greenhouse gases. Its percentage of the world output is anticipated to decrease over the foreseeable future as newly industrializing countries, particularly China, India, and Brazil, produce a proportionately larger share of the world's emissions. According to the U.S. Energy Information Administration, U.S. emissions are economically efficient, with emissions per dollar

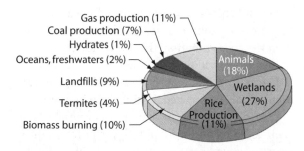

Gas production (11%)
Coal production (7%)
Hydrates (1%)
Oceans, freshwaters (2%)
Landfills (9%)
Termites (4%)
Biomass burning (10%)
Animals (18%)
Wetlands (27%)
Rice Production (11%)

Figure 12.3   Sources of net global emission of atmospheric methane. *Data from*: Mikaloff Fletcher, S. E., Tans, P. P., Bruhwiler, L. M., Miller, J. B., and Heimann, M. (2004), "CH4 sources estimated from atmospheric observations of CH4 and its C-13/C-12 isotopic ratios: 1. Inverse modeling of source processes," *Global Biogeochemical Cycles*, 18(4). UC Los Angeles: Retrieved from: http://escholarship.org/uc/item/7fm946gq.

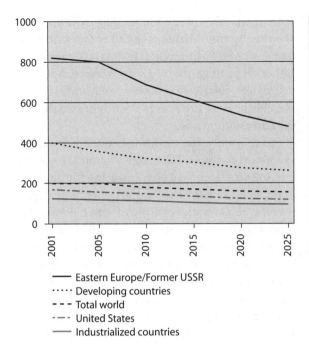

Figure 12.4 Current and projected carbon emissions standardized by gross national product (metric tons per million dollars). *Data from*: *International Energy Outlook, 2003*. Energy Information Administration.

of gross national product below the world average and far below the developing world's average (**Figure 12.4**). Fossil fuel emissions within the United States are concentrated in regions nearest to the deposits, with coal emissions concentrated in the Appalachian states and northern Plains and oil and gas in the Gulf Coast region (**Figure 12.5**).

Indirect Effects of Increasing Greenhouse Gases
In addition to higher atmospheric temperatures, other effects from the increasing concentrations of greenhouse gases are likely to result. Increased air temperatures are likely to cause some melting and *sublimation* in the *cryosphere*. Such melting would decrease the *albedo* of polar regions as darker surfaces replace the snow and ice, exacerbating the situation in a *positive feedback system*. The meltwater would then induce sea level increases, which would endanger coastal cities, increase coastal erosion problems, and cause *saltwater intrusion* in coastal zones.

The oceans are also likely to expand slightly in volume as the *density* of the ocean waters decreases under warming conditions. This factor alone is estimated to contribute perhaps 25 percent of the sea level rise that we might see. When combined with

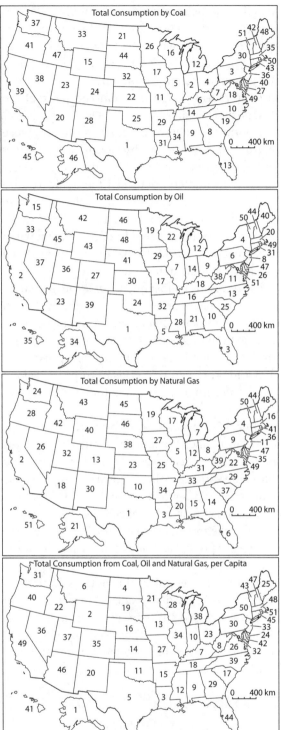

Figure 12.5 Fossil fuel consumption ranked by state, 2007. *Data from*: U.S. Energy Information Administration [http://www.eia.doe.gov/emeu/states/sep_sum/plain_html/rank_use_per_cap.html]. Accessed April 2010.

the effect of the addition of meltwater, sea level is expected to rise substantially in the next century. Some have suggested that a 1- to 2-mmyear^{-1} increase in sea level rise has occurred throughout the twentieth century, with perhaps another 0.1- to 0.2-mmyear^{-1} increase from the peak of the Wisconsin Glacial Phase until 1900.

Another possible indirect effect is an increase in *tropical cyclone* activity. Because *hurricanes* and other tropical cyclones require warm water for their formation, increased temperatures are likely to increase the surface area of Earth where tropical cyclones might form and track, in addition to increasing the length of the season when they can exist. The additional energy from global warming would then be available to drive the *evaporation* and subsequent *latent heat* release that would energize tropical cyclones. Although it is possible that tropical cyclones may be more intense than in the past because of global warming, some scientists say that the effect is exaggerated and other natural cycles in tropical cyclone activity dominate any trends induced by global warming.

Other effects may not even be known or predictable to us today. For example, circulation changes may occur with changes in global temperatures that could shift midlatitude storm tracks and cause relatively rapid regional climatic shifts. Some of these changes may be beneficial to some economic sectors, whereas others may benefit other sectors. One rather predictable outcome is that any shifts in surface water availability are likely to be accompanied by geopolitical problems as nations struggle for control of the water. Temperature- and moisture-induced shifts in agricultural and bioclimatic zones could likely be accompanied by severe and unpredictable ecological, economic, and political impacts.

Urban Heat Island

A second anthropogenic cause of climatic change also contributes to global warming. This effect is associated with urbanization. Cities are generally 1 to 5 C° (2 to 9 F°) warmer than the surrounding countryside, for several reasons. First, the urban environment has less evaporation and *transpiration* (*evapotranspiration*) of water from soil and plants than rural areas because of the relatively sparse vegetation found in urban areas. This leaves more available energy for *sensible heat* and less shading by trees. Urban rainfall is more likely to become *runoff* immediately after a rainstorm rather than lingering near the surface, and this also causes relatively less latent heating and more sensible heating to occur in the urban area than in the surrounding area.

Additionally, the buildings and streets themselves are composed of materials that are conducive to extra storage of energy during the daytime and reemission of that stored energy at night, thereby delaying the cooling of urban areas. Materials such as concrete, asphalt, and brick have greater **heat capacity** and **thermal conductivity** than moist, open soil and vegetation. Heat capacity refers to the amount of energy that would need to be added to a unit volume of a substance to raise its temperature by 1 K in the *Kelvin temperature scale*. Thus, the units of heat capacity are *Joules* m^{-3} K^{-1}. This term is related to *specific heat*, except that specific heat refers to the amount of heat needed to increase a unit mass (not volume) by 1 K. Thermal conductivity refers to the number of Watts of power required to increase the temperature at a point 1 m into the ground by 1 K. The units of thermal conductivity are W m^{-1} K^{-1}. Heat capacity and thermal conductivity of urban materials allow them to have an ability to absorb a tremendous amount of heat and also conduct that heat (energy) more readily than materials such as open soil. In warm climates without air conditioning systems, building walls tend to be thick enough to absorb the high quantities of afternoon heat without allowing the heat to penetrate to the interior part of the building. Yet the conductivity must be high enough to allow that heat to flow back outward from the walls during the nocturnal hours. This "building material" effect is a major contributor to the urban heat island at night. The dark and irregular surfaces of buildings also have an impact on retaining heat. Vertical walls permit multiple reflections of *insolation* (**Figure 12.6**) and delay and reduce the amount of insolation reflected back toward space. This increases the energy available to heat the air over cities.

A final reason for the heat island is that urban areas have greater amounts of waste heat released to the atmosphere than rural areas, because of the intense concentration of human activity. This energy includes that used for industrial and do-

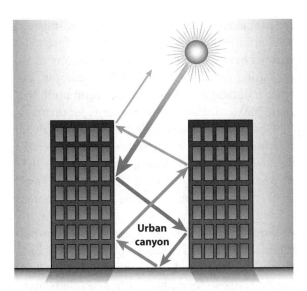

Figure 12.6 Effect of building geometry on the urban heat island.

mestic purposes, transportation, heating, and illumination. Any object with a light or a motor releases waste heat. Even air conditioning units give off waste heat to the atmosphere!

The urban heat island tends to be magnified under certain conditions. The effect is greatest for large cities that are surrounded by rural areas or are near oceans. The heat island becomes more prominent under clear skies rather than under cloudy conditions because clear skies allow more longwave energy to escape from rural terrain, but urban-produced **smog** (a combination of smoke and fog), fog, and cloud cover still prevent some longwave energy from escaping to space over the city unimpeded. Finally, the heat island is best developed under calm conditions. Stronger winds hinder longwave energy from escaping to space over the surrounding rural area because as the radiation encounters more matter from the atmosphere advected by the wind, there is an increased chance of it being absorbed by the matter. The presence of storms or strong pressure gradients diminishes the heat island effect because of the winds associated with them.

Diurnally, the urban heat island usually reaches its magnitude in late evening. At that time of day the industrial, transportation, and domestic waste heat from the day's activities have already warmed the atmosphere. By late evening this extra energy is supplemented by the *conduction* of stored energy in the concrete and walls of buildings to the ambient air. Sometimes the extra warmth at night decreases the atmospheric *stability* to such an extent that a nocturnal *temperature inversion* is prevented from forming locally. Such a scenario allows for a deeper mixing layer over the city and more vertical distance for pollution to be dispersed.

Cities also typically display other climatic contrasts with their surrounding rural environments. *Relative humidity* tends to be reduced in urban environments. Although this modification is partially caused by the increased urban temperatures—recall that temperature and relative humidity are inversely related—part of the difference is also caused by the relative lack of transpiration and evaporation in the urban environment. Although relative humidity is generally decreased, fogs are still common in cities because of the city's abundance of **hygroscopic nuclei**. These are solid *aerosols* that act as such efficient *condensation nuclei*—encouraging the conversion of water vapor to liquid water around them—that fog can occur even at relative humidities far below 100 percent.

Winds are also modified in urban areas. Prevailing winds are lighter than in surrounding rural areas, because buildings act as obstructions and increase *friction*. However, local winds within the city are often very strong, as a *wind channeling* effect around and between buildings—the so-called **urban canyon**—can create locally windy conditions. **Figure 12.7** shows this channeling effect. As rural winds enter the urban environment, they get "squeezed" by the buildings such that the surface area upon which their force is applied must decrease. When the winds become funneled into

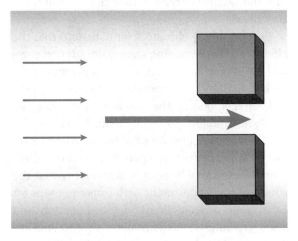

Figure 12.7 Effect of channeling in the urban canyon on wind speed.

a smaller area, the wind speed must increase to compensate for the "squeezing."

Studies of precipitation enhancement by urban areas are somewhat inconclusive. Most suggest that cities cause an increase in rainfall downwind—perhaps up to 10 percent. However, it is difficult to attribute the precipitation enhancement to the urban area because of the nonquantifiable influence of topographical sites, distance to water bodies, and other factors.

Global Warming: The Great Debate

Mainstream Perspective Few climate scientists today question that the Earth–ocean–atmosphere system has displayed a significant warming trend over the last century. However, debate still persists over the cause and the severity of the problem. Most climate scientists believe that increasing greenhouse gas concentrations are the primary cause of the warming and that global warming is a major problem that needs immediate attention. Several valid assertions are used to support this claim.

First, those who are concerned about global warming note that experiments demonstrate that greenhouse gases are efficient absorbers and radiators of energy of the relevant wavelengths. We know from ice cores and the instrumental record that we are increasing greenhouse gas emissions. Because of that evidence, most scientists state that there is no reason to doubt the problem is real. Likewise, the indisputable fact that urban surfaces and waste heat cause warming, combined with the increase in urbanization over time, give these scientists even more evidence to conclude that the warming trend is tied closely to human activities.

Second, many climate scientists note that the effects of global warming are possibly detrimental and are largely unknown. Unlike a chemistry experiment, there is no laboratory for analyzing possible effects, finding the "answer," and using the result to establish policy. Instead, Earth itself is the "laboratory," and we have to live with the result of the experiment, regardless of its outcome. Along these same lines, it is noteworthy that the effects of the "experiment" will linger perhaps for generations even if fossil fuel combustion and other greenhouse gas emissions were to end immediately. Thus, any further delay in addressing the problem will cause even more drastic impacts. This argument implies that the more-developed nations in the world should be the leaders in establishing policy to curtail global warming, because climatic conditions in the world up to this point have coincided with prosperity for them.

Finally, those who are concerned about global warming state that we should be moving toward a world where fossil fuels are phased out, regardless of whether global warming is a problem. Fossil fuels represent a nonrenewable resource, whereas other sources of energy such as solar power, hydropower, and wind power are renewable. The pollutants emitted into the atmosphere by fossil fuels bring environmental, epidemiological, and economic consequences, but the renewable sources of energy listed above cause no such problems. Additionally, fossil fuels are distributed unevenly and lead to international geopolitical tensions, whereas the renewable sources of energy are more evenly distributed and potentially more accessible to more of the world's people, especially as technological developments reduce the cost of constructing such power-generating facilities. Proponents of this argument note that if global warming is the catalyst for changes in energy policy that lead us to more sustainable, inexpensive, and environmentally friendly sources of energy, then so be it.

Skeptical Perspective Some climate scientists believe that global warming is a less serious problem, claiming that many statements purported by the more popular view are unrealistic and sensationalized. Though most climate scientists disagree with the skeptical perspective, the argument is an important, healthy, and central theme in the development of any scientific body of knowledge. Such skepticism has led to greater depth of research into climatic systems, the development of better climatic simulation computer models, and ultimately a heightened insight into the interrelationships of the Earth–ocean–atmosphere system. Thus, healthy debate has an important place in science, including climatology.

One popular "skeptical" argument is that there are too many uncertainties to establish drastic changes in policy. For example, we do not know whether *negative feedback systems* might curtail the warming. Increased temperature would likely cause increased evaporation, which then would cause increased *condensation* and cloud cover. This extra cloudiness may then decrease the amount of shortwave energy reaching Earth's surface, thereby keeping temperatures within a reasonable range of normal conditions. A similar kind of negative feedback prevents temperatures in tropical rain forests from rising too high. Another possible negative feedback system is the potentially increasing role of the oceans as a CO_2 sink, through both natural and anthropogenic enhancement. This mechanism is seen by the skeptics as offering hope for solving the problem before it becomes more serious.

Skeptics also suggest that even if negative feedbacks do not curtail the process, the warming may simply be part of a larger natural cycle that will swing back toward cooling again sometime in the future. In support of this sentiment, they point to the fact that climatologists in the 1970s and 1980s were expressing concern about Earth slipping into a new *glacial phase*. They also state that much of the observed warming of the past century may simply be part of the natural cycle that pulled Earth out of the *Little Ice Age* and the Wisconsin Glacial Phase. Finally, they note that a 1 percent change in the natural solar output could cause more climatic change than all anthropogenic effects combined.

Some paleoclimatological evidence suggests that temperature trends may precede trends in CO_2. In other words, CO_2 may respond to temperature variation rather than cause it. One explanation for this sequence that fuels the skeptics' perspective is that the solubility of CO_2 in oceans is inversely proportional to ocean temperature. So a slight increase in temperature would result in less CO_2 dissolved in the ocean, and a higher concentration of atmospheric CO_2. A further complication is the slow response of oceans to atmospheric temperature changes; the changes in atmospheric CO_2 caused by changes in oceanic assimilation may occur decades or even centuries after the atmospheric temperature change occurred. The entire process is very complicated, and the suggestion that increasing CO_2 is an effect rather than a cause of global warming remains very controversial today.

The skeptical view of the global warming problem also includes the notion that higher temperatures may provide more benefits than costs. For instance, agriculture may become more efficient if warming lengthens the growing season, particularly if the warming is concentrated at night (when killing frosts would occur at the beginning and end of the growing season). Furthermore, the extra CO_2 released by fossil fuel emissions is shown to be favorable to the development of many of the world's major crops, because additional carbon acts as a fertilizer. Higher temperatures may also cause fewer economic losses in some other industries such as transportation, tourism, and recreation.

Another source of doubt regarding the global warming problem involves the data. Skeptics wonder whether erroneous data may be providing a misleading impression of the amount of warming that has occurred. Specifically, they point to the fact that the reliable long-term weather stations upon which the global temperature curve is calculated are overwhelmingly biased toward land-based sites (only 30 percent of Earth's surface) in the developed world (less than 20 percent of Earth's land surface). Although the curve shown in Figure 12.2 suggests that sudden warming has occurred over the last 120 years, changes in monitoring procedures and instrumentation have caused concerns among some that the data may be at least somewhat biased. Concern is also generated by the fact that many stations became closer to urbanized areas as the urban areas grew over time. The skeptics claim that the "global" temperature record may be created as an average of a series of locally enhanced increasing urban heat island signals.

Another concern with the data record involves the global cooling trend shown from about 1945 to approximately 1980 (see Figure 11.5). The skeptics point out that this cooling should not have occurred if emission of greenhouse gases is to blame. This period corresponds to the time when the world's industrial output increased tremendously after World War II, with little or no concern for the environmental impacts of the smokestack industries. Skeptics claim that either the data record is flawed or negative feedbacks must have occurred that suppressed the warming at that time. If negative feedbacks or other compensating factors did occur, similar feedbacks or phenomena may curtail future warming.

For many years the skeptics argued that because the long-term satellite-based records disagreed with meteorological station-based records by showing that the daytime atmosphere was cooling over time, the surface temperature record may be biased. At a minimum, they pointed out that the disagreement between surface and satellite-based atmospheric temperatures shows that the data are unreliable. However, research published in 2005 suggests that the atmosphere is indeed warming along with the surface. The research shows that sensors on satellites and weather balloons had not been properly shielded from direct sunshine early in the data record. As a result, daytime temperatures were overestimated in the early years of satellite observations and (especially) *radiosondes*. This error led to the false impression that atmospheric temperatures have cooled over time.

The debate over global warming has been divisive, with scientists on each side sometimes accusing the other of propagating weak and faulty science at the expense of fairness and objectivity, to pursue their own agendas. A recent manifestation of this debate involves the **hockey stick** temperature curve (**Figure 12.8**). The hockey stick refers to the shape of the global temperature curve reconstructed by proxy evidence such as tree

rings, coral, ice cores, and historical and instrumental records. In the curve, temperatures are relatively stable but fall very slowly from about AD 1000 to about 1900. This forms the shaft of the hockey stick. The last 100 years show abrupt warming that is largely attributable to greenhouse gas emissions.

Although evidence for the validity of the hockey stick graph is convincing, the research remains controversial for three major reasons. First, it suggests that climatic phenomena such as the Little Ice Age and the *Little Climatic Optimum* did not influence global temperature records substantially. Second, the methods of deriving the temperature records and the certainty involved in the calculations have been called into question by skeptics, though some of those skeptics are not atmospheric scientists. Third, the graph is controversial because it was an important part of a report published in 2001 by the **Intergovernmental Panel on Climate Change** to show that global warming is a major concern. Interestingly, the graph was not included in the group's 2007 summary report to policymakers. The next Intergovernmental Panel on Climate Change report will be released in 2014.

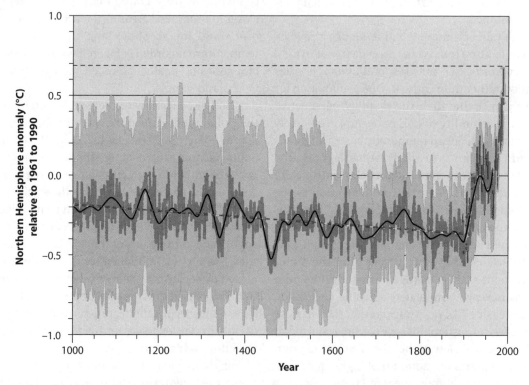

Figure 12.8 "Hockey stick" global temperature curve. *Adapted from*: Mann, M.E., R.S. Bradley, and M.K. Hughes. Northern hemisphere temperatures during the past millennium: inferences, uncertainties, and limitations. *Geophysical Research Letters,* 26, 759–762 (1999).

Most atmospheric scientists support the notion that anthropogenic influences have, and are, significantly impacting the climate system. But the degree to which people are contributing to the problem is not known and probably cannot be known. Our understanding of the anthropogenic global warming problem has been strengthened by examining issues expounded upon by the skeptics. This has probably led to more realistic assessments of future climatic conditions, because many past assessments predicted unrealistically large increases in global temperatures and resulting impacts. Unfortunately, some still use these unrealistic assessments in public forums in an attempt to advance political agendas.

There are still many unresolved anthropogenic global warming issues among scientists regarding how societies should implement policy related to the issues. One seemingly fair approach is for climatologists and other earth and environmental scientists to assess the economic costs and benefits of human activities contributing to global warming. If the benefit exceeds the cost, then a "business as usual" approach should be taken to environmental problems. However, the "cost" of our activities is often inherently nonquantifiable, as in the case of human health and mortality. In addition, there are often "hidden costs"—costs that we are unaware of until after they occur. As climatologists and other environmental scientists expand their knowledge of the climate system, these hidden costs may be understood better. Finally, the issue is complicated by the fact that the costs and benefits will not occur evenly across Earth, so some nations have more at stake than others. The nature of the warming and whether it will occur primarily in one part of Earth or another, the distribution of warming in the seasonal and daily cycles, and the "side effects" of the warming must be understood better.

■ Atmospheric Pollution

The anthropogenic greenhouse effect and the urban heat island are both non-natural phenomena that lead to surface warming. Likewise, both are considered to be permanent in that their effects cannot be reversed easily on human time scales. Other anthropogenic impacts on climate do not cause significant warming and are not necessarily permanent. For instance, aside from its other effects, polluting the atmosphere allows more insolation to be reflected and scattered back out to space before it reaches Earth. Pollution may be masking the thermal effects of increased greenhouse gas emissions to some extent. But atmospheric pollution is not necessarily a permanent feature of the atmosphere.

An extreme example of the effect of atmospheric pollutants would occur in a **nuclear winter**—the hypothesis that a large-scale detonation of nuclear weapons could create enough soot to effectively prevent insolation from reaching much of Earth's surface for up to several weeks. Global temperatures could be reduced by as much as 20 C° (36 F°) in the process. Such a notion is similar to the theory that the dinosaurs and many other creatures may have succumbed to the *K-T extinction* 65 million years ago because an asteroid or comet struck Earth. The impact is believed to have stirred up enough dust in the atmosphere to reflect most of the insolation for some time, thereby damaging the food supply of the organisms that became extinct.

Recent calculations of the climatic impacts of nuclear winter suggest a less drastic scenario than the one originally developed in the early 1980s, in part because of improvements in atmospheric models that allow for oceanic exchange of energy with the atmosphere. The term **nuclear autumn** has been used in recent years to describe this somewhat less dramatic version of a nuclear winter. Nevertheless, the climatic effects would be devastating.

Global Dimming

Perhaps the biggest "secret" among major anthropogenic climatic change issues is that of **global dimming**—the observed decreasing trend in insolation striking Earth over recent decades. Estimates generally suggest that a 2 to 3 percent rate of decrease in insolation occurred each decade from about 1961 through 1990. Interestingly, recent research suggests that the trend toward dimming may have reversed since the 1990s.

The global dimming phenomenon received little acceptance from many climatologists until very recently, in part because the dimming seemed to contradict the observed increase in global temperatures during the 1980s. It is now widely believed that the decreasing trend in shortwave radiation receipt between 1961 and 1990 is real and that increasing atmospheric pollution and aerosols from 1961 through 1990 caused at least some of this trend. The additional global cloud cover resulting from the increasing condensation nuclei from pollution and aerosols seems to have reflected an increasing amount of insolation over that period. The result was a suppression of daytime high temperatures, somewhat offsetting the warming effects of greenhouse gas emissions and the urban heat island. This pollution may also partially explain the global cooling trend from about 1945 through 1980. Alternately, temperature may continue to climb in some cases even under increasing pollutants because cloud cover from pollution and aerosols may increase nocturnal temperatures by impeding longwave radiation loss to space. In the days after September 11, 2001, studies showed that the absence of jet contrails (condensation trails) associated with the grounding of all U.S. flights was associated with an increase in the daily temperature ranges (the difference between maximum and minimum temperature) of approximately $1\,C°\,(2\,F°)$ in the United States. These results support the notion that aerosols, including jet contrails, have been masking the effects of global warming, at least during the daylight hours, and perhaps reducing the diurnal temperature range.

The implication is that pollution abatement in the last 15 to 20 years may be at least partially responsible for the reversal of the global dimming trend. This may mean that as skies around the world become less polluted, the effect of greenhouse gas emissions may magnify. The amount of magnification is unclear and the potential for renewed global dimming from additional pollution in rapidly expanding economies in the developing world exists.

Although many effects of atmospheric pollution are global, most of the direct effects are primarily local or regional. It is important to examine the major causes, types, distribution, and effects of atmospheric pollution.

Atmospheric Factors Affecting Pollution Concentrations

Several atmospheric factors affect the magnitude of the air pollution problem. First, atmospheric stability is an important consideration. During times when the atmosphere is stable, vertical motion in the atmosphere is suppressed, and air pollution remains near the source. Dispersion of atmospheric pollution occurs more easily when the atmosphere is unstable, because *unstable atmospheres* are associated with rising motions of *adiabatically* moving parcels of air.

Because *anticyclones* are often associated with a *stable atmosphere*, they are linked to poor dispersion capability and poor air quality. A particularly good example of anticyclonic effects on stability occurs in the case of the eastern sides of the *subtropical anticyclones*. The eastern sides of these highs support strong sinking motion, largely because the cold ocean currents on the eastern sides provide strong temperature inversions and very stable atmospheres. The California coast is one example of a place affected by these conditions. In summer the nearby *Hawaiian high* strengthens and expands over the region, partially supported by the cold *California current*, which provides for a persistent inversion near the coast. The coastal mountain ranges also contribute by confining sinking motions in the atmosphere to the coastal zone, where more than 20 million southern Californians live. Automobile exhaust and other pollutants are also confined to this narrow zone. Even the afternoon onshore breezes provide little relief because the coastal mountains prevent eastward dispersion of pollutants. In winter the Hawaiian high weakens, shrinks, and migrates southward with the direct rays of the Sun, and California is affected more by *midlatitude wave cyclones* that provide rainfall and relief from the poor air quality through enhanced vertical and horizontal mixing. *Cyclones*, then, often facilitate dispersal of pollutants, both because of the upward motion characteristic of unstable atmospheres that often accompanies cyclones and also because of the characteristically strong pressure gradients (and, therefore, winds) associated with storminess.

A second atmospheric factor that affects air pollution is wind. Obviously, windy conditions disperse atmospheric pollutants and dilute them with distance. But this may cause pollution problems downwind from the pollution source. For example, Canadians often state that pollution generated in the midwestern United States drifts in the westerlies toward Ontario, Quebec, and the Maritime Provinces. Air pollution does not respect international borders.

A final important atmospheric factor is insolation. Although solar radiation may seem to be harmless, it may trigger chemical reactions that produce poor air quality in a form known as **photochemical smog**. Sunlight can cause preexisting pollutants that are only somewhat harmful to be transformed into more dangerous pollutants. Again, California provides an effective example of the problem of photochemical pollution. Recall from Chapter 9 that most of California has a *Mediterranean climate*—characterized by sunny summers. The abundant sunshine in summer combines with the stability associated with the eastern side of the Hawaiian high to make photochemical pollution particularly troublesome.

Atmospheric dispersion modelers use principles of atmospheric stability, radiation, and dynamics to forecast the days on which air quality is likely to be a problem. By understanding such principles, an appropriate balance can be reached such that the benefit of human activities (e.g., industry, economic development, transportation) can be achieved for the minimal cost (i.e., air quality problems).

Air Quality Legislation in the United States

Before 1955 no federal laws governing air quality existed in the United States. Instead, air pollution was regulated only at the state and local levels. This meant that one community might have had drastically different air quality standards from a community nearby. Of course, this also meant that pollution from a community that did not regulate air quality could drift to a nearby community without any recourse from the federal government.

In 1955 Congress passed the **Air Pollution Control Act**, which was the first attempt at federal oversight in the United States. However, this legislation was more effective in making people aware that air pollution is a problem than it was in solving the problem. The **Clean Air Act of 1963**, with several subsequent amendments, became the first major effective legislation for regulating air quality and monitoring both stationary sources, such as power plants, and mobile sources, such as automobiles.

One amendment that was monumental in its influence was the **Clean Air Act of 1970**, passed during the Nixon administration. The Clean Air Act has been modified several times since 1970, most notably in 1977 and 1990. One of the most sweeping reforms created by the Clean Air Act of 1970 was the creation of a new federal agency called the **Environmental Protection Agency (EPA)**. The EPA was required to generate a list of **criteria pollutants**—those that are sufficiently widespread and dangerous in the United States that they must be monitored at the federal level. The EPA was also required to determine the **National Air Quality Standards (NAQS)** of those criteria pollutants. The NAQS determined the "acceptable" concentration of each pollutant (in ppm of the atmospheric molecules). Obviously, this was a difficult job because the same concentration that may be harmless for most of the population may be life threatening to an elderly person, a newborn infant, or a person with existing respiratory disease.

In 1971 the EPA produced the list of criteria pollutants. The list focused on sulfur oxides (SO_x), carbon monoxide (CO), *particulates* (a family of tiny suspended solid aerosols), lead, nitrogen oxides (NO_x)—**nitric oxide (NO)** and **nitrogen dioxide (NO_2)** in particular—and photochemical oxidants (primarily **tropospheric ozone**). Any areas that persistently exceed the NAQS of a pollutant are said to be **nonattainment zones**. Today, concentrations of all criteria pollutants have been reduced sharply in the United States, including in most nonattainment zones, in large part because of the success of the EPA's guidance over the years. It is important to note that the EPA makes no attempt to set standards on pollutants that are not widespread nationwide, regardless of their danger. The decision on whether to allow obscure pollutants to be emitted into the air is left to the local and state governments, even if they are dangerous.

■ Classifying Air Pollutants

By Response

Threshold pollutants are contaminants not known to be harmful in sufficiently small concentrations but that elicit a response after the concentration exceeds a particular threshold. Such pollutants may be considered to have a concentration–response curve like those shown in **Figure 12.9**. Note that the vertical line in this figure represents the threshold concentration. All criteria pollutants except one—tropospheric ozone—are considered to be

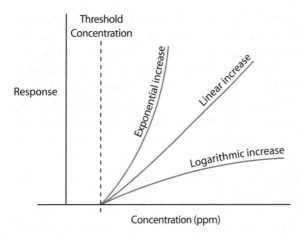

Figure 12.9 Example of a concentration–response curve for a threshold pollutant.

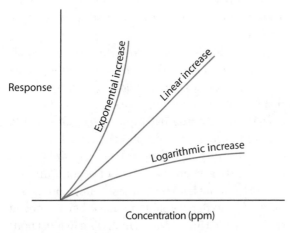

Figure 12.10 Example of a concentration–response curve for a nonthreshold pollutant.

threshold pollutants. **Nonthreshold pollutants** evoke a response as soon as they exist at any concentration, meaning that the threshold concentration would be 0 ppm. Some examples of concentration–response curves for nonthreshold pollutants are shown in **Figure 12.10**. The only criteria pollutant that is a nonthreshold pollutant is tropospheric ozone.

If a decision is to be made about whether a pollution-generating facility is to be allowed to move into a community, several questions relating to the response need to be addressed. First, is the pollutant to be released a threshold or nonthreshold pollutant, and if it is a threshold pollutant, what is the threshold concentration? Second, will the emission of the pollutant exceed this threshold? Third, does the concentration–

response curve look more like the exponential curve or the logarithmic curve in Figures 12.9 and 12.10? Finally, what is the response? For example, if the concentration to be emitted does not exceed the threshold, or if the response increases only slightly with increased concentrations and the response is only a minor ailment, then perhaps the benefit of the factory or other emitting source may exceed the cost. However, if the threshold concentration is to be exceeded greatly, the response increases considerably with increased concentrations, and if the response is a serious illness, then the benefit probably would not exceed the cost. Unfortunately, scientists cannot always be certain about the nature of the threshold concentration or the response.

By Source

Primary Pollutants We can also classify pollutants by the mechanism by which they appear in the atmosphere. **Primary pollutants** are those solids, liquids, or gases emitted directly by human activities. In general, solids and liquids are "washed out" of the atmosphere by precipitation more easily than gases. Among the criteria pollutants, all except tropospheric ozone and some types of SO_x and particulates are primary.

Each of the primary criteria pollutants has impacts that can be harmful to humans, animals, and plants. Because the United States is the largest producer of pollution, we focus on pollutants and trends in that country.

Sulfur Oxides It has been estimated that about half of the global atmospheric SO_x emissions are emitted naturally. However, in industrialized areas most atmospheric SO_x is anthropogenic in origin, resulting primarily from fossil fuel combustion. Diesel fuel combustion, steel mills, and oil refineries are particularly large producers of atmospheric SO_x. Of the three major types of fossil fuels, coal releases far more SO_x per energy unit than oil, and oil releases far more than natural gas.

The most prevalent and dangerous form of SO_x is SO_2. As illustrated in **Figure 12.11**, U.S. counties in nonattainment of SO_2 standards are concentrated in the Intermontaine Basin region of the western United States. SO_2 tends to destroy plant hormones, stunting growth and reducing reproduction rates. As a result, the economic cost to

Figure 12.11   Nonattainment areas for SO$_2$ in the United States as of July 2009. *Modified from: The Green Book, Non-Attainment Areas for Criteria Pollutants*, U.S. EPA/Office of Air Quality Planning and Standards. http://www.epa.gov/air/oaqps/greenbk/.

agriculture and forestry in areas with high atmospheric SO$_2$ content can be significant. As we see shortly, SO$_x$ can also have important impacts on creating **secondary pollutants**, through the generation of acid precipitation, which impacts vegetation and groundwater.

Humans are also impacted directly by SO$_2$. Excessive exposure to SO$_2$ is associated with respiratory illnesses, including asthma, bronchitis, emphysema, and lung cancer. It is unclear whether SO$_2$ can cause these illnesses or whether it only accelerates and intensifies these respiratory problems, but sharp increases in respiratory disease have been recorded over the past century in industrialized regions.

Carbon Monoxide   Carbon monoxide (CO) is a dangerous pollutant even in very small concentrations, even to healthy people. Most CO is produced by automobiles. Like monatomic oxygen, CO is unstable in the free atmosphere, so when it is emitted it immediately attempts to bond with another atom or molecule. This may hinder the ability of hemoglobin—a protein in red blood cells—to deliver O$_2$ to body organs and tissues because inhaled CO may outcompete hemoglobin for this oxygen, thereby starving the brain, heart, and other organs of oxygen.

Perhaps you have walked briskly in a downtown area of a major city and felt tired unusually quickly. It is very likely that in that situation the sleepy sensation occurred because your brain was receiving a bit less oxygen than usual after you inhaled CO. Malfunctioning furnaces can release

CO to a sealed-up home in winter, causing fatalities. Another means of inhaling CO is through smoking. Only one county, in southern Nevada, is currently not in compliance with the U.S. EPA CO standard.

Particulates   Two major classes of particulates have been recognized by the EPA. **PM$_{10}$** is the category of particulates with diameters of less than 10 μm, whereas **PM$_{2.5}$** (also known as **fine particles**) refers to those particulates that have diameters smaller than 2.5 μm. Recent research suggests that even though PM$_{10}$ can be harmful, PM$_{2.5}$ is of particular concern, because cilia in the bronchiole airways have great difficulty in removing it from the human body. These cilia normally move larger foreign substances, including inhaled pollutants and large bacteria, upward to the throat, where they can be expelled from the body via the digestive system. Particulates are believed to cause health problems in people with heart and lung disease. They also reduce visibility and can disrupt the balance of nutrients in soils when they settle to the ground. Finally, they damage buildings and monuments by abrasion as they continuously strike these surfaces.

Although many particulates enter the atmosphere via natural processes, some of these solid and liquid aerosols are emitted directly from vehicles, soil plowing, wood burning, and other human activities that generate dust. Other anthropogenic particulates arrive indirectly when fuel combustion combines with water vapor and sunlight. PM$_{2.5}$ nonattainment areas are widespread across the mid-Atlantic states, the Appalachian region, the area surrounding the Great Salt Lake, and California (**Figure 12.12**).

Lead   Lead was listed as a criteria pollutant in 1978, and since that time it has been rapidly phased out of the atmosphere in the United States. Formerly an ingredient in gasoline and paints, it was shown to be linked to numerous brain and neurological disorders (collectively termed **plumbism**), and has been known to harm fish and wildlife. Today, the primary non-natural source of lead in the atmosphere is metal processing plants. Although concentrations are still high near smelters and incinerators, lead is no longer a widespread problem in the United States. As of June 2009 only two counties in the United States—one in Montana and one in Missouri—were in nonattainment of the lead standard.

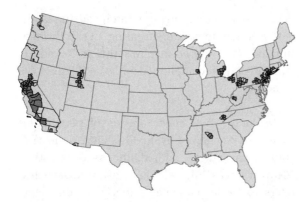

Figure 12.12 Nonattainment areas for PM$_{2.5}$ in the United States as of July 2009. When a portion of a county is in color, only that part of the county is within a nonattainment area. *Modified from: The Green Book, Non-Attainment Areas for Criteria Pollutants,* U.S. EPA/Office of Air Quality Planning and Standards. http://www.epa.gov/air/oaqps/greenbk/.

Nitrogen Oxides Nitrogen oxides are produced naturally through various processes, including bacterial action in the soil, lightning strikes, and volcanic eruptions, but they are also produced by fossil fuel combustion and forest fires (the latter of which may be natural or human induced). Automobile exhaust is a very important component of NO$_x$ emissions in heavily populated areas. However, with the recent attainment of the standard in four counties surrounding Los Angeles, all counties in the United States are currently in attainment for NO$_x$.

As a primary pollutant, NO$_x$ is harmful because of its ability to erode away objects in contact with it, such as building stone, tombstones, and lungs. As we saw earlier in this chapter, one form of NO$_x$, nitrous oxide, is a minor greenhouse gas. Nitrogen oxides are also associated with poor visibility. However, the most serious impacts of NO$_x$ tend to be associated with its role as a secondary pollutant because of its ability to react with other substances to produce different photochemical pollutants. These are presented shortly in the discussion of secondary pollution and acid precipitation.

Secondary Pollutants Secondary pollutants arrive in the atmosphere as a byproduct of chemical reactions between primary pollutants and other matter or energy in the atmosphere. Two primary pollutants can interact to create another pollutant, which may be even more hazardous than the two original primary pollutants combined.

Industrial Smog One type of secondary pollutant is **industrial smog** (sometimes known as **gray smog**). Industrial smog results from chemical reactions between two primary pollutants—*sulfur dioxide (SO$_2$)* and particulates—particularly in the presence of fog. The biggest danger of industrial smog is that the particles are too small to be removed by cilia, making them quite damaging to the respiratory system.

Perhaps the most widely recognized industrial smog episode in the United States occurred in Donora, Pennsylvania, a small industrial town nestled in a narrow, steep valley. On October 30, 1948, 17 people perished after an especially persistent anticyclone produced a strong temperature inversion while SO$_2$ and particulates continued to be released into the atmosphere by local industrial activities. Limited vertical mixing in the stable atmosphere caused lethal point source pollution concentrations. Such an event, when deadly concentrations of a pollutant build near the emitter, is termed a **fumigation event**.

A far more deadly fumigation event occurred in London in December 1952, when death rates were over three times the normal rate for a period of 3 weeks of persistently stable atmospheric conditions. Over 2000 deaths were associated with this event that saw dense coal smoke combining with fog to produce smog. The fear of other deadly episodes helped to inspire the first federal clean air legislation in the United States.

Photochemical Smog A second family of secondary pollutants is termed photochemical smog (sometimes known as **brown smog**), which results when sunlight triggers a chemical reaction that creates pollution. These very complicated reactions occur relatively easily in the presence of nitrogen oxides, **hydrocarbons**, and oxygen. Unlike industrial smog, photochemical smog usually occurs in relatively dry air.

One type of photochemical smog involves the *photodissociation* of NO$_2$ into NO and O in the presence of sunlight. Because monatomic oxygen is very unstable in the atmosphere, it is likely to bond with a diatomic (O$_2$) molecule (which comprises 21 percent of the mass of the dry atmosphere) to form tropospheric ozone (O$_3$). Although this O$_3$ is chemically identical to the O$_3$ that protects us from harmful *ultraviolet radiation* in the *stratosphere*, O$_3$ near the ground is very toxic. This feature is exemplified by the EPA's snappy rhyme about O$_3$—"good up high, bad nearby." The NO from the photodissociated NO$_2$ may then combine with certain hydrocarbons in a series of complicated reactions to form a family of secondary pollutants called **peroxyacetyl nitrates**.

Like NO_2, O_3 is known for wearing away objects in contact with it. These objects may include leaf cuticles, building stone, and human lungs. No conclusive evidence exists to blame O_3 for causing respiratory illness, but it is evident that O_3 worsens existing respiratory problems. Plants and trees that have been affected by O_3 are more susceptible to disease and insect infestation. When trees on slopes succumb to the disease or infestation, the slope becomes more susceptible to mudslides. In California, this problem has resulted in the construction of huge and expensive debris basins to protect homes in adjacent valleys. Any cost–benefit analysis for pollution must include hidden costs such as the purchase, installation, and maintenance of these basins. Further hidden costs are associated with degradation of the natural environment. Hydrocarbons are important in the production of some types of photochemical smog, including peroxyacetyl nitrates, and they also act as O_3 precursor compounds (a substance that must already be in place in the environment before a following chemical reaction can occur). They consist of a wide variety of chemicals, including some familiar ones such as benzene, formaldehyde, and methyl chloroform. Some hydrocarbons, such as isoprene, are emitted directly into the atmosphere naturally by vegetation; a familiar result is the hazy color that gives the Smoky Mountains their name. Some hydrocarbons are emitted by human activities as primary pollutants during fuel combustion (particularly by automobiles) and in the use of solvents, paints, and other industrial products.

Currently, many U.S. counties exceed the EPA's standard for O_3 (**Figure 12.13**). The primary reason for these high numbers is that the EPA's requirements are becoming more stringent over time as more is learned about the harmful effects of O_3 rather than by an actual deterioration of the air quality. The counties in most serious violation of the standard are concentrated along the northeast coast, California, western Lake Michigan coast, and some urban/industrial areas in the Deep South.

Acid Precipitation   A third family of secondary pollutants is **acid precipitation**, which occurs when certain primary pollutants are deposited on the surface. Though far less common than acid rainfall, **acid fog** is a much more dangerous phenomenon,

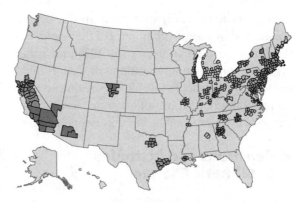

Figure 12.13 Nonattainment areas for ozone in the United States as of July 2009. *Modified from: The Green Book, Non-Attainment Areas for Criteria Pollutants*, U.S. EPA/Office of Air Quality Planning and Standards. http://www.epa.gov/air/oaqps/greenbk/.

because of the potential for humans and animals to breathe the acidic concoction. Although all precipitation is naturally acidic due to the incorporation of CO_2 into falling precipitation, precipitation exceeding natural values of acidity are primarily anthropogenic in origin (although some natural processes, such as volcanic eruptions, may create acid precipitation, they are infrequent). The most common form of acid precipitation results from the emission of SO_2 in humid environments. When SO_2 reacts with water, it produces H_2SO_3, which can then react with oxygen to produce **sulfuric acid** (H_2SO_4), a very corrosive acid. As we saw earlier, coal combustion releases SO_2 to the atmosphere, and because most early coal-intensive industries in the United States and western Europe were located in humid environments, this form of acid precipitation has been a major problem historically.

Another type of acid precipitation results through complicated processes involving reactions between NO_x and water to form **nitric acid (HNO_3)**, another very corrosive acid. This form of acid precipitation is more problematic in the "freeway cities" of the southern and western United States. It is important to note that either H_2SO_4 or HNO_3 can reach the surface through precipitation—or **wet deposition**—or the SO_2 and NO_x may reach the surface and interact with soil water to produce the acids in a **dry deposition** process.

The good news is that acid precipitation has not been as severe a problem in the United States as was initially anticipated. This outcome is similar to that for many other pollutants that were initially forecasted to increase in concentration far more than has actually occurred. The fact that air quality is generally improving in the United States, and in several other parts of the world, is a testament to the success of the EPA and other organizations in improving air quality. Additional engineering developments have also allowed us to increase the benefit of pollution-generating machines while reducing the environmental cost.

■ Reactions and Attitudes to Climatic Change

Many of the human-induced forcing mechanisms discussed above cause global climatic change by shifting prominent atmospheric circulation features, which create different impacts across space. They may cause a warming trend in one location and a cooling trend in another. At this time the complexity of the climate system prohibits us from understanding all the potential impacts. Even so, several strategies can be used in policies designed to deal with the threat of climatic change. These include prevention, mitigation, and adaptation.

Prevention

The tactics designed to reduce a potential threat are referred to as **prevention**. One technique of prevention in the global warming debate is to prevent (or at least minimize) the emission of greenhouse gases. The most recent significant attempt at prevention of global warming was the Kyoto Treaty that was signed by over 150 nations and took effect in February 2005. This agreement requires that the wealthy nations contributing most of the world's greenhouse gases reduce their collective greenhouse gas emissions to 5.2 percent below the 1990 levels. Each nation has a target reduction under the treaty. Although several of the world's most industrialized nations must reduce their emissions, some (such as Russia) are currently below their emissions quota; these nations may sell the rights to emit greenhouse gases to others that cannot meet their quota. This idea may be used by countries, states, cities, and industrial companies in a "cap and trade" policy. Although some may view cap and trade as a tax on economic development, such a policy may represent the only realistic means of holding or reducing emissions across the global economic landscape.

The United States refused to ratify the Kyoto treaty for two major reasons: (1) It would create too much risk of economic instability in the United States, especially considering the fact that the economic impacts of global warming and of cuts to fossil fuel emissions are not clearly understood, and (2) it does not require fossil fuel emissions cuts from 14 of the top 20 fossil fuel-emitting nations, including developing nations such as China, Brazil, and India, which are all projected to be producing a greater share of the world's fossil fuel emissions than the United States within 25 to 30 years. Currently, other than the United States, Australia is the largest economic power to refuse to ratify the treaty.

In July 2005 the United States entered into a plan to reduce human-based climatic change with five other nations (Australia, India, Japan, China, and South Korea). This plan focuses more on the development of clean energy sources instead of capping emissions, as in the Kyoto Treaty. Such a strategy may be warranted as recent analysis into Kyoto compliance shows that all except one of the countries that signed the treaty have actually increased greenhouse emissions since agreeing to the reductions.

Mitigation

The attempts to decrease the risk associated with a threat, rather than decreasing the threat itself, are referred to as **mitigation**. Whereas prevention strategies attempt to eliminate or reduce the problem of global warming, mitigation might involve building sea walls to combat the risk of coastal land loss from sea level rise resulting from global warming. Or mitigation may involve the introduction of new hybrid crops that would be more resistant to extreme weather variations caused by global warming. Or perhaps mitigation may simply involve purchasing fans for senior citizens to reduce their risk of death during a hot summer. Although mitigation strategies may be successful, they are sometimes criticized for being expensive and not "getting to the root of the problem."

Adaptation

A strategy that involves simply living with a threat and the consequences associated with the threat is referred to as **adaptation**. It follows a "business as usual" approach, despite the problem at hand. The advantage of adaptation is that the preparation cost is zero, but the obvious disadvantage is that after the damage occurs, the full brunt of the impact must be borne. People living in hurricane- or earthquake-prone areas practice adaptation as a strategy for dealing with the problems. Until recently, the world had largely used an adaptation strategy in dealing with global warming.

Continued Research

For complicated environmental problems, such as global warming, policy must be enacted even while the problem is only partially understood. Regardless of whether prevention, mitigation, or adaptation strategies (or some combination of the three) are used, research on such problems must continue. Sometimes, the policy must be changed as research reveals more about the nature and impact of the problem. Not surprisingly, the general public becomes impatient when policy must be changed in response to the increased knowledge.

Research can take one of two forms, or some combination of the two. First, the modeling approach involves the generation of mathematical equations to predict the future severity of the problem. For example, complicated computerized climate models are used to predict the future climate under different magnitudes and rates of the input of CO_2 into the atmosphere. Unfortunately, different models project different results for temperature and especially precipitation for similar locations. In some cases even the sign of the change in a predicted variable is opposite that indicated by other models. For example, one model may predict future desertification for a location, whereas another indicates above-normal precipitation for the same location and time. These inaccuracies are largely confined to regional or local spatial scales, but modeled results at larger scales are more uniform. All major models suggest that global warming will continue and will have potentially destructive impacts. Policymakers must decide whether this model output warrants serious action. Many policymakers are now answering "yes" to this question.

Alternatively, the **empirical approach** involves the use of climatic data to uncover historical trends in variables and relationships between variables. This approach is advantageous because high-quality data represent actual observed conditions rather than a hypothetical, derived condition that may or may not actually happen. However, it is dangerous and generally inappropriate to extrapolate trends in atmospheric variables out into the future, especially if the mechanism causing the trend in the variable being studied is not fully understood. For example, during the late 1970s many scientists were discussing future cooling trends as evidenced by the prevalence of severe winters and cooler-than-normal summers. This climatic trend quickly reversed itself in the 1980s. Simple extrapolation of any trend is unwarranted, because positive feedbacks could cause an acceleration of the observed trend, whereas negative feedbacks could cause a data-driven trend to reverse itself at some point in the future.

In reality, both modeling and empirical approaches should be, and are, used to facilitate understanding. For example, climate models use empirical data to handle some processes for which our understanding of the physics is poor. The modeling and empirical approaches can complement each other or at least provide a "cross-check" to determine the degree of confidence of results from either approach.

■ Summary

Human activities have had a substantial impact on climate. Two broad categories of activities have led to a global warming effect on climate over the past century. First, the release of certain gases, particularly carbon dioxide, methane, and nitrous oxide, as a result of human activities has intensified the natural process of retention of longwave radiation near Earth's surface—the greenhouse effect. Oceans, soils, and vegetation effectively remove the most important anthropogenic greenhouse gas—carbon dioxide—from the atmosphere, but the rate of removal is being exceeded by the rate of production. The measurable warming over the past century potentially caused by this anthropogenic greenhouse effect is expected to continue and have major consequences. Some of these anticipated consequences include rising sea levels, melting in the cryosphere, circulation shifts, changes in the characteristics of severe storms, and shifts in agricultural and bioclimatic zones.

A major activity that contributes to warming on a local as well as possibly a regional level is urbanization. Local heating in urban areas is enhanced by the increased retention and slow release of heat in construction materials and streets. The lack of vegetation and surface water storage in urban areas also contributes to this urban heat island by causing a proportionately higher amount of surface energy to be used for sensible heating, while less latent heating occurs compared with rural areas. Finally, waste heat generated by domestic and industrial activities also contributes to this urban heat island. At least some of this excess energy adds to local- as well as regional-scale heating. The degree to which urban-derived energy infiltrates upward in scale to contribute to the global scale warming is not yet fully understood.

Although climatologists generally agree that a global warming trend is present in the climatic record over the past century, the cause and importance of this trend is still debated by some. Most believe that global warming should be curtailed because the trend is reliably measurable. They contend that the effects are real and may be disastrous, and that further delay will magnify the problems even more. They suggest that greenhouse gas emissions should be reduced regardless of their impact on the greenhouse effect. However, some believe there is too much uncertainty about the causes and impacts of the warming, feedbacks that may reverse the trend, and the data revealing the warming to enact strict and drastic policy to limit greenhouse gas emissions.

Some anthropogenic problems may be offsetting global warming. In particular, air pollution may reduce the amount of insolation incident on Earth's surface, thereby cooling the planet. An observed global dimming from 1961 to 1990 supports this notion and suggests that the impacts of greenhouse gas emissions may have been masked during this time by the reflection and scattering of some daytime shortwave radiation to space.

Atmospheric stability, wind, and sunshine all govern the severity of pollution episodes. Stable atmospheres allow less pollution dispersal than unstable atmospheres. Windy conditions disperse and dilute pollutants, but light winds can cause pollution to drift long distances from its source. Sunshine can trigger the formation of photochemical pollutants such as ozone and peroxyacetyl nitrates, particularly when in the presence of hydrocarbons.

Landmark legislation on federal air quality in the United States occurred with the passage of the Clean Air Act of 1970. This legislation required that a newly created Environmental Protection Agency (EPA) create a list of widespread problematic pollutants, known as criteria pollutants, and set standards on acceptable concentrations of these pollutants. The criteria pollutants were sulfur oxides, carbon monoxide, particulates, lead, nitrogen oxides, and photochemical oxidants (primarily tropospheric ozone). Although localized concentrations of obscure pollutants may be very hazardous, concentrations of all criteria pollutants have decreased in the past 35 years nationally.

Pollutants can be categorized by response and by source. Threshold pollutants elicit a response (whether it be respiratory illness, decreased crop yield, or any other type of impact) only after a certain threshold concentration is exceeded. Nonthreshold pollutants elicit a response as soon as they exist in any concentration. The source of primary pollutants is direct emission by human activities. Most criteria pollutants are primary pollutants. Secondary pollutants arrive in the atmosphere as a result of chemical reactions between primary pollutants or between a primary pollutant and some other matter or energy, such as water or sunshine. Three forms of secondary pollutants are industrial smog, photochemical smog, and acid precipitation.

Policy enacted in response to climatic change takes similar forms to reactions to other forms of hazard or crisis. Prevention measures seek to eliminate a threat. Mitigation strategies are aimed at minimizing the impact of a threat. Adaptation policies are designed to cope with a threat and its possible consequences. Regardless of which strategy or strategies are used in dealing with the hazard posed by anthropogenic climatic change, continued research to understand the problem and its potential impacts is critical to minimizing impacts to human health and the environment.

▶ **Key Terms**

Absolute zero	Acid precipitation	*Advection*
Absorption	Adaptation	*Aerosol*
Acid fog	*Adiabatically*	Air Pollution Control Act

Albedo
Anaerobic
Anthropogenic
Anticyclone
Atmospheric window
Biosphere
Brown smog
California current
"Cap and Trade" Policy
Carbon monoxide (CO)
Carbonic acid (H_2CO_3)
Clean Air Act of 1963
Clean Air Act of 1970
Condensation
Condensation nuclei
Conduction
Convection
Counter-radiation
Criteria pollutant
Cryosphere
Cyclone
Deforestation
Density
Desertification
Dry deposition
Empirical approach
Environmental Protection Agency
    (EPA)
Evaporation
Evapotranspiration
Fine particles
Fluorocarbon
Fossil fuel
Friction
Fumigation event
Glacial phase
Global dimming
Global warming
Gray smog
Greenhouse effect
Greenhouse gas
Hawaiian high
Heat capacity

Hockey stick
Humus
Hurricane
Hydrocarbon
Hygroscopic nuclei
Ice core
Industrial smog
Insolation
Intergovernmental Panel
    on Climate Change
Joule
Kelvin temperature scale
K-T extinction
Kyoto Treaty
Latent heat
Lithification
Little Climatic Optimum
Little Ice Age
Longwave radiation
Mediterranean climate
Methane
Methane hydrates
Methanogen
Micrometer
Midlatitude wave cyclone
Mitigation
National Air Quality Standards
    (NAQS)
Negative feedback system
Nitric acid (HNO_3)
Nitric oxide (NO)
Nitrogen dioxide (NO_2)
Nitrogen oxide (NO_x)
Nitrous oxide (N_2O)
Nonattainment zone
Nonthreshold pollutant
Nuclear autumn
Nuclear winter
Particulates
Peroxyacetyl nitrate
Photochemical smog
Photodissociation
Photosynthesis

Plumbism
$PM_{2.5}$
PM_{10}
Positive feedback system
Prevention
Primary pollutant
Proxy evidence
Radiosonde
Relative humidity
Reservoir
Runoff
Saltwater intrusion
Secondary pollutant
Sensible heat
Shortwave radiation
Sink
Smog
Specific heat
Stability
Stable atmosphere
Stratosphere
Sublimation
Subtropical anticyclone
Sulfur dioxide (SO_2)
Sulfuric acid (H_2SO_4)
Temperature inversion
Thermal conductivity
Threshold pollutant
Transpiration
Tropical cyclone
Tropospheric ozone
Tundra
Ultraviolet radiation
Unstable atmosphere
Urban canyon
Urban heat island
Wavelength
Wet deposition
Wind channeling
Wisconsin Glacial Phase
Terms in italics have appeared
    in at least one previous
    chapter.

▶ Review Questions

1. Why is anthropogenic climatic change an important issue in contemporary climatology?

2. How does the greenhouse effect work and why is it not solely a natural process?

3. What similarities and differences exist between the heating in the lower atmosphere in the greenhouse effect and the heating of the inside of an actual greenhouse?

4. What is the most abundant greenhouse gas and at what wavelengths is it most effective?

5. What are the sources and sinks of atmospheric carbon dioxide?

6. What are the sources of atmospheric methane?

7. What are methane hydrates and what role do they play in atmospheric methane concentrations?

8. What country is the world's leading emitter of greenhouse gases, and which are projected to be the leading countries in the future?

9. How might positive feedback systems in polar regions accelerate global warming?

10. Why are urban areas warmer than their surrounding rural environments?

11. Other than increased temperatures, what climatological differences exist between cities and their surrounding rural environments?

12. What are the major arguments that environmentalists use to defend their position that the anthropogenic greenhouse effect is a major problem that must be solved?

13. What arguments do skeptics give to justify their position that the anthropogenic greenhouse effect is not as serious as some would suggest?

14. How has the hockey stick temperature curve shaped the global warming debate in recent years?

15. What factors are assumed to have caused global dimming and why has the global dimming trend apparently reversed in recent years?

16. What did studies of the atmosphere in the absence of jet contrails reveal about global dimming in the days after September 11, 2001?

17. How does atmospheric stability affect pollution concentrations?

18. What factors cause California to have severe problems from photochemical pollution?

19. Why was the National Air Quality Standard for criteria pollutants difficult to determine?

20. What are the criteria pollutants? Of these, which are primary pollutants?

21. What are the major anthropogenic sources of atmospheric sulfur oxides?

22. What danger does carbon monoxide pose if inhaled?

23. What are the primary classes of particulates? Of these, which is the more harmful class? Why?

24. How does industrial smog form?

25. How does tropospheric ozone form?

26. What primary pollutants are the major contributors to acid rain, and how does the distribution of these pollutants differ?

27. What are the advantages and disadvantages of mitigation as a strategy in combating anthropogenic climatic change?

▶ Questions for Thought

1. Compare and contrast the history of federal environmental legislation and/or the national federal agency that governs air quality in a country other than the United States to the history and development of U.S. air quality legislation by the EPA.

2. The text discussed prevention, mitigation, and adaptation as strategies to deal with the effects of global climatic change. Think of examples of each of these three strategies in dealing with another threat besides climatic change. Some examples include earthquake risk, bioterrorism, and floodplain development.

3. Other than the topics discussed in this chapter (anthropogenic greenhouse effect, urban heat island, pollution, nuclear winter/autumn, and global dimming), what human activities contribute to altering local and regional climatic environments?

4. Watch the 2006 movie *An Inconvenient Truth*. Does this film portray the global warming problem accurately? Why or why not?

---

http://physicalscience.jbpub.com/climatology

Connect to this book's Web site: http://physical science.jbpub.com/climatology. The site provides chapter outlines, further readings, and other tools to help you study for your class. You can also follow useful links for additional information on the topics covered in this chapter.

13

Linking Spatial and Temporal Aspects of Climate Through Quantitative Methods

Chapter at a Glance

Computerized Climate Models
   Types of GCMs
   Representing the Earth–Ocean–Atmosphere
     System in GCMs
   Data for GCMs
Seven Basic Equations
   Navier-Stokes Equations of Motion
   Thermodynamic Energy Equation
   Moisture Conservation Equation
   Continuity Equation
   Equation of State
   Similarities and Differences Between GCMs and
     Weather Forecasting Models
Statistical Techniques
Atmospheric Teleconnections
   Extratropical Teleconnections in the Pacific:
     The Pacific Decadal Oscillation
   Extratropical Teleconnections in the Atlantic
     Ocean
   Teleconnections over North America
Summary
Review

The first seven chapters focused on the processes that contribute to climate. Chapters 8 through 10 synthesized the information about climatic processes by describing how those processes act to produce climates that vary spatially across Earth. Chapters 11 and 12 discussed the variability and changes known to occur in climates across time because of natural and *anthropogenic* processes. The purpose of this chapter is to describe the major mathematical tools available to climatologists to assess the differences in climates across space

and time simultaneously. These extremely powerful and important tools represent the state of the art in modern climatology.

■ Computerized Climate Models

Computer models are important for simulating atmospheric processes. Some models may serve their purpose quite satisfactorily even though they may be simple. For example, a *surface water balance* model may provide useful information on the impact of a forecasted precipitation event on the hydrology and soil moisture of a particular location. Or an **energy balance model** might be useful for estimating the changes in *insolation* if the solar output were increased or decreased by a certain percentage, or if the *Milankovitch cycles* changed Earth's orbital *eccentricity* or direction or degree of tilt on its axis.

Other climate models are by necessity more complicated. Some models attempt to predict the motion of storms or other patterns of atmospheric behavior through **analog forecasting**—identifying cases from the past and assuming that the atmosphere will behave in an analogous way in a similar situation in the future. Of course, analog models require abundant and accurate historical data to be useful.

Other, more complicated models generally involve the prediction of the global atmospheric circulation based on dynamic conditions rather than by analog comparisons. These models usually operate at the planetary scale because the behavior of the atmosphere is inherently "continuous" around the world, such that processes occurring in one location will be felt elsewhere. These types of models are collectively known as **general circulation models (GCMs)**.

Several research groups around the world are leaders in the generation of GCMs. In the United States the National Aeronautics and Space Administration's Geophysical Fluid Dynamics Laboratory (and associated scientists at Princeton University) and the Goddard Institute of Space Sciences are both world leaders. Other leading institutions include the Hadley Centre in England, the Max Planck Institute for Meteorology (Germany), the Canadian Climate Centre, the Japanese Meteorological Agency, and the Australian Bureau of Meteorology. In recent years several other nations have developed their own GCMs, partially in response to an increased need to understand potential impacts of global climatic change on their territories.

A critical feature of any GCM is the **resolution**, an attribute that refers to the coarseness, or precision, of observations over space (**spatial resolution**) and time (**temporal resolution**). Spatial resolution is reminiscent of pixels on a computer monitor or digital image. Finer spatial resolution provides more pixels per unit area or volume and more detail in the model output across space, but at the expense of a higher number of calculations and greater data handling capacity. It is tempting to consider spatial resolution in only two-dimensional space, but actually GCMs must consider spatial resolution in three dimensions. A GCM that divides the atmosphere into small three-dimensional "boxes" simulates the behavior of the atmosphere in greater spatial detail than one that uses larger boxes. A typical GCM may have a spatial resolution of perhaps $2 \times 2°$ of *latitude* and *longitude*, or, because the lines of longitude converge toward the poles, an equally spaced grid may have a spatial resolution of perhaps $300 \times 300 \, km$ ($185 \times 185 \, mi$), with perhaps 30 vertical levels.

Temporal resolution refers to the degree of detail in time that a model attempts to forecast. For example, does the model attempt to forecast hourly or daily temperatures (as weather forecasting models may) or monthly, seasonal, annual, or even decadal mean temperatures only (as GCMs may attempt)? A related concept is the **time step**—the length of time into the future that is forecasted by each "run" of the model. If you were asked to forecast your university football team's record 5 years from now, you might proceed by forecasting its record next year. Then you might base your forecast of the following year's record on an as-

sumption that what you predicted this year actually did occur. Then you might proceed, 1 year at a time, until you reached the 5-year forecast. In this example your time step is 1 year.

Similarly, a weather forecasting model that uses a time step of 5 minutes makes a forecast for 5 minutes into the future, then uses that model output as the input for the next 5-minute forecast, and so on. The model is run **iteratively**—that is, over and over again—to a forecasting period such as 24 hours. Smaller time steps require more calculations but perhaps provide better accuracy than longer time steps. Regardless, for the model to be of any use, the time step must be small enough to allow for any disturbance (even those that travel fastest) to appear within the domain of analysis. If a fast-traveling wave can enter the "real world" and then leave it in less time than the time step, the model could never "catch" this feature and take its effects into account.

Types of GCMs

Atmospheric general circulation models consider only those processes that occur in the atmosphere. At very short time scales, these may provide adequate simulation of atmospheric motion, so atmospheric GCMs may be adequate for weather forecasting purposes. However, these may not be applicable for climatological applications because of the importance of the interactions between the atmosphere and the *lithosphere, hydrosphere, cryosphere*, and *biosphere* at longer time scales. Climatological forecasting models usually require more sophistication, as we see later in this chapter.

Like atmospheric GCMs, **oceanic general circulation models** simulate circulation, but in this case the circulation remains in the hydrosphere. The same equations govern motion in the ocean as in the atmosphere, because both media are fluids, but oceanic circulation is complicated by the fact that the circulation is confined to the ocean basins rather than being "free" to circulate anywhere as the atmosphere does. Furthermore, ocean circulation involves a fluid that is 800 times denser than the fluid circulating in the atmosphere. Assumptions that can be made for the atmosphere cannot be made for the ocean, and vice versa, because the higher *density* of the oceans causes a much longer reaction time to changing forces than in air.

Coupled atmosphere–ocean general circulation models include an ability to simulate not only the circulations of both the atmosphere and the ocean but also the feedbacks and energy transfer between those two fluids. Some crude atmosphere-ocean GCMs may treat the ocean as a **swamp model**—a computerized model in which the water is assumed to have a different *albedo* than land but no capability to store heat. A more advanced type of approach uses a **slab model**, in which a constant depth and ability to store heat are assumed. Nevertheless, even these models may not provide the most reliable projections of climatic change and its impacts. Other atmosphere–ocean GCMs include far more detailed calculations involving physical processes, feedbacks, and energy exchanges. Some may even include mechanisms for accounting for changes to the global *carbon cycle*, but these details come at the expense of spatial and temporal resolution. Spatial resolution of a few hundred kilometers is a reasonable goal in such sophisticated models.

In some cases output from a global-scale GCM may be used as input to a **regional climate model (RCM)**. Such models zoom in on the particular region of interest and contain much greater spatial resolution because the geographical area of interest is smaller. Specifically, RCMs may contain a spatial resolution as small as 50 km (30 mi). This feature enables them to mimic the effects of local surface features and short waves in the atmosphere better than global-scale GCMs. However, because of the increasing likelihood of errors magnifying over time when only a small region is represented in the model, RCMs are usually used only to simulate climates of the near future (perhaps a few decades at most). The **National Center for Atmospheric Research** and Penn State University collaborated to produce an RCM called the MM5 (*Mesoscale* Model version 5) model that is run for the eastern United States.

As an alternative to using RCMs, some climatologists have used **downscaling** to improve upon the spatial resolution of current GCMs. **Spatial downscaling** is a process that involves deriving a statistical relationship between GCM output and atmospheric conditions at a finer grid scale than the GCM can provide. For example, if a GCM outputs a temperature of 10°C in 2100 for a 300×300 km surface area, downscaling may fine-tune that projection for different land covers within that surface area based on derived statistical relationships between temperature and land cover.

Temporal downscaling projects output at one temporal scale to a finer temporal scale using statistical relationships. For example, if a GCM outputs the annual temperature in 2025, temporal downscaling techniques might be used to resolve that projection down to the monthly level. In some cases output may even be temporally downscaled to the daily level, particularly for projections with shorter lead times.

Representing the Earth–Ocean–Atmosphere System in GCMs

Two major families of GCMs can be identified, based on the strategy used for modeling geophysical processes. **Gridpoint models** represent atmospheric phenomena by dividing Earth into equidistant points with imaginary three-dimensional "boxes" centered on those points representing the smallest area in the atmosphere for which motion can be calculated by the model (**Figure 13.1**). The data at those points are interpolated from available atmospheric data and observations. The models are then run by using **finite differencing**—a technique that determines the values of the atmospheric variables at some point into the future. These procedures calculate the value of each atmospheric variable at a gridpoint based on the rate of *advection* of calculated values at each adjacent point, including the points above and below the point of interest.

Spectral models adopt the strategy of expressing atmospheric motion in terms of a series of waves. A schematic of the process of applying spectral techniques is shown in **Figure 13.2**. In spectral methods the collected data are interpolated to a grid system with multiple vertical layers, as they are in gridpoint models. Once the gridded variables are collected, each variable is plotted by latitude and (separately) by longitude, for a grid of latitude and longitude lines that are as fine as the spatial resolution of the model. For example, for 80°W longitude each value of temperature is plotted from the South Pole to the North Pole, with latitude along the *x*-axis and temperature along the *y*-axis. Mathematical techniques are then used to identify curves that approach as many of the points in the graph as possible, so that the temperature distribution from the South Pole to the North Pole

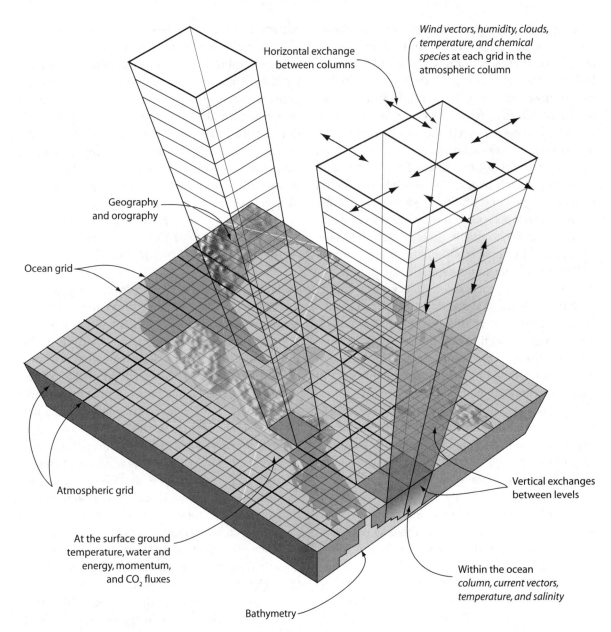

Figure 13.1 Schematic of a gridpoint model. *Adapted from*: McGuffie, K., Henderson-Sellers, A. *A Climate Modeling Primer*, 3rd Edition, John Wiley & Sons, (New York, NY, 2005).

along the 80°W meridian of longitude can be expressed mathematically. After the mathematical expression of the distribution of all variables at all vertical levels, the "results" of the forecast at the time step of interest are projected back to the gridded "data" for input to the next iteration of the model.

The primary advantage of spectral techniques is the representation of atmospheric motion as wavelike, which is a very good assumption in the real world. Any motion that occurs on one side of Earth "connects" after that motion propagates around the world back toward the original loca-

tion. Spectral analysis lends itself well to mimicking this type of atmospheric process. Even in spectral analysis, however, vertical transfers of energy, matter, and *momentum* are still calculated as they are in gridpoint models.

Data for GCMs

Because meteorological data are not collected at equally spaced locations around Earth (except in the case of satellite data), the data at the network of gridpoints are most likely to be interpolated computationally using spatial interpolation meth-

1. Data collected at several altitudes for a given variable (e.g., temperature) are quality controlled…

2. …and then interpolated to a grid system with multiple layers.

3. The data at each latitude and longitude line are plotted at each vertical level, and the spectral transformation plots the curve that fits the distribution best.

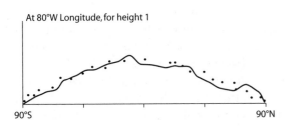

4. The data points are shifted to meet the graphs generated by the spectral transformation, at all latitude and longitude lines.

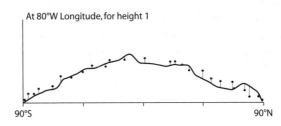

5. Fluxes are then computed in three-dimensional space, as in the gridpoint models.

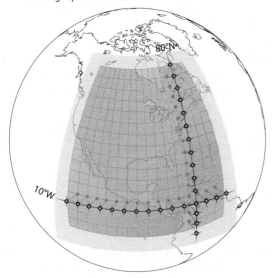

Figure 13.2 Schematic of a spectral model. *Adapted from*: McGuffie, K., Henderson-Sellers, A. *A Climate Modeling Primer*, 3rd Edition, John Wiley & Sons, (New York, NY, 2005).

ods from the available, collected data. Therefore, the data that are input into the gridpoint models are likely to be generated rather than actually observed. Of course, any errors present in the data generation process would appear in the model even before any forecasting calculations were made. Note that because upper-air observations are far sparser than surface-based observations, the upper-air data must be interpolated based on far fewer "actual" data points than at the surface. But because upper-level observations are affected less by the land cover and local conditions, the sparse network of points is less problematic than a lack of adequate data at the surface.

■ Seven Basic Equations

All weather and climate result from three fundamental entities: motion (both horizontal and vertical), energy, and moisture. Without any of these three, weather and climate do not exist. Any attempt to simulate the behavior of the atmosphere dynamically must include a representation of the fundamental equations that govern motion, energy, and moisture in the atmosphere. Seven equations accomplish this task. These equations are a set of nonlinear, partial differential equations. To gain an appreciation for the complexity of solving such equations simultaneously, imagine a system of two equations with only two unknowns, such as the following:

$$\begin{cases} 2x + 7y = 23 \\ x + 2y = 10 \end{cases}$$

The solution of this system of equations is simple, because both are linear (i.e., the variables do not have exponents other than 1) and because there are only two unknowns.

By contrast, each of the *seven* basic equations is tied together and must be solved simultaneously. To solve seven nonlinear partial differential equations simultaneously is an incredibly intensive computational task. Indeed, no branch of mathematics has yet been invented to accomplish this task efficiently. Instead, the equations must be solved iteratively, repeatedly "guessing" the solution and using each guess to focus in on the solution.

Navier-Stokes Equations of Motion

All GCMs and all weather forecasting models use the seven basic equations (in some form) to ana-

lyze and forecast atmospheric flow and behavior. The first three equations are collectively known as the *Navier-Stokes equations of motion*, named after Claude Navier, a French physicist whose equations for vibrational motion were generalized by several scientists, including Sir George Stokes, for use in a wide variety of applications. If you studied calculus you may remember Stokes' theorem, an important principle in three-dimensional vector calculus. Stokes was interested in the mathematics of motion, and his work has been applied to all areas of fluid mechanics, including large-scale ocean circulation, hydrodynamics, aerodynamics, and atmospheric circulation.

One of the Navier-Stokes equations governs velocity in the west-to-east direction (the positive u direction), another is for south-to-north velocity (positive v), and the third is for down-to-up velocity (positive w). East-to-west, north-to-south, and up-to-down velocities are simply represented as the negative u, v, and w directions, respectively. Thus, all motion (advection and *convection*) can be described by some combination of the u, v, and w equations. The equations themselves describe the rate of change of u, v, and w over time at each gridpoint, and the rate of change of a velocity over time is actually acceleration. The first Navier-Stokes equation may be written in the form of **Equation 13.1**, in which the symbols in the first five terms are simply read as "the partial derivative of (*variable in the numerator*) with respect to (*variable in the denominator*)." The "partial derivative" is the rate of change of the variable in the numerator over the variable in the denominator, *while holding the effect of all of other variables constant.* For example, term 1 in the equation below is expressed as the rate of change of westerly-to-easterly component of wind velocity (u) over time (an acceleration) at the gridpoint of interest, expressing *how much has the west-to-east component of wind changed since the last observation at this gridpoint?* Notice that because time is in the denominator, this equation has forecasting implications—how much has u changed at this point since the last time step?

$$\frac{\partial u}{\partial t} = -u\frac{\partial u}{\partial x} - v\frac{\partial u}{\partial y} - w\frac{\partial u}{\partial z} - \alpha\frac{\partial p}{\partial x} + fv + F_x$$

$$\quad 1 \qquad 2 \qquad 3 \qquad 4 \qquad 5 \qquad 6 \quad 7$$

13.1

The terms on the right side of the equation represent the factors that can influence the rate of change in u over time. Terms 2 and 3 represent the rate of advection of u in the west-east (x) and south-north (y) directions, respectively. As the wind velocity (u and v, respectively) increases, the rate of advection increases. Terms 2 and 3 also show that as the gradient of wind speed increases (in the x and y directions, respectively), the rate of advection increases because of the tendency to equal out imbalances of energy, matter, and momentum. Similarly, term 4 shows the vertical transfer (through convection) of u in the down-to-up (z) direction.

Term 5 describes the effect of the west–east *pressure gradient force*, where a is the **specific volume** (the reciprocal of density)—the amount of space occupied by a solid, liquid, or gas divided by its mass. The pressure gradient force only matters in the x direction because only the change in u with respect to time is sought in this equation. Term 6 represents the *Coriolis effect*, whereby the *Coriolis parameter* is

$$f = 2\,\Omega\,\sin\varphi$$

with Ω representing the angular velocity of Earth ($7.27 \times 10^{-5}\ \text{s}^{-1}$) and φ expressing the latitude. Finally, F_x in term 7 represents the impact of *friction* in the x-direction. Friction is very difficult to quantify, so different models treat the effect of friction differently.

Likewise, the second Navier-Stokes equation can be expressed as shown in **Equation 13.2**. This equation differs from the first equation because it determines the rate of change of the v component of atmospheric velocity over time. Correspondingly, the terms on the right side of the equation are modified to reflect this change.

$$\frac{\partial v}{\partial t} = -u\frac{\partial v}{\partial x} - v\frac{\partial v}{\partial y} - w\frac{\partial v}{\partial z} - \alpha\frac{\partial p}{\partial y} + fu + F_y$$

$$\qquad 1 \qquad 2 \qquad 3 \qquad 4 \qquad 5 \qquad 6 \quad 7$$

13.2

The third Navier-Stokes equation is for vertical motion (**Equation 13.3**).

$$\frac{\partial w}{\partial t} = -u\frac{\partial w}{\partial x} - v\frac{\partial w}{\partial y} - w\frac{\partial w}{\partial z} - a\frac{\partial p}{\partial z} + (2\Omega\cos\varphi)u + F_z - g$$

$$\quad 1 \qquad 2 \qquad 3 \qquad 4 \qquad 5 \qquad\qquad 6 \qquad 7 \quad 8$$

13.3

Interestingly, in the vertical direction, the Coriolis term (term 6) contains a cosine function rather than the sine function as in the first two equations. This is because the Coriolis effect depends slightly on vertical motion (rather than just latitude), but this is such a small term compared with the others that it can usually be ignored (except when analyzing the acceleration in the vertical direction). Additionally, note the presence of the gravity term (term 8). Unlike in the first two equations, gravity must be considered when evaluating vertical motion.

Several of the terms in the third Navier-Stokes equation are very small compared with the others, particularly for motion at the *synoptic scale* and *planetary scale*. Terms 5 and 8 tend to dominate, because for broad-scale atmospheric motion w does not change much in the x, y, or z directions—the air moves horizontally to a much greater extent than vertically—and because the Coriolis and frictional terms are minimal. As a result, for ease and speed of calculation, the small terms are often ignored. The third Navier-Stokes equation for vertical motion is often simplified to

$$-\alpha\frac{\partial p}{\partial z} = g$$

This simplified form is the *hydrostatic equation* that was discussed in Chapter 5. It expresses the balance between the upward-directed pressure gradient force (i.e., the buoyancy term) and the downward-directed gravitational force. The negative sign is necessary because *pressure* decreases as height (z) increases.

Simply stated, the hydrostatic equation shows that buoyancy and gravity are the two major factors that govern vertical motion. Of course, in some cases where vertical acceleration is important—for example, in local-scale thunderstorms or in situations where advection of w is important—the atmosphere is not behaving as a hydrostatic fluid and the neglect of the six terms in the "full" Navier-Stokes equation for w is unacceptable. It is no surprise, then, that *local-scale* thunderstorms, which have significant vertical acceleration, are difficult or impossible to predict precisely. Numerical models generally do not have adequate input data, computational power, and spatial/temporal resolution to allow for the full consideration of all of the terms in the third Navier-Stokes equation.

Likewise, the first two equations of motion may also be simplified to include only those terms that tend to dominate. The advection terms (that is, those terms that govern the rate of change of the velocity terms u, v, and w across space) are often much smaller (particularly for synoptic and planetary scales of motion) than the effects of acceleration, the pressure gradient force, the Coriolis effect, and friction. This simplification reduces the first two equations to

$$\frac{\partial u}{\partial t} = -\alpha \frac{\partial p}{\partial x} + fv + F_x \quad \text{and}$$

$$\frac{\partial v}{\partial t} = -\alpha \frac{\partial p}{\partial y} + fu + F_y$$

Collectively, these two equations and the hydrostatic equation are termed the **primitive equations of motion**. If a model is described as "primitive," it assumes that there is no advection of wind velocity in the west-east, south-north, or down-up direction. The benefit of such a simplification is that computational resources can be devoted toward improving the temporal and/or spatial resolution but at a cost of providing a cruder representation of the atmospheric flow. Local atmospheric flow may be modeled poorly when making the assumptions required to use the primitive equations, particularly in disturbed weather.

Thermodynamic Energy Equation

The fourth of the seven basic equations is the **thermodynamic energy equation**. It expresses the sum of the **internal energy** of the individual molecules comprising the atmosphere and the *kinetic energy* that drives the synoptic- and planetary-scale motion (i.e., the wind). The internal energy is the *"thermo"* part of the term—temperature actually represents the internal energy present in matter. Kinetic energy is the *"dynamic"* part of the term.

The temperature in this and other equations is always represented using the *Kelvin temperature scale*. The reason for this is that only in the Kelvin system is internal energy proportional to the temperature (i.e., doubling the Kelvin temperature of something means that the internal energy is doubled). This is not true in the Celsius or Fahrenheit scale—when the temperature increases from 10 to 20 °C, the energy content of the air does not double. Likewise, 0 K—*absolute zero*—actually represents the situation in which zero internal energy exists and all molecular motion ceases, unlike the situation for 0 °C or 0 °F.

The *"thermo"* part of the equation can occur through either **diabatic** or *adiabatic* heating/cooling. Diabatic heating of the atmosphere at a gridpoint refers to heating with an energy change and comes from outside sources, such as insolation, *latent heat*, or *sensible heat* in the boundary layer. The diabatic heating changes each minute of the day, as the solar angle changes, cloud cover increases or decreases, and water changes phase. A small time step is, therefore, necessary to capture the changes in diabatic heating effectively.

As we saw in Chapter 5, adiabatic heating/cooling—a process without an energy exchange—comes from expanding/contracting motion as an air parcel rises/sinks. In the case of heating (either diabatically, adiabatically, or some combination of the two), the internal energy of the molecules, and, therefore, the temperature, increases at the gridpoint. Likewise, cooling the atmosphere (either diabatically or adiabatically, or both) decreases molecular speed and decreases the internal energy component of the total thermodynamic energy.

The *"dynamic"* part of the equation occurs through advection of energy from elsewhere (different x, y, and z coordinates). These terms are reminiscent of those in the Navier-Stokes equations of motion. The complete thermodynamic energy equation can be expressed as shown in **Equation 13.4**. In this equation term 1 expresses the rate of change of temperature over time at a gridpoint. This is an important quantity because once we know how much temperature change exists at a point for some time step into the future we can then use that temperature to calculate the energy that exists to fuel further vertical and horizontal motion in the atmosphere even farther into the future. Term 2 represents a diabatic heating term, where q is the heat (or energy) added (in *Joules* kg^{-1}) and C_p is the **specific heat at constant pressure** for air (1005 J kg^{-1} K^{-1}). The latter quantity refers to the number of Joules of heat/energy that must be added to 1 kg of a substance (in this case, air) to raise its temperature by 1 C° (or 1 K). Heat is gained or lost continuously, as the solar angle changes or clouds move in front of the Sun at a point.

$$\frac{\partial T}{\partial t} = \frac{1}{C_p}\frac{dq}{dt} + \frac{\alpha}{C_p}\frac{\partial p}{\partial t} - u\frac{\partial T}{\partial x} - v\frac{\partial T}{\partial y} - w\frac{\partial T}{\partial z}$$

$$\quad\; 1 \qquad\; 2 \qquad\quad 3 \qquad\; 4 \qquad\; 5 \qquad\; 6$$

13.4

The third term represents the rate of temperature change due to adiabatic vertical motion, with α representing specific volume. The units of α are in $m^3 \, kg^{-1}$ (volume divided by mass). Pressure (p) has units of *Pascals* (Pa, which is force divided by area), and *Newton's second law* states that force can be expressed in units of mass times acceleration. Pressure, therefore, really has fundamental units of mass times acceleration divided by area, or $kg \, m \, s^{-2} \, m^{-2}$, or $kg \, m^{-1} \, s^{-2}$. This gives term 3 units of $K \, s^{-1}$. You can verify these units at the end of this chapter in the Questions for Thought section. The last three terms represent the advection of temperature in the west–east (x), south–north (y), and down–up (z) directions, respectively.

Moisture Conservation Equation

The fifth of the basic equations in all GCMs is the **moisture conservation equation**. Atmospheric moisture is usually represented by *specific humidity*, which is normally abbreviated as q, but here we use m because q is also usually used to represent the heat (energy) added in the thermodynamic energy equation. The moisture conservation equation states that the change in moisture content at a gridpoint is equal to the sum of the contribution of moisture via advection and the contribution via phase changes. One form is shown in **Equation 13.5**.

$$\frac{\partial m}{\partial t} = -u\frac{\partial m}{\partial x} - v\frac{\partial m}{\partial y} - w\frac{\partial m}{\partial z} + S \qquad \text{13.5}$$

$$\quad 1 \qquad\quad 2 \qquad\quad 3 \qquad\quad 4 \quad\; 5$$

The first term is the temporal rate of change of moisture (represented by specific humidity). The second, third, and fourth terms represent the advection of specific humidity in the west–east, south–north, and down–up directions, respectively. Finally, term 5 denotes changes of moisture caused by phase changes, such as *evaporation*, which would add moisture to the air. Often, this term is difficult to quantify. When the physics governing certain processes (such as friction or phase changes) cannot be quantified easily, models sometimes represent such effects with **parameterizations**—statistically based "fudge factors" that are used when the precise equations to simulate the process are either unknown or impossible to include because the input data are not available to solve those equations.

Continuity Equation

The **continuity equation**, sometimes known as the **mass conservation equation**, is another vital component of all GCMs and weather forecasting models. Total mass must be conserved (i.e., remain constant) in any model, because the *first law of thermodynamics* states that energy must be conserved. Because energy is related to matter according to Einstein's theory of relativity, mass must also be explained. The amount of mass entering any three-dimensional box centered on a gridpoint must be balanced by the amount of mass leaving adjacent boxes to enter the box. Likewise, the amount of mass leaving the imaginary box of interest must be balanced by mass entering adjacent boxes from it.

In the atmospheric sciences mass is an inconvenient variable to measure and monitor. When a "box" of the atmosphere is considered, volume becomes the variable of interest, because the atmosphere is a compressible fluid whose volume changes when mass changes. It is convenient to divide all quantities in an equation by mass. Notice that entities such as C_p tend to have units of mass in the denominator ($J \, kg^{-1} \, K^{-1}$), because the quantity has been divided by mass. In the continuity equation it is convenient to believe that the volume divided by the mass must be conserved. But volume divided by mass is simply the reciprocal of density. Thus, one form of the continuity equation states that the temporal rate of change of density in the atmosphere is produced by some combination of advection of more (or less) density from the west–east, south–north, and down–up direction and by **velocity convergence**—the pile-up of atmospheric density (or mass) from the west–east, south–north, and down–up directions.

To understand velocity convergence better, imagine the box of air in the atmosphere again, as shown in **Figure 13.3a**. If more air enters the box than leaves it from a particular direction, because the air flowing into the box is moving faster than the air leaving the box, velocity convergence is occurring. Even if advection does not bring in air of different density, if *more* air of the same density is approaching the gridpoint than is moving away from it, the density of the air in the box must increase.

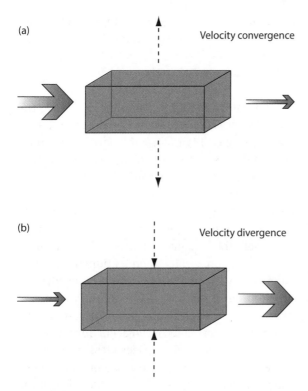

(a)

Velocity convergence

(b)

Velocity divergence

Figure 13.3   Velocity convergence and divergence.

$$\frac{\partial \rho}{\partial t} = -u\frac{\partial \rho}{\partial x} - v\frac{\partial \rho}{\partial y} - w\frac{\partial \rho}{\partial z} - \rho\frac{\partial u}{\partial x} - \rho\frac{\partial v}{\partial y} - \rho\frac{\partial w}{\partial z}$$
$$\qquad 1 \qquad 2 \qquad 3 \qquad 4 \qquad 5 \qquad 6 \qquad 7$$

13.6

In most analyses at synoptic (or larger) scales, the velocity convergence/divergence terms far exceed the advection terms. It is relatively unusual to find a situation in which air of significantly different density moves into a place *from a broad area*. But it is not so unusual to have air piling up or spreading out at a location. At mesoscales and local scales both the advection and the convergence/divergence components can be important. For example, when a strong frontal boundary is present, density advection becomes more important. In convective thunderstorms, local convergence/divergence as well as advection of air with great density differences—whether they occur between colder versus warmer, or wetter versus drier air—can create even greater effects of local convergence.

Equation of State

The final equation is also the simplest. It is the "glue" that binds the other equations because it relates pressure, density (or specific volume), and temperature. All of the previous six equations (except the moisture conservation equation) explicitly express either temperature, density, or pressure. Even the moisture conservation equation implicitly expresses these quantities in its parameterization of moisture transfer via phase changes. All three quantities, however, are not represented together in any of the equations despite the fact that they are all related to each other. The *equation of state*, also known as the *ideal gas law,* expresses the relationship between these three fundamental entities so that they can be used in the other equations.

As we saw in Chapter 3, the equation of state for dry air is

$$P\alpha = R_d T$$

where R_d is the dry gas constant (287 J kg^{-1}K^{-1}). In this equation pressure is expressed in Pascals and temperature is expressed in the Kelvin temperature scale to make the units cancel. When dealing with moist air (i.e., air with any water vapor in it), a few adjustments must be made. The pressure and

Of course, it is also possible that more mass (or density) could be leaving the box than is entering (from one or more directions), even if advection is not changing the density of the air approaching the box. Such a situation is termed **velocity divergence** (Figure 13.3b). When velocity convergence or divergence occurs at a gridpoint, the extra (or reduced) mass (or density) affects all of the first four equations (including the calculation of the rate of motion in the x-, y-, and z-directions and the rate of temperature change). This is because density is represented by its reciprocal, specific volume (α), in each of these equations. The extra mass also affects moisture conservation because specific humidity is a function of mass. As we are about to see, the extra mass also affects the final result of the other basic equations.

Quantitatively, the continuity equation can be expressed as shown in **Equation 13.6**, where the first term represents the rate of change of density from one time interval to the next. Terms 2 through 4 represent the advection of density in the west–east, south–north, and down–up directions, respectively. Terms 5 through 7 represent the velocity convergence/divergence in the same three directions.

density would then represent the sum of that due to dry air and that caused by water vapor. R_d must then be adjusted to take the moisture into account. This adjustment creates the **equation of state for moist air**, which is expressed as

$$P\alpha = \frac{R_d\left(1+\dfrac{r}{\varepsilon}\right)}{1+r}T$$

where r is the **mixing ratio**—a *dimensionless quantity* expressing the ratio of the mass of water vapor present in the air to the mass of dry air—and ε is the dimensionless ratio of the molecular weight of water vapor to the molecular weight of dry air (0.622). Mixing ratio is closely related to specific humidity (m) by

$$m = \frac{r}{r+1}$$

Furthermore, m itself is a dimensionless quantity that can be derived based on known quantities by using

$$m = \frac{e\varepsilon}{p}$$

Vapor pressure (e) can be calculated from *dewpoint temperature* by using a variant of the *Clausius-Clapeyron equation*:

$$e = 611\text{Pa}\ \exp^{\frac{L_v M_v}{R^*}\left(\frac{1}{273\text{K}} - \frac{1}{T_d}\right)}$$

where L_v is the latent heat of condensation (2.5008×10^6 J kg^{-1}), M_v is the molecular weight of water vapor (0.018015 kg mol^{-1}), R^* is the universal gas constant (8.314 J mol^{-1} K^{-1}), and T_d is dewpoint temperature in Kelvins. It should be noted that over an ice surface, L_v is substituted by the latent heat of sublimation (L_s, or 2.8345×10^6 J kg^{-1} at 0°C).

The terms multiplied by R_d in the equation of state for moist air are sometimes lumped together to form a new quantity called **virtual temperature** (T_v), where

$$T_v = \left(\frac{1+\dfrac{r}{\varepsilon}}{1+r}\right)T$$

to give

$$P\alpha = R_d T_v$$

Virtual temperature is the temperature that dry air would have if its pressure and volume were equal to that of a sample of moist air. Because moist air adds pressure (i.e., vapor pressure) and pressure is directly proportional to temperature (according to the equations of state for dry and moist air), T_v cannot be less than T.

Similarities and Differences Between GCMs and Weather Forecasting Models

Weather forecasting models and GCMs are similar in many ways but different in others. Both types of models use the same general equations to simulate atmospheric motion and energy, moisture, and momentum exchanges, because the same physical processes affect the circulation of the atmosphere whether it is at "weather" or "climate" time scales. The most fundamental differences between GCMs and weather forecasting models involve the difference in temporal scale. Unlike a weather forecasting model, a GCM would not attempt to project the temperature of a specific day, even if prediction of daily conditions could somehow be made accurately to the distant future. Similarly, because of the longer distance of the prediction, the time step in GCMs is generally longer than that for weather forecasting models. A typical time step on GCMs may range perhaps from half a day to a month, compared with only a few minutes for a weather forecasting model.

The importance of accuracy during **initialization**—the setting of initial conditions in a model in preparation for a forecast—is also somewhat relaxed in GCMs. Of course, some errors in initial conditions can magnify over time to create dubious model output. To understand this better, imagine an arrow that hits within a bull's-eye but is only slightly off from dead center at a target 25 m away. Although a bull's-eye occurred on a close target, the same arrow might be several centimeters away from the center of a target located 50 m away even when the archer aimed the arrow in the exact same manner. The same shot extended to a target located 100 m away may miss the target completely. This example demonstrates that even a tiny error embedded in a model magnifies in time (as did the arrow) to produce unrealistic results (i.e., miss the target completely). Meteorologist Edward Lorenz named this magnification of errors the **butterfly effect**, a theory based on the idea that a butterfly that flaps its wings in one part of the world could create a tiny perturbation that could cause a chain of magnified events that ultimately alters the planetary-scale atmospheric flow. This idea is at the core of **chaos theory**, which suggests an inherent lack of predictability (sometimes known as a **dynamic instability**) in some systems despite the fact that an order underlies them.

Another fundamental difference is the importance of representing interactions and feedbacks between the atmosphere and the lithosphere, hydrosphere, and biosphere. At meteorological time scales such influences are often negligible compared with other errors that result from erroneous input data, poor parameterization techniques, and/or poorly understood physical processes, but these interactions and feedbacks can be extremely important over longer time periods. For example, a weather forecasting model would not need to consider increased *desertification* in the *Sahel* in predicting differences between tomorrow's weather and today's in the United States; likewise, increased *deforestation* at distant locations by tomorrow would be inconsequential for weather prediction. However, if a GCM is to be run to simulate the climate in 2100, the effects of Sahelian desertification and deforestation between now and 2100 must be considered because over such long periods of time it is likely to influence global circulation.

■ Statistical Techniques

In addition to the mathematical methods in GCMs, climatologists use a number of statistical methodologies. Several of these are simple statistical methods that are used across the spectrum of natural and social sciences. These are not discussed here because they are addressed in introductory statistics textbooks. Modern climatic analysis includes other, more advanced statistical techniques that can be useful. For example, as introduced in Chapter 8, *eigenvector analysis* provides a useful suite of information because these techniques are

the only ones that allow for simultaneous examination of variability and changes in climate across space and time.

As we saw in Chapter 8, it is useful to keep in mind that in climatological data sets, three entities are usually present: (1) atmospheric variables (such as temperature, precipitation, pressure), (2) spatial variables (e.g., weather stations), and (3) temporal variables (e.g., months since 1900). If any one of these three entities contains only one variable, eigenvector analysis can identify relationships between the other two sets of variables simultaneously. In analyses of synoptic circulation, only one atmospheric variable, *geopotential height* or sea level pressure, is usually necessary, because geopotential height or pressure gradients cause atmospheric motion in response to the height or pressure differences represented. Synoptic analysis can often use eigenvector techniques to identify spatiotemporal variability and change in one aspect of climate, such as 500-mb heights.

Although the nuances of the differences between each eigenvector technique listed below are not addressed here, the methods all share some common characteristics. A data set is assembled that contains variables representing the two varying entities under consideration (for example, months from 1900 through 2010 along the rows and weather stations along the columns) for only one variable of the third entity (sea level pressure, for example). The eigenvector technique then uses matrix algebraic procedures to transform that matrix of data (in this case the weather stations vs. months) into two other matrices. The first of these matrices has the first entity's variables (in this example, the set of all weather stations in the study area) along the rows and the *loading* for each eigenvector along the columns. The first eigenvector explains the maximum amount of variance in the data set, and each loading on that eigenvector describes how strongly the particular observation (each weather station, in this example) is represented by that eigenvector. The total sum of the loadings for all eigenvectors must equal the number of observations (in this case, weather stations) and is termed the *eigenvalue*. So if the data set contains 27 weather stations, then the sum of the eigenvalues for each eigenvector must be 27. The first eigenvector always has the highest eigenvalue, the second has the second-highest eigenvalue, and so on.

If the variable in the loadings matrix is a spatial variable (as in this example), then the loadings for each *component* can be shown using isoline maps. These maps reveal the spatial component to the explained variance. In our example, a loadings map for the first component reveals the geographical areas where variability in sea level pressure is highest during the 1900–2010 period.

In addition to the loadings matrix, a second output matrix is produced. That matrix contains the second entity's variables (in this example, all monthly weather observations) along the rows. Each variable has a *score* for each eigenvalue, which indicates the magnitude of the influence on the explained variability on that eigenvector (mapped by the loadings matrix) during each time unit of the time series. We can examine the trends over time for the scores in that eigenvector to determine whether the influence has increased or decreased temporally. Furthermore, scores for the eigenvectors, which represent a time series for a particular component, may be correlated against some other phenomenon's time series. This correlation could then be used to reach a conclusion about the relationship between the flow pattern represented by the mapped isolines of loadings for a given eigenvector and the environmental variable, such as dissolved oxygen content in Lake Champlain over time. Which atmospheric flow pattern is coincident with the most oxygenated water in Lake Champlain? Which is associated with the lowest?

Two main variants of eigenvector analysis have been used in climatological studies. The first is *principal components analysis*, which is also essentially the same as two techniques known as singular value decomposition and empirical orthogonal function analysis. A slight variant is **common factor analysis**. Although both techniques work on the same basic principles described above, the principal components analysis model has proved to be the favorite in most climatological applications. The reason is that the matrix transformations in principal components analysis assume no underlying uniqueness among the variables. In other words, nothing unique at weather station A makes station A's data differ from that at station B. Because all stations are influenced by the same Sun, with the same atmospheric composition, the same physical laws, and the same interactions with other "spheres," this assumption is reasonable. By contrast, the common factor analysis model tends to be preferred in social scientific research when dealing with human subjects, because it allows for some "unique" element to contribute to the data. For example, one person may have had a unique life experience that made her answers on a survey differ from the results that would be expected for another female in her peer group.

Eigenvector analysis, and particularly principal components analysis, has been used in *synoptic climatology* to delineate the major modes of variability in the atmospheric flow. This application is useful because atmospheric flow is known to be associated with a wide range of other variables. Because flow patterns are represented by the geographical distribution of geopotential height or pressure, the mapped loadings reveal the locations where geopotential height (or pressure) varies the most. And when the scores matrix is compared with some other environmental impact, we get a picture of where and when the pressure (and, therefore, the circulation) feature is occurring, and an assessment of the impact of that circulation feature on the environmental problem of interest.

■ Atmospheric Teleconnections

Collectively, results of eigenvector-based studies (and other studies that were not based on eigenvector analysis) identified the major atmospheric *teleconnections*—the correlations or "see-saws" in geopotential height or sea level pressure patterns across a large area. Because the total atmospheric mass is conserved, increases in one location (i.e., higher pressure) must be compensated elsewhere by equal decreases in mass (i.e., lower pressure). Geopotential height or pressure anomalies in certain geographical locations tend to act in tandem—when geopotential height or pressure is abnormally high in one location, it is abnormally low simultaneously in the other, teleconnected location. The see-saw or set of *action centers* is the teleconnection, and the teleconnection can be a source of important regional- to planetary-scale climate impacts. It is, therefore, important to identify these teleconnections. Because the three important tropical climatological phenonoma—the

Southern Oscillation (SO), *Madden-Julian Oscillation*, and *Quasi-Biennial Oscillation*—have already been discussed in Chapter 10 for their role in shaping low-latitude climates, we consider only extratropical teleconnections here.

Extratropical Teleconnections in the Pacific: The Pacific Decadal Oscillation

In 1996 fisheries scientist Steven Hare and other scientists, including Nathan Mantua, noticed a climatic oscillation that seemed to be associated with periods of productive and nonproductive years in the Alaskan salmon industry. This discovery led to the description of the *Pacific Decadal Oscillation* (PDO). Like the *El Niño* phase of the SO (**Figure 13.4**), the "positive" or "warm" phase of the PDO is characterized by a periodic warming of the tropical central and eastern Pacific Ocean (**Figure 13.5a**). This warming tends to propagate northward along coastal North America, weakening the cold *California current*, but unlike the SO this warming accompanies abnormally low sea surface temperatures in the north central Pacific Ocean, in the vicinity of the *Aleutian low*. During this "positive" or "warm" phase of the PDO, westerly winds on the southern end of the Aleutian low are stronger than usual (Figure 13.5a).

During the "negative" or "cool" phase of the PDO the opposite conditions occur (Figure 13.5b). Waters in the tropical central and eastern Pacific are anomalously cold (as in the *La Niña* phase of the SO), and these cold waters tend to propagate poleward along the west coast of the Americas, leaving the cold currents even colder than usual. At the same time, warm waters tend to exist in the northern central and western Pacific Ocean, and the Aleutian low is weakened.

The PDO, then, is similar to the SO, but with two major exceptions. First, the primary "signature" of the PDO is apparent not only in the tropical Pacific but also in the extratropical North Pacific Ocean. Second, the PDO fluctuates much more slowly than the SO. A PDO regime may last 20 to 30 years, during which one phase of the oscillation tends to dominate, whereas the SO fluctuates back and forth over periods of 2 to 7 years. During times when the SO and the PDO both favor the same conditions (i.e., a "warm phase" or a "cold phase" in the central and eastern tropical

Pacific), the impacts tend to be magnified. When one of these Pacific teleconnections is in the "warm" phase and the other is in the "cold" phase, the effects tend to be dampened.

The PDO has undergone a few major "regime shifts" over the last 120 years (**Figure 13.6**). From 1925 to about 1946 warm phase conditions were prevalent. Cool-phase conditions dominated from 1947 until 1976, when the warm-phase reemerged. The period since 1999 is uncertain, as both warm- and cool-phase conditions show some signs of dominance.

Extratropical Teleconnections in the Atlantic Ocean

Atlantic Multidecadal Oscillation   A phenomenon related to the PDO exists in the North Atlantic Ocean basin. The **Atlantic Multidecadal Oscillation (AMO)** refers to a cycling of warm and cold surface waters over time scales ranging from about 15 to 30 years. But the Atlantic Ocean is not large enough to develop a see-saw pattern in sea surface temperatures as well as in the Pacific. Variations in the AMO impact climate in various ways. For example, the AMO is largely responsible for the cyclical nature of *tropical cyclone* frequency and intensity in the basin. Currently, the AMO is in a warm phase, and the associated high sea surface temperatures contributed to the very unusual characteristics of the 2005 Atlantic hurricane season when a record number of tropical cyclones occurred. Additionally, intensities were anomalously high for many of the 2005 storms, particularly Hurricanes Katrina, Rita, and Wilma. Although many point to global warming factors as the cause of the 2005 Atlantic hurricane season, the fact that the 2009 North Atlantic hurricane season was relatively quiet indicates that many factors combine to determine hurricane frequencies and associated intensities.

North Atlantic Oscillation   In addition to the AMO, two teleconnections in the Atlantic Ocean cause well-known impacts on weather and climate in North America and Europe. Of these, the first to be identified was the **North Atlantic Oscillation (NAO)**, a see-saw in pressure patterns between the *Bermuda-Azores high* and the *Icelandic low*. At times the sea level pressure associated with the Bermuda-Azores high is even higher than usual, while at the same time the central pressures of the

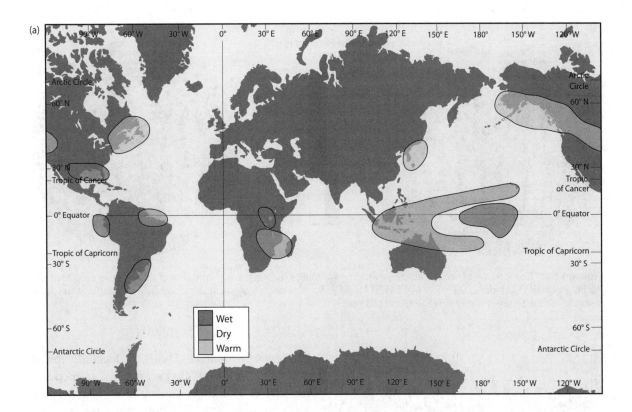

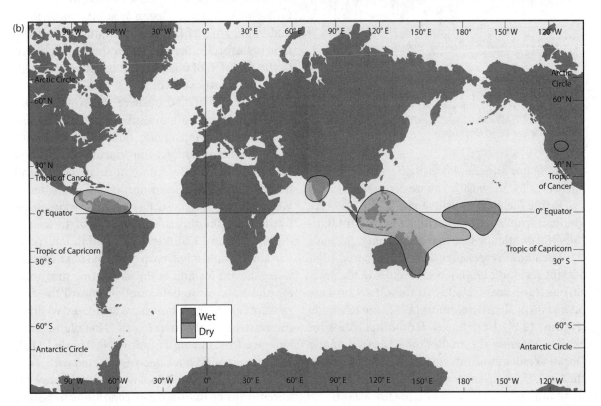

Figure 13.4   Impacts of the El Niño phase of the Southern Oscillation in (a) January and (b) July. *Modified from*: "Global Consequences of El Niño, Impacts of El Niño and Benefits of El Niño Prediction," El Niño Theme Page, NOAA, http://www.pmel.noaa.gov/tao/elnino/impacts.html.

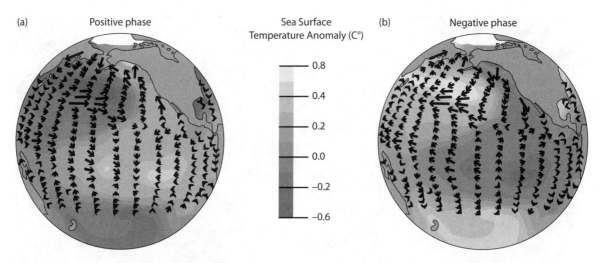

Figure 13.5 Comparison of sea surface temperature anomalies during the (a) positive and (b) negative phases of the Pacific Decadal Oscillation. (See color plate 13.5) *Modified from*: Nathan Mantua at the University of Washington's Joint institute for the Study of the Atmosphere and Oceans.

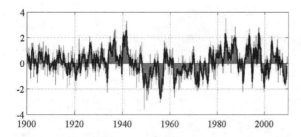

Figure 13.6 Time series of the monthly Pacific Decadal Oscillation index between 1900 and September 2009. (See color plate 13.6) *Modified from*: Nathan Mantua at the University of Washington's Joint institute for the Study of the Atmosphere and Oceans.

Icelandic low are even lower than usual. This scenario is called the positive phase of the NAO.

During the NAO's positive phase, the pressure gradient from south to north across the midlatitude North Atlantic Ocean is very strong. Because the Bermuda-Azores high is a warm-cored high and the Icelandic low is a cold-cored low, the pressure features associated with these two systems persist vertically in the atmosphere to great heights (**Figure 13.7**). Upper-level latitudinal pressure gradients across the midlatitude North Atlantic Ocean are also strong during the positive phase of the NAO. This causes very strong zonal geostrophic winds and strong westerlies across the Atlantic.

At other times the see-saw is "tilted" in the opposite direction. During these periods the Bermuda-Azores high has central pressures that are not as high as usual, while simultaneously the

low pressure associated with the Icelandic low is not as low as usual. This scenario is termed the negative NAO pattern. During times of a negative NAO the westerly circulation across the midlatitude North Atlantic Ocean is weak, with *ridges* and *troughs* more likely to appear (in a *meridional flow* pattern). Unlike the SO no rhythmic pattern in the duration of positive and negative NAO patterns has been observed.

One of the first to observe the NAO was Sir Gilbert Walker, the namesake of the *Walker circulation*—the atmospheric flow associated with the SO. Walker noticed that during times when Iceland was warmer than normal, Greenland tended to be colder than normal, and vice versa. We now know that these observations were caused by the strengthening and weakening of the Icelandic low (**Figure 13.8**). A strong Icelandic low (positive NAO) would advect warm air and near-surface water toward Iceland at the same time that polar air and water were being pulled toward the east coast of Greenland. Some have wondered whether the settlement and naming of "Greenland" may have occurred during a negative NAO period. Such conditions would have been associated with a relatively weak Icelandic low, which would have allowed warmer-than-normal conditions to persist over Greenland while below-normal air and water temperatures affected Iceland (Figure 13.8).

The impacts of the NAO are strong, particularly in winter, for Europe and the Mediterranean

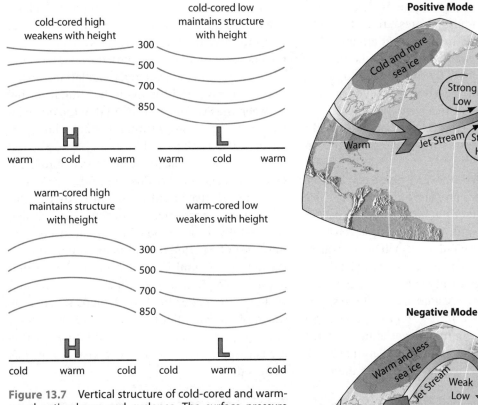

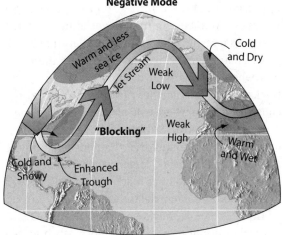

Figure 13.7 Vertical structure of cold-cored and warm-cored anticyclones and cyclones. The surface pressure values work with the thermal structure of the surface. A high-pressure area at the surface appears as a "dome" in the constant pressure field above it, whereas a low-pressure system appears as a "bowl" in the constant pressure field above it. The thermal structure, however, may eventually overcome this near-surface pressure pattern. Warm-cored systems cause the pressure level to bulge upward and cold-cored systems cause the pressure height to be deflected downward, in fulfillment of the hydrostatic equation. The result is the pressure pattern shown here for the various pressure/thermal combinations.

Figure 13.8 North Atlantic Oscillation.

Basin. Positive NAO winters usually bring dry conditions to southern Europe and mild, wet conditions to northern Europe, coincident with the primary storm track. Some impacts are also felt in eastern North America. For example, positive NAO winters are usually mild and wet in New England with relatively warm waters in the adjacent Atlantic Ocean, as the strengthened westerlies discourage the southward movement of bitter cold air toward the region and tropical moisture can penetrate farther north than usual. By contrast, negative NAO winters usually see more *continental polar (cP) air masses* with less precipitation over eastern North America and northern Europe, as the ridge-to-trough sides of the *Rossby*

waves tend to spill cold air southward. Atlantic waters are generally colder than normal in the Atlantic Ocean adjacent to North America during these times.

Arctic Oscillation (Annular Mode) In 1998 meteorologists David Thompson and John Wallace used eigenvector analysis to reveal the presence of a NAO-like oscillation in which the see-saw in pressure exists throughout most of the troposphere between the north polar region and the midlatitudes. The NAO is considered by some to be the regional expression of this hemispheric-scale phenomenon termed the **Arctic Oscillation (AO)** or the **Annular mode**, as it is sometimes known. In one phase (the "warm" or "positive" phase), the AO is characterized by lower-than-normal pressure in the polar region and above-normal pressure around 45°N. This results in a strengthening and "tightening" of the *circumpolar vortex*. By contrast, the opposite phase (the "cold" or "negative" phase) of the seesaw is associated with relatively high pressure at the North Pole and relatively low pressure at the midlatitudes. This results in a weakening of the circumpolar vortex.

The "warmth" in the positive phase results from the weakened surface pressure gradients that allow relatively warm water from the subtropical oceans to propagate northward. At the same time, relatively strong polar westerly winds exist aloft in the "warm" phase because the lower pressure at the poles drives a strong pressure gradient aloft over the *polar cell* (where surface *polar easterlies* occur). These strong upper-level polar westerlies tend to prevent air masses of Arctic origin from penetrating far to the south. As a result, conditions in midlatitude North America and Eurasia tend to be warmer than normal when the AO is in its positive phase. Storm tracks are usually farther north than normal because the region where cold air and warm air meet retreats poleward. Subtropical locations then tend to be warmer and drier than normal, and the *trade winds* blow over a larger region and are often strengthened, without as much nearby influence of midlatitude circulation features.

In the "cold" or "negative" phase of the AO, different conditions prevail. The abnormally high sea level pressure at the pole causes an expansion of the circumpolar vortex, which tends to push the warm water away from polar regions. At the same time, the relatively high geopotential heights aloft over the pole act to weaken the normal westerlies aloft over the polar easterlies. This is because high pressure exists to the right of the flow in the northern hemisphere, thereby working against the normal westerly circulation aloft in the polar cell. With the weaker westerlies aloft and an expanded circumpolar vortex, Arctic air can spill southward relatively easily. This results in colder-than-normal conditions in midlatitude North America and Eurasia and storm tracks that are displaced southward. Subtropical locations tend to be cooler and wetter than normal, and the tropical trade winds are weakened as midlatitude circulation features are displaced anomalously equatorward.

The AO moves from a dominant positive phase to a dominant negative phase and back again over long time periods. The periodicity is far longer than the fluctuations associated with the NAO. Because of this, some have considered the relationship between the NAO and the AO to be similar to that between the SO and the PDO in that the two teleconnection pairs show similar behavior but operate on different time scales. It has been said that Adolph Hitler decided to invade Russia in 1941 because he did not believe Russia could experience a third consecutive bitterly cold winter, thereby ensuring that his army could continue the invasion successfully through the winter. As it turned out, the onset of a period of cold-phase dominance of the AO caused a miscalculation in Hitler's thinking. The cold phase also dominated the 1960s and 1970s, whereas the 1980s and 1990s saw mostly warm-phase conditions. Some recent signs point to the return of cold-phase conditions, but it is too early to tell whether a true phase change has occurred.

Teleconnections over North America

Pacific-North American Pattern The **Pacific-North American (PNA) pattern** has regional importance for North American weather and climate. This teleconnection represents the oscillation between geopotential height in the midtroposphere of northwestern North America, with opposing heights upwind in the northern Pacific and downwind in the southeastern United States. The PNA index is positive when the midtropospheric ridge is amplified over western North America and a concurrent trough is amplified over the eastern portions of the United States (**Figure 13.9**). Likewise, the PNA pattern is said to be in its negative mode when the ridge–trough pair over midlatitude North America is deamplified (**zonal flow**), or even inverted, with a trough over western North America and a ridge over the east. The PNA pattern describes the condition of the Rossby wave flow across midlatitude North America.

The PNA pattern is known to exert a major impact on atmospheric variability in the United States. In its positive mode the southeastern United States tends to be colder than normal because it lies on the poleward side of the *polar front jet stream*, which dips southward in the trough

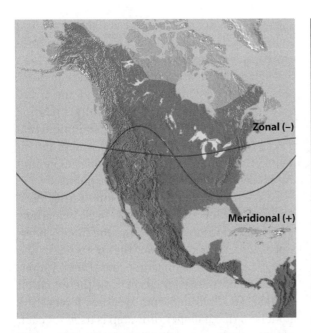

Figure 13.9 Rossby wave flow associated with the Pacific-North American teleconnection pattern. *Modified from*: Vega, A. J. 1994. The Influence of Regional-Scale Circulation on Precipitation Variability in the Southern United States. A dissertation submitted to the Faculty of Louisiana State University, Department of Geography and Anthropology, 300 pp.

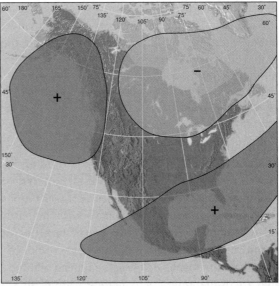

Figure 13.10 Regions of pressure variability associated with the positive mode of the Tropical-Northern Hemisphere teleconnection. *Adapted from*: NOAA/NWS/Climate Prediction Center, Northern Hemisphere Teleconnection Patterns. http://www.cpc.ncep.noaa.gov/data/teledoc/ tnh_map.shtml. Last accessed July 5, 2005.

situated over the eastern United States. Simultaneously, the northwestern states are warmer than normal, and they lie south of the polar front wave train, beneath the ridge. In the negative mode the Northwest is colder than normal, whereas the Southeast is relatively warm. The PNA pattern is also associated with a more complicated, but still distinguishable, set of precipitation conditions across the United States. For example, the Ohio Valley often experiences wetter-than-normal conditions during negative PNA periods, particularly in winter and spring, as midlatitude storms form and track along the nearby cold-warm boundary.

Tropical-Northern Hemisphere Teleconnection
The **Tropical-Northern Hemisphere (TNH) teleconnection** has been shown by climatologists Kingtse Mo and Robert Livezey to represent another important component of atmospheric variability over North America, particularly in winter. The nodes of this midtropospheric see-saw are in the Aleutian low area and the subtropical Mexico–southeastern U.S. region, with a center of opposite sign over northeastern Canada (**Figure 13.10**). Sometimes a

ridge is positioned to the north over the North Pacific with an accompanying trough over the Hudson Bay area (describing the "positive" mode). At other times the trough is over the North Pacific with a ridge near Hudson Bay (the "negative" mode). These extremes of the TNH pattern represent variability in the location and intensity of the polar front jet stream over North America. The TNH pattern is important for explaining westerly flow over North America. It also affects the position of the trough-to-ridge side of the Rossby waves (which is the side that supports stormy weather) and the ridge-to-trough sides of the Rossby waves (which can advect stable, polar air southward into the United States).

There is some evidence to suggest that the teleconnections themselves are teleconnected. In addition to the NAO and the AO demonstrating many of the same features, the negative phase of the SO (El Niño) often triggers a positive PNA pattern and a negative TNH pattern. Furthermore, a positive SO (La Niña phase) is often associated with enhanced convection from the Madden-Julian Oscillation. Work by climatologist Jeff Rogers indicates that the SO and NAO also show some signs of correlation. These "teleconnected teleconnections" provide some evidence that the global distribution of pressure features resembles gears in a machine in that changes in one location affect changes elsewhere. This demonstrates that the global climate is intricately balanced, with perturbations in one area likely to be felt far away.

■ Summary

Contemporary climatology is quantitative and analytical, and the current trend is toward an increasing emphasis on quantitative and analytical methods. Aside from basic statistical techniques that have become commonplace across the natural and social sciences, two broad families of quantitative techniques can be recognized in climatology. The first involves mathematical methods that are used in numerical models of atmospheric behavior. Some of these models may be simple, but they may serve their purpose adequately. Other numerical models are inherently more complex because they are designed to simulate and predict atmospheric dynamics and thermodynamics at the planetary scale. These general circulation models are the most sophisticated tools available to predict atmospheric motion at climatological time scales. They may be designed to simulate atmospheric or oceanic motion exclusively, or they may be "coupled" to incorporate exchanges of energy, matter, and momentum between the atmosphere and the ocean, land, and ice surface that may be important at climatological time scales. Output from GCMs can be used as input to a more focused analysis of a particular region through either further dynamical modeling or a downscaling of the results. The input data are projected to a grid, and the GCMs may run by making calculations at each gridpoint and basing forecasts on the conditions at adjacent gridpoints.

These models are relying increasingly on spectral techniques in which atmospheric motion is represented as a series of sinusoidal waves rather than through exclusive use of gridpoint representations. Several research institutions around the world have developed and are continually updating GCMs to make use of increasing computational power, better surface observations, and improvements in our understanding of the physics behind the processes influencing the atmosphere.

GCMs are similar to weather forecasting models in that they contain seven basic equations governing atmospheric behavior: the three Navier-Stokes equations of motion, the thermodynamic energy equation, the moisture conservation equation, the continuity equation, and the equation of state. GCMs differ from weather forecasting models in that a longer time step is used for each successive projection, initial conditions are less important, and representation of interactions and feedbacks between the atmosphere and the hydrosphere, lithosphere, cryosphere, and biosphere is emphasized.

Statistical methods also have been used extensively in climatology to provide information about both spatial and temporal variability in climates simultaneously. In this regard eigenvector techniques have been especially useful. Such techniques have revealed various atmospheric teleconnection patterns that exert a significant influence on a wide array of atmospheric and environmental features at time scales from weeks to decades. These teleconnections include the Southern Oscillation, Madden-Julian Oscillation, Pacific Decadal Oscillation, North Atlantic Oscillation, Arctic Oscillation, Pacific-North American pattern, and Tropical-Northern Hemisphere teleconnection.

▶ Key Terms

Absolute zero	Arctic Oscillation (AO)	*Carbon cycle*
Action center	Atlantic Multidecadal Oscillation	Chaos theory
Adiabatic	(AMO)	*Circumpolar vortex*
Advection	Atmospheric general circulation	*Clausius-Clapeyron equation*
Albedo	model	Common factor analysis
Aleutian low	*Bermuda-Azores high*	*Component*
Analog forecasting	*Biosphere*	Continental polar air mass
Annular mode	Butterfly effect	Continuity equation
Anthropogenic	*California current*	*Convection*

Coriolis effect
Coriolis parameter
Coupled atmosphere–ocean
  general circulation model
Cryosphere
Deforestation
Density
Desertification
Dewpoint temperature
Diabatic
Dimensionless quantity
Downscaling
Dynamic instability
Eccentricity
Eigenvalue
Eigenvector analysis
El Niño
Energy balance model
Equation of state
Equation of state for moist air
Evaporation
Finite differencing
First law of thermodynamics
Friction
General circulation model (GCM)
Geopotential height
Gridpoint model
Hydrosphere
Hydrostatic equation
Icelandic low
Ideal gas law
Initialization
Insolation
Internal energy
Iterative
Joule
Kelvin temperature scale

Kinetic energy
La Niña
Latent heat
Latitude
Lithosphere
Loading
Local-scale
Longitude
Madden-Julian Oscillation
Mass conservation equation
Meridional flow
Mesoscale
Milankovitch cycle
Mixing ratio
Moisture conservation equation
Momentum
National Center for Atmospheric
  Research
Navier-Stokes equations of motion
Newton's second law
North Atlantic Oscillation (NAO)
Oceanic general circulation model
Pacific Decadal Oscillation (PDO)
Pacific-North American (PNA)
  pattern
Parameterization
Pascal
Planetary scale
Polar cell
Polar easterlies
Polar front jet stream
Pressure
Pressure gradient force
Primitive equations of motion
Principal components analysis
Quasi-biennial oscillation
Regional climate model (RCM)

Resolution
Ridge
Rossby wave
Sahel
Score
Sensible heat
Slab model
Southern Oscillation
Southern Oscillation index
Spatial downscaling
Spatial resolution
Specific heat at constant
  pressure
Specific humidity
Specific volume
Spectral model
Surface water balance
Swamp model
Synoptic climatology
Synoptic scale
Teleconnection
Temporal downscaling
Temporal resolution
Thermodynamic energy equation
Time step
Trade winds
Tropical cyclone
Tropical-Northern Hemisphere
  (TNH) teleconnection
Trough
Velocity convergence
Velocity divergence
Virtual temperature (T_v)
Walker circulation
Zonal flow
Terms in italics have appeared in at
  least one previous chapter.

▶ Review Questions

1. What are the advantages and disadvantages of increasing the spatial and temporal resolution in a GCM?
2. What is a "coupled" GCM?
3. Why is downscaling sometimes necessary when evaluating GCM output?
4. What is the main advantage of a spectral model over a gridpoint model?
5. What are the primitive equations of motion and what is their purpose?
6. What two kinds of energy are represented as components of the thermodynamic energy equation?
7. What is the difference between diabatic and adiabatic heating?
8. What is the purpose of the continuity equation?
9. What is virtual temperature?
10. Are initial conditions more important in weather forecasting models or in GCMs? Why?
11. What is the primary advantage of using eigenvector techniques to identify climatic patterns?

12. Why is principal components analysis generally preferred over common factor analysis in climatological studies?
13. Compare and contrast the Pacific Decadal Oscillation and the Southern Oscillation.
14. What impact does the Pacific-North American teleconnection have on North American climate?

▶ Questions for Thought

1. Verify that the units cancel to the appropriate units in each of the seven basic equations that appear in all GCMs.
2. Why are numerical models of atmospheric motion sometimes inaccurate even though the equations representing the atmospheric dynamics and thermodynamics involved are known?
3. Find and use some form of climatic data (such as temperature) for a town along with teleconnection indices posted online to examine the influence of one of the teleconnections described in this chapter on the local climate.

---

http://physicalscience.jbpub.com/climatology

Connect to this book's Web site: http://physical science.jbpub.com/climatology. The site provides chapter outlines, further readings, and other tools to help you study for your class. You can also follow useful links for additional information on the topics covered in this chapter.

Relationships Between Climate and Other Endeavors

Chapter 14—Applied Climatology, Climate Impacts, and Climatic Data
Climate Impacts
Climatological Data Sources
Summary

Chapter 15—Future of Climatology
Relationship Between Climatology and Meteorology
Interdisciplinary Work With Other Scientists
Interaction Between Climate Scientists and Nonscientific Professionals
Improved Atmospheric Data Availability and Display
Recognition of the Possibility for Rapid Climate Change
Summary

14 | Applied Climatology, Climate Impacts, and Climatic Data

Chapter at a Glance

Climate Impacts
   Impacts on Natural Systems
   Impacts on Societal Systems
   Impacts on Human Health and Comfort
Climatological Data Sources
   Data Collection Agencies
   Primary Versus Secondary Data
   Secondary Data Sources for Applied Synoptic
     and Dynamic Climatological Studies
   Secondary Sources for Applied Studies of
     the Climate System
   Secondary Sources Well Suited to Studies
     in Paleoclimatology and Climate Change
   Secondary Sources Well Suited to Studies
     in Physical Climatology
Summary
Review

In each of the previous chapters, climatology has been discussed from a "process" approach. Emphasis was placed on the factors that cause climates to be the way they are, from the *microscale* to the *planetary scale*. The first seven chapters concentrated on the processes themselves. Chapters 8 through 10 explained the spatial distributions of climates resulting from those processes, and Chapters 11 and 12 emphasized temporal variability and changes in the climates resulting from those factors. Chapter 13 described analytical and quantitative methods of identifying patterns resulting from those processes.

This chapter takes a completely different perspective. It examines the *effects* of climate rather than the *causes*. For example, imagine that a 5-cm (2-in) rainstorm occurred over a 3-hour period. A hydroclimatologist or a synoptic, dynamic, or tropical climatologist may be concerned with the atmospheric conditions that caused the event to occur. However, it is also perfectly legitimate to wonder not why the event occurred, but why it is important and what its impact is. How often do events of such magnitude occur at this location within a 3-hour period at this time of year? What impact did the event have on the local corn crop?

Applied climatology is the study of the effects of climate on natural and social systems. These effects range from the direct to the indirect. Both numerical and statistical models can be used to estimate these effects. As we saw in the previous chapter, output from climate models can undergo *downscaling* to provide more localized estimates of the modeled conditions. The model output can then be used to plan and prepare accordingly for impacts in various sectors of society.

A different approach to the assessment of impacts is to examine the relationship between "actual" past atmospheric conditions and observed features of the natural or social systems that climate is believed to impact. According to the *empirical approach*, climatic data are used to uncover historical trends. The scope of applied climatology, then, includes the impacts and applications of atmospheric data (either modeled or measured) to address some type of practical research question that falls outside of the atmospheric sciences.

The applied climatologist may also use these data to estimate atmospheric conditions at some specific time of interest in the past. For example, a legal consultant may need to know whether the Sun was positioned in the line of sight of a traffic light at the time of a traffic accident. An entomologist may want to determine whether atmospheric conditions may have played a role in the migration of a certain insect pest during a period of time. An engineer may be interested in the frequency that 5 cm (2 in) of precipitation falls within a 3-hour period at a location, because a culvert may need to be designed to withstand frequent flooding. An architect may need to know the weight of snow that can be expected to accumulate on a roof, so that it may be designed accordingly. A technician working with a scientific instrument may need to know the atmospheric pressure at a certain location so that a sensitive piece of equipment can be calibrated accurately. A farmer may want to understand the climatological *surface water balance* in a location before deciding whether to plant a certain crop that may need to be irrigated. These are just a few of the many types of questions that applied climatologists address.

In some cases data sources are available to answer the question at hand, but a more likely scenario is that no formal data exist for the place and time necessary. In such cases the applied climatologist must act as a detective by trying to determine the most likely conditions at the time of interest. Sometimes this detective work may involve the interpolation of data at nearby (or sometimes not-so-nearby) places, when the location of interest does not have a weather station. At other times a nearby station may exist, but it may not measure the variable that is needed, which means that the applied climatologist may need to make an educated guess based on the data that are available. Finally, in some cases the variable of interest is measured at a nearby site, but the data may be missing on a day or on a series of days. In that case the applied climatologist may estimate the values on the days with missing observations. In short, the applied climatologist must be able to supply the most appropriate data, or at least provide reliable estimates when "actual" data do not exist. Applied climatologists are often called to give expert testimony in court, both to provide their best guess of the atmospheric conditions at the time of interest and to comment on the reliability of the data upon which their opinions are based.

This chapter begins by presenting specific examples of the impacts of climate on various components of the natural environment and sectors of the economy. It then examines some sources of climatological data for use in applied climato-

logical problems and discusses some of the applications and limitations of certain types of data.

■ Climate Impacts

Few entities affect the array of natural and social systems to the extent that climate does. **Primary climate impacts** are natural or social systems affected most directly by climate, and **secondary climate impacts** are those affected only indirectly. In reality, some effects fall somewhere between primary and secondary, but for simplicity we consider all impacts to be either "primary" or "secondary." Both types of impacts occur in natural and human systems.

Impacts on Natural Systems

Processes in the atmosphere are inherently tied to those in the *cryosphere, hydrosphere, biosphere*, and *lithosphere*. These systems influence the climate system, which is also impacted by processes occurring in all the other "spheres" through many interrelated feedback loops. Despite continual interaction between the "spheres" climate may be appropriately considered to be the most overriding influence on the natural environment.

Perhaps the most direct impact of climate is felt in the cryosphere. Snow- and ice-covered surfaces reflect as much as 90 percent of the *shortwave radiation* incident upon them. Relatively modest changes in polar temperatures can, therefore, melt a small amount of ice or freeze a small amount of water. As was discussed previously, these changes alter *albedo* slightly. The changes, however, tend to magnify over time, because small increases in albedo make further cooling more likely, as more snow and ice exist. Slight increases in temperature likewise melt a small amount of the cryosphere, which decrease polar albedo and allow further warming to take place as the snow and ice are replaced by darker surfaces. These changes represent a *positive feedback system* because relatively small impacts magnify over time.

In the hydrosphere, temperature and precipitation directly affect the amount, seasonality, and distribution of water available at the surface—the surface water balance. Changes to the global climate system inherently affect water on Earth's surface because, as pointed out in Chapter 6, *evapotranspiration* is the only variable that appears in both the energy balance and the water balance equations. Changes in energy input at the surface, therefore, inherently change the water balance. These primary impacts of climate cause direct impacts in human systems, such as agriculture, fisheries, recreation, and environmental planning.

The hydrosphere is also affected by increases or decreases in salinity resulting from changes to the input of fresh water at the surface. These salinity changes can then affect the biosphere directly by impacting fisheries, coastal vegetation, and coastal and marine ecosystem productivity. In addition, changes to the atmosphere impact the deep-water circulation in the oceans by altering the conveyor belt that is part of the *thermohaline circulation* described in Chapter 4.

Within the biosphere, climatological impacts are felt in numerous other ways. The effects of predator–prey relationships, disease and insect infestations, and ecosystem stability are inherently tied to climatic conditions. Magnitude and timing of precipitation affect all these ecological features, winds affect pollen distribution and dispersal, and solar radiation relates to rates of *photosynthesis* and ecosystem productivity.

Climatological impacts are felt directly in the lithosphere. The field of climatic geomorphology is emerging as an important and interesting area of research inquiry. Most notably the field investigates how temperature and moisture affect rates of weathering, soil formation, and subsurface freeze–thaw. Precipitation events, glaciers, and winds also shape landforms, and coastal landforms are shaped by both acute events such as individual storms and the chronic process of everyday weather phenomena. And at both short and long time scales, volcanic activity causes feedbacks on climate. These regions are likely topics of climatic geomorphology research.

Impacts on Societal Systems

Climate affects some societal systems in a primary manner and others to a secondary extent. The most direct impact may be on agriculture. The intensity, magnitude, duration, and timing of precipitation; the length of time between the last spring frost and the first autumn frost (the **growing season**); the severity of storms; and temperature all impact agriculture directly. These factors affect agricultural commodities and futures, and many commodities brokerage firms provide advice and assess weather-related risk for the agricultural sector.

Climate also affects many forms of recreation. The ski industry, for example, relies on subfreezing temperatures and snowfall. In many cases perception of climate is as important as the climate itself. For example, during active periods of hurricane activity, tourist activity may decrease, both from the meteorological event itself and from the climatological perception. Eight hurricanes hit Florida between August 2004 and October 2005 (Charley, Frances, Ivan, Jeanne, Dennis, Katrina, Tammy, and Wilma), and there are strong signs that tourism was affected so detrimentally that the surging economy in that state was interrupted. How many more hurricanes will strike before the collective mindset of vacationers begins to remove Florida from their late summer and early autumn travel plans?

Transportation is another societal sector directly affected by climate. Extreme events play an even greater role in creating impacts to the transportation industry than to other sectors. A snowier-than-usual winter will obliterate the budget of municipalities for snow removal before the end of the snow season, and impacts will thereafter be even greater. In other cases forecasts of heavy snow that never materialize can also be very costly, because municipal workers that are held for overtime have no work to do. This is an especially difficult problem in heavily populated areas that experience occasional strong winter snowstorms. In the northeastern United States, for example, *nor'easters* that track up the Atlantic coast may dump 45 cm (18 in) of snow on Boston, New York, or Philadelphia. But a slightly more eastward track may provide no snow at all for these cities. In this case an incorrect forecast can be very costly. The American architect/inventor/engineer Buckminster Fuller believed that the construction of a large geodesic dome over midtown Manhattan would pay for itself after only 10 years from the savings in snow removal alone.

Climate also affects energy supply and demand directly. Power companies must often buy additional power from other companies during unseasonably cold winters and hot summers in their service areas. In general, the farther in advance such purchases are made, the lower the cost of the power. Consulting climatologists are employed to provide long-lead climatic forecasts to minimize additional expense to their companies. Climatologists are also consulted to determine feasibility of solar and wind power. Some recent evidence suggests that coastal winds may be sufficiently strong and persistent to provide for a reasonable share of the nation's energy needs in the near future, especially if the turbines are very high. Another area of potential wind power development is the northern Great Plains of the United States. This region represents one of the best locations on the planet for harvesting wind power and has been called the "Saudi Arabia of wind power" by many environmentalists.

The insurance and risk management industry is also directly tied to weather and climate. Two major factors contribute to increasing risks to life and property caused by weather and climate. First, more people are living in areas vulnerable to atmospheric (and other) natural hazards now than ever before. Development along hurricane-prone coastlines and construction in floodplains represent two examples of this increasing hazard. It is well known that the bulk of hurricane storm surge damage occurs within 100 m (330 ft) of the high tide mark, yet development within that zone has exploded in recent decades.

Second, a major concern is that global climatic change will cause more severe weather events and more chronic climatic problems (and, therefore, higher insurance premiums), particularly until people can adapt better to the changing environmental conditions. A 2005 report by Allianz Group—an international financial services provider—indicated that climatic change is increasing the potential for property damage at a rate of 2 to 4 percent per year and that insurance premiums must increase to meet this increasing threat. Similarly, risk management trading (typically structured as a "climate contract" or "derivative" and operating on the principle of an insurance policy) has mushroomed into a major industry with the recognition that weather/climate may affect as much as 70 percent of all business activity. These transactions provide assistance when weather conditions exceed certain thresholds that cause hardship for the contracted business.

Secondary societal impacts are felt in numerous sectors. For example, food prices are a result of both primary atmospheric influences on agriculture and other nonatmospheric factors. Likewise, economic and political conditions can be affected indirectly by climate, particularly in smaller political units that might be overwhelmed by extreme weather and climate events. In his 2001 book, *The Little Ice Age*, Brian Fagan argues that the shortage of grain and bread in the 1780s was one factor that contributed to the timing of the French Revolution. Even sectors such as fashion, art, and literature have been linked indirectly to atmospheric features.

Impacts on Human Health and Comfort

Atmospheric conditions exert a primary impact on many types of bacteria, viruses, and vector-borne diseases. Extremes of heat and cold cause numerous deaths each year. Recent research by climatologists at Arizona State University suggests that databases vary widely in reporting the total number of heat- and cold-related deaths in the United States, depending on, among other factors, the method of determining whether or not a death was considered to be a result of atmospheric conditions. Regardless, it appears that heat waves may cause more massive weather mortality events than cold waves. On the other hand, cold waves are more likely to cause a larger number of isolated deaths—people stranded alone in a deadly situation—and on the whole, these deaths seem to make cold weather more deadly than hot weather, even if increased severity of diseases from cold weather are not included. It is feared that long-term climatic change will increasingly threaten human health. For example, global warming may already be causing the mosquito-borne dengue fever and other diseases to spread to locations beyond their traditional tropical domains.

■ Climatological Data Sources

Data Collection Agencies

Compared with most other regions of the world, the United States is well served by historical records of climatological data. This has been attributed to the need to keep accurate records so that settlers would know what to expect from a relatively unfamiliar environment. Weather records in the 1700s and 1800s were kept by the more educated professionals such as surgeons, statesmen, and engineers. The **Federal Weather Service** was established by Congress as a component of the U.S. Army in 1870, then moved in 1891 to the U.S. Department of Agriculture and renamed the **United States Weather Bureau.** In 1940, in response to the increasing needs of weather data in the transportation and newly evolving aviation industry, the Weather Bureau was moved to the Department of Commerce. The demands on the Weather Bureau became so heavy that in 1970, coincident with the 100th anniversary of organized reporting, forecasting, and dissemination of atmospheric data to the public, the U.S. Weather Bureau was renamed the *National Weather Service.* The National Weather Service was established as a branch of the *National Oceanic and Atmospheric Administration* (NOAA), a new agency that oversaw other organizations as well, whose missions ranged from fisheries management to environmental data collection to oversight of the newly launched weather satellite program (**Figure 14.1**). NOAA grew quickly, as its budget tripled in its first 10 years of existence. From the beginning its mission has been to protect life and property and to recognize the integrative aspects of the Earth–ocean–atmosphere system.

The data collection arm of what is now NOAA is the **National Environmental Satellite, Data, and Information Service (NESDIS),** headquartered in Silver Spring, Maryland. Within NESDIS are 11 offices that collect geophysical data of all types. These include the **National Climatic Data Center**

(NCDC), which is headquartered in Asheville, North Carolina. NCDC was formed in 1951 (even predating NOAA) when several offices were merged. Today, NCDC archives weather data obtained by National Weather Service, Military Services, Federal Aviation Administration, U.S. Coast Guard, and voluntary observers.

With the increasing collection, availability, and demand for atmospheric data over the years, many additions and improvements to the NCDC program have been made. For instance, nearly every U.S. state has a state climatologist who is responsible for the archival and dissemination of atmospheric data to the public and private sectors for climatological applications. In some states the **State Climatology Program** is funded more heavily than in others, so the scope of the program varies widely from state to state. In addition, six **Regional Climate Centers** (**Figure 14.2**) distributed around the United States specialize in the archiving of atmospheric data and monitor atmospheric events in their respective regions. These were established in the 1980s, as a result of the passage in 1978 of the **National Climate Program Act,** which greatly increased the scope and mission of atmospheric data collection and archiving in the United States. The demand for data from NCDC and other sources has increased tremendously in recent years, coinciding with the growth of the applied climatology industry as well as the growth of sources of data for the United States and world from agencies outside of NOAA.

Recent climatological studies have benefited from the widespread availability of data. Most data

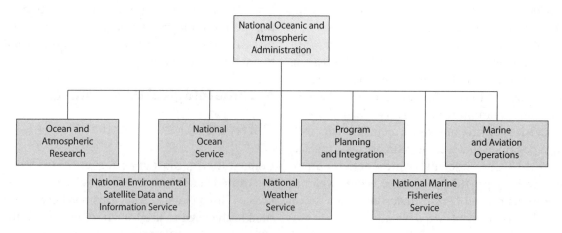

Figure 14.1   A simplified organizational chart for the National Oceanic and Atmospheric Administration. Courtesy of NOAA/Workforce Management Office.

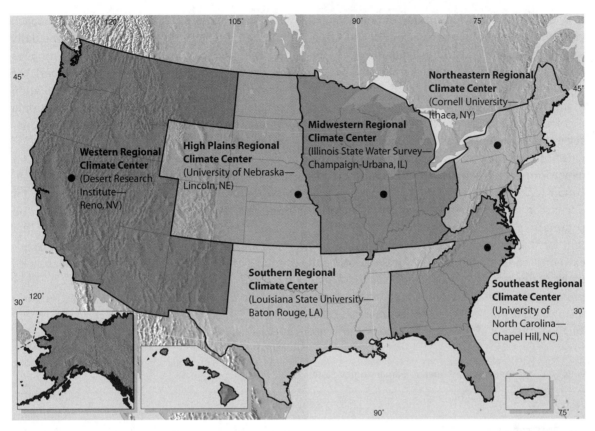

Figure 14.2 Regional climate centers and their areas of coverage. *Adapted from*: NOAA/NESDIS/National Climatic Data Center/Regional Climate Centers.

are now accessible via the Internet, with many data sets provided free of charge or for only a nominal fee. This ease of availability has increased the number of independent studies on climate, leading to a heightened understanding of climatic processes.

Primary Versus Secondary Data

Primary data are defined as data collected by an investigator using thermometers, *radiometers*, moisture sensors, or other equipment in the field. **Secondary data** are data used by researchers after being collected, quality controlled, and compiled elsewhere, usually in digital format. Because of the nature of this research, those working at *local scales* and microscales, such as boundary layer climatologists, are more likely than synoptic and dynamic climatologists to use primary data. Likewise, synopticians and dynamicists are more likely to use secondary data sources. Primary data can be collected using methods such as those de-

scribed in Chapter 5. Some important secondary sources are discussed below. No attempt is made here to describe all major sources of secondary data—there are many!

Secondary Data Sources for Applied Synoptic and Dynamic Climatological Studies

In recent years the availability of high-quality climatological data sets for research has blossomed. Currently, the **reanalysis data set** produced by the *National Center for Atmospheric Research* in cooperation with the **National Centers for Environmental Prediction** is among the most widely used resources of global climatic data for dynamic and synoptic studies. The "reanalysis" involves the compilation and assimilation of observed data from a wide variety of sources, such as surface meteorological stations, buoys, satellites, aircraft, and ships, into a single data set for each variable analyzed. This collection of data sets, interpolated to a grid, provides a multitude of atmospheric variables at several vertical levels in the atmosphere. Much of the data are available on the Internet at four observation times per day, back to 1948. Other data are aggregated into daily or monthly totals.

In the United States NCDC/NOAA produces several publications on an ongoing basis that provide atmospheric data on the Internet and also in paper form. A particularly useful monthly publication is *Climatological Data (CD)*, which is produced for each state and contains a monthly summary of statewide areally weighted precipitation for that month, along with both means and extremes of temperature and the locations and dates during which the extremes occurred. It also provides these data and more for all official monitoring stations around the state of interest (**Table 14.1**). At the end of the year *Climatological Data:*

Annual Summary (CDAS) is also published for each state. Although *CDAS* does not replicate the daily data that were already published in each month's *CD* for that year, it does report monthly totals for each month in that year and annual compilation statistics (**Table 14.2**). In general, if monthly data are required, *CDAS* provides a "shortcut" in analyzing station data.

For more detailed information, NCDC/NOAA provides other resources. For example, *Local Climatological Data (LCD)* is published monthly for more than 270 sites around the United States and Pacific Islands. All large cities and many smaller cities are represented in this publication. An advantage of *LCD* is that it provides some data with temporal resolution as high as every 3 hours rather than just the daily data included in *CD*. Furthermore, *LCD* provides additional variables. **Table 14.3** shows the data that are available in that

Table 14.1 Information Contained in Climatological Data

Section	Data Included
Statewide summary	Monthly precipitation total compared against other years
	Temperature and precipitation extremes for the month
Monthly station and divisional temperature	Average daily maxima and minima
	Monthly mean and departure from normal
	Extreme high and low and date of occurrence
	Heating and cooling degree days
	Number of days with maximum $\geq 90°F$ or $\leq 32°F$
	Number of days with minimum $\leq 32°$ or $\leq 0°F$
Monthly station and divisional precipitation	Monthly total and departure from normal
	Greatest day and date of occurrence
	Snowfall total
	Maximum snowfall depth on ground and date of occurrence
	Number of days with snowfall of $\geq 0.10, 0.50,$ or 1.00 in.
Daily station and divisional information	Precipitation
	Temperature
	Soil temperature
	Pan evaporation
	Wind (accumulated daily totals, measured in miles on an "odometer")
Station index	Identifying index number
	Climate division
	County
	Latitude/longitude
	Elevation
	Time of observation
	Observer name or company/agency name
	Reference notes
	Map of state with stations labeled

Table 14.2 Information Contained in *Climatological Data: Annual Summary*

Section	Data Included
Statewide summary	Line graph of statewide annual precipitation plotted against that of other years in the historical record
Station and divisional precipitation	Monthly total and departure from normal
Average station and divisional temperature	Monthly average and departure from normal
	Date of last spring minimum at or below 16°F, 20°F, 24°F, 28°F, and 32°F and temperature
	Date of first autumn minimum at or below 32°F, 28°F, 24°F, 20°F, and 16°F, and temperature
	Number of consecutive days between dates with temperatures at or below 16°F, 20°F, 24°F, 28°F, and 32°F (if less than 365)
Other	Monthly and seasonal cooling degree days*
	Monthly average soil temperatures and extremes
	Monthly total pan evaporation (in inches) and wind movement (in miles)

*Monthly and seasonal heating degree days are contained in the July issue of *Climatological Data*.

Table 14.3 Information Contained in *Local Climatological Data*

Section	Data Included
Daily totals	Temperatures: maximum, minimum, average, departure from normal
	Average daily dewpoint temperature
	Average daily wet bulb temperature
	Heating and cooling degree days
	Type of weather observed; rain, drizzle, thunderstorms, haze, fog, dust storms, tornadoes, blowing snow, etc.
	Snow/ice depth and water equivalent (at a fixed time each day)
	Average station pressure, and adjusted sea-level pressure
	Mean wind speed, resultant direction
	Maximum 5-second and 2-minute wind speed and direction
	Monthly averages/totals
	Visibility in miles
	Sunrise and sunset times
Hourly precipitation	Precipitation totals ending each hour
	Maximum 5-, 10-, 15-, 20-, 30-, 45-, 60-, 80-, 100-, 120-, 150-, and 180-minute precipitation total and date of occurrence
Observations at 3-hourly intervals	Sky cover (clear, scattered, broken, overcast)
	Cloud ceiling height
	Visibility
	Type of weather observed; rain, drizzle, thunderstorms, haze, fog, dust storms, tornadoes, blowing snow, etc.
	Temperature, dewpoint temperature, wet bulb temperature
	Relative humidity
	Wind speed and direction
	Station pressure and adjusted sea-level pressure

resource. The disadvantage of *LCD* is that it is available for far fewer sites than *CD* provides. In general, if data are needed at a particular site, *LCD* should be used if available for that site or for a nearby city, if data are required for variables that are not reported by *CD*, and/or if higher temporal resolution is necessary. On the other hand, if no *LCD* exists near the site of interest or if high spatial resolution of data that are provided by *CD* is required, then *CD* is used.

For occasions when 3-hourly data do not provide sufficient temporal resolution to solve the problem at hand, surface data collected on an hourly basis can be used. Such hourly data are available from NCDC or the regional climate centers for more than 500 stations around the United States and 12,000 stations worldwide via the **surface airways observations**. Many, but certainly not all, of the surface airways sites in the United States are also sites published in *LCD*. These data usually include the variables shown in **Table 14.4**.

Precipitation may vary tremendously in small spatial and temporal periods and is a factor in many applied climatological problems. NOAA's

Hourly Precipitation Data (HPD) is another publication designed to provide precipitation data at temporal and spatial resolution as fine as possible. Like *CD*, *HPD* is also published monthly by state. A typical state may have approximately one *HPD* site for every two counties, on average. The specific precipitation data types available in *HPD* are described in **Table 14.5**. Among the most widely used information in *HPD* is that which relates to the maximum intensity of precipitation to fall within specific time durations. To document such heavy precipitation events and severe weather, *Storm Data* is a monthly publication that provides a chronological summary, by state, of severe and unusual weather that occurred during that month, including data on the storm trajectories, damage, and casualties resulting from each event.

A final important secondary data source for the United States is NOAA's *Daily Synoptic Series*, a booklet of seven daily surface and 500-mb weather maps. Applied climatologists can often use these to determine the most likely time of passage of a frontal system or to provide other clues that might identify the most likely time that a temperature or precipitation event occurred, when using daily data.

NCDC offers a variety of other products that are not described in detail here. A few examples include global climatological normals, engineering weather data, archived hourly radar images of the United States, rainfall frequency atlas of the United States for various durations and magnitudes of precipitation events, and satellite imagery of significant weather events in the past.

The **World Meteorological Organization (WMO)** of the United Nations also dispenses a variety of climatological data products for use in large-scale synoptic and dynamic climatological studies, including some of the same data sets produced by NOAA/NESDIS/NCDC. Other data sets in the WMO archives are contributed by organizations around the world, such as the U.K.'s Hadley

Table 14.4 Data Available Through Surface Airways Observations

Wind speed and direction
Ceiling height
Horizontal visibility
Sky cover
Altimeter setting
Station pressure
Sea level pressure
Air temperature
Wet bulb temperature
Dewpoint temperature
Relative humidity
Solar radiation
Present weather conditions (snow, thunderstorm, haze, etc.)

Table 14.5 Information Contained in *Hourly Precipitation Data*

Section	Data Included
Daily precipitation totals by station	Precipitation by station, including the type of gauge used
Hourly precipitation totals by station	Precipitation totals at each hour, ending on the hour
Monthly precipitation maxima by station	Maximum for measurement periods of 15, 30, and 45 minutes and 1, 2, 3, 6, 12, and 24 hours, with the date and time of occurrence

Centre and the German and Japanese Weather Services.

Considerations Regarding Instrumentation Used
One important consideration when using *HPD* is the type of rain gauge used at the station of interest. Although the **standard nonrecording rain gauge** (see Figure 6.10) is used at most of the 13,000 precipitation gauges in the United States (including all of the sites reported in *CD*), hourly stations must have more sophisticated equipment for measuring rain automatically. In most of the United States two types of gauges are most prevalent for hourly recordings.

The **Fischer-Porter rain gauge** (**Figure 14.3**) is a 1.52-m (5-ft) tall tube with a 0.6-m (2-ft) diameter. The Fischer-Porter gauge reports precipitation collected in a bucket based on the water's weight. Every 15 minutes a special tape is punched mechanically to report a running total of the precipitation (in tenths of an inch) since the bucket was emptied. Network connections allow instant access to the data.

An important consideration of the Fischer-Porter gauge is that it is only precise to 2.5 mm (0.10 in) of precipitation. In other words, 2.5 mm (0.10 in)

Figure 14.3 Fischer-Porter rain gauge.

of rainfall must occur before any rainfall is reported. This can be problematic for some applications. For instance, many rainfall-related traffic accidents occur in the first few minutes of precipitation because the slight amount of water causes roads to become slick from dissolving oils previously deposited on the roadway. After several minutes of moderate precipitation, the oils are washed from the road, making them somewhat safer. If precipitation data are needed to determine whether rain may have played a role in an automobile accident, the time of precipitation onset may not be readily apparent from the Fischer-Porter readings, because a significant amount of time may be required for as much as 2.5 mm (0.10 in) to be accumulated. Some light rain showers may go totally unreported. Likewise, if an amount of less than 2.5 mm (0.10 in) accumulates in the gauge, a precipitation total of 2.5 mm (0.10 in) will be reported relatively quickly after the beginning of the next rain shower (which may occur days later) because some precipitation will have accumulated in the gauge from the previous shower. For these reasons readings from Fischer-Porter gauges should be interpreted with caution, particularly in certain applications.

The other common type of automated precipitation recorder is the **tipping bucket rain gauge**, an instrument consisting of two small buckets that alternate in the collection of precipitation (**Figure 14.4**). When .25 mm (0.01 in) of precipitation occurs, the active bucket dumps its water while the other bucket quickly moves into its place to begin collecting the precipitation. The number of "tips" is monitored electronically to provide a continuous measure of precipitation. The tipping bucket gauge is advantageous in that it measures with precision of 0.25 mm (0.01 in), but a major disadvantage is that some precipitation is not "captured" in the brief moment when the buckets are alternating. This *undercatch* can be significant during heavy rainstorms. Problems related to undercatch in standard nonrecording rain gauges were described in Chapter 6.

In addition to precipitation, temperature data are also routinely collected and reported at some 5000 "official" stations around the United States and published in secondary data sources such as *CD* and *LCD*. Because data at most stations are collected by volunteers and are not automated (i.e., computerized) stations, the thermometers and/or

Figure 14.4 A tipping-bucket rain gauge.

precipitation gauges are only checked once daily. For this, a special type of instrument, called a **maximum/minimum thermometer,** is used. Although there are several varieties of maximum/minimum thermometers, the basic principle is that the instrument leaves a mark at the highest temperature that had occurred since the last resetting of the instrument. Likewise, the base of the mercury allows for the lowest temperature since the last resetting to be reported. At a designated time each day, the observer notes the maximum and minimum temperatures and then resets the instrument, usually by twirling it. The maximum/minimum thermometers are necessary because volunteers cannot be asked to check their thermometers at multiple times of the day and night; with this instrument at least the high and low temperatures for the previous 24 hours can be noted.

Time of Observation   When using data collected daily by maximum/minimum thermometers or by standard nonrecording rain gauges, the time of ob-

servation is a critical consideration and is reported in *CD*. To illustrate how the time of observation can influence the data set, imagine a station that reports at 3:00 PM. Because 3:00 PM is near the typical time of maximum temperature for the day, it is possible that on a hot day the temperature at 2:59 PM was 37°C (99°F) and was the high temperature for the day. When the observer notes the maximum temperature at 3:00 PM, she reports the temperature that occurred at 2:59 PM, and then she resets the instrument. At 3:01 PM the temperature is still likely to be 37°C (99°F), but this temperature is not counted until the following calendar day. So even though only 1 day with a 37°C temperature may have occurred, *CD* may report it as two separate days of high temperatures of 37°C. Thus, in cases of heat waves, afternoon-reporting stations are likely to "double-count" the maximum temperatures.

The opposite situation can occur for morning-reporting stations. For example, if a station's time of observation is 7:00 AM, a cold wave that occurred on a single morning could be reported in the daily data as having 2 "days" with low temperatures hovering at the temperature of the single cold day.

This situation is known as **time of observation bias.** Many stations in the United States have changed from afternoon-reporting to morning-reporting stations over the past 40 to 50 years, because more of the general population is employed during the afternoon hours and has less time at home to monitor weather conditions. The bias toward lower temperatures has, therefore, increased over time. The fact that temperatures have risen substantially in the last 30 years, despite this general trend toward more morning-reporting stations, provides further evidence that the observed warming is significant.

Time of observation is also critical when analyzing daily precipitation data. It is important to remember that a "daily" precipitation total really refers to the 24-hour total terminating at the time of observation on the calendar day that the precipitation is reported. Precipitation reported on a certain day at a morning-reporting station is very likely to have fallen on the previous calendar day, particularly in places and times of the year when afternoon *convective precipitation* supplies a large proportion of the precipitation. No adjustment of temperature or daily precipitation totals is made before publication to account for the "most likely" time of the precipitation or the high or low temperatures—it is up to the applied climatologist to determine the most likely time/day that a temperature or precipitation event occurred. This could be achieved by comparing the data of interest with data measured at a nearby site that reports at intervals of 1 or 3 hours, such as *HPD* or *LCD*.

Other Variables Although most variables reported in the various secondary data sources are self-explanatory, a few need additional explanation. One such set is the "degree-day" variables—a concept that can be used to extract additional meaning from the daily temperature data for specific applications. For example, in energy applications the **heating degree-day (HDD)** can be used to index the amount of energy needed to heat or cool a building. Heating is usually assumed to be required when the daily mean temperature of a location is 18°C (65°F) or below. Each degree of daily mean temperature below this base is counted as one HDD. For example, if the maximum temperature on a given day was 22°C (70°F) and the minimum temperature was 10°C (50°F), 2 Celsius or 5 Fahrenheit HDDs would have accumulated on that day because the average temperature was 16°C (60°F). Monthly and seasonal cumulative totals of HDDs for the heating season (usually considered to be July 1 through June 30, in the northern hemisphere) are published in *CD* and *LCD*. Energy planners devise mathematical relationships between HDDs and energy consumption, and long-lead seasonal climatic outlooks are used to estimate energy demands for the upcoming season.

Likewise, the **cooling degree-day (CDD)** operates on the same principle as the HDD, but it is used to identify the energy demand for cooling a building whenever the daily mean temperature exceeds the base of 18°C (65°F). So for instance, if the maximum and minimum temperatures were 32°C (90°F) and 20°C (68°F), respectively, on a given day, then 8 Celsius or 14 Fahrenheit CDDs were accumulated. Because cooling is generally not necessary in the winter months, in the northern hemisphere the CDD season is generally considered to run from January 1 through December 31. Factors other than temperature may determine energy consumption, including cultural behavior and meteorological factors. For example, many consumers in the southern United States tend to use their air conditioners on some days even when daily temperatures fall below 18°C (65°F), because humidity and/or lack of wind might lead to discomfort if they opened their windows. HDDs and CDDs are published in *CD*, *CDAS*, and *LCD*.

For agricultural applications of temperature data, a similar concept is the **growing degree-day (GDD)**, a variable that is accumulated for each degree that the daily mean temperature exceeds some base value. The base value is determined experimentally, varies by crop, and is the theoretical minimum daily temperature required to provide enough energy for that crop to grow. For example, a typical base value for alfalfa is 5°C (41°F). On a given day, if the high temperature is 18°C (64°F) and the minimum temperature is 4°C (39°F), then 6 Celsius GDDs are accumulated. If the daily mean temperature falls below the threshold, then 0 GDDs are accumulated. It is impossible to have "negative" GDDs, just as it is impossible to accumulate negative HDDs or CDDs. Because each crop theoretically requires a certain number of GDDs before harvest, a running total of GDDs for a growing season tells a farmer whether the crop is on schedule to be harvested on time. Of course, the GDD concept assumes that other conditions are met adequately, such as precipitation totals and timing, protection from disease, and nutrient availability.

A related agricultural application is the **crop heat unit (CHU)**, which actually calculates two values: One base temperature is for the daily minimum temperature and the other is for the daily maximum temperature. As for HDDs, CDDs, and GDDs, if the daily minimum and/or maximum temperature is below the base, zero CHUs are accumulated for that calculation, and for each degree above the base, 1 CHU is reported for that calculation. The daily maximum temperature also has an "optimal" temperature, which is the temperature at which maximum growth occurs for a given crop. If the daily maximum temperature exceeds this optimal temperature, the CHU for the daily maximum begins to fall. The total CHU for the day is simply the mean of the CHU calculation for the minimum and maximum temperatures on that day. Although the CHU is not published in the sources described above, it is used more commonly in Canada.

Wind data are often tricky to decipher in secondary data sources. In *LCD* wind is reported in tens of degrees using 360° compass headings from north. Thus, a wind direction reported as "9" represents winds coming from 90°—an easterly wind (i.e., wind blowing from east to west). Likewise, "18" represents a southerly wind, "27" is a westerly wind, and "30" would be a wind from the westnorthwest. In this notation, "0" is usually reserved for a calm wind (which, by definition, has no direction) and "36" is used for a northerly wind. In some applications wind speed and direction are transformed into west–east and south–north vector components—u and v, as described in Chapter 13.

Some secondary sources report wind according to the **Beaufort wind scale**, a simple scale that measures wind in units of 0 to 12 "Beauforts," which convey information about the observed effect of the wind's force rather than the wind speed itself. In this system, "0" represents calm winds and "12" corresponds to hurricane-force winds, as shown in **Table 14.6**. Beauforts are typically used in situations where wind measurement with an

Table 14.6 Beaufort Wind Scale, for Use at Sea and on Land

Class	Speed (mph)	WMO Classification	Appearance on Sea Surface	Appearance on Land Surface
0	0–1	Calm	Smooth and mirror-like	Calm, smoke rises vertically
1	1–3	Light air	Scaly ripples, no foam crests	Smoke drift indicates wind direction, still wind vanes
2	4–7	Light breeze	Small wavelets, crests glassy, no breaking	Wind felt on face, leaves rustle, vanes begin to move
3	8–12	Gentle breeze	Large wavelets, crests begin to break, scattered whitecaps	Leaves and small twigs constantly moving, light flags extended
4	13–18	Moderate breeze	Small waves 1–4 ft becoming longer, numerous whitecaps	Dust, leaves, and loose paper lifted, small tree branches move
5	19–24	Fresh breeze	Moderate waves 4–8 ft taking longer form, many whitecaps, some spray	Small trees in leaf begin to sway
6	25–31	Strong breeze	Larger waves 8–13 ft, whitecaps common, more spray	Larger tree branches moving, whistling in wires
7	32–38	Near gale	Sea heaps up, waves 13–20 ft, white foam streaks off breakers	Whole trees moving, resistance felt walking against wind
8	39–46	Gale	Moderately high (13–20 ft) waves of greater length, edges of crests begin to break into spindrift, foam blown in streaks	Whole trees in motion, resistance felt walking against wind
9	47–54	Strong gale	High waves (20 ft), sea begins to roll, dense streaks of foam, spray may reduce visibility	Slight structural damage occurs, slate blows off roofs
10	55–63	Storm	Very high waves (20–30 ft) with overhanging crests, sea white with densely blown foam, heavy rolling, lowered visibility	Seldom experienced on land, trees broken or uprooted, "considerable structural damage"
11	64–72	Violent storm	Exceptionally high (30–45 ft) waves, foam patches cover sea, visibility more reduced	
12	73+	Hurricane	Air filled with foam, waves over 45 ft, sea completely white with driving spray, visibility greatly reduced	

Data from: National Weather Storm Prediction Center, NOAA [http://www.spc.noaa.gov/faq/tornado/beaufort.html]. Accessed April, 2010.

anemometer is not feasible, such as on the ocean, in military combat zones, and in remote terrain.

Secondary Sources for Applied Studies of the Climate System

Because of the local nature of the processes, secondary data sources for analyzing vegetation–climate relationships are more difficult to find. Usually, the researcher must use instrumentation to gather primary data but some secondary data sources are available. For example, the **Long-term Ecological Research (LTER) Program**, which began in 1980, provides bioclimatological data (along with other environmental information such as hydrological measurements, soil composition and quality, land-use change, and even demographic data) for 26 sites across North America, the Caribbean, and Antarctica (**Figure 14.5**). These data reveal information about surface–atmosphere interactions of climatological importance, including broad-scale trace gas composition, which would be difficult to monitor without such a sophisticated network of sensors. Similarly, the **AmeriFlux** program began in 1996 as part of a wider network

of stations including **EuroFlux**. These stations provide data on energy, matter, and *momentum* exchanges in the biosphere, with a particular emphasis placed on carbon *fluxes*. The AmeriFlux and EuroFlux stations are shown in **Figures 14.6** and **14.7**, respectively.

Other agencies collect, maintain, and archive cryospheric and hydrologic data of interest to climatologists. For example, a major clearinghouse for cryospheric data, including snow cover, glaciers, continental ice sheets, avalanches, permafrost, and sea ice, is located in Boulder, Colorado, at the **National Snow and Ice Data Center**. This agency was formed in 1982 as a means of consolidating operations from several preexisting locations. The **National Data Buoy Center**, located at Stennis Space Center in Mississippi, is responsible for dispatching and maintaining a network of buoys and coastal stations around North America

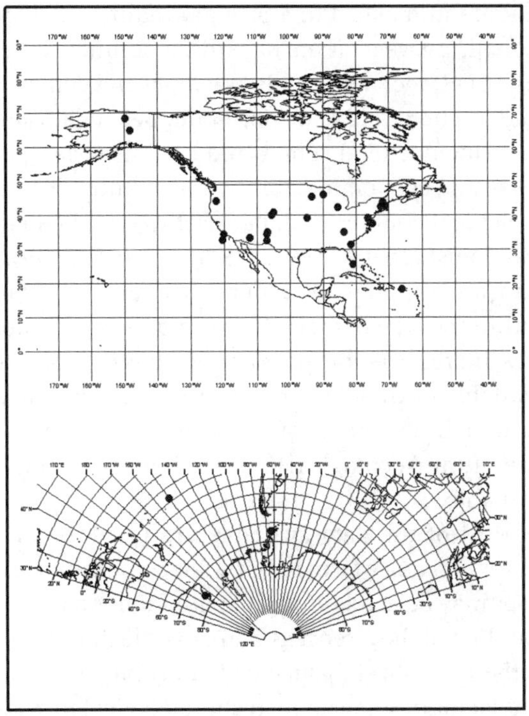

Figure 14.6 The Ameriflux research stations. Courtesy of Ameriflux, Oak Ridge National Laboratories, U.S. Department of Energy.

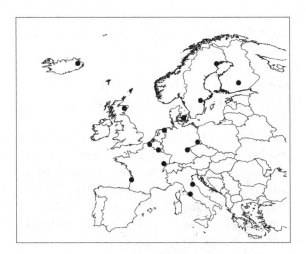

Figure 14.7 The Euroflux research stations. Courtesy of Robert V. Rohli.

Figure 14.5 The locations of stations in the Long-term Ecological Research Program. Courtesy of Robert V. Rohli.

and in some other parts of the world that collect atmospheric data. The **Global Energy and Water Cycle Experiment (GEWEX)**, under the auspices of the **World Climate Research Programme**, provides data on fluxes of energy, water, and momentum between the surface and atmosphere, both on land and in the oceans. The goal is to support improvements in the forecasting of global and regional climatic change.

Secondary Sources Well Suited to Studies in Paleoclimatology and Climate Change

An often-used source of data well suited to global change detection as well as atmospheric composition, *fossil fuel* consumption, and the carbon cycle, is provided by the **Carbon Dioxide Information Analysis Center (CDIAC)** at Oak Ridge National Laboratory in Tennessee. CDIAC also produces the global **Historical Climatology Network**—a set of mean monthly temperatures, precipitation, and sea level pressure for thousands of stations worldwide for time periods of up to 50 years. The U.S. component of this data set consists of daily and monthly records. A major advantage of the HCN over *CD* and other sources is that much effort has been invested in ensuring that only high-quality, accurate data are included and that station moves are minimized.

The **World Data Center for Paleoclimatology**, headquartered in Boulder, Colorado, is a branch of NCDC that acts as a repository for paleoclimatological data sets at thousands of locations worldwide. These data include both historical and proxy sources described in Chapter 11. Researchers who are willing to share their findings with the public contribute the data.

Secondary Sources Well Suited to Studies in Physical Climatology

Although the LTER, GEWEX, and CDIAC projects provide free global data on atmospheric gas concentrations, other data sets are widely used for other types of studies in physical climatology. For instance, the **Global Hydrology Center** of the **National Aeronautics and Space Administration** provides a wide array of satellite-based observations for climatological research, including such variables as lightning strikes, aerosol loadings,

and data for studies of cloud formation and physics. NOAA's **Climate Monitoring and Diagnostics Laboratory** maintains data related to stratospheric ozone concentrations and other atmospheric phenomena. The U.S. Department of Energy's **Renewable Resource Data Center** provides *insolation* and wind data, for a series of sites around the United States.

■ Summary

Applied climatology investigates the impacts of climate on other phenomena. Although computer models can be used to estimate these impacts, empirical (historical data-driven) approaches are also important. Applied climatologists must be knowledgeable about both the available data sources and any shortcomings in answering the research question at hand. They must also be adept at making educated guesses about the condition of the atmosphere in cases when formal, reliable atmospheric data are unavailable.

Primary impacts of climate include those systems that climate affects directly, whereas secondary impacts are felt only indirectly. Both types of impacts occur in both natural and social systems and also affect human health. Some sectors that are influenced in a primary manner include agriculture, vegetation, hazards, insurance, water resources, recreation, energy supply and demand, construction, and transportation. Secondary impacts include food supply and cost, fisheries, political and economic stability, and health.

Several sources of data assist the applied climatologist. In the United States most of these data are supplied by federal agencies that fall within the Department of Commerce, and a rapidly increasing proportion of these data are available online. Other data are being supplied by non–Department of Commerce agencies, as satellite and ecological data become more widely available.

Most of the data available are provided in a self-explanatory format, but a few important variables may need additional explanation. An understanding of the strengths and possible limitations of the type of rain gauge used may be an important factor in some applied climatological investigations. Likewise, the time of observation may be important in reconstructing events using data from daily-reporting stations. Finally, it is important to note that wind data are often supplied in vector format, with the direction listed in tens of degrees using a 360° compass heading. For energy and agricultural applications, the degree-day and crop growth unit concepts are also important. Several data sources for studies on the relationships between the atmosphere and the other "spheres," along with paleoclimates and physical climatology, are also available.

▶ Key Terms

Albedo
AmeriFlux
Applied climatology
Beaufort wind scale
Biosphere
Carbon Dioxide Information Analysis Center (CDIAC)
Climate Monitoring and Diagnostics Laboratory
Climatological Data (CD)
Climatological Data: Annual Summary (CDAS)
Convective precipitation
Cooling degree-day (CDD)
Crop heat unit (CHU)
Cryosphere
Daily Synoptic Series
Downscaling
Empirical approach
EuroFlux
Evapotranspiration
Federal Weather Service
Fischer-Porter rain gauge
Flux
Fossil fuel
Global Energy and Water Cycle Experiment (GEWEX)
Global Hydrology Center
Growing degree-day (GDD)
Growing season
Heating degree-day (HDD)

Historical Climatology Network
Hourly Precipitation Data (HPD)
Hydrosphere
Insolation
Lithosphere
Local Climatological Data (LCD)
Local scale
Long-term Ecological Research (LTER) Program
Maximum/minimum thermometer
Microscale
Momentum
National Aeronautics and Space Administration
National Center for Atmospheric Research
National Centers for Environmental Prediction
National Climate Program Act
National Climatic Data Center (NCDC)
National Data Buoy Center
National Environmental Satellite, Data, and Information Service (NESDIS)
National Oceanic and Atmospheric Administration (NOAA)
National Snow and Ice Data Center
National Weather Service
Nor'easter
Photosynthesis

Planetary scale
Positive feedback system
Primary climate impact
Primary data
Radiometer
Reanalysis data set
Regional Climate Center
Renewable Resource Data Center
Secondary climate impact
Secondary data
Shortwave radiation
Standard nonrecording rain gauge
State Climatology Program
Storm Data
Surface airways observation
Surface water balance
Thermohaline circulation
Time of observation bias
Tipping bucket rain gauge
Undercatch
United States Weather Bureau
World Climate Research Programme
World Data Center for Paleoclimatology
World Meteorological Organization (WMO)
Terms in italics have appeared in at least one previous chapter.

▶ Review Questions

1. How does the empirical approach differ from the modeling approach in solving atmospheric problems?

2. The text compares applied climatologists to detectives. In what way do you believe this comparison is valid?

3. Name five social or economic sectors of activity that are affected by climate and determine whether these impacts are primary or secondary.

4. How does the total of heat-related deaths compare with that of cold-related deaths in the United States?

5. What is the main advantage of the National Center for Atmospheric Research reanalysis data set?

6. What advantages do *HPD* provide that are not provided by *LCD*?

7. How does a Fischer-Porter rain gauge work? What caveats should be heeded when using Fischer-Porter data? Why?

8. Why is the time of observation a critical variable in applied climatological analysis?

9. What is a cooling degree-day and why is it an important variable?

10. What is the difference between a crop heat unit and a growing degree-day?

11. What are the advantages and disadvantages of using the Beaufort wind scale?

12. A traffic accident occurred on the corner of Final and Exam Streets at a particular day and time. What sources would you consult to identify the weather conditions at that place and time?

13. Name three data sets that you might recommend to a friend who is doing research on long-term changes in atmospheric carbon dioxide.

▶ Questions for Thought

1. A homeowner calls you, a climatologist at the Greater Good Environmental Consulting Company, because he believes there may be a crack in the foundation of his built-in swimming pool. He indicates that it seems to him the water is disappearing from the pool faster than it should be, given that it is April. Using what you learned in this chapter and in Chapter 6, how would you go about determining whether he is justified in his accusation that the construction work was faulty?

2. Consider the following statement: "Applied climatology represents the best hope for climatologists to provide a contribution to knowledge to others outside the field." To what extent do you agree with this?

3. How would you go about finding out the last day when hail fell in your hometown?

---

http://physicalscience.jbpub.com/climatology

Connect to this book's Web site: http://physical science.jbpub.com/climatology. The site provides chapter outlines, further readings, and other tools to help you study for your class. You can also follow useful links for additional information on the topics covered in this chapter.

15

Future of Climatology

Chapter at a Glance

Relationship Between Climatology and
  Meteorology
Interdisciplinary Work With Other Scientists
Interaction Between Climate Scientists and
  Nonscientific Professionals
Improved Atmospheric Data Availability and
  Display
Recognition of the Possibility for Rapid Climate
  Change
Summary
Review

Our understanding of atmospheric processes, models to simulate those processes, and data that show the results of those processes, as described in the first 14 chapters, has propelled climatology to a position of respect among the natural sciences. But what trends and challenges will characterize climatology in the coming years? What lines of inquiry will be pursued? Which ideas about Earth's climate system will be debated? This chapter offers some perspectives on possible answers to these questions.

■ Relationship Between Climatology and Meteorology

In the years ahead we can expect that the distinction between the traditionally defined domains of climatology and meteorology will continue to blur. Well-trained climatologists will increasingly be expected to have an understanding of the physical principles usually taught in meteorology courses. These perspectives include a firm grounding in dynamic, synoptic, and physical meteorology. Likewise, well-trained meteorologists will benefit from the broad and integrative perspective traditionally offered by climatology courses. The result is a cross-fertilization of ideas that strengthens both of the atmospheric sciences.

■ Interdisciplinary Work With Other Scientists

Climatologists will increasingly find themselves as parts of interdisciplinary teams whose goal is to solve complex climate-related problems. Already, paleoclimatological research has reached this interdisciplinary sphere. As we saw in Chapter 11, biological, geological, geochemical, and historical evidence is often used in tandem to decipher climates of the past, and the interpretation of such evidence requires expertise in a range of disciplines. In addition, the increased emphasis on interactions between the atmosphere and various other physical "spheres," such as Earth–ocean–atmosphere interactions and feedbacks, will necessitate increasing interdisciplinary work between climatologists and oceanographers, vulcanologists, geophysicists, biological scientists, and geochemists. In short, climatology will move closer to the "hub" of scientific environmental analysis and take on more of an all-encompassing (holistic) scientific perspective. This is logical because climate is the most fundamental limiting factor in the environment. To a large degree, this "coming of age" means that climatologists will "see the forest through the trees" while the other natural scientists focus more specifically on "seeing the trees" in examining the relationship between climate and their specialization.

Attention to these interactions between the various spheres of the Earth–ocean–atmosphere system will be focused increasingly on interactions between processes at various scales. This concept has been described as a "cascade of linkages" by climatologist Werner Terjung over 30 years ago. One manifestation of this cascade is the trend toward understanding the influence of global atmospheric *teleconnections* on local, near-surface atmospheric features (see Chapter 5), the *surface water balance* (see Chapter 6), and circulations that impact the climate system (see Chapter 7) to produce distinctive local and regional climates (see Chapters 8 through 10). Others are the generation of regional climate models (see Chapter 13) to simulate impacts of the broad-scale atmospheric circulation and the use of climatic data sources (see Chapter 14) to quantify this impact in a local area. In the future, climate models will have increasingly finer resolution, and replacement of **downscaling** techniques at a given scale with model output at that scale is inevitable.

Interaction Between Climate Scientists and Nonscientific Professionals

As the human influence on global climate becomes even more indisputable, climate scientists not only will collaborate with other natural scientists but also will interact increasingly with policymakers, social scientists, and the mass media. Policymakers will come under increased pressure to make decisions based on sound science. Versatile specialists who have background in both climate science and climate policy will be in increasing demand in both the political and environmental consulting arenas and at local, regional, national, and international levels. At the local and regional levels, climate consultants will increasingly work with urban and regional planners and representatives of local industry to ensure compliance with federal clean air standards.

In the United States and many other parts of the world, emission standards are becoming stricter over time as technology allows for improved energy efficiency and alternative sources of energy as a means of curtailing *global warming* and enhancing the local quality of life. An improved understanding of the climate system allows for increasing the benefits of industry while reducing the cost of that industry in terms of environmental impact. Likewise, improved understanding of the local climate system will aid in design, engi-

neering, and agricultural applications. These **greening initiatives** are currently being set in place to mitigate future environmental problems. In the national and international spheres, efforts to understand climatic change through modeling and the *empirical approach* will guide policy designed to prevent, mitigate, and adapt to those changes.

Future national and international debate will focus on devising fair and efficient means of ensuring a safe global atmospheric environment while minimizing economic costs. Professional climatologists, such as those who participate in the *Intergovernmental Panel on Climate Change* sanctioned by the United Nations and the *World Meteorological Organization*, will play a central role in these discussions. The strengthening link between climate science and policy will also have repercussions in climatology because of the increased interactions between climatologists and social scientists. The realization that human behavior is likely to affect and be affected by climate will draw such attention. Climatic implications on energy consumption, migration patterns, recreation, disaster planning and management, and political and agricultural systems are only a few of the areas that are likely to receive increasing attention.

As a result of increased awareness of the importance of climate, climatologists are appearing in the local and national media more and more over time. For example, seasonal hurricane forecasts by William Gray, a professor at Colorado State University, generate much public interest. Other documentaries and movies about climatologists have been made in recent years.

Improved Atmospheric Data Availability and Display

The increasing availability of high-quality atmospheric data for general use will continue to revolutionize the study of climate. Already, atmospheric data are more widely available free of charge than at any previous time. These newly available data sets will facilitate the study of climate by the private sector and by researchers outside of the major research institutions around the world. Furthermore, as future satellite-derived data sets reach lengths of a few decades, new climatic cycles and features will be discovered. In the coming years the influence of sophisticated techniques for analysis and display in climatology will become even stronger. Powerful statistical and other quantitative techniques will continue to characterize the study of climate. Included among these is the increasing implementation of the techniques of **artificial neural networks**—a method of modeling processes computationally in a manner analogous to the processing of information by the human brain, with simultaneous calculations and an ability to "learn" from previously discovered relationships between variables. These statistical methods are especially well-suited to climatology because they can be used to model feedbacks and nonlinear processes—two important and complex features of climate that we have seen throughout this book.

Visualization—a tool of increasing importance for displaying data in climatology—has traditionally referred to the use of maps and other graphics, but in recent years the scope of visualization has broadened with advances in mapping and display technology. For instance, three-dimensional maps showing temperature or pressure surfaces are now quite common. Also, animations that show hurricane structure, pollution advection, and other phenomena across space and time simultaneously have revolutionized the way that we can convey information in climatology. Because climate inherently varies across both space and time, the trend toward the increasing use of visualization techniques is likely to continue.

Other tools that will be used with rapidly increasing frequency in climatology involve *geographic information systems* (GIS)—the use of computer hardware and software to assemble, store, update, analyze, and display spatial information and to link nonspatial information to the spatial features. Although some applications may use GIS only for mapping (visualization), GIS can also be used for many more powerful analytical purposes. For example, GIS can measure areas of enclosed polygons, distances between points, or optimal routes on a road map. If you have ever used Mapquest to find a location, you have already used a GIS. GIS can also be used to provide accurate and high-resolution surface and topographic information, two critical controls in climatic processes. As data become more widely available and digital file sizes become more manageable by improvements in technology, an increased need for GIS to analyze that data will be apparent. GIS is already used to link disparate entities important to climatology, such as combining maps of cities

with elevation information to determine flood potential given a certain quantity of precipitation. Analytic features of GIS will be integrated more effectively into climatological analysis, particularly in combination with image processing software that will store data collected from satellites in a GIS. Many scientists view GIS as an important analytical link between a vast array of cultural elements and the scientific disciplines of climatology, remote sensing, geography, biology, and geology.

Recognition of the Possibility for Rapid Climate Change

One recently accepted phenomenon uncovered with new data and modeling techniques is that of rapid climate change. There is already an increasing recognition that "rapid" climate change is not only possible but perhaps even likely, as positive feedbacks within the climate system can accelerate climatic changes, whether or not they are *anthropogenic* in origin. Recent evidence suggests that the abrupt warming of the present *interglacial phase*, which began about 10,000 years ago, took place in a much shorter time period than was previously believed. Some suggest that the warming may have occurred in only a few decades. Changes in the oceanic *thermohaline circulation* are believed to trigger a *positive feedback system* that could plunge Earth into significantly colder conditions. Perhaps such changes could end the present interglacial phase as quickly as that phase arrived. Regardless of the type of climatic change that actually occurs, we do know that climate will change, and our understanding of the processes involved in the changes will only benefit society.

■ Climatology as Part of the Ultimate Goal of Sustainability

The notion that human practices can, and do, alter the atmosphere has filtered to the general public over the recent past. Increased general awareness of anthropogenically altered natural environments has led to a rethinking of many human practices, especially those that produce significant amounts of waste. Climatologists must be at the forefront of efforts toward achieving a more productive and healthy environment.

Public pressures have called for more environmental awareness both in the domestic and business sectors. Domestic practices such as recycling programs have been instituted in most communities, and increasing pressure has been placed on those communities that do not currently support such programs. Other practices have also been instituted to reduce waste in the domestic domain. Some, such as buying local goods over those transported from long distances, seek to reduce transportation costs and associated waste and pollution. Further practices, such as reducing the use of chemical fertilizers, seek to reduce human impacts on the local environment, thereby better preserving natural environments. Overall, a greater sensitivity toward human impacts has permeated through the domestic area.

However, domestic waste is responsible for only about 30 percent of the total human waste produced in the United States. The remainder comes from the industrial and agricultural sectors. Many businesses originally believed that shifting to "green" practices would be overly costly, and, therefore, many sought to discredit such practices. Over time many have realized that implementing green initiatives can be profitable.

Such approaches move humans toward **sustainability**—development in such a way as to eliminate negative impacts, including both resource depletion and environmental degradation. Sustainable practices seek to restore balance to natural ecosystems while still making use of natural resources for human use. The UN Brundtland Commission refers to sustainable development as "Development that meets the needs of the present without compromising the ability of future generations to meet their own needs."

At the broadest scale, Earth itself operates in a sustainable manner. Nutrients are continuously recycled through biological, chemical, physical, and geological processes over time. All healthy ecosystems exist in a sustainable manner. Humans are the only animals whose practices at broad spatial and temporal scales have stood in the way of a sustainable earth. It is currently realized that humans cannot continue to exist in a nonsustainable lifestyle.

Because business and industry produce so much human-related waste, it is important to involve those sectors in sustainable practices. The notion of sustainable industry practices led to the **Triple Bottom Line (TBL) concept**. Before the notion of sustainability gained acceptance, the bottom line simply meant economic gain. If a company made money, the mission was complete. The TBL concept underscores the importance of not only economic performance (the first bottom line) but also social (second bottom line) and environmental (third bottom line) performance. TBL practices show greater respect for all living things (including humans) as well as the wise management of resources.

Many companies have used the TBL approach to maximize economic performance. For example, a nonsustainable practice is payment for food waste to be carried to landfills by a waste removal provider. A more sustainable, TBL-based approach involves selling the food waste to a composting company, which can then use it in the production of an organic fertilizer, or to sell the food waste directly to farmers, who can either compost it into fertilizer or use it directly as feedstock. The sustainable practice turns an expense into a money-making venture while simultaneously reducing the amount of material moving to landfills.

As an additional benefit, companies can enjoy increasing public-perceived good will through advertisement of their "green" practices, because culture continues to place increasing value on green initiatives. As public pressures mount, such practices are becoming required of many industries. Lists are now generated which rank major companies and industries according to their levels of "greenness."

The government sector can play a leadership role in determining the direction of sustainability, not solely through legislation and education but also through direct practice. For example, universities may mandate that only recycled paper be purchased. Through this simple action, paper companies are forced to shift their practices to accommodate the demand for recycled products. Government agencies may also require vendors to meet some standard relative to the generation of waste. Companies vying for government contracts are then forced to eliminate waste or face large reductions in business. These practices streamline the production of goods and services that are in demand, making the involved industries more profitable.

Such incentives must go much further for the entirety of humans to become truly sustainable. For example, a government could require that a minimum of 20 percent of the energy needs of its vendors be provided through renewable sources. Such an incentive would spur the development of solar, wind, geothermal, and other potential energy industries. These energy sources would then become increasingly feasible for public consumption. The resulting decrease in reliance on fossil fuels for energy production would lead to less overall waste and environmental degradation and a trajectory toward sustainability.

The ultimate goal of the human condition should be to evolve to an entirely sustainable state of existence. Implementation of policy in support of sound climate science should support this goal. Only in a truly sustainable global environment will the ecosystems that further support the human condition be preserved. Technology has progressed to a point whereby true sustainability is possible. We are currently awaiting the proper implementation incentives. Sustainability is the only path to ensure the health of the only planetary body known to be capable of supporting life, and climatologists should play a significant role in achieving sustainability.

■ Summary

Climatology will continue to be a vibrant, interesting, and integrative field of inquiry in the foreseeable future. Its importance beyond simply understanding climate "for its own sake" is already being recognized and appreciated by meteorologists and other scientists as well as nonscientists. Improved data quality and availability, along with improvements in model resolution and increasing use of quantitative methods, will allow climatologists to play an expanding role in understanding the Earth–ocean–atmosphere system and its relationship to human activities. Most importantly, climatologists should contribute toward the goal of achieving sustainability—the only goal that would ensure that continued habitability of our planet into the distant future.

▶ Key Terms

Anthropogenic
Artificial neural network
Downscaling
Empirical approach
Geographic information systems
Global warming
Greening initiatives

Interglacial phase
Intergovernmental Panel on Climate Change
Positive feedback system
Surface water balance
Sustainability
Teleconnection

Thermohaline circulation
Triple Bottom Line (TBL) concept
Visualization
World Meteorological Organization
Terms in italics have appeared in at least one previous chapter.

▶ Review Questions

1. The text suggests that climatology will become increasingly interdisciplinary in the future. Is this an advantage or a disadvantage for the future of climatology? Why?
2. How does the "systems" approach to the study of climate affect the way that climatologists of the future will need to be trained?
3. What advantages do artificial neural networks offer for the study of climate?
4. What advantages do geographic information systems offer for climatologists?

▶ Questions for Thought

1. Imagine yourself applying for a job as a climatologist with the World Health Organization upon graduation. What kinds of skills would you like to be able to tell the interviewer you have acquired?

2. Think back to the types of proxy evidence for paleoclimates that were discussed in Chapter 11. Name as many different types of scientists as you can that would participate in each of those types of investigations.
3. What kinds of future changes might occur in higher education and in society in general that would make climatology a more important or less important topic of study?

http://physicalscience.jbpub.com/climatology

Connect to this book's Web site: http://physical science.jbpub.com/climatology. The site provides chapter outlines, further readings, and other tools to help you study for your class. You can also follow useful links for additional information on the topics covered in this chapter.

Glossary

Note: Numbers in parentheses indicate chapters in which the term was used.

30–60 Day Oscillation *See* Madden-Julian Oscillation.

40–50 Day Oscillation *See* Madden-Julian Oscillation.

Ablation The combined effect of melting and sublimation that causes a glacier to retreat. (4, 10, 11)

Absolute dating Any means of assessing the age (in number of years since formation) of rocks, sediments, or fossils by radiometric techniques. (11)

Absolute humidity The mass of water vapor per cubic meter of air. (5)

Absolute vorticity The spin of a body that results from the combined effect of relative vorticity and planetary vorticity. (7, 8)

Absolute zero The theoretical temperature at which all molecular motion ceases and no internal energy is present. (3, 5, 12, 13)

Absorption The process by which radiant energy is retained by matter and converted to internal energy of that matter. (5, 12)

Acacia A type of spiny tree that is characteristic of Tropical Savanna climates. (10)

Accumulated potential water loss In a decreasing availability water balance model, a running total of the P-PE value, used to decrease the modeled loss of water from soil with increasing cumulative aridity. (6)

Accumulation The expansion of a glacier due to the effect of precipitation, condensation, and deposition. (4)

Acid fog A secondary pollutant formed when a primary pollutant interacts with fog droplets to produce a chemical change. (12)

Acid precipitation A secondary pollutant formed when certain primary pollutants interact with water from precipitation, either before or after the precipitation reaches the surface. (12)

Action center A pair of locations where ridging and troughing or pressure anomalies of opposite sign frequently occur simultaneously. (7, 13)

Adaptation A strategy for dealing with climate change that refers to simply living with the threat and the consequences associated with the threat. (12)

Adiabat A plot on a thermodynamic diagram that shows the temperature profile of an adiabatically moving air parcel at a given place and time. (5)

Adiabatic motion A type of thermodynamic process that occurs without any gain or loss of energy from outside the system. (5, 9, 12, 13)

Advection A type of convection characterized by the horizontal transfer of matter and/or energy. (3, 4, 5, 6, 8, 9, 10, 12, 13)

Advection fog A type of fog that occurs when air flows across a colder surface and is chilled to its dewpoint temperature from below. (9, 10)

Aerosol A tiny suspended solid or liquid particle in the air, including soot, volcanic ash, soil particles, salt crystals, and fog and cloud droplets. (2, 3, 4, 5, 12)

Aghulas current A warm, surface ocean current of the subtropical Indian Ocean that moves southward along the southeastern coast of Africa, guided by the winds on the western side of the Indian Ocean anticyclone. (10)

Agricultural climatology The branch of bioclimatology concerned with the impact of atmospheric properties and processes on living things of economic value, including crops and livestock. (1)

Air mass A body of air that is relatively uniform in its characteristics for distances of hundreds to thousands of square kilometers. (8, 9, 10, 11)

Air Pollution Control Act The 1955 legislation that represented the first attempt at federal oversight of air quality in the United States. (12)

Albedo The ratio of the shortwave energy reflected from a mass divided by the total shortwave energy incident on that mass. (4, 5, 7, 9, 10, 11, 12, 13, 14)

Alberta clipper A midlatitude wave cyclone with cyclogenesis on the eastern side of the Canadian Rockies and a generally east-northeastward track. (9)

Aleutian low The semipermanent subpolar low pressure system over the north Pacific Ocean. (7, 9, 13)

AmeriFlux A network of stations, funded by the U.S. Department of Energy and other federal agencies that provides data on energy, matter, and momentum exchanges in the biosphere, with particular emphasis placed on carbon fluxes. (14)

Amplitude The degree of meridionality of a Rossby or other type of wave. (4, 7)

Anaerobic Any process that occurs in the absence of oxygen. (12)

Analemma A graph in the general shape of a figure eight that shows the equation of time adjustment as a function of solar declination. (5)

Analog forecasting Predicting atmospheric behavior based on the outcome of a similar synoptic situation in the past. (13)

Angiosperm A flowering plant. (11)

Annular mode *See* Arctic Oscillation.

Anomaly A departure from the expected atmospheric conditions at a place. (10, 11)

Antarctic Bottom Water (AABW) The deepest global ocean current, circumnavigating Antarctica in all the southern oceans and leading to the Atlantic Deep Water. (4)

Antarctic Circle The 66.5°S parallel of latitude; the most equatorward latitude that experiences 24 consecutive hours of daylight on the December solstice and 24 consecutive hours of night on the June solstice. (3)

Antarctic Deep Water (ADW) A moderately deep and salty part of the global ocean conveyor belt originating near Antarctica that flows northward and upward, eventually flowing beneath the North Atlantic Deep Water. (4)

Antarctic Intermediate Water (AAIW) Part of the global ocean conveyor belt that flows northward beneath South Atlantic and South Pacific waters and above the North Atlantic Deep Water until it converges with northern ocean currents. (4)

Anthropogenic Human induced. (4, 10, 12, 13, 15)

Anticyclone Any enclosed area of relatively high pressure, rotating generally clockwise in the northern hemisphere and counterclockwise in the southern hemisphere. (3, 7, 9, 10, 12)

Aphelion The day on which Earth is at its farthest orbital extent from the Sun, corresponding approximately to July 4. (3, 10, 11)

Applied climatology The subfield of climatology concerned with the effects and impacts of climate on various sectors, such as agriculture, forestry, architecture, insurance, engineering, health, and transportation. (1, 14)

Archimedes' principle The notion that the buoyancy of any mass immersed in a fluid is equal to the weight of the fluid that the mass displaces. (11)

Arctic (A) air mass A body of air that is relatively uniform in its very cold and very dry characteristics for hundreds or thousands of square kilometers, emanating from high-latitude source regions that are either inland or over sea ice. (8, 9)

Arctic Circle The 66.5°N parallel of latitude; the most equatorward latitude that experiences 24 consecutive hours of daylight on the June solstice and 24 consecutive hours of night on the December solstice. (3, 5, 9)

Arctic Oscillation (AO) A teleconnection in which pressure anomalies see-saw between the polar region and around 45°N, resulting in variability of the circumpolar vortex. (13)

Artificial neural network A method of modeling processes in a manner analogous to the processing of information by the human brain, with simultaneous calculations and an ability to "learn" from previously discovered relationships between variables. (15)

Asian monsoon The seasonal reversal of continental-scale circulation between offshore and onshore in Asia, caused by the winter dominance of the Siberian high and the summer dominance of a thermally driven low-pressure system over southern Asia. (9, 10, 11)

Atlantic Multidecadal Oscillation A cycling of warm and cold surface waters over time scales ranging from about 15 to 30 years. (13)

Atmosphere The thin layer of gases suspended above the surface of a celestial body. (1, 2)

Atmospheric general circulation model A complicated computer-based model that simulates and/or predicts global atmospheric circulation based on dynamic conditions but that does not take processes from the hydrosphere, lithosphere, biosphere, or cryosphere into account. (13)

Atmospheric window The range of the electromagnetic spectrum from about 8 to 13 micrometers, in which the atmospheric absorption of terrestrial radiation occurs only when water vapor is abundant. (12)

Atomic mass The total mass of an atom, essentially equal (in atomic mass units) to the sum of protons and neutrons contained in that atom. (11)

Attenuation The depletion of the solar rays as they pass through the atmosphere through the combined effect of absorption and scattering by atmospheric particles. (3, 5)

Autumnal equinox The day of the year midway between the summer and winter solstices, when the solar declination is at the equator, neither hemisphere is tilted toward or away from the Sun, and day length is 12 hours over the entire Earth; currently corresponds roughly to September 22 (the September equinox) for a northern hemisphere observer and March 21 (the March equinox) for a southern hemisphere observer. (3)

Axial tilt The angle between the vertical and Earth's axis; corresponds approximately to 23.5° currently in geological history. (3)

Azimuth angle A coordinate indicating the number of degrees that the Sun's position deviates laterally from true north relative to the observer, ranging from 0° to 360°, with 180° representing a position in the southern sky. (5)

Azores high *See* Bermuda-Azores high.

Backing A counterclockwise change in wind direction with increasing height. (3)

Baroclinic zone Any region (at the surface or aloft) where warm air is laterally adjacent to much colder air. (7)

Baroclinicity The intermixing of radically different air masses. (7, 9)

Beaufort wind scale A simple scale from 0 to 12, based on the observed effect of wind's force rather than the wind speed itself. (14)

Benguela current A cold, surface ocean current of the South Atlantic Ocean that moves northward along the southwestern coast of Africa, guided by the winds on the eastern side of the South Atlantic high. (10)

Bergeron Climatic Classification System A genetic climatic categorization of climate at a place, determined by the frequency with which it is dominated by certain types of weather. (8)

Bermuda-Azores high The semipermanent subtropical high-pressure system over the North Atlantic Ocean. (7, 9, 10, 13)

Bernoulli effect A phenomenon whereby an increase in wind speed above a friction-generating object causes a decrease in air pressure on one side of that object; allows lift of airplanes and undercatch in rain gauges. (6)

Berson westerlies Lower-stratospheric winds that blow from west to east as one phase of the quasi-biennial oscillation. (10)

Bioclimatology The subfield of climatology concerned with the relationship between atmospheric properties and processes and the well-being of humans and other life forms. (1)

Biome A natural vegetation community. (8)

Biosphere The component of the Earth–ocean–atmosphere system containing life, including parts of the atmosphere, hydrosphere, and lithosphere. (1, 2, 5, 11, 12, 13, 14)

Biotemperature In the Holdridge Life Zones Climatic Classification System, an adjusted temperature that considers temperatures below freezing or above 30°C to provide no contribution to the proliferation of life (i.e., to have a biotemperature of zero), because plants will be dormant whether the temperature is freezing, far below freezing, or above 30°C. (8)

Blaney-Criddle model A simple method of estimating monthly evapotranspiration as a function of a crop-specific coefficient, monthly mean temperature, and monthly percentage of daytime hours of the year. (6)

Blocking anticyclone The positioning of a surface anticyclone or ridge, or an upper-level ridge over the same location for a period of many days, resulting in an extended period of synoptic-scale subsidence and clear, dry conditions. (9)

Bolide A luminous meteor. (11)

Bora wind Winds forming in the Balkan highlands and spilling toward lower elevations through gaps in the Dinaric Alps toward the Adriatic Sea. (9)

Boreal forest The extensive, slow-growing coniferous forest of subarctic climates of North America. (9)

Bottleneck A significant drop in population that triggers rapid genetic divergence in surviving individuals. (4)

Boundary-layer climatology The subfield of climatology concerned with atmospheric properties and exchanges of energy, matter, and momentum in the near-surface atmosphere. (1)

Brown smog *See* Photochemical smog.

Budyko Climatic Classification System A method of categorizing climates based on the ratio of energy received to energy required to evaporate local precipitation. (8)

Budyko ratio In the Budyko Climatic Classification System, the ratio of net radiation received to the energy required to evaporate a unit volume of local precipitation; a ratio of 1.0 separates dry from humid climates. (8)

Buoyancy The tendency of a fluid to float on top of a denser fluid. (5)

Butterfly effect The magnification of errors over time as described by chaos theory caused by erroneous or unknown initial conditions. (13)

California current A cold ocean current of the eastern North Pacific Ocean that moves southward near the west coast of North America, guided by the winds on the eastern side of the Pacific high. (4, 9, 12, 13)

Canary current A cold, surface ocean current of the North Atlantic Ocean that moves southward along the northwestern coast of Africa, guided by the winds of the eastern side of the Bermuda-Azores high. (4, 9, 10)

Carbon cycle The continuous movement of carbon through the Earth–ocean–atmosphere system. (2, 13, 14)

Carbon Dioxide Information Analysis Center (CDIAC) An agency within the U.S. Department of Energy that produces the often-used global Historical Climatology Network and disseminates data related to atmospheric composition, fossil fuel consumption, and the carbon cycle. (14)

Carbon monoxide (CO) A colorless, odorless gas monitored by the U.S. Environmental Protection Agency as a criteria pollutant. (12)

Carbonic acid (H_2CO_3) A weak acid formed by the bonding of carbon dioxide and water. (12)

Catastrophism The notion that cataclysmic events reshape Earth's environment to a greater extent than gradual, persistent forces. (11)

Cenozoic Era The present era of geological time, which began with the K-T extinction approximately 65 million years ago and contains the Tertiary and Quaternary Periods. (11)

Centrifugal force (CF) An apparent force directed radially outward from the center of curvature in a rotating system, caused by the inertia of the rotating mass. (3, 5)

Chaos theory An idea suggesting an inherent lack of predictability in some systems despite the fact that an order underlies them, because small occurrences that may be difficult or impossible to model can create large impacts to apparently unrelated phenomena. (13)

Charles' Law The property of an ideal gas that temperature and density are inversely related if pressure is held constant. (2)

Chimu floods Ancient periodic flooding in the Moche Valley of Peru, believed to be linked to extreme El Niño events. (4)

Chinook wind A warm, dry wind caused as air descends the leeward sides of the western mountain ranges in North America (except southern California). (9)

Chlorofluorocarbon (CFC) A family of compounds consisting of carbon, hydrogen, oxygen, and fluorine that are used for various domestic and industrial purposes; CFCs bond with monatomic oxygen in the atmosphere, thereby preventing the formation of ozone, and also act as minor greenhouse gases. (2, 10)

Circle of illumination The imaginary line, as viewed from space, that separates the illuminated and dark halves of Earth at any given time. (3)

Circumpolar vortex An area of strong winds aloft encircling the hemisphere and surrounding the high surface atmospheric pressures with low geopotential heights aloft that form over the poles. (7, 9, 10, 13)

Class A Evaporation Pan An instrument used for measuring evaporation, consisting of a metal pan 4 ft in diameter that sits on a short wooden platform and is exposed to the elements. (5)

Clausius-Clapeyron equation The equation that expresses saturation vapor pressure as a function of temperature. (5, 6, 13)

Clean Air Act of 1963 The first major effective legislation in the United States for regulating air quality and monitoring both stationary pollution sources, such as power plants, and mobile pollution sources, such as automobiles. (12)

Clean Air Act of 1970 Sweeping legislation that created the Environmental Protection Agency and required it to identify criteria pollutants to be monitored, determine the National Air Quality Standards of those criteria pollutants, and set up state-level environmental protection agencies in each state. (12)

Climate The long-term, overall state of the atmosphere at a given place, both at the surface and aloft, including not only the mean conditions but also variability and seasonality of those conditions and factors that cause those conditions. (1)

Climate division A subset of a state of the United States into which point-based climatic data are aggregated for analysis and display. (6)

Climate Monitoring and Diagnostics Laboratory An agency within the U.S. National Oceanic and Atmospheric Administration that maintains data related to stratospheric ozone concentrations and other atmospheric phenomena. (14)

Climatological Data (CD) A monthly publication of the National Climatic Data Center for each state that contains a monthly summary of statewide precipitation for that month, along with both means and extremes of temperature, as well as other daily data for all official monitoring stations in the state. (14)

Climatological Data Annual Summary (CDAS) A monthly publication from the National Oceanic and Atmospheric Administration for each state that contains monthly totals by site in that state for each month in that year, as well as annual compilation statistics. (14)

Climatology The scientific study of climate; the branch of the atmospheric sciences that analyzes the long-term state of the atmosphere, explains the causes of those conditions, and forecasts possible changes in those conditions at seasonal time scales and beyond. (1)

Cloud seeding The intentional introduction of substances into the atmosphere in an attempt to cause or enhance precipitation. (1)

Cluster analysis A quantitative technique in which the scores resulting from an eigenvector analysis are grouped together based on their position in n-dimensional space, where n is the number of components derived by the analysis. (8)

Coastal Kelvin wave Poleward-moving ocean waves along the western coast of the Americas, resulting from the splitting of the eastward-moving equatorial Kelvin wave at the coast. (4)

Cold air damming A condition whereby low-level cold air becomes trapped as a midlatitude surface anticyclone or cyclone forces surface air toward a mountain range, often along the eastern sides of both the Rocky and the Appalachian mountains when synoptic-scale systems track eastward. (9)

Cold air drainage Any situation in which cold, dense air is pulled to lower elevations by gravity. (9)

Cold front A narrow zone separating two air mass types in which the colder air mass is actively displacing the warmer air mass. (8, 9)

Cold Steppe (BSk) climate A semiarid climate in which the mean annual temperature is below 18°C (64°F). (9, 10)

Colorado wave cyclone A midlatitude wave cyclone with cyclogenesis on the eastern side of the Rockies in the vicinity of Colorado and tracking generally northeastward. (9)

Common factor analysis A type of eigenvector analysis that assumes some underlying uniqueness among the variables that contributes to the eigenvector. (13)

Common Water (CW) The deepest large-scale component of the global ocean circulation in the Pacific Ocean, originating near Antarctica as a branch of the Antarctic Bottom Water. (4)

Component A mode of variability in a data set revealed in principal components analysis or another form of eigenvector analysis, analogous to a regression line in bivariate regression analysis. (8, 13)

Condensation The phase change of water molecules from a gaseous to a liquid state. (2, 5, 6, 12)

Condensation nuclei Solid aerosols that encourage the conversion of water vapor to liquid water around them. (12)

Conditional instability An atmospheric condition in which rising motion is encouraged if the atmosphere is saturated but discouraged if the atmosphere is unsaturated. (5)

Conduction The transfer of energy by one molecule to another molecule touching it, and then to the next molecule, etc. (5, 12)

Confluence The horizontal convergence of air streams moving in the same direction at a given altitude. (7)

Conservation of angular momentum The property of curved motion that states that the product of mass, velocity, and radius of curvature is constant. (7)

Constant absolute vorticity trajectory The characteristic movement of air in a manner such that the total amount of vorticity is conserved, resulting in the wavelike shape of Rossby wave flow across the midlatitudes. (7)

Constant gas A component of the atmosphere that maintains its concentration across space and time. (2)

Content analysis An objective method of converting qualitative text into discrete categories for inferring climatic conditions from historical writings. (11)

Continental Polar (cP) air mass A body of air that is relatively uniform in its cold and dry characteristics for hundreds or thousands of square kilometers, emanating from inland, high-latitude source regions. (8, 9, 13)

Continental Tropical (cT) air mass A body of air that is relatively uniform in its hot and dry characteristics for hundreds or thousands of square kilometers, emanating from inland, low-latitude source regions. (8, 9, 10)

Continentality The degree to which a location is distant from a large body of water. (3, 4, 7, 9, 10)

Continuity equation The equation found in all general circulation models and weather forecasting models, also known as the mass conservation equation, that ensures that the amount of mass entering any imaginary three-dimensional "box" centered on a gridpoint must be balanced by the amount of mass leaving adjacent boxes to enter the box of interest. Likewise, the amount of mass leaving the imaginary box of interest must be balanced by mass entering adjacent boxes from it. (13)

Convection The vertical transfer of matter and the energy that accompanies it through a fluid. (3, 4, 5, 7, 9, 10, 12, 13)

Convective precipitation Precipitation resulting from atmospheric instability, particularly on warm, sunny afternoons; associated with cumuliform cloudiness. (10, 14)

Cooling degree-day (CDD) An index of the amount of cooling that is required for human comfort in a building; 1 CDD is accumulated for each degree that the daily mean temperature rises above the base temperature of 65°F (18°C). (14)

Coriolis effect (CE) An apparent deflective force that causes moving objects to appear to be deflected to the right in the northern hemisphere and to the left in the southern hemisphere, resulting from the rotation of Earth. (3, 4, 5, 7, 9, 10, 11, 13)

Coriolis parameter The product of twice the angular velocity of Earth and the sine of the latitude; proportional to the magnitude of the Coriolis deflection, it represents the magnitude of planetary vorticity. (7, 9, 13)

Counter-radiation Longwave radiation emitted by the atmosphere downward to the surface. (5, 11, 12)

Counter-trade winds Surface winds immediately north of the equator that blow from southwest to northeast, and immediately south of the equator that blow from northwest to southeast, resulting from Coriolis deflection in the opposite direction from the trade winds as the air crosses the equator. (10)

Coupled atmosphere–ocean general circulation model A complicated computer-based model that simulates and/or predicts global atmospheric and oceanic circulation based on dynamical conditions, including effects of feedbacks and energy transfer between these fluids. (13)

Crest The magnitude of peak streamflow following a precipitation event, expressed either as the water height in the stream or as the volume of stream discharge. (6)

Cretaceous Period The last period in the Mesozoic Era, terminating about 65 million years ago with the K-T extinction. (11)

Criteria pollutant Pollutants that are deemed to be sufficiently widespread and dangerous in the United States that they should be monitored at the federal level by the Environmental Protection Agency; includes sulfur oxides, carbon monoxide, particulates, lead, nitrogen oxides, and photochemical oxidants. (12)

Crop Heat Unit (CHU) An index of the exposure of crops to the warmth required for growth, commonly used in Canada; for each degree above the base (for both minimum and maximum daily temperature), 1 CHU is reported for that calculation, but for each degree that the maximum temperature surpasses the optimal temperature, 1 CHU is lost. (14)

Crop Moisture Index (CMI) A water-balance-based measure of drought severity in which moisture in the uppermost layers is emphasized, for suitability in many agricultural applications. (6)

Cross-cutting relationships A principle used in relative dating techniques that any igneous rock formation must be younger than sedimentary rock layers through which it cross-cuts. (11)

Cryogenian Period The last period of the Proterozoic Eon, from about 800 to 600 million years ago, which coincided with perhaps the most impressive glaciation in geological history. (11)

Cryosphere The region of the Earth–ocean–atmosphere system comprising frozen water in all its forms (glaciers, sea ice, surface ice, permafrost, and snow) seasonally or "permanently." (1, 4, 5, 6, 11, 12, 13, 14)

Cumuliform cloud A set of liquid droplets and ice suspended in the atmosphere oriented vertically, characterized by localized cloudiness and perhaps short periods of relatively intense precipitation. (8, 9, 10)

Cumulonimbus cloud A set of liquid droplets and ice suspended in the atmosphere oriented vertically in very unstable atmospheric conditions; severe weather clouds. (8, 9)

Cyclogenesis The beginning stage of storm formation. (7, 9)

Cyclone Any enclosed area of relatively low pressure, rotating generally counterclockwise in the northern hemisphere and clockwise in the southern hemisphere. (3, 4, 7, 8, 9, 10, 12)

Daily mean temperature The average temperature at a given place over the course of a day, calculated simply by averaging the maximum and minimum temperature for that day. (1, 14)

Daily Synoptic Series A booklet of 7 daily surface and 500-mb weather maps, produced and circulated by the U.S. National Oceanic and Atmospheric Administration. (14)

Daughter isotope The form of an element that remains after the radioactive decay of the parent isotope. (11)

December solstice The day of the year when the solar declination is at the Tropic of Capricorn, the northern hemisphere is tilted maximally away from the Sun, the southern hemisphere is tilted maximally toward the Sun, and day length is minimized in the northern hemisphere and maximized in the southern hemisphere; currently corresponds to December 22 and the winter solstice for a northern hemisphere observer and the summer solstice for a southern hemisphere observer. (3)

Decreasing availability water balance model A method of accounting for water in the Earth–ocean–atmosphere system at a location over a period of time; assumes that as soil-moisture withdrawal increases, the water becomes increasingly more difficult to be moved upward via evapotranspiration. (6)

Deficit (D) The amount of water (in centimeters or inches of equivalent precipitation) needed by the atmosphere over a given time period but not supplied to the atmosphere either through precipitation or soil moisture withdrawal. (6, 8, 9, 10)

Deforestation The systematic and widespread clearing of forested landscapes. (4, 6, 10, 11, 12, 13)

Dendroclimatology The science of using annual tree rings to infer paleoclimates. (11)

Density A physical property of matter represented by the ratio of mass to volume. (2, 5, 7, 12, 13)

Deposition The transformation of water molecules from a gaseous state directly to a solid state. (4, 5, 6, 9, 10)

Depression storage Water that is stored in puddles, gutters, and other natural or human-made "depressions" on the landscape. (6)

Desertification The conversion of a climate toward increasing aridity, either through natural or anthropogenic causes. (4, 10, 12, 13)

Dewpoint temperature (T_d) The temperature at which saturation occurs at a given place and time. (5, 6, 9, 10, 13)

Dextral An individual of a species of foraminifera whose shell developed with a coiling on the right side, thereby suggesting that the ocean was relatively warm at the time that the individual lived. (11)

Diabatic A type of heating or cooling that occurs from external sources through energy exchange in fulfillment of the second law of thermodynamics; includes solar radiation, latent heating or cooling, and sensible heating in the boundary layer. (13)

Diffluence The horizontal spreading of air streams moving in the same direction at a given altitude. (7)

Diffuse radiation Solar radiation that is scattered in the atmosphere down to the surface. (5)

Dimensionless quantity A ratio or constant that has no units associated with it. (5, 13)

Direct radiation Solar radiation that is neither absorbed, scattered, nor reflected in the atmosphere but is instead transmitted to Earth's surface. (5, 9)

Diversity The variety in life forms found in a location. (10)

Dixie Alley The region of high tornado frequencies from Texas to Florida. (9)

Doldrums A nickname given to the belt of calm winds between the northeast trades and southeast trades. (7)

Downdraft A small-scale downward rush of air associated with a cumulonimbus cloud. (3)

Downscaling The process of using statistical techniques to derive a finer spatial and/or temporal scale than is output by a general circulation model. (13, 14, 15)

Downwelling An oceanic condition commonly found on the western sides of ocean basins in which surface airflow pushes water against a coastline, forcing it to sink. (4)

Dry deposition A process by which acid precipitation is produced only after the precipitation reaches the surface and interacts with soil water to produce the acids. (12)

Dryline The line of contact between a maritime tropical air mass and a continental tropical air mass in the warm sector of a midlatitude wave cyclone. (9)

Dryness index (DI) In the Thornthwaite Climatic Classification System, the ratio of deficit to potential evapotranspiration in the surface water balance at a location. (8)

Dust Veil Index (DVI) A scale used to estimate the total amount of particulate matter in the atmosphere based on surface air temperatures and the amount of solar radiation reaching the surface. (4)

Dynamic climatology The physics of broad-scale atmospheric motion at long time scales. (1)

Dynamic instability An inherent lack of predictability in a complicated process, despite the presence of order within that process. (13)

East African low-level jet A fast-moving river of east-to-west-flowing air that develops during the day-time hours across the East African. (10)

East-coast occlusion An occluded front characterized by a cold, dry air mass moving into warmer, wetter maritime air, inducing strong vertical lift, cumulonimbus cloud development, and heavy precipitation, similar to the case of a cold front. (9)

Easterly wave A "kink" in the trade wind flow that represents a location where low pressure associated with the Intertropical Convergence Zone can extend slightly farther poleward than elsewhere. (10)

Eccentricity The degree to which Earth's orbit deviates from being circular at any given time in geological history. (11, 13)

Ecotone A transition zone between two biomes. (8, 9)

Eddy covariance The measurement of turbulent fluxes of energy, matter, and momentum based on measured covariance of fluctuations between vertical wind velocity (for vertical fluxes) and changes in temperature, specific humidity, and horizontal wind speed, respectively. (5)

Eigenvalue In eigenvector analysis, the total variance accounted for by a component. (8, 13)

Eigenvector analysis A family of quantitative techniques based on algebraic transformation of a data matrix, to allow for relationships between two different sets of variables to be analyzed simultaneously; in climatology, these techniques are most often used for the simultaneous examination of variability and changes in climate across space and time. (8, 13)

Ekman spiral The idealized depiction of the change in wind vector with height in the atmosphere, or the flow vector in the oceanic boundary layer, caused by changes in the impact of friction and Coriolis deflection with distance from the surface. (3, 4, 10)

Electromagnetic spectrum The full assemblage of all possible wavelengths of energy of the form that advances as disturbances in electrical and/or magnetic fields. (2)

El Niño An oceanic phenomenon during which the normal accumulation of warm tropical waters in the western tropical Pacific migrates eastward, resulting in a decrease in the upwelling of cold water along the tropical South American coast. (4, 9, 10, 11, 13)

El Niño–Southern Oscillation (ENSO) event The combined oceanic and atmospheric alterations that result from the migration of warm tropical waters from the western to the eastern tropical Pacific Ocean. (4, 10)

Empirical approach The use of climatic data to uncover historical trends in variables and relationships between variables. (12, 14, 15)

Empirical climatic classification system Any method of categorizing climates based on the data that represent the effects of the forcing mechanisms rather than the causes of the climates. (8)

Energy balance The relationship between the radiant flux received at the surface and the convective and conductive fluxes. (5)

Energy balance model A representation of the (usually global) availability of energy to account for effects of changes to inputs, reservoirs, or surface features such as global cloud cover, soil moisture, or deforestation. (13)

Environmental lapse rate (γ) The rate at which temperature decreases with increasing height in a static atmosphere; the global annual average is 6.5 C°/km but varies widely across space and time. (2, 5, 9)

Environmental Protection Agency The office of the U.S. government created by the Clean Air Act of 1970, designed to promote healthy environmental conditions in the United States. (12)

Eon The broadest division of geological time, lasting for hundreds of millions or billions of years. (11)

Epoch A unit of geological time within a period. (11)

Equal availability water balance model A method of accounting for water in the Earth–ocean–atmosphere system at a location over a period of time, by assuming that soil water is equally available to be evapotranspired regardless of the amount of storage in the soil. (6)

Equation of state The relationship of pressure to temperature and density (or specific volume) in an ideal gas; also known as the *ideal gas law*. (3, 5, 7, 13)

Equation of state for moist air The relationship of pressure to temperature and density (or specific volume) in an ideal gas, when the effect of water vapor in the air is taken into account. (13)

Equation of time An adjustment made when computing solar time from zone mean time to account for the slight variations in Earth's orbital velocity throughout the year; a function of time of year. (5)

Equatorial (E) air mass A body of air that is relatively uniform in its very warm and moist characteristics for hundreds or thousands of kilometers, emanating from low-latitude, oceanic source regions. (8, 10)

Equatorial Kelvin wave A near-surface wave that propagates from west to east across the tropical Pacific Ocean, particularly during El Niño events, in opposition to the normal trade wind–induced waves. (4, 10)

Equatorial Rossby wave A near-surface wave that propagates from east to west across the tropical Pacific Ocean in response to the equatorial Kelvin wave, signaling the demise of an El Niño event. (4)

Equilibrium line The line at which the rate of accumulation equals the rate of ablation in a glacier. (4)

Equinox A day on which 12 hours of daylight occur worldwide; corresponds approximately to March 21 and September 22. (5)

Era The broadest division of geological time within an eon. (11)

Eukaryote Multicelled organisms that evolved in Earth's early ocean and released carbon dioxide to the atmosphere through fermentation. (2)

EuroFlux A network of stations established in 1995 that provides data on energy, matter, and momentum exchanges in the biosphere across a wide range of climatic types in Europe, with particular emphasis placed on carbon fluxes. (14)

Eustatic Globally uniform, as in sea level. (10, 11)

Evaporation The transformation of water molecules from a liquid to a gaseous state directly, without being routed through vegetation in the process. (1, 2, 3, 4, 5, 6, 7, 8, 9, 10, 11, 12, 13)

Evapotranspiration (ET) The movement of water from the soil to the atmosphere either directly or after moving upward through vegetation. (5, 6, 8, 9, 10, 12, 14)

Extreme An unusually high or low value of a measured atmospheric feature at a given place and time. (1, 8, 11)

Faint young Sun paradox The apparent contradiction that solar output was weaker early in Earth's history than today, despite the lack of convincing evidence that the Earth was ever significantly colder than today. (2, 11)

Federal Weather Service The forerunner of the National Weather Service in the United States, established in 1870 as a component of the U.S. Army. (14)

Fermentation The process by which simple organisms acquire energy through the breakdown of food in the absence of oxygen, releasing carbon dioxide to the atmosphere. (2)

Fick's law The principle that the flux of any entity is proportional to the gradient of that entity in the fluid through which the flux occurs. (5)

Field capacity In the surface water balance, the maximum amount of precipitated water (in centimeters of equivalent precipitation) capable of being held in a soil against the force of gravity; also known as *soil moisture storage capacity*. (6)

Filchner Ice Shelf One of two thick, vast sheets of ice protruding from Antarctica into the Weddell Sea, which opens to the Atlantic Ocean. (10)

Findlater jet *See* Somali jet.

Fine particles *See* $PM_{2.5}$.

Finite differencing In computer models, a technique for calculating the value of each atmospheric variable at a gridpoint based on the calculated values at each adjacent point (including the points above and below the point of interest). (13)

First law of thermodynamics The principle that energy can never be simply created or destroyed in nature; it may only change from one form to another. (5, 11, 13)

Fischer-Porter rain gauge An instrument that collects precipitation in a bucket, measures it based on the water's weight, and punches a special tape mechanically every 15 minutes to report a running total of the precipitation (in tenths of an inch) since the bucket was emptied. (14)

Fluorocarbon A family of chemicals emitted to the atmosphere by industrial activities; although very minute in concentration, an extremely efficient absorber of terrestrial energy. (12)

Flux A flow. (2, 3, 5, 6, 10, 14)

Föehn wind Warm, dry winds caused when air flows down the leeward side of the Alps and undergoes adiabatic warming. (9)

Foraminifera An order of mostly marine protozoans that can provide evidence of paleoclimatic conditions. (11)

Forced convection *See* Mechanical turbulence.

Fossil fuel Hydrocarbon deposit (primarily in the form of coal, oil, and natural gas) derived from formerly living matter that releases carbon dioxide, methane, and other compounds to the atmosphere when burned. (2, 4, 11, 12, 14)

Free atmosphere The layer above the transition layer of the troposphere beginning at approximately 1 km above the surface; characterized by the negligible impact of friction. (3, 5)

Free convection *See* Thermal turbulence.

Frequency The rate of occurrence of a particular phenomenon at a particular place over a long period of time. (1)

Friction A force caused by one body resisting the forward motion of an adjacent mass. (1, 3, 5, 10, 12, 13)

Frigid Zone The area of Earth having the coldest climatic type in the climatic classification system of the ancient Greeks. (8)

Front A narrow zone between two air masses with distinct temperature and moisture conditions. (8, 10)

Frostpoint temperature The temperature to which the air would have to be cooled for saturation to be reached and frost to form at a given place and time. (5)

Fumigation event A situation that exists when limited vertical mixing and a highly stable atmosphere cause deadly concentrations of a pollutant to accumulate near the emission source. (12)

General circulation The interconnected mechanisms by which energy is transferred across the atmosphere at the planetary scale. (7)

General circulation model (GCM) Any complicated computer-based model that simulates and/or predicts global atmospheric circulation based on dynamic conditions. (13)

Genetic climatic classification system Any method of categorizing climates based on the causes of the climates rather than the data that represent the effects of the forcing mechanisms. (8)

Geographic information systems (GIS) Computer hardware and software used to assemble, store, update, analyze, and display spatial information, and to link nonspatial information to the spatial features. (6, 15)

Geopotential height The altitude (in meters above sea level) of a given pressure surface over a particular location at a certain time. (7, 8, 9, 10, 13)

Geostrophic balance An equilibrium established between the pressure gradient force and the Coriolis effect, when the effects of other forces such as friction are negligible; results in air flowing parallel to isobars with higher pressure to the right of the flow in the northern hemisphere and to the left of the flow in the southern hemisphere. (3, 5, 7, 10)

Geostrophic flow In the oceans, the flow resulting from a balance between the tendency of the Coriolis effect (manifested as the Ekman spiral) and gravity. (4)

Geostrophic wind The airflow that results when the pressure gradient force balances the Coriolis effect and other forces are negligible; results in air flowing parallel to isobars. (3, 9)

Glacial advance Any time during an ice age when "permanent" ice is migrating equatorward and downslope. (4, 10)

Glacial erratic A rock that differs in composition from the surrounding geology, suggesting that a glacier must have moved the rock from a location near other rocks similar to its composition to its present location. (11)

Glacial phase Any interval of time within an ice age characterized by an equatorward and downslope displacement of ice. (11, 12)

Global dimming The long-term, slight decrease in solar radiation striking Earth's surface over recent decades. (12)

Global Energy and Water Cycle Experiment (GEWEX) An initiative as part of the World Climate Research Programme that promotes an improved understanding of the global energy and hydrologic budgets by providing data on fluxes of energy, water, and momentum between the surface and the atmosphere, both on land and in the ocean. (14)

Global hydrologic cycle The constant cycling of water (in all of its forms) throughout the Earth–ocean–atmosphere system. (1, 6)

Global Hydrology Center A U.S. agency within the National Aeronautics and Space Administration that provides a wide array of satellite-based observations for climatological research, including such variables as lightning strikes, aerosol loadings, and data for studies of cloud formation and physics. (14)

Global warming Irreversible increases in Earth's surface atmospheric temperatures caused by increasing greenhouse gas concentrations. (4, 10, 12, 15)

Global water balance equation The expression of precipitation as the sum of evapotranspiration and runoff. (6)

Gondwanan Ice Age The period of "permanent" ice on Earth's surface at the end of the Paleozoic Era, occurring after the breakup of Pangaea. (11)

Gradient Richardson number (Ri) An index for describing the relative importance of thermal and mechanical turbulence and for assessing atmospheric stability in the local, near-surface atmosphere. (5)

Gradient wind equation The equation that relates the speed of air moving in curved flow to geostrophic wind speed. (9)

Gray smog *See* Industrial smog.

Greenhouse effect The phenomenon by which Earth's lower atmosphere is warmed by the presence of gases that absorb longwave radiation from the Earth and emit radiation downward to the Earth. (2, 9, 10, 11, 12)

Greenhouse gas An atmospheric pollutant that contributes to the greenhouse effect by enhancing the efficiency of absorption of longwave radiation emitted from the surface and reemitting radiation back downward effectively as counter-radiation. (2, 4, 9, 11, 12)

Greening initiatives The use of improved technology and understanding of the Earth–ocean–atmosphere system to enhance the quality of the environment. (15)

Gridpoint model A representation of circulation created by dividing Earth into equidistant points with imaginary three-dimensional "boxes" centered on those points representing the smallest area in the atmosphere for which motion can be calculated. (13)

Gridpoint technique A spatial analytic method for converting point data to areal data, in which an artificial grid is superimposed over the study region, point totals are calculated for each of the resulting gridpoints based on an inverse-distance-weighted function of the value at each of the known stations, and the mean of the derived values represents the areal mean. (6)

Groundwater Water stored below the soil layer. (6)

Growing degree-day (GDD) An index of the exposure of crops to the warmth required for growth; one unit is accumulated for each degree that the daily mean temperature exceeds the theoretical minimum daily temperature required to provide enough energy for that crop to grow. (14)

Growing season The interval of time between the last spring frost and the first autumn frost. (14)

Guard cell A cell on a leaf that works as part of a pair to control the opening of a stomate. (5, 6)

Gulf Coast cyclone A midlatitude wave cyclone with cyclogenesis in the northwestern Gulf of Mexico and a general northeastward track. (9)

Gulf Stream A warm ocean current flowing northward along the eastern coast of North America and becoming the North Atlantic Drift near northwestern Europe. (3, 4, 9)

Gust front *See* Outflow boundary.

Gymnosperm Plants such as conifers that do not have seeds enclosed within an ovary. (11)

Gyre A circular current in the near-surface ocean, caused by a coupling of the atmospheric and surface oceanic circulation around a subtropical anticyclone. (4, 9, 10)

Hadley cell A convection cell, with rising motion near the equator at the Intertropical Convergence Zone and sinking motion near 30° latitude in each hemisphere, most prominent at the longitude experiencing diurnal heating at any given time. (7, 9, 11)

Half-life The amount of time required for one-half of the parent atoms to decay into daughter product in a radioactive element. (11)

Harmattan winds Northeasterly surface winds that blow hot, dry, Sahara Desert winds toward the Gulf of Guinea region in association with the northeast trade winds. (10)

Hawaiian high The semipermanent subtropical high-pressure system over the northern subtropical Pacific Ocean; also known as the *North Pacific high*. (7, 9, 10, 12)

Heat capacity The amount of heat (energy) necessary to be added to a unit volume of a substance to raise its temperature by one Kelvin; distinguished from specific heat by the fact that specific heat refers to a unit mass rather than a unit volume. (12)

Heat index A contrived "temperature" based on how the temperature "feels" for people by taking into account the ease with which evaporative cooling can occur. (3, 9)

Heating degree-day (HDD) A variable that indexes the amount of heating that is required for human comfort in a building; 1 HDD is accumulated for each degree that the daily mean temperature falls below the base temperature of 65°F (18°C). (14)

Heterosphere The layer of the atmosphere corresponding to the thermosphere, characterized by a layering of gases according to density. (2)

Highland (H) climate In the Köppen system, a climate type in rugged terrain characterized by great climatic contrasts across space caused by elevation differences. (9, 10)

Historical Climatology Network A data set of mean monthly temperature, precipitation, and sea level pressure for thousands of stations worldwide for time periods of up to 50 years. (14)

Hockey stick The shape of the global temperature curve for the past 1000 years reconstructed by proxy evidence; characterized by a very slow decline from about AD 1000 to about 1900, followed by abrupt warming in the last 100 years. (12)

Holdridge Life Zones Climatic Classification System A global categorization system of climate, most commonly used in tropical areas, that incorporates annual precipitation and the ratio of a derived PE-to-precipitation. (8)

Holocene Epoch The current epoch of geological time; embedded within the Quaternary period, corresponding to the current interglacial phase and the last 10,000 years. (4, 11)

Holocene Interglacial Phase The most recent epoch of time in geological history, consisting of the last 9000 years, characterized by the poleward and upslope retreat of the "permanent" ice on Earth's surface. (4, 11)

Homosphere The layer of the atmosphere corresponding to the troposphere, stratosphere, and mesosphere, where atmospheric gases are continuously mixed by winds. (2)

Horizontal convergence The collision of two streams of air moving laterally in opposite directions, or from the same direction but with one stream moving faster and "catching up with" the other. (10)

Horse latitudes Generic term for the latitudinal belt around 30°N and 30°S. (7)

Hot Steppe (BSh) climate A semiarid climate in which the mean annual temperature is above 18°C (64°F). (9, 10)

Hourly Precipitation Data (HPD) A publication issued monthly by the U.S. National Oceanic and Atmospheric Administration that provides hourly precipitation data for sites in each state, along with the maximum intensity of precipitation to fall within specific time durations during that month. (14)

Human bioclimatology The study of the relationship between people and the atmospheric environment immediately surrounding them and affecting their health, comfort, and performance. (1)

Humboldt (Peru) current A cold, surface ocean current of the South Pacific Ocean that moves northward along the coast of South America, guided by the winds of the eastern side of the South Pacific high. (4, 10)

Humid Continental (Dfa, Dfb) climate In the Köppen system, a climatic type of the midlatitudes generally found inland and on the east coasts of continents, characterized by warm to hot summers, cold winters, and wet conditions year-round. (9)

Humid Continental (Dwa, Dwb) Winter Dry climate In the Köppen system, a climatic type of the midlatitudes generally found inland and on the east coasts of continents, characterized by warm to hot summers, cold winters, and wet conditions only in the warm season. (9)

Humid Subtropical (Cfa) climate In the Köppen system, a climatic type of the near-tropical midlatitudes characterized by hot summers, mild winters, and wet conditions year-round. (9, 10)

Humid Subtropical Winter Dry (Cwa) climate In the Köppen system, a climatic type of the near-tropical midlatitudes characterized by hot summers, mild winters, and wet conditions except in the cold months. (9)

Humidity index (HI) In the Thornthwaite Climatic Classification System, the ratio of monthly surplus to potential evapotranspiration in the monthly surface water balance at a location. (8)

Humus Partially decomposed organic matter in soil. (12)

Hurricane Regional name given to the strongest type of tropical cyclone in the Atlantic Ocean basin. (9, 10, 11, 12)

Hybrid climatic classification technique Any method of climatic categorization in which investigator-identified "prototype" atmospheric circulation patterns are used as input into an automated cluster analysis, so that other maps can be matched automatically to the prototype map that they most resemble. (8)

Hydrocarbon A type of compound that contains only hydrogen and carbon. (12)

Hydroclimatology The study of the interaction between the atmosphere and near-surface water at long time scales; the atmospheric component of the global hydrologic cycle. (1)

Hydrosphere The component of the Earth–ocean–atmosphere system consisting of liquid water, including the oceans, groundwater, surface streams and lakes, atmospheric water, and soil water. (1, 2, 4, 5, 6, 11, 13, 14)

Hydrostatic equation The equation that relates changes in pressure to changes in height and density in an atmosphere characterized by hydrostatic equilibrium. (5, 13)

Hydrostatic equilibrium A balance between the downward-directed gravitational force and the upward-directed buoyancy force in the atmosphere. (2, 3)

Hygrometer An instrument that measures relative humidity. (5)

Hygroscopic nuclei Solid aerosols that act as very efficient condensation nuclei. (12)

Hypsithermal A period of time from about 8000 to 5000 years ago, characterized by global average temperatures that were perhaps 2 C° (4 F°) above today's levels. (11)

Ice age Any time in geological history during which ice exists continuously for a long period of time over some section of Earth. (4, 11)

Ice Cap (EF) climate In the Köppen system, a climatic type of the high latitudes characterized by mean temperatures below freezing in every month of the year. (9, 10)

Ice core A sample drilled into an ice surface and brought to a laboratory for subsequent analysis of composition of air bubbles, pollen grains, and other proxy evidence of paleoclimates. (11, 12)

Icelandic low The semipermanent subpolar low-pressure system over the North Atlantic Ocean. (7, 9, 13)

Ideal gas law *See* Equation of state.

Indian Ocean high The subtropical anticyclone of the Indian Ocean. (7, 10)

Industrial smog A secondary pollutant that results from chemical reactions between sulfur dioxide (SO_2) and particulates; also known as *gray smog* and *London smog*. (12)

Inertia The property of matter that causes it to retain its velocity (or lack thereof) until acted upon by a force. (3)

Inertial period The length of time required for a fluid's inertia to be balanced by the Coriolis effect. (3)

Infiltration The downward movement of water from the surface into the soil and plant rooting zone. (6, 9)

Initialization The setting of initial conditions in a model in preparation for a forecast. (13)

Insolation Incoming solar (shortwave) radiation; radiation emitted by the Sun and intercepted by the Earth–ocean–atmosphere system. (2, 3, 4, 5, 6, 7, 9, 10, 11, 12, 13, 14)

Interception Water held on surfaces other than the soil, such as tree leaves and rooftops, until it eventually evaporates. (6)

Interglacial phase Any interval of several thousand years within an ice age when "permanent" ice is present but not very extensive on Earth's surface. (4, 11, 15)

Intergovernmental Panel on Climate Change An international group sanctioned by the United Nations and the World Meteorological Organization to address issues and policies related to global climate change. (12, 15)

Internal energy The capacity for individual molecules to move. (13)

International date line The imaginary zigzagging line roughly corresponding to the 180° meridian of longitude; by international agreement, calendars are advanced by one calendar day when moving westward across this line and calendars are set backward one day when moving eastward across the line. (5, 9)

Intertropical Convergence Zone (ITCZ) The belt of low surface pressure around the equatorial region of Earth, caused by rising motion in the general circulation. (7, 10)

Inverse square law The axiom stating that the intensity of radiation received by a body is inversely proportional to the square of the distance between the emitting and the receiving body. (11)

Inverted trough An elongated zone of low pressure in the tropics that extends poleward (in either hemisphere). (10)

Isobar A line on a map that connects locations having the same atmospheric pressure. (3)

Isohyet A line on a map that connects locations having the same precipitation total over a given period of time. (6)

Isohyetal method A spatial analytic method for converting point data to areal data, in which isolines are drawn based on the point data and the mean value between each isoline is weighted by the percentage of area enclosed between those isolines to determine the overall areal mean value. (6)

Isohypse On upper-level atmospheric maps, a line connecting locations having the same geopotential height. (7, 9, 10)

Isostatic rebound A local rising in continental crust as a reaction to the weight of ice being removed; results in a decrease in local sea level. (11)

Isotherm A line on a map that connects locations having the same temperature. (3, 6, 9)

Isothermal layer Any zone in the atmosphere in which temperature changes relatively little with height, as in the lower stratosphere. (2)

Isotope A form of an element having the same number of protons but different numbers of neutrons from another form of that element. (11)

Iterative Involving repetition, as in the repetitive automated calculation of solutions to a set of differential equations until one is found. (13)

Jet streak The region of greatest wind speed within a jet stream. (7)

Jet stream A fast-moving river of air that moves generally from west to east across the upper troposphere. (4, 7)

Joule The metric unit of energy or work, equivalent to one kilogram times one square meter divided by one squared second. (3, 5, 6, 12, 13)

June solstice The day of the year when the solar declination is at the Tropic of Cancer, the northern hemisphere is tilted maximally toward the Sun, the southern hemisphere is tilted maximally away the Sun, and day length is maximized in the northern hemisphere and minimized in the southern hemisphere; currently corresponds to June 21 and the "summer solstice" for a northern hemisphere observer and the "winter solstice" for a southern hemisphere observer. (3)

K-T extinction A relatively sudden mass elimination of approximately 85 percent of the species existing on Earth that occurred about 65 million years ago. (11, 12)

Katabatic wind A local circulation in which air moves downslope, typically at night when the air adjacent to the surface increases in density as it cools; also known as *mountain wind*. (9, 10)

Keeling curve The graph of increasing atmospheric carbon dioxide through time, measured continuously since 1957 at the Mauna Loa Observatory in Hawaii. (2)

Keetch-Byram Drought Index A measure of moisture conditions designed to assess the risk of fire potential by examining the relationship of evapotranspiration to precipitation in the organic matter on a forest floor and in the uppermost soil layers. (6)

Kelvin temperature scale A direct system of representing the internal energy content of matter by maintaining proportionality between temperature and internal energy. (3, 5, 6, 7, 12, 13)

Kinetic energy Energy possessed by a body as a consequence of the motion of that body (rather than the molecular motion of individual particles within that body). (3, 5, 7, 13)

Köppen Climatic Classification System An empirical method of categorizing climates so that they correspond to major biomes, based on monthly temperature and precipitation data and the seasonality of those data. (8)

Krakatau A volcano in Indonesia that erupted in 1883, causing massive damage and fatalities. (4, 10)

Krakatau easterlies Lower-stratospheric winds that blow from east to west as one phase of the quasi-biennial oscillation. (10)

Kyoto Treaty An international agreement that requires wealthy nations contributing most of the world's greenhouse gases to reduce their collective greenhouse gas emissions to 5.2 percent below the 1990 levels. (12)

Labrador current A cold surface ocean current that moves southward along Canada's Atlantic coast. (9)

Lake effect snow Snow that forms immediately downwind of a large lake, with totals enhanced by the moisture and atmospheric instability provided by the lake. (3, 9)

Laki Fissure A volcano that erupted in Iceland in 1783, causing devastating impacts. (4)

Lamb Weather Types A climatic classification system based on the position of a location relative to a midlatitude wave cyclone, designed for application in the British Isles. (8)

Laminar layer The few millimeters of the atmosphere nearest to the surface or elements on the surface, such as crops, trees, buildings, or waves on the sea. (5)

La Nada A condition of the tropical Pacific Ocean characterized by neither El Niño nor La Niña conditions. (4)

La Niña An oceanic phenomenon during which the normal accumulation of warm tropical waters in the western tropical Pacific is even more exaggerated than usual, associated with a strengthening of the trade winds and Walker circulation, and an increase in the upwelling of cold water along the tropical South American coast. (4, 10, 11, 13)

Lapse rate A rate of change of temperature with height. (5)

Latent energy (heat) Radiant energy that evaporates water in the Earth–ocean–atmosphere system rather than heating the atmosphere or surface. (3, 4, 8, 9, 10, 12, 13)

Latent heat flux *See* Turbulent flux of latent heat.

Latitude A set of imaginary lines that runs east–west and measures distances north or south of the equator; 0° latitude corresponds to the equator, with 90°N and 90°S corresponding to the North and South Poles, respectively. (3, 4, 5, 6, 7, 8, 9, 10, 13)

Leeward Downwind; the side of a topographic feature facing away from the wind. (3, 7, 8, 9, 10)

Level of nondivergence The height in the atmosphere at which neither convergence nor divergence is occurring, despite having convergence above it and divergence below it (or vice versa). (7)

Lifting condensation level The height in the atmosphere at which the air becomes saturated and an adiabatically rising air parcel no longer cools at the unsaturated adiabatic lapse rate but begins cooling instead at the saturated adiabatic lapse rate; corresponds to a cloud base. (5)

Lithification The process by which sedimentary rock is formed from compaction and cementing together of unconsolidated sediments. (12)

Lithosphere The uppermost part of the solid Earth, from the surface to approximately 100 km of depth; the solid-Earth component of the Earth–ocean–atmosphere system. (1, 2, 5, 6, 11, 13, 14)

Little Climatic Optimum (LCO) The period from about AD 900 to about 1450, characterized by relatively warm conditions in Europe; also known as the Medieval Warm Period. (11, 12)

Little Ice Age (LIA) The period from approximately AD 1450 to 1850, characterized by colder global conditions than at present. (11, 12)

Loading In eigenvector analysis, a coefficient that is proportional to the degree to which the eigenvector model represents the variability in each observation of one set of variables collectively for all observations in another set of variables simultaneously; in climatology, the loadings often represent the correlation between a component and the atmospheric variable of interest at each station in the data set, collectively across all times of observation. (8, 13)

Local apparent time *See* Solar time.

Local Climatological Data (LCD) A monthly publication of the National Climatic Data Center of the National Oceanic and Atmospheric Administration for about 270 sites around the United States; contains data with temporal resolution as high as every 3 hours and data for more variables than *Climatological Data* provides. (14)

Local scale A spatial scale of analysis between about 0.5 and 5 km (0.3 and 3 mi), including atmospheric phenomena that generally persist from a few minutes to a few hours. (1, 13, 14)

Loch Lomond Stadial The European name for the Younger Dryas. (11)

Loess A fine, angular-grained soil or wind-blown deposit formed by the grinding of glaciers against bare rock until it forms a powdery consistency. (11)

Logarithmic wind profile The theoretical variation of wind speed with height under neutral stability conditions, in which wind speed varies linearly with the natural logarithm of height. (5)

Longitude A set of imaginary lines that runs north–south and measures distances east or west of the prime meridian; 0° longitude corresponds to the prime meridian, and 180° corresponds approximately to the international date line. (3, 5, 7, 9, 10, 13)

Long-term Ecological Research Program An initiative that collects and disseminates bioclimatological data (along with other environmental information such as hydrological measurements, soil composition and quality, land-use change, and even demographic data) for 26 sites across North America, the Caribbean, and Antarctica. (14)

Long wave *See* Rossby wave.

Longwave radiation Electromagnetic radiation at wavelengths greater than 4.0 micrometers, generally considered in the atmospheric sciences to be synonymous with radiation emitted by Earth and the atmosphere. (5, 9, 10, 11, 12)

Loop current A surface ocean circulation that branches away from the Gulf Stream into the Gulf of Mexico immediately west of the Florida peninsula. (9)

Low-level jet stream (LLJ) A river of air at approximately the 800- to 900-mb level, moving northward from the western Gulf of Mexico at perhaps 40–110 km hr^{-1} (25–70 mi hr^{-1}). (9)

Lysimeter An instrument that is installed underground and weighs the soil and vegetation on top of it to measure volumetric water content. (5, 6)

Madden-Julian Oscillation (MJO) A system of continual intra-annual variations of upper- and lower-level winds, sea surface temperatures, and associated cloudiness and precipitation in the tropical Pacific Ocean. (10, 13)

Magma Molten rock material beneath Earth's surface. (11)

March equinox The day of the year midway between the December and June solstices, when the solar declination is at the equator, neither hemisphere is tilted toward or away from the Sun, and day length is 12 hours over the entire Earth; currently corresponds roughly to March 21. (3)

Marine Dry Winter (Cwb) climate A variant of the Marine West Coast climate found in a small area of southeastern Africa and Madagascar, characterized by warm summers, mild winters, and abundant precipitation except in winter. (10)

Marine West Coast climate In the Köppen system, a climatic type of the mid to high latitudes characterized by warm to mild summers, cool winters, and wet conditions year-round. (9)

Marine West Coast Cool Summer (Cfc) climate A Marine West Coast climate with only 1 to 3 months having average temperatures above 10°C (50°F). (9)

Marine West Coast Warm Summer (Cfb) climate A Marine West Coast climate with more than 3 months having average temperatures above 10°C (50°F). (9, 10)

Maritime Continent Australia and the islands southeast of continental Asia including Indonesia, the Philippines, New Guinea, and thousands of other islands occupying the tropical and southwestern Pacific Ocean basin. (4, 10)

Maritime effect A moderating influence on temperature by bodies of water, caused primarily by the high specific heat of water, circulation within water, relatively high transparency of water, and latent heating over water bodies. (3, 4, 9)

Maritime polar (mP) air mass A body of air that is relatively uniform in its cool and humid characteristics for hundreds or thousands of square kilometers, emanating from high-latitude oceanic source regions. (8, 9)

Maritime tropical (mT) air mass A body of air that is relatively uniform in its warm and humid characteristics for hundreds or thousands of square kilometers, emanating from low-latitude oceanic source regions. (8, 9, 10)

Mass conservation equation *See* Continuity equation.

Mass spectrometry An instrumental laboratory technique that can determine the chemical composition of a substance. (11)

Maunder Minimum The interval of time from about AD 1645 to 1715 in which few sunspots were observed. (11)

Maximum/minimum thermometer A special type of instrument that measures temperature by leaving a mark at the highest temperature that had occurred since the last resetting of the instrument and having the base of the mercury sit at the lowest temperature since the last resetting. (14)

Mean free path The distance that a molecule must travel in the atmosphere before encountering another molecule. (2)

Mechanical turbulence Microscale convective motion resulting from vertical gradients of horizontal momentum. (5)

Medieval Warm Period *See* Little Climatic Optimum.

Mediterranean (Cs) climate In the Köppen system, a climatic type of the midlatitudes characterized by dry and warm to hot summers and by mild, wet winters. (9, 10, 12)

Mediterranean Hot Summer (Csa) climate In the Köppen system, a Mediterranean climate characterized by a warmest month averaging above 22°C (72°F). (9, 10)

Mediterranean Warm Summer (Csb) climate In the Köppen system, a Mediterranean climate characterized by a warmest month averaging below 22°C (72°F). (9, 10)

Meltemi wind A katabatic wind system that develops when high pressure occurs over the Balkan region near Hungary and a cyclonic center lies to the southeast near or over Turkey, resulting in an offshore wind blowing over the Aegean Sea. (9)

Meridional flow Rossby waves with high amplitudes (i.e., pronounced ridges and troughs), characterized by strong north–south and south–north components of atmospheric motion. (7, 13)

Mesopause The boundary separating the mesosphere from the thermosphere. (2)

Mesoscale A spatial scale of analysis between about 5 and 100 km (3 and 60 mi), including atmospheric phenomena that typically persist from a few hours to a few days. (1, 13)

Mesosphere A layer of the atmosphere in which temperature decreases with height, from about 48 to 80 km (29 to 50 mi) above the surface. (2)

Mesothermal Any of the mild, midlatitude climatic types in the Köppen classification system, including Humid Subtropical, Mediterranean, Marine West Coast, and Marine Dry Weather climates. (8, 9, 10)

Mesozoic Era The era of geological time with generally warm conditions, beginning at the end of major extinctions associated with the Gondwanan Ice Age approximately 250 million years ago and ending with the K-T extinction about 65 million years ago. (11)

Meteorology The scientific study of weather; the atmospheric science that analyzes the properties and causes of atmospheric conditions at a specific place and time. (1)

Methane (CH_4) A compound consisting of one atom of carbon and four of hydrogen that acts as a greenhouse gas despite a very low atmospheric concentration. (2, 11, 12)

Methane hydrates Frozen lattices of water molecules surrounding CH_4 molecules. (12)

Methanogen Any of various bacteria that are intolerant of atmospheric oxygen and give off methane. (2, 11, 12)

Micrometer One millionth of a meter, represented by μm. (2, 12)

Microscale A spatial scale of analysis smaller than about 0.5 km (0.3 mi), including atmospheric phenomena that typically persist from a few seconds to a few hours. (1, 5, 13, 14)

Microthermal Either of the cold, midlatitude climatic types in the Köppen classification system, including Humid Continental and Subarctic. (8, 9)

Midlatitude Cold Desert (BWk) climate In the Köppen system, a climatic type characterized by potential evapotranspiration that far exceeds precipitation and a climatological mean annual temperature that is below 18°C (64°F). (9, 10)

Midlatitude (frontal) wave cyclone The largest storm systems on the planet, consisting of a low-pressure center with cyclonic circulation and fronts extending from that low-pressure center. (3. 8, 9, 12)

Midlatitude westerlies The prevailing general circulation feature over the middle latitudes in both hemispheres, both at the surface and aloft, in which flow occurs primarily from west to east. (7, 9, 10)

Milankovitch cycles Any of the three axial and orbital variations (eccentricity, precession, tilt) that cause Earth's climate to vary on time scales of thousands of years. (11, 13)

Millibar A unit of pressure in the metric system equivalent to 100 Pascals; average sea level pressure is 1013.25 mb. (3, 5)

Mississippian Period A period in the latter part of the Paleozoic Era largely within the Gondwanan Ice Age, from about 350 to 325 million years ago. (11)

Mistral wind Air that flows downslope from the Alps and spills though the Rhone Valley, where it spreads along the Mediterranean coast of France. (9)

Mitigation A strategy for dealing with climate change that attempts to decrease the risk associated with the threat rather than decreasing the threat itself. (12)

Mixing ratio The dimensionless ratio of the mass of water vapor present in the air to the mass of dry air. (13)

Moisture conservation equation The equation in general circulation models that expresses the change in moisture content at a gridpoint as the sum of the contribution of moisture via advection and the contribution via phase changes. (13)

Moisture index (MI) In the Thornthwaite Climatic Classification System, a variable calculated as the ratio of the difference between total surplus and deficit to potential evapotranspiration at a location. (8)

Momentum A property of a body in motion equivalent to the product of mass and velocity. (1, 3, 4, 5, 6, 7, 11, 13, 14)

Monsoon A seasonal reversal of wind caused by synoptic-scale pressure changes, often accompanied by seasonal changes in moisture. (9, 11)

Moraine A large pile of rock and soil debris deposited by a glacier. (11)

Mountain wind *See* Katabatic wind.

Mount Pinatubo A volcano in the Philippines that erupted on June 15, 1991 and has been linked to a temporary reduction in subsequent global temperatures. (4)

Mount Toba A volcano that erupted in Sumatra about 71,000 years ago, accelerating the last glacial advance. (4)

Muller Weather Types A climatic classification system based on the position of a location relative to a midlatitude wave cyclone, designed for application on the U.S. Gulf Coast. (8)

National Aeronautics and Space Administration An independent agency of the federal government that oversees the U.S. space flight program. (14)

National Air Quality Standards The maximum concentration of a pollutant (in parts per million of the atmospheric molecules) that is deemed by the U.S. Environmental Protection Agency to be acceptable. (12)

National Center for Atmospheric Research A federally funded research center in Boulder, Colorado, that provides research opportunities and data, including the reanalysis data set. (13, 14)

National Centers for Environmental Prediction A branch of the U.S. National Weather Service that provides specialized forecasts through nine offices, including the Climate Prediction Center, Hydrometeorological Prediction Center, Aviation Weather Center, Tropical Prediction Center, and Storm Prediction Center. (14)

National Climate Program Act The 1978 legislation that greatly increased the scope and mission of atmospheric data collection and archiving in the United States and included the establishment of the Regional Climate Centers. (14)

National Climatic Data Center (NCDC) A U.S. federal agency within the National Environmental Satellite, Data, and Information Service (NESDIS) that archives and disseminates atmospheric data. (14)

National Data Buoy Center (NDBC) A U.S. federal agency within the National Oceanic and Atmospheric Administration that is responsible for dispatching and maintaining a network of buoys and coastal stations around North America and in some other parts of the world that collect atmospheric data. (14)

National Environmental Satellite, Data, and Information Service (NESDIS) The office within the National Oceanic and Atmospheric Administration in the United States that collects and disseminates atmospheric and other geophysical data. (14)

National Oceanic and Atmospheric Administration (NOAA) The federal agency in the United States that includes the National Weather Service in addition to managing the nation's fisheries, and collecting and disseminating environmental data. (6, 14)

National Snow and Ice Data Center A U.S. federally funded center that provides cryospheric data, including snow cover, glaciers, continental ice sheets, avalanches, permafrost, and sea ice. (14)

National Weather Service The present federal agency that oversees government-sponsored weather forecasting in the United States, established in 1970 when the U.S. Weather Bureau was renamed and moved to the Department of Commerce. (1, 3, 5, 6, 14)

Navier-Stokes equations of motion The three equations found in all general circulation models that express motion in the west–east, south–north, and down–up directions. (3, 13)

Nebula A vast cloud of interstellar dust, from which stars and planets form. (2)

Negative feedback system An input to a system that decreases the likelihood of further changes of the same type to the system. (4, 12)

Negative vorticity Rotation that is clockwise in the northern hemisphere or counterclockwise in the southern hemisphere; anticyclonic rotation. (7)

Net radiation (Q*) The difference between radiant energy absorbed by matter and that emitted by matter. (5, 6, 8)

Neutral atmosphere An atmospheric condition in which the buoyancy force balances the gravitational force, and a hypothetical air parcel will neither rise nor sink once a lifting force is removed. (5)

Newton (N) The metric unit of force. (3)

Newton's laws of motion The physical laws that govern flow upon which the Navier-Stokes equations are based. (3)

Newton's second law of motion The notion that force on a moving body can be expressed as the product of mass and acceleration. (2, 7, 13)

Nitric acid (HNO$_3$) A very strong acid formed by complicated processes involving interactions between nitrogen oxides and water. (12)

Nitric oxide (NO) A type of nitrogen oxide photodissociated from NO$_2$ in the presence of ultraviolet radiation in the troposphere; monitored by the U.S. Environmental Protection Agency as a criteria pollutant. (12)

Nitrogen dioxide (NO$_2$) A type of nitrogen oxide that is noted for wearing away objects in contact with it; monitored by the U.S. Environmental Protection Agency as a criteria pollutant. (12)

Nitrogen oxide (NO$_x$) Any compound formed by the bonding of a nitrogen atom with one or more atoms of oxygen; monitored by the U.S. Environmental Protection Agency as a criteria pollutant. (12)

Nitrous oxide (N$_2$O) A non-natural greenhouse gas that contributes a tiny percentage to the anthropogenic effect. (12)

Noise *See* Random variability.

Nonattainment zone Any area that persistently exceeds the satisfactory concentration threshold of a pollutant. (12)

Nonthreshold pollutant A contaminant that evokes a response as soon as it exists at any concentration. (12)

Nor'easter A midlatitude wave cyclone with cyclogenesis over the mid-Atlantic coast near Cape Hatteras and a generally north-northeastward track in the Atlantic Ocean. (9, 14)

Normal The average measurement of an atmospheric feature at a given place and time. (1, 8)

North American summer monsoon The tendency for onshore flow associated with a thermal low centered over Arizona in summer, resulting in more precipitation in late summer than in the rest of the year. (9)

North Atlantic Deep Water (NADW) A part of the global ocean conveyor belt originating from the remnant of the North Atlantic Drift; flows southward near the coast of North America in the Atlantic Ocean and eventually above the Antarctic Deep Water. (4)

North Atlantic Drift The northernmost part of the Gulf Stream current; moderates the climate of northwestern Europe. (3, 4, 9)

North Atlantic Oscillation (NAO) A teleconnection in which pressure patterns see-saw between the Bermuda-Azores high and the Icelandic low. (13)

North Equatorial current An east-to-west flow of surface water in the northern hemisphere immediately north of the equator. (4, 9)

North Pacific high *See* Hawaiian high.

North Pacific Intermediate Water (NPIW) A part of the global ocean conveyor belt in the Pacific Ocean that overlies the Pacific Subarctic Water. (4)

Northeast trade winds The dominant surface circulation pattern in the tropical northern hemisphere, with consistent winds blowing from northeast to southwest. (4, 7, 10)

Nuclear autumn A somewhat less dramatic scenario than "nuclear winter" of the impact of large scale detonation of nuclear weapons on the Earth–ocean–atmosphere system. (12)

Nuclear fusion A process in star formation in which lighter elements (principally hydrogen) are converted into heavier elements (primarily helium). (2)

Nuclear winter The notion that a large-scale detonation of nuclear weapons could create so much soot that it would effectively block the Sun's energy from reaching much of Earth's surface for up to several weeks, leading to widespread surface cooling. (12)

Occluded front A surface front that results in the late stage of a midlatitude wave cyclone's life cycle after the cold front has overtaken the warm front; results in warm sector being wedged upward between cold air behind the cold front and cool air ahead of the warm front. (8, 9)

Occlusion The process by which a cold front overtakes a warm front in a midlatitude wave cyclone. (9)

Oceania A broad geographic region of the Pacific Ocean, including the Maritime Continent (except for Australia) and the islands comprising Melanesia, Micronesia, and Polynesia. (10)

Oceanic general circulation model A complicated computer-based model that simulates and/or predicts global hydrospheric circulation based on dynamic conditions. (13)

Oort cloud A collection of icy comets, dust, and debris surrounding the outer edges of our solar system. (2)

Optical air mass The ratio of the distance of the path length of a beam of insolation to the minimum distance (perpendicular to the surface) that it could possibly have to travel to reach the surface. (5)

Ordovician Period A period of geological time from about 490 through 440 million years ago, within the Paleozoic Era, during which a massive extinction coincided with a major but short-lived glaciation. (11)

Orogeny The process by which mountains form. (11)

Orographic effect The enhancement of precipitation on the windward sides of slopes. (3, 6, 9, 10)

Oroshi wind A katabatic wind system in Japan, similar to the bora of Europe. (9)

Outflow boundary A line of contact between a cold downdraft that hits the surface in a thunderstorm and warmer air in advance of the system. (9)

Outgassing The release of gases from the lithosphere to the atmosphere—for example, as through volcanic activity. (2, 4)

Ozone (O_3) A molecule containing three oxygen atoms that absorbs ultraviolet radiation in the stratosphere but represents a toxic pollutant to humans in the troposphere. (2, 10, 12)

Pacific Decadal Oscillation (PDO) A teleconnection in sea surface temperatures of the north Pacific Ocean, with the action centers located in the tropical central to eastern Pacific and the northwestern Pacific. (4, 13)

Pacific-North American (PNA) pattern A teleconnection that represents the oscillation between geopotential height in the midtroposphere of northwestern North America, with opposing heights upwind in the northern Pacific and downwind in the southeastern United States. (13)

Pacific Subarctic Water (PSW) Part of the deep water conveyor belt emanating from the Common Water and moving northward in the Pacific Ocean. (4)

Paleoclimatology The scientific study of climates of the preinstrumental period and the causal mechanisms that produced those climates. (1, 11)

Paleotempestology The science of identifying signatures of ancient storms primarily from sediment records. (11)

Paleozoic Era The first era within the current Phanerozoic Eon, lasting from about 540 to 250 million years ago, beginning with the first fish and shellfish and ending with the massive Permian extinction. (11)

Palmer Drought Severity Index (PDSI) A water-balance-based measure of drought severity in which relatively deep soil-moisture conditions are emphasized, for assessing long-term moisture conditions. (6)

Palmer Hydrological Drought Index A water-balance-based measure of drought severity used for even longer-term analysis than the Palmer Drought Severity Index, used to assess conditions relating to groundwater availability and reservoir supplies. (6)

Palynology The scientific study of pollen, particularly as an indicator of paleoenvironmental conditions. (11)

Pan adjustment coefficient A factor used to adjust measured Class A pan evaporation downward to provide an estimate of potential evapotranspiration. (5)

Pangaea The supercontinent that existed during the late Paleozoic Era. (11)

Parallelism The property of Earth's axis of remaining tilted at the same fixed angle throughout its revolution about the Sun. (3)

Parameterization A statistically based "correction factor" used when the precise equations to simulate a process are either unknown or impossible to include in a model because the input data are not available to solve the equation. (13)

Parent isotope The form of an element that exists before radioactive decay begins. (11)

Particulates A criteria pollutant monitored by the U.S. Environmental Protection Agency consisting of a family of tiny suspended solid aerosols. (4, 12)

Pascal A metric unit of pressure equal to one Newton per square meter. (3, 5, 7, 13)

Path length The distance through the atmosphere that a ray of insolation travels. (3, 5)

Penman-Monteith potential evapotranspiration equation A theoretical, physically based method of computing potential evapotranspiration, requiring the measurement of net radiation, substrate heat flux, vapor pressure (or some humidity measurement that can be converted to vapor pressure), temperature, and resistance. (6)

Pennsylvanian Period A period in the latter part of the Paleozoic Era following the Mississippian Period and largely within the Gondwanan Ice Age, from about 325 to 300 million years ago. (11)

Percolation The gradual downward movement of water through the lithosphere to replenish the groundwater supply. (6)

Perihelion The day on which Earth is at its nearest distance from the Sun, corresponding approximately to January 3. (3, 10, 11)

Period The largest subdivision of geological time within an era, usually lasting tens of millions of years. (11)

Periodic variability A fluctuation from the mean of any atmospheric variable that recurs with some relatively constant regularity. (11)

Permafrost Permanently frozen subsoil. (9)

Permeability The property that describes the ability of water to flow through soil (or rock, in the case of groundwater recharge). (6)

Permian Period The last period in the Paleozoic Era, at the end of the Gondwanan Ice Age, from about 300 to 250 million years ago, coinciding with perhaps the most complete mass extinction other than the one occurring presently. (11)

Peroxyacetyl nitrate A family of secondary pollutants formed when nitric oxide from photodissociated nitrogen dioxide combines with volatile organic compounds. (12)

Peru current *See* Humboldt current.

Phanerozoic Eon The current eon of geological time, beginning approximately 540 million years ago with the beginning of the Paleozoic Era and the evolution of the first fish and shellfish. (11)

Photochemical smog A secondary pollutant formed when sunlight interacts with a primary pollutant, triggering a chemical change; also known as *brown smog* and *L.A.-type smog*. (12)

Photodissociation The splitting apart of molecules caused by exposure to radiation. (2, 12)

Photosynthesis The process by which green plants derive energy through the breakdown of food, releasing oxygen to the atmosphere. (2, 4, 5, 10, 11, 12, 14)

Physical climatology The study of the nature of atmospheric energy and matter at long time scales. (1)

Pineapple Express A type of midlatitude wave cyclone moving from near Hawaii across California, common during El Niño winters. (9)

Plane of the ecliptic The imaginary plane bisecting Earth and the Sun, on which Earth and other planets revolve about the Sun. (3)

Planetary albedo The component of solar radiation reaching the top of Earth's atmosphere reflected by cloud tops, scattered by atmospheric particles back to space, or reflected from the surface all the way out to space. (5)

Planetary boundary layer The lowest 500 to 1000 meters of the atmosphere, including the laminar, roughness, and transition layers. (5)

Planetary scale The broadest spatial scale of climatological analysis, including phenomena on the order of 10,000 to 40,000 km (6,000 to 24,000 mi) and persisting from weeks to months. (1, 5, 13, 14)

Planetary vorticity The spin acquired by a body as a consequence of its location on another rotating body, such as Earth. (7)

Planetesimal Small celestial bodies of condensed debris moving over wildly eccentric orbits about the Sun during the formation of the solar system. (2)

Pleistocene Epoch The interval of geological time from about 1.8 million years ago until about 10,000 years ago. (4, 11)

Plumbism Any brain or neurological disorder caused by lead poisoning. (12)

Pluvial lake An ancient lake that existed in a wetter and/or colder period than today. (11)

PM$_{2.5}$ The category of particulates with diameters less than 2.5 µm; also known as *fine particles*. (12)

PM$_{10}$ The category of particulates with diameters less than 10 µm. (12)

Polar cell The feature of the general circulation from the pole to about 60° latitude, including the polar high, surface polar easterlies, subpolar lows, and upper-level westerlies in each hemisphere. (7, 9, 13)

Polar climate The coldest of the broad categories of climate types in the Köppen system. (9, 10)

Polar easterlies The prevailing general circulation feature from the pole to about 60° latitude in both hemispheres at the surface, in which flow occurs primarily from east to west. (7, 13)

Polar front jet stream A large river of fast-moving air in the upper troposphere over the midlatitudes in each hemisphere that transports energy and moisture via the Rossby waves; exists at zones where cold air is adjacent to much warmer air. (4, 7, 9, 13)

Polar high The semipermanent surface anticyclone that sits over or near the North and South Poles. (7)

Porosity The ratio of the volume of all subsurface pore spaces in soil or rock to the volume of the whole. (6)

Positive feedback system An input that creates change to a system in such a way that additional, similar changes in the system occur as a result. (4, 10, 11, 12, 14, 15)

Positive vorticity Rotation that is counterclockwise in the northern hemisphere or clockwise in the southern hemisphere; cyclonic rotation. (7, 9)

Potential evapotranspiration (PE) The maximum amount of evapotranspiration that would occur at a place over a given time if water were not a limiting factor. (5, 6, 8, 9, 10)

Potential temperature The temperature that a parcel of air would have if moved dry adiabatically from its height in the atmosphere to the 1000-mb level. If the lapse rate of potential temperature is positive, the atmosphere is stable; if it is negative, the atmosphere is unstable. (5)

Potential vorticity The ratio of the sum of relative vorticity and planetary vorticity to the height of a column of mass; tends to remain constant. (7, 9, 11)

Precession The direction of Earth's axial tilt at any given time in geological history. (11)

Precipitable water The depth of water (in centimeters) in a column of atmosphere if all water molecules in that column condensed and precipitated. (6)

Pressure The amount of force exerted on a given area. (1, 3, 13)

Pressure gradient force (PGF) The force that initiates movement of air from areas of higher pressure to areas of lower pressure at right angles to the isobars, in fulfillment of the second law of thermodynamics. (3, 5, 7, 10, 13)

Prevention A strategy for dealing with climate change that implements tactics designed to reduce the threat. (12)

Primary climate impact A natural or social system that is affected most directly by climate. (14)

Primary data Data collected by an investigator using thermometers, radiometers, moisture sensors, or other equipment in the field. (14)

Primary pollutant Any contaminant that enters the atmosphere as a direct emission by human activities. (12)

Prime meridian The meridian of longitude that runs through Greenwich, England, and corresponds to 08 longitude. (3)

Primitive equations of motion A simplification of the Navier-Stokes equations of motion in which acceleration of motion in the west–east, south–north, and down–up directions is assumed to be zero. (13)

Principal components analysis (PCA) The type of eigenvector technique most often used in climatology. (8, 13)

Principal meridian The lines of longitude spaced evenly at 15° intervals, beginning at 0° (the prime meridian), along which zone mean time corresponds to solar time. (5)

Principle of Superposition The axiom that older geological materials will be found successively lower in sedimentary rock layers than younger materials; used in most relative dating techniques. (11)

Principle of Uniformitarianism The axiom that the processes that occurred in the past are the same ones that occur in the present, at the same rates, and that the same environmental conditions required by a given species today were also favorable for that or similar species in the past. (11)

Prokaryote Single-celled ancestors of bacteria and blue-green algae that existed in Earth's primordial ocean. (2)

Proterozoic Eon The eon of geological time prior to the current eon, lasting from approximately 2.5 billion years ago to approximately 550 million years ago. (11)

Proxy evidence Indirect clues about past climatic conditions, used when direct measurements are not available; includes geological, biological, geophysical, and historical clues. (1, 11, 12)

Pyranometer A type of radiometer that measures the flux of shortwave radiant energy. (5)

Pyrgeometer A type of radiometer that measures the flux of longwave radiant energy. (5)

Pyrradiometer A type of radiometer that measures the combined flux of longwave and shortwave radiant energy. (5)

Quasi-biennial oscillation (QBO) A continual reversal of wind direction in the lower stratosphere with an average period of 26 months, alternating between the Krakatau easterlies and the Berson westerlies. (10, 11, 13)

Quaternary Period The current period of geological history, encompassing the last 1.8 million years and coinciding with the intensification of the present ice age. (11)

Radiant energy Radiation transmitted by means of electromagnetic waves. (3)

Radiation balance equation The expression of net radiation as the sum of the net shortwave radiant energy at the surface and the net longwave radiant energy at the surface. (5)

Radioactive decay The rate at which spontaneous nuclear disintegration of certain isotopes of certain elements occurs. (11)

Radiometer Any instrument that measures the intensity of radiation (in Watts) over a unit area (in square meters) of the surface. (5, 14)

Radiometric dating A laboratory technique used to determine the approximate age of a specimen, based on the principle of radioactive decay. (11)

Radiosonde An instrument package that ascends with a weather balloon and measures the vertical profile of atmospheric properties such as temperature, humidity, wind speed and direction, and pressure. (5, 11, 12)

Rain machine Nickname given to tropical rain forest climates where a large percentage of the evaporated and transpired water falls locally as precipitation. (4, 10)

Rain shadow effect The dryness experienced on the leeward side of a mountain, both because the moisture and precipitation is left on the windward side where the air ascends the slope and because of the adiabatic warming associated with descending air on the leeward side. (9, 10)

Random variability A fluctuation from the mean of any atmospheric variable that does not appear to be caused systematically; also known as *noise*. (11)

Reanalysis dataset A widely used source of global climatic data for dynamic and synoptic studies produced by the U.S. National Center for Atmospheric Research and the National Centers for Environmental Prediction. (14)

Reflection A change of direction of a beam of radiant energy that is not scattered in a multitude of different directions. (5)

Refraction The bending of light when it encounters a medium of different density. (3)

Regional Climate Center One of the six offices distributed around the United States that specializes in the archival and dissemination of climatic data and monitoring weather and climate events in their respective regions. (14)

Regional climate model (RCM) A representation of circulation that zooms in on the particular region of interest and usually contains much greater spatial resolution than general circulation models because the geographical area of interest is smaller. (13)

Regional climatology A description of the climate of a particular part of Earth's surface, sometimes including an explanation of the processes generating that climate. (1)

Regression line A line drawn across a scatterplot of data points designed to pass as closely as possible to as many points as possible, so as to minimize the squared sum of the distances between each point and the line in the y direction. (8)

Relative dating Any means of assessing the age of rocks, sediments, or fossils only in comparison with (i.e., younger than or older than) that of some other rock, sediment, or fossil based on the principle of superposition. (11)

Relative humidity The ratio of vapor pressure divided by saturation vapor pressure. (5, 6, 9, 12)

Relative vorticity The spin of a body that occurs because the object itself is turning rather than because the object on which it rests is turning. (7)

Renewable Resource Data Center An agency within the U.S. Department of Commerce that provides solar radiation and wind data for a series of sites around the United States. (14)

Reservoir A component of a system that effectively stores matter and/or energy for a certain period of time, after which it allows for the movement of that matter and/or energy to another component of the system. (2, 6, 12)

Residence time The mean length of time that an individual molecule remains suspended in the atmosphere. (2)

Residual The variance in a data set that remains unexplained by a regression line or a component. (8)

Resistance A component of force exerted by air that works parallel and in the opposite direction to the turbulent flux; in general, relatively high during times of static stability and lower during times of static instability. (6)

Resolution The sharpness of observations either in space or time. (13)

Revolution The path of a planet around the Sun; one complete Earth revolution requires approximately 365.25 days. (3)

Ridge Any elongated area of high atmospheric pressure on a map. (4, 7, 8, 9, 10, 13)

Riming The deposition of water vapor directly onto a solid surface. (10)

Rock striation A groove cut into bedrock by rocks embedded within the base of moving glaciers. (11)

Ronne Ice Shelf One of two thick, vast sheets of ice protruding from Antarctica into the Weddell Sea, which opens to the Atlantic Ocean. (10)

Ross Ice Shelf A thick, vast sheet of ice protruding from Antarctica into the Ross Sea, which opens to the Pacific Ocean. (10)

Rossby wave Upper-tropospheric, midlatitude waves with wavelengths of hundreds to thousands of kilometers. (7, 8, 9, 10, 13)

Rotation The spin of a body on its axis; one complete rotation of Earth requires approximately 24 hours. (3, 5)

Roughness layer A layer of very strong mechanical turbulence above the laminar layer, extending to approximately 50 to 100 m above Earth's surface. (5)

Roughness length The height above Earth's surface at which the air theoretically begins to move laterally in the logarithmic wind profile. (5)

Runoff Precipitated water that runs downhill to feed streams. (6, 10, 12)

Sahel A semiarid area of north central Africa south of the Sahara Desert, where widespread desertification has occurred in the last 30 years. (4, 10, 13)

Saltwater intrusion Movement of saltwater upstream into a freshwater estuary or stream. (6, 12)

Santa Ana wind A warm, dry wind resulting from descent of air on leeward sides of the mountains of southern California. (9)

Sargasso Sea The region of the North Atlantic Ocean where the Bermuda-Azores high is located. (7)

Saturated adiabatic lapse rate The rate of decrease of temperature as an adiabatically moving air parcel with a relative humidity at 100 percent rises; always less than 10 C°/km. (5)

Saturation vapor pressure (e_s) The atmospheric pressure exerted by water vapor when the air is at saturation. (5, 6, 7)

Scattering The deflection of electromagnetic radiation is in various directions upon interacting with matter in the atmosphere. (5)

Score In eigenvector analysis, a coefficient that is proportional to the degree to which the eigenvector model represents the variability in each observation of one set of variables collectively for all observations in another set of variables in the data set simultaneously; in climatology, it is common for the scores matrix to represent the relationship between each component and each time of observation, collectively across all stations simultaneously. (8, 13)

Scree plot A graph depicting the eigenvalue for each successive component or factor in eigenvector analysis. (8)

Second law of thermodynamics The fundamental principle that energy moves from areas of higher concentration to areas of lower concentration, with the result that disorder, or entropy, increases in a system. (2, 7)

Secondary circulation A smaller circulation feature that is embedded within the larger general circulation, such as a midlatitude wave cyclone or migratory anticyclone. (7, 8)

Secondary climate impact A natural or social system that is affected only indirectly by climate. (14)

Secondary data Data used by researchers after being collected, quality-controlled, and compiled elsewhere, usually in digital format. (14)

Secondary pollutant A contaminant that arrives in the atmosphere as a byproduct of a chemical reaction between primary pollutants and/or other matter or energy in the atmosphere. (12)

Sensible energy (heat) Radiant energy that heats the Earth–ocean–atmosphere system rather than evaporates water. (3, 4, 7, 9, 10, 12, 13)

Sensible heat flux *See* Turbulent flux of sensible heat.

September equinox The day of the year midway between the June and December solstices, when the solar declination is at the equator, neither hemisphere is tilted toward or away from the Sun, and day length is 12 hours over the entire Earth; currently corresponds roughly to September 22. (3)

Shearing stress *See* Vertical flux of horizontal momentum.

Shortwave radiation Electromagnetic radiation at wavelengths shorter than 4.0 μm, usually considered synonymous with radiation emitted by the Sun. (2, 5, 9, 12, 14)

Siberian high The strong surface anticyclone that persists through winter over the high latitudes of continental Asia. (7, 9, 10)

Sinistral An individual of a species of foraminifera whose shell developed with a coiling on the left side, thereby suggesting that the ocean was relatively cold at the time that the individual lived. (11)

Sink A route or mechanism by which matter or energy is removed from circulation in a system by being held in a reservoir for an extended period of time. (2, 4, 12)

Skew T–Log P Diagram A type of thermodynamic diagram that shows temperature on lines slanted upward emanating from the x-axis and the logarithm of pressure along the y-axis. (5)

Slab model A computer model in which water is assumed to have a different albedo than land and a constant depth and ability to store heat. (13)

Slash and burn agriculture A type of migratory farming in which forests are continually cleared and burned. (4)

Smog A combination of smoke and fog. (12)

Snowball Earth hypothesis The controversial idea that conditions were so cold during the Cryogenian Period that ice even existed toward the equatorial land masses and oceans. (11)

Soil heat flux plate An instrument for measuring the substrate heat flux. (5)

Soil moisture recharge In the surface water balance, the amount of precipitated water (in centimeters or inches of equivalent precipitation) during a given time period that replenishes the storage in the soil. (6, 10)

Soil moisture storage capacity *See* Field capacity.

Soil moisture withdrawal In the surface water balance, the amount of soil water (in centimeters or inches of equivalent precipitation) used by the atmosphere during a given time period to fulfill or partially fulfill the atmospheric demand. (6)

Solar declination The parallel of latitude on which the direct rays of the Sun fall on a given day; varies between the Tropic of Cancer and the Tropic of Capricorn throughout the year. (3, 5, 7, 10)

Solar noon The time when the Sun is at its highest point in the sky for that day at that location. (3, 5, 7)

Solar time The "true" time at a location based on the location's position with respect to the Sun; incorporates the zone mean adjustment and equation of time adjustment; also known as *local apparent time*. (5)

Solar wind Radioactive particles from the Sun moving through space at the speed of light. (2)

Somali jet A low-level, fast-moving river of air that provides southwesterly flow paralleling the Somali coast and advects moisture into southern Asia in association with the Asian summer monsoon system; also known as the *Findlater jet*. (10)

Source area Locations in the ocean near the surface that provide water to the depths. (4)

Source region The place where an air mass forms and obtains its characteristics. (8, 9)

South Atlantic high The semipermanent anticyclone in the subtropical South Atlantic Ocean. (7, 10)

South Pacific Convergence Zone (SPCZ) A band of convection, most active during the summer months when surface heating is maximized, that branches southward or southeastward from the Intertropical Convergence Zone in the western South Pacific Ocean. (10)

South Pacific high The semipermanent anticyclone in the subtropical South Pacific Ocean. (7, 10)

Southeast trade winds The dominant surface circulation pattern in the tropical southern hemisphere, with consistent winds blowing from southeast to northwest. (4, 7, 10)

Southern Oscillation A see-saw effect (teleconnection) of surface atmospheric pressure between the eastern and western equatorial Pacific Ocean in which higher-than-normal pressure in one of these two regions is coincident with lower-than-normal pressure in the other. (4, 10, 11, 13)

Southern Oscillation Index (SOI) A measure of the condition of the Southern Oscillation at a given time, derived from sea level pressure differences between the eastern Pacific (represented by Tahiti) and western Pacific (represented by Darwin, Australia); positive values represent La Niña conditions, whereas negative values represent El Niño conditions. (4, 13)

Southwesterly monsoon The continental-scale, prevailing surface airflow over much of Asia in summer, characterized by onshore flow. (9, 10)

Spatial downscaling The process of deriving a statistical relationship between general circulation model output and atmospheric conditions at a finer grid scale than the general circulation model can provide. (13)

Spatial resolution The distance between observations of a variable at a given location. (13)

Spatial Synoptic Classification An automated method of categorizing climates in which eigenvector analysis is conducted on atmospheric variables over time at a single site, the analysis is repeated for many other sites, and locations where similar characteristics of the atmosphere tend to occur simultaneously are clustered in the same category. (8)

Specific heat The amount of heat (energy) required to raise the temperature of a 1 g mass by 1 C° or 1 K; distinguished from heat capacity by the fact that heat capacity refers to a unit volume rather than a unit mass. (3, 7, 12)

Specific heat at constant pressure The amount of energy that must be added to 1 kg of a substance (in this case, air) to raise its temperature by 1 C° (or 1 K) assuming that pressure does not change; approximately equivalent to 1005 J kg^{-1} K^{-1}. (13)

Specific humidity The ratio of the mass of moist air to the total mass of the air. (5, 10, 13)

Specific volume The amount of space occupied by a solid, liquid, or gas divided by its mass; the reciprocal of density. (13)

Spectral model A general circulation model that expresses motion in terms of a series of waves. (13)

Squall line A line of thunderstorms within the warm sector of a midlatitude cyclone in advance of a cold front. (9)

Stability A condition of the local atmosphere at a given place and time that describes the likelihood of a hypothetical parcel of air to rise or sink spontaneously. (5, 6, 7, 9, 10, 12)

Stable atmosphere An atmospheric condition in which rising motion is suppressed because the relevant adiabatic lapse rate exceeds the environmental lapse rate, thereby making the hypothetical adiabatically moving air parcel cooler than its surrounding environment and decreasing its buoyancy. (4, 5, 12)

Stable stratification Any situation in which colder waters underlie warmer water; analogous to a temperature inversion in the ocean. (4, 8)

Standard nonrecording rain gauge The nonautomated instrument for measuring rainfall that is used at the vast majority of the 13,000 precipitation gauges in the United States, including all sites reported in *Climatological Data*. (14)

State Climatology Program Offices in each U.S. state that are responsible for the archival and dissemination of current and historical atmospheric data from that state to the public and private sectors. (14)

Stationary front A narrow zone separating two air mass types in which neither is actively displacing the other at a given time. (8)

Step change An abrupt shift to a different mean, without a slow "drift" toward that mean. (11)

Steppe (BS) climate In the Köppen system, a climatic type characterized by potential evapotranspiration that moderately exceeds precipitation. (8, 9, 10)

Stomate A tiny pore in leaf tissue, through which transpiration occurs. (5, 6)

Stommel model A simplified model of the general circulation of the deep ocean. (4)

Storm Data A monthly publication by the National Climatic Data Center of the National Oceanic and Atmospheric Administration that provides a chronological summary, by U.S. state, of severe and unusual weather that occurred in that month, including data on storm trajectories, damage, and casualties resulting from each event. (14)

Stratiform cloud A set of liquid droplets or ice crystals suspended in the atmosphere oriented in a horizontal sheet, characterized by widespread cloudiness and perhaps long periods of precipitation of weak or moderate intensity. (8, 9, 10)

Stratopause The thin layer of the atmosphere separating the stratosphere from the mesosphere. (2)

Stratosphere The layer of the atmosphere between the troposphere and mesosphere, containing the ozone layer and characterized by a temperature inversion; begins at approximately 8 to 20 km (5 to 13 mi) above the surface and extends to approximately 48 km (29 mi) above the surface. (2, 4, 10, 11, 12)

Streamline analysis A form of atmospheric analysis that uses lines to represent the flow at any given time and level of the atmosphere. (10)

Subarctic (Dfc, Dwc, Dwd) climate In the Köppen system, a climatic type of the high latitudes characterized by at least one month with a climatological mean temperature above 10°C (50°F) but extremely cold winters, very high annual temperature ranges, and adequate moisture. (9)

Subduction zone A location where a crustal plate is displaced beneath an adjacent, less-dense plate. (11)

Sublimation The transformation of water molecules from a solid directly to a gaseous state. (4, 5, 12)

Subpolar low An area of semipermanently low surface pressure centered on the oceans at about 60° latitude in each hemisphere. (7)

Substrate heat flux (Q_G) The flow of energy from the surface into the ground (and vice versa) via conduction. (5, 6)

Subtropical anticyclone A semipermanent surface high-pressure cell at latitudes of approximately 20 to 30° in each hemisphere over the oceans. (4, 7, 8, 10, 12)

Subtropical Hot Desert (BWh) climate In the Köppen system, a climatic type characterized by potential evapotranspiration that far exceeds precipitation and a climatological mean temperature above 18°C (64°F). (9, 10)

Subtropical jet stream A large river of fast-moving air generally from west-to-east in the upper troposphere near 30° latitude in each hemisphere; exists because of the increasing Coriolis deflection of accelerating poleward-moving air as angular momentum is conserved. (4, 7)

Sulfur dioxide (SO_2) A colorless, irritating gas that is emitted from volcanic eruptions and combustion of fossil fuel. (4, 12)

Sulfuric acid (H_2SO_4) A very corrosive acid that forms when sulfur dioxide (SO_2) reacts with water to produce H_2SO_3 (sulfurous acid), which can then react with oxygen to form H_2SO_4. (12)

Summer monsoon The continental-scale circulation feature, particularly in Asia, whereby surface air moves inland because of a thermally driven low-pressure area over the continent. (9)

Summer solstice The day of the year when the solar declination is at its farthest poleward extent in the observer's hemisphere, the observer's hemisphere is tilted maximally toward the Sun, and day length is maximized in the observer's hemisphere; currently corresponds to June 21 for a northern hemisphere observer and December 22 for a southern hemisphere observer. (3, 5)

Summer thermal efficiency concentration In the Thornthwaite Climatic Classification System, the ratio of potential evapotranspiration in the summer months to total annual potential evapotranspiration at a location. (8)

Sun path diagram A graph that shows the position in the Sun using zenith and azimuth angles, relative to an observer at a given parallel of latitude, throughout the course of a day and year. (5)

Sunspot A huge magnetic storm that appears as a darker region on the Sun's surface, varying in number and intensity over time. (11)

Surface airways observation A collection of hourly atmospheric data for more than 500 stations around the United States and 12,000 stations worldwide. (14)

Surface boundary layer (SBL) The lowest 100-or-so meters of the troposphere, characterized by constant fluxes of energy, matter, and momentum with height (not across space) at a given instant in time; includes the laminar and roughness layers. (5, 9)

Surface tension The attraction of molecules to each other within water. (6)

Surface water balance The representation of the hydrological cycle at a local scale, accounting for the input, output, and storage of water and interactions with available surface energy and moisture. (6, 9, 10, 13, 14, 15)

Surplus In the surface water balance, the amount of water (in centimeters of equivalent precipitation) that is supplied to the surface as precipitation in a given time period in excess of that which can be evapotranspired or stored in the soil. (6, 8, 9, 10)

Sustainability Development without depleting resources or degrading the environment to an extent that would compromise the ability of future generations to meet their own needs. (15)

Swamp model A computer model in which water is assumed to have a different albedo from land but no capability to store heat. (13)

Synoptic climatology The study of the relationship between regional-scale atmospheric circulation and environmental features at the surface. (1, 13)

Synoptic scale A spatial scale of analysis between about 100 and 10,000 km (60 and 6000 mi), including atmospheric phenomena that typically persist from days to weeks. (1, 13)

Taiga The Russian term for the boreal forest. (9)

Tambora An Indonesian volcano that erupted in 1815 and is alleged to have caused the "Year without a Summer" of 1816. (4, 11)

Teleconnection A set of locations that tend to have a "see-saw" pattern of pressure and ridging/troughing with each other at some height in the troposphere. (7, 11, 13, 15)

Temperate Zone The area of Earth having a moderate climatic type in the climatic classification system of the ancient Greeks. (8)

Temperature inversion Any situation in which the temperature of the static atmosphere increases as height increases. (2, 5, 9, 10, 12)

Temporal downscaling The process of projecting output from a general circulation model at one temporal scale to a finer temporal scale using statistical relationships. (13)

Temporal resolution The amount of time between observations of a variable at a given location. (13)

Temporal trend A slow drift over time toward an increasing or decreasing mean in some atmospheric variable. (11)

Tertiary Period The first of two periods in the Cenozoic Era, from about 65 million years ago until about 2 million years ago. (11)

Thermal conductivity The ability of a solid to allow heat (energy) to propagate through it; the number of watts per square meter of energy required to raise the temperature at a point 1 meter into the ground by 1 K. (12)

Thermal efficiency index (T/ET) In the Thornthwaite Climatic Classification System, the ratio of temperature to calculated evapotranspiration; increasing values represent increasingly arid climates. (8)

Thermal low Any area of relatively low surface pressure because of the ascent of relatively warm air. (9, 10)

Thermal turbulence Microscale convective motion resulting from unstable atmospheric conditions. (5)

Thermocline The boundary between relatively warm surface waters and colder deep waters. (4, 10)

Thermodynamic diagram Any type of graph that shows a vertical profile of various meteorological variables and can assist with the determination of stability conditions. (5)

Thermodynamic energy equation The equation that expresses the rate of change of temperature over time as the sum of the diabatic heating, adiabatic temperature change due to vertical motion, and temperature advection. (13)

Thermohaline circulation The global, three-dimensional oceanic circulation system that results from inequalities of heat and salinity across large distances. (3, 4, 14, 15)

Thermohaline current Any of the various subcomponents of the global thermohaline circulation. (4)

Thermosphere The highest layer of the atmosphere, beginning at approximately 80 km (50 mi) above Earth's surface and gradually extending to outer space. (2)

Thiessen polygon method A spatial analytic method for converting point data to areal data in which perpendicular bisectors for the lines connecting each adjacent station in the study area are connected, and the percentage of the total area enclosed within each resulting subarea is calculated so that the point-based totals can be weighted by these percentages to determine an overall areal mean value. (6)

Thornthwaite Climatic Classification System A method of categorizing climate types based on the interplay between temperature and moisture as represented by the surface water balance. (8)

Thornthwaite potential evapotranspiration A set of empirical equations based only on the temperature and latitude used to estimate the maximum combined amount of evaporation and transpiration of the location of interest. (6)

Threshold pollutant A contaminant that is not known to be harmful in sufficiently small concentrations but that elicits a response after the concentration exceeds a particular threshold. (12)

Time of observation bias The bias in weather records caused by a permanent change in the meteorological observing station's official time at which observations are measured and recorded. (14)

Time step The length of time into the future that is forecasted by each "run" of a computer model. (13)

Time zone A longitudinally based region across Earth's surface theoretically centered on a principal meridian, in which, by convention, clocks are set to the same time. (3, 5)

Tipping bucket rain gauge An instrument for the automated measurement and recording of precipitation consisting of two small buckets that alternate in the collection of precipitation. (14)

Tornado Alley The swath from south-central Canada southward through central Texas where tornado frequency is the highest on Earth. (9)

Torrid Zone The area of Earth having the warmest climatic type in the climatic classification system of the ancient Greeks. (8)

Trade surge A sudden strengthening in the trade winds that resumes the flow of moisture inland and the wet season in tropical monsoon climates. (10)

Trade winds The dominant near-surface atmospheric circulation systems between the subtropical and equatorial latitudes in each hemisphere, consisting of the northeast trades, southeast trades, and counter-trade winds. (4, 13)

Transition layer The layer in the lower troposphere that extends from the top of the surface boundary layer to approximately 500 to 1000 m above the surface, characterized by strong turbulent fluxes and also a notable decrease in friction with height. (5)

Transpiration The process by which vegetation loses water from its leaf surface that had moved from the soil through its roots and stem. (4, 5, 6, 8, 12)

Transverse wind shear The shearing motions that result from variations in horizontal wind speed across a horizontal surface. (7)

Treeline The line that separates the subarctic from the polar climates, either latitudinally or altitudinally, in the Köppen system. (8, 9, 11)

Triple Bottom Line (TBL) concept The doctrine that social and environmental performance must be considered in addition to economic performance in assessing corporate success. (15)

Tropic of Cancer The 23.5°N parallel of latitude; represents the solar declination on the June solstice—the most poleward extent of the solar declination in the northern hemisphere. (3, 7, 10, 11)

Tropic of Capricorn The 23.5°S parallel of latitude; represents the solar declination on the December solstice—the most poleward extent of the solar declination in the southern hemisphere. (3, 7, 10, 11)

Tropical (A) climate In the Köppen system, any climatic type in which the mean monthly temperature of the coldest month exceeds 18°C (64°F) and moisture is generally adequate through at least half of the year; includes tropical rain forest, tropical monsoon, and tropical savanna climates. (10)

Tropical cyclone Any storm of tropical origin. (3, 9, 10, 11, 12, 13)

Tropical Monsoon (Am) climate In the Köppen system, a climatic type in which the coldest month has a mean temperature above 18°C (64°F) and a short dry season. (10)

Tropical-Northern Hemisphere (TNH) teleconnection A see-saw in geopotential height patterns between the Aleutian low area and the subtropical Mexico–southeastern U.S. region, with a center of opposite sign over northeastern Canada. (13)

Tropical Rain Forest (Af) climate In the Köppen system, a climatic type in which the coldest month has a mean temperature above 18°C (64°F) and precipitation generally exceeds potential evapotranspiration year-round. (10)

Tropical Savanna (Aw) climate In the Köppen system, a climatic type in which the coldest month has a mean temperature above 18°C (64°F), abundant precipitation exists in the high-sun season, and deficits dominate the low-sun season. (10)

Tropopause The boundary separating the troposphere and the stratosphere. (2, 5, 7, 10)

Troposphere The lowest 8 to 20 km of the atmosphere, characterized by temperatures that usually decrease with height, abundant convection, most of the atmosphere's mass, and nearly all of the atmosphere's water—the weather and climate layer. (2, 3, 5, 7, 10)

Tropospheric ozone A secondary, nonthreshold pollutant formed primarily by photodissociation of NO_2 in the presence of ultraviolet radiation that seeps into the troposphere; monitored by the U.S. Environmental Protection Agency as a criteria pollutant. (12)

Trough Any elongated area of low atmospheric pressure on a map. (4, 7, 9, 10, 13)

True Desert (BW) climate In the Köppen system, a climatic type characterized by potential evapotranspiration that far exceeds precipitation in most months. (9, 10)

Tundra Any of various types of polar shrubs and successively shorter polar vegetation. (12)

Tundra (ET) climate In the Köppen system, a climatic type of the high latitudes characterized by mean monthly temperature below 10°C (50°F) in each month but with at least one month having a climatological mean temperature above freezing. (8, 9, 10)

Turbulence Any irregular fluctuation of flow in a fluid. (5)

Turbulent flux of latent heat (latent heat flux) The convective transport of energy absorbed during evaporation and sublimation, usually from the surface upward. (5)

Turbulent flux of sensible heat (sensible heat flux) The convective transport of energy that can be felt as heat, usually from the surface upward. (5)

Turc method A procedure for estimating daily evapotranspiration as a function of mean daily temperature, incoming shortwave radiation, and a humidity-based value. (6)

Turkana jet stream An easterly or southeasterly low-level, fast-moving river of air that exists between the Ethiopian and East African highlands, peaking in intensity between February and March. (10)

Typhoon Regional name given to the strongest type of tropical cyclone in the western Pacific Ocean. (10)

Ultraviolet (UV) radiation Electromagnetic radiation having wavelengths between approximately 0.005 and 0.4 micrometers. (2, 10, 12)

Undercatch The reduced measurement of precipitation in a rain gauge compared with actual precipitation fallen, as a result of factors such as the lateral wind movement across the top of the gauge, water remaining on the side of the tube rather than at the bottom, and evaporation from inside the tube. (6, 14)

United States Weather Bureau The forerunner of the National Weather Service, established in 1891 when the Federal Weather Bureau was renamed and moved to the U.S. Department of Agriculture. (14)

Unsaturated adiabatic lapse rate (UALR) The rate of change of temperature as an adiabatically moving air parcel with a relative humidity below 100 percent rises or sinks; always 10 C°/km. (5, 9)

Unstable atmosphere An atmospheric condition in which rising motion occurs because the environmental lapse rate exceeds the relevant adiabatic lapse rate, thereby making the hypothetical adiabatically moving air parcel warmer than its surrounding environment and increasing its buoyancy. (4, 5, 12)

Updraft A small-scale upward thrust of air in an unstable atmosphere, mainly associated with cumuliform clouds. (3)

Upwelling An ascending current of usually cold water caused by the advection of surface waters. (3, 4, 10)

Urban canyon The narrow zone between buildings in an intensely urbanized area. (12)

Urban heat island An isolated zone of relatively high temperatures in built-up areas; caused by the lack of vegetation, decreased evaporative cooling, waste heat from domestic and industrial processes, and thermal properties of construction materials. (1, 3, 12)

Vapor pressure (e) The contribution of the total atmospheric pressure exerted by water vapor molecules. (5, 6, 10)

Variable gas A component of the atmosphere that varies in its concentration across space and/or time. (2)

Variability change A temporal drift to larger or smaller fluctuations about the mean value of an atmospheric variable, with or without a change in the mean. (11)

Varve An annual set of layers of silt and clay deposited on the bottoms of lakes and ponds that freeze in winter and thaw in summer, thereby giving proxy evidence of the length of the frost-free season. (1, 11)

Veering A clockwise change in wind direction with increasing height. (3)

Velocity convergence A situation in which more mass is entering an imaginary "box" of atmosphere than is leaving (from one or more directions); accounted for by the *continuity equation*. (13)

Velocity divergence A situation in which more mass is leaving an imaginary "box" of atmosphere than is entering (from one or more directions); accounted for by the *continuity equation*. (13)

Vernal equinox The day of the year midway between the winter and summer solstices, when the solar declination is at the equator, neither hemisphere is tilted toward or away from the Sun, and day length is 12 hours over the entire Earth; currently corresponds roughly to March 21 (the March equinox) for a northern hemisphere observer and September 22 (the September equinox) for a southern hemisphere observer. (3)

Vertical flux of horizontal momentum (shearing stress) The downward propagation of horizontal momentum caused by vertical wind shear. (5)

Vertical wind shear The change in horizontal wind speed and/or direction with height. (5)

Vertical zonation Climatic variety caused by local-scale elevation differences. (9)

Virtual temperature The temperature that dry air would have if its pressure and volume were equal to a sample of moist air; always greater than or equal to temperature. (13)

Visualization Any graphical technique that uses sequences of maps to show spatial features across time. (15)

Vog Volcanic smog. (4)

Volcanic Explosivity Index (VEI) A scale between 1 and 8 used to describe the magnitude, intensity, dispersion, and destructiveness of individual volcanic events. (4)

Vorticity The rotation or spin of any object. (7)

Walker circulation The "normal" atmospheric circulation system in the equatorial Pacific Ocean, characterized by high surface pressure and concurrent sinking motion in the east and relatively low surface pressure and concurrent rising motion in the west, with easterly trade winds at the surface and westerlies aloft. (4, 13)

Warm front A narrow zone separating two air mass types in which the warmer air mass is actively displacing the colder air mass. (8)

Warm sector The wedge-shaped area of a midlatitude wave cyclone between the cold front and the warm front. (8)

Wavelength The distance between two successive crests or ridges in any wave. (2, 5, 7, 12)

Weather The overall instantaneous condition of the atmosphere at a certain place and time. (1)

West African monsoonsystem A seasonal reversal of winds in tropical western Africa. (10)

West-coast occlusion An occluded front associated with the movement of maritime air intruding upon continental air as the system migrates inland in the midlatitude westerlies; characterized by conditions similar to a warm front, with gentle uplift of moist cool air, stratiform clouds, and drizzle. (9)

Western-boundary intensification Any very deep, fast-moving western ocean basin current, such as the Gulf Stream. (4)

Wet bulb depression The difference between the dry bulb temperature and the wet bulb temperature when measuring relative humidity. (6)

Wet deposition A situation in which acid precipitation is produced by reactions between precipitation and primary pollutants before the falling precipitation reaches the surface. (12)

Whiteout An atmospheric condition caused primarily by wind-blown ice that makes the sky appear to blend with the snow-covered surface, eliminating the horizon. (10)

Wilting point The minimum amount of water (in centimeters or inches of precipitation equivalent) that is necessary in the rooting zone to allow extraction by plants. (6)

Wind Air in motion; the transfer of atmospheric mass from one location to another. (1, 3)

Wind channeling The constriction of air as it flows between narrow obstructions, triggering a velocity increase. (9, 10, 12)

Wind-chill factor A contrived "temperature" that represents human comfort by taking both temperature and wind speed into account. (10)

Windward Upwind; the side of a mountain facing the wind. (3, 9, 10)

Winter solstice The day of the year when the solar declination is at its farthest poleward extent in the opposite hemisphere, the observer's hemisphere is tilted maximally away from the Sun, and day length is minimized in the observer's hemisphere; currently corresponds to December 22 for a northern hemisphere observer and June 21 for a southern hemisphere observer. (3, 5)

Wisconsin Glacial Phase The last major glacial advance, peaking at about 18,000 years ago. (4, 10, 11, 12)

World Climate Research Programme (WCRP) An initiative sponsored by the World Meteorological Organization to address climatic research questions, especially those that promote the successful prediction of global and regional climatic variations. (14)

World Data Center for Paleoclimatology A branch of the U.S. National Climatic Data Center that acts as a repository for paleoclimatological data sets at thousands of locations worldwide. (14)

World Meteorological Organization (WMO) An agency of the United Nations that dispenses a variety of climatological data products for use in large-scale synoptic and dynamic climatological studies, in addition to promoting the protection of the international atmospheric environment and protecting people from atmospheric hazards. (14, 15)

Younger Dryas A brief but abrupt and notable "lapse" back to cold conditions after the ice had retreated following the last glacial maximum, lasting from about 13,000 years to 11,500 years before the present time; known as the *Loch Lomond Stadial* in Europe. (11)

Zenith angle A coordinate indicating the number of degrees that the Sun is located away from the vertical relative to an observer's location, ranging from 0° to 90°. (5)

Zonal flow Rossby waves with low amplitudes (i.e., deamplified ridges and troughs), characterized by strong west–east and weak south–north and north–south components of atmospheric motion. (7, 13)

Zone mean time A commonly used system of timekeeping in which the "clock time" at a location is synchronized across a range of longitudes so that no adjustment is necessary except when crossing from one time zone to another. (5)

Index

Italicized page locators indicate a figure or photo; tables are noted with a *t*.

A

ablation, 74
  in Antarctica, 233
  snow/ice, 73–74
absolute dating, 271
absolute humidity, 91
absolute vorticity, 141–142, 166
absolute zero, 36, 287
  radiation and, 87
  thermodynamic energy equation and, 316–317
absorption, 81, 85–86
acacia trees, 239, *239*
acceleration, 11
Accra, Ghana, 246
  climograph for, *247*
  water balance diagram for, *247*
accumulated potential water loss, 120
accumulation, snow/ice, 73–74
acid fog, 303
acid precipitation, 303–304
action centers, 142
adaptation, 304–305
adiabatic heating, 316
adiabatic lapse rate
  saturated, 96, *97*
  unsaturated, 95–96, *97,* 177
adiabatic motion, 95–96, 316–317
adiabats, 98
advection
  in air masses, 164
  in baroclinic zones, 143–144
  continentality and, 34
  on land vs. water, 132–133
  radiation and, 87
  vorticity, 141
  water flux via, 105–106
advection fogs, 200, 230
aerosols, 11
  general circulation of, 126
  in global dimming, 297–298
  hygroscopic nuclei, 293
  indices of, 70–71
  insolation absorbed by, 86–87
  insolation scattered by, 27–28
  from volcanic activity, 70
Africa
  climatic setting of, 222–225
  desertification in, 72–73, *73*
  Marine West Coast climates in, 251
  physiographic features of, *222*
  temperature range in, 51
Aghulas current, 223
agricultural climatology, 6, 7
agriculture
  climate impacts on, 334
  in deforestation, 71–72
  in desertification, 72–73
  droughts and, 223
  insolation levels and, 81
  methane in, 290
  slash and burn, 71–72
  temperature data and, 343
Agung, Indonesia, eruption, 70*t*
airflow
  convergence and divergence in, 144, *144*
  in cyclones and anticyclones, 43–45, *44, 45*
  diffluence and confluence in, 145, *145*
  meridional, 137–138, *138*
  over mountainous terrain, 142–143, *143*
  upper-level, 137–147
air masses, 162–164
  in African climates, 223–225
  Rocky Mountains and, 177
  in tropical atmospheres, 220–221
air pollution, 6, 297–299. *See also* pollution
  classification of pollutants in, 299–304
Air Pollution Control Act (1955), 299
air quality legislation, 299
Alaska
  Subarctic climate in, 210
  Tundra climates in, 213
  volcanic activity in, 69, 70*t*

albedo
   in Antarctica, 233
   definition of, 70, 86
   desertification and, 73
   ice and, 74–75
   insolation and, 86–87, 87*t*
   planetary, 86
   polar cells and, 128–129
   in Tundra climates, 214
   volcanic activity and, 70
Alberta clippers, 177, 178
Albuquerque, New Mexico
   climate in, 187
   climograph for, 187, *188*
   water balance diagram for, 187, *188*
Aleutian Islands
   latitude of, 174
   volcanic activity in, 69
Aleutian low, 136
Alice Springs, Australia
   climograph for, *247*
   water balance diagram for, *247*
Alps, 180–181
   Highland climates in, 216
   Mediterranean climates and, 201
altitude
   in Antarctica, 232
   geopotential height and, 131–132
   temperature and, 47, 95–97
Amazon basin
   Atacama Desert and, 244
   climatic effects of, 230, *230*
   deforestation of, 231
   in ENSO events, 66
   orographic influences of, 245
   runoff in, 106
AmeriFlux program, 345, *345*
ammonia (NH_3), 18
amplitude
   jet stream, 67, 68
   of upper-level airflow, 137
anaerobic decomposition, 290
analemmas, 84, *84*
analog forecasting, 309
Andes Mountains, 230, *230,* 231, 244
angiosperms, 265, 275
angular momentum, conservation of, 130–131
animal migrations, 67, 240–241, 276
annual climatic normals, 262
Annular mode, 325–326
anomalies, 226
   climatic change and, 260–261, 262
Antananarivo, Madagascar
   climograph for, *252*
   water balance diagram for, *252*

Antarctica
   climatic setting of, 232–233
   Ice Cap climate in, 253–254
   ice sheet melting in, 75, *266,* 267
   July isotherms in, 49, *49*
   permanent ice in, 74
   physiographic features of, *232*
   Polar climate in, 253
   temperature range in, 51
Antarctic Bottom Water, 60
Antarctic Circle, day length in, 32
Antarctic Deep Water, 60
Antarctic Intermediate Water, 60
anthropogenic changes, 286–308
   adaptation to, 304–305
   in Africa, 223
   in Antarctica, 233
   atmospheric pollution, 297–304
   carbon cycle, 14–16
   climate modification, 4, 8, 262
   continued research on, 305
   deforestation and desertification, 71–73, *73*
   global dimming, 297–298
   global warming, 68–69, 286–297
   mitigation of, 304
   ozone layer damage, 20–21
   preventing, 304
   rapid climate change and, 351
   reactions and attitudes to, 304–305
   sustainability and, 351–353
   variable gases concentrations, 16–17
anticyclones
   airflow in, 43–45, *44,* 45, *45*
   blocking, 181, *181*
   definition of, 44
   Mediterranean climates and, 201
   negative vorticity in, 139, 141
   permanent, 136
   pollution concentrations and, 298
   ridges compared with, 138
   as secondary circulations, 166
   semipermanent subtropical, 47
   subtropical, 54
aphelion, 29, 222
   orbit eccentricity and, 278
   precession and, *279,* 279–280
applied climatology, 6–7, 332–348
   on climate impacts, 333–335
   data sources in, 335–346
   definition of, 332
   on human health and comfort, 335
   scope of, 332
   secondary sources for, 345–346
Arabian Desert, 224
Arabian Peninsula, 224, 242

Aral Sea, 184, 187
Archimedes' Principle, 267
architecture, 81, *81*
Arctic (A) air masses, 164, *164*, 175
Arctic Circle
    day length in, 31–32
    Sun path diagram for, 82
Arctic Ocean, 75, 215
Arctic Oscillation (AO), 325–326
Arequipa, Peru
    climograph for, *256*
    water balance diagram for, *256*
argon (Ar), 12, 13, 13*t*, 16
Arid (B) climates
    forcing mechanisms in, 184–185
    geographic extent of, 184
    Midlatitude Cold Desert, 185–188
    northern hemisphere, 184–191
    southern hemisphere, 242–247
    Steppe, 157, 184, 188–191, 246–247
    Subtropical Hot Desert, 185
    True Desert, 157, 183, 184, 185, 242–246
arid climatic classification, 155, 155*t*, 156
Arizona, humidity in, 35
Arizona State University, 335
artificial neural networks, 351
Asia
    climatic setting of, 181–184
    coastal zones in, 183–184
    continentality effects in, 34, 136
    general characteristics of, 181–182
    Highland climates in, 216
    Humid Subtropical (Cfa) climates in, 192
    January isotherms in, 50, *50*
    July isotherms in, 49, *49*
    katabatic winds in, 216–217
    Mediterranean climate in, 197
    mountain ranges in, 183
    Steppe climates in, 184, 246
    taiga in, 210–211
    temperature range in, 51
    Tundra climates in, 213
Asian monsoons, 182–183, 187, 238
    in Africa, 225
    in India, 241
    katabatic winds in, 216–217
    Quasi-Biennial Oscillation in, 229, 283
    Subtropical Dry Winter climates and, 195
Atacama Desert, 56, 63, 230
    climatic classification of, 242
    cold currents in development of, 244–245
    ENSO events and, 67
Atlantic Multidecadal Oscillation (AMO), 322
Atlantic Ocean
    circulation in, 57

climate effects of on North America, 174
currents in, 47
extratropical teleconnections in, 322–326, *325*
geostrophic flow in, *57*, 57–58
hurricane season in, 230
as source area, 59
subtropical anticyclones in, 128
thermohaline currents in, *59*, 59–60
tropical cyclones in, 248
atmosphere, 11–23
    carbon cycle and, 13–16
    composition of, 12–17, 13*t*, 69
    constant and variable gases in, 16–17, 16*t*
    definition of, 2
    faint young Sun paradox and, 17–18
    free, 42
    interactions of with other "spheres," 3*t*
    local, moisture in, 90–94, *91, 93*
    local, stability in, 97–99
    mass of, 11
    neutral, 95
    origin of, 11–12
    path length in, 27–28, *28*
    spatial-temporal relationships in, 4–5, 5*t*
    stable, 55
    structure of, 18–22, *19*
    unstable, 55
    weather and, 3
atmospheric general circulation models, 310
atmospheric moisture, 90–94, *91, 93*
atmospheric stability, 95–97
    assessing, 97–99
atmospheric statics, 94–99
atmospheric teleconnections, 321–328
atmospheric waves, 138
atmospheric window, 287, 287*t*
atomic mass, 270
attenuation, *27*, 27–28, *28*, 81
aurora australis, 21
aurora borealis, 21, *21*
Australia
    Arid climates in, 242
    Bureau of Meterologocy, 310
    climatic setting of, 225–230
    deserts in, 244
    El Niño-Southern Oscillation events in, 63
    ENSO events in, 66, 226–227
    general characteristics of, 225–226
    Madden-Julian Oscillation in, 228–229
    Mesothermal climates in, 248
    physiographic features of, *225*
    Quasi-Biennial Oscillation in, 229–230
    Steppe climates in, 246
    temperature range in, 51

autumnal equinox, 32
axial precession, *279*, 279–280
axial tilt, *29*, 29–30, *278*, 278–279
    day length and, 30–33, *31*
azimuth angle, 81, *82*

B
backing, 44
Baghdad, Iraq
    climograph for, *189*
    water balance diagram for, *189*
Baltic Sea, 201
baroclinicity
    in Asia, 184
    in North America, 175
baroclinic zones, 143–144, *144*, 146–147
    polar front jet stream and, 145–146
barometers, 37–38, *38*
Barranquilla, Colombia
    climograph for, *244*
    water balance diagram for, *244*
Barringer Crater, Arizona, 280, *280*
bathtubs, water flow in, 41
Baton Rouge, Louisiana
    climograph for, *193*
    monthly water balance in, 119–120, 119*t*
    water balance diagram for, *193*
Bay of Bengal, 182
Beaufort wind scale, 344–345, 344*t*
Beijing, China
    climograph for, *210*
    water balance diagram for, *210*
Belize City, Belize
    climograph for, *236*
    water balance diagram for, *236*, 237
Benguela current, 223
Bentley Subglacial Trench, Antarctica, 232,
    232
Bergeron, Tor, 161
Bergeron Climatic Classification System,
    161–162
Bering Strait, 266, 277, *277*
Berlin, Germany
    climograph for, *203*
    water balance diagram for, *203*
Bermuda-Azores high, 128, *135*, 135–136
    effects of on Africa, 223
    Mediterranean climates and, 201
    North Atlantic Oscillation and, 324
    Rocky Mountains and, 177
Bernoulli effect, 114
Berson westerlies, 229
between-group variability, 153, 171
bias, in data, 8, 342
bioclimatology, 6

biodiversity, 72
biological evidence of climate change,
    274–276
biomes, 154
biosphere
    applied climatology on, 334
    in atmosphere evolution, 13
    as carbon reservoir, 13–14, *14*, 16
    definition of, 2
    hydroclimatology and, 6
    interactions of with atmosphere, 3*t*
biotemperature, 159
Bjerknes, Jacob, 62, 162
Blaney, H. F., 110
Blaney-Criddle model, 110
"Blizzard of the Century" (1993), 179
blocking anticyclones, 181, *181*
Bogotá, Colombia
    climatic classification of, 152
boiling point, 48
Boise, Idaho
    climograph for, *191*
    water balance diagram for, *191*
bolide, 265, 270
bora winds, 216
boreal forests, 210, 212
bottlenecks, evolutionary, 71
boundary-layer climatology, 5–6, 100
bouyancy
    of the atmosphere, 11, 18
    atmospheric stability and, 95
    hydrostatic equation and, 94
    moisture movement and, 94
    vertical motion and, 43
    Walker circulation and, 63
Brazil, rain forest destruction in, 72. *See also*
    Amazon basin
Brazilian Highlands, 230, *230*
Brisbane, Australia
    climograph for, *249*
    water balance diagram for, *249*
bromine, 20
Broome, Australia
    climograph of, 246–247, *248*
    water balance diagram for of, *248*
brown smog, 299, 302–303
Brundtland Commission, 352
Budyko, M. I., 159
Budyko Climatic Classification System, 159
Budyko ratio, 159
Buenos Aires, Argentina
    climograph for, *249*
    water balance diagram for, *249*
buoys, 345–346
butterfly effect, 320

C

Cairns, Australia
  climograph for, *243*
  water balance diagram for, *243*
Cairo, Egypt
  climatic classification of, 152
  climograph for, *246*
  water balance diagram for, *246*
calcium carbonate ($CaCO_3$), 276, 289
Calgary, Alberta
  climograph for, *191*
  water balance diagram for, *191*
California, photochemical pollution in, 299, 303
California current, 67, 179, 230–231
  pollution concentrations and, 298
Canada
  ice pack in, 75
  ocean current effects on, 47
  Subarctic climate in, 210
  Tundra climates in, 213
Canadian Climate Centre, 310
Canary Current, 57, 180, 223
  western-boundary intensification and, 58, *58*
Cantabrian Mountains, 188
cap and trade policies, 304
Capetown, South Africa
  climograph for, *250*
  forcing mechanisms in, 159, 161
  water balance diagram for, *250*
carbon cycle, 13–16, 72
carbon dioxide (CO_2), 12
  in atmosphere composition, 13, 13*t*, 14–16, 15*t*
  emission trends in, 290, *291*
  faint young Sun paradox and, 17–18
  global warming skeptics on, 295
  as greenhouse gas, 287–289
  ice core data on, 274
  increase in atmospheric, 14–16
  rain forests and, 72
  reservoirs of, 13–14, *14*
  sinks for, 14, 288–289
  U.S. emissions of, *288*
  as variable gas, 16, 16*t*
  volcanic activity and, 69
Carbon Dioxide Information Analysis Center
    (CDIAC), 346
carbonic acid (H_2CO_3), 289
carbon monoxide (CO), 16, 16*t*, 290, 301
Caribou, Maine
  climograph for, *207*
  water balance diagram for, *207*
Carlsbad Caverns, New Mexico, 271
Cascade Mountains, 161, *163*, 184
  Highland climates in, 216
cascade of linkages, 350

catrastrophism, 269–270
Cedar City, Utah
  climograph for, *190*
  water balance diagram for, *190*
Cenozoic Era, *264*, 265
Central America
  rain forest destruction in, 72
centrifugal force (CF), 41–42
  in the free atmosphere, 80
  friction and, 42
CFCs. *See* chlorofluorocarbons (CFCs)
chaos theory, 320
Charles' law, 18–19
Cherrapunji, India, rainfall record in, 182–183,
    196–197
Chesapeake Bay, 267
Chile
  maritime effects on, 56
Chimu Floods, 63
China
  cyclones in, 183–184
  greenhouse gas emissions in, 290
  Mesothermal climates in, 191, 192
  Steppe climate in, 184
Chinook winds, 177
chlorine, 20
chlorofluorocarbons (CFCs), 16–17, 20, 253
Christchurch, New Zealand
  climograph for, *251*
  water balance diagram for, *251*
Cincinnati, Ohio, time zone of, 84
circle of illumination, 29, 30–32, *31*
circulation, 35–47
  cyclones and anticyclones in, 43–45
  general, 126–150
  oceanic, 46–47, 54–61
  pressure in, 36–38, *38*
  prototype patterns of, 171
  secondary, 126–127
  Sun as driver of, 80–81
  thermohaline, 47
  vertical motion in, 43
  Walker, *62*, 62–63
  wind in, 38–43
circumpolar vortex, 145, *145*, 253, 326
Class A Evaporation Pans, 92–93, *93*
classification
  climatic, 152–173
  ineffective, 152–153
  uses of, 152
Clausius-Clapeyron equation, 90–91, *91*, 92, 110
Clean Air Act of 1963, 299
Clean Air Act of 1970, 299
climate
  controls on, 26–53

definition of, 3
    impacts of on human health and comfort, 335
    latitude effects on, *28*, 28–29
    longitude effects on, 26–28, *27, 28*
    regionalization of, 153
climate divisions, 121–122
climate impacts, 333–335
    on natural systems, 333–334
    on societal systems, 334–335
Climate Monitoring and Diagnostics Laboratory, 346
Climate Prediction Center, 121, 122
climate system, 26–53
    circulation, 35–47
    continentality in, *33*, 33–35
    cryospheric changes and, 73–75
    deforestation and desertification and, 71–73
    effects on, 54–77
    El Niño-Southern Oscillation events and, 61–69
    energy in, 80–90
    latitude in, 26–29, *27, 28*
    local features in, *48*, 48–49
    ocean circulation in, 54–61
    spatial and seasonal variation in, 49–51
    topography in, 47–48
    volcanic activity and, 69–71
climatic boundaries, 112
climatic change and variability, 260–285
    continental drift and landforms in, 277, *277*
    definition of, 261–262
    in geological history, 262–269
    information sources on historical, 269–276
    lithospheric and cryospheric evidence of, 271–274
    Milankovitch cycles and, 277–280
    natural causes of, 273–283
    radiometric dating and, 270–271
    rapid, recognition of, 351
    secondary sources on, 346
climatic classification, 152–173
    air masses and fronts in, 162–167
    classical age of, 153–159
    early attempts at, 153
    etymology of, 153–154
    genetic, 150, 161–167
    local and regional, 167–168
    quantitative analysis to derive, 168–171
climatic normals, 3–4
climatic sites, 26, 29
climatic situation, 26
Climatological Data (CD), 337, 338*t*, 340
Climatological Data: Annual Summary (CDAS), 337, *338*
climatologists, 3

climatology, 3–4
    applied, 332–348
    definition of, 2
    derivation of the word, 2–3
    future of, 349–353
    improved atmospheric data and display in, 350–351
    interdiscipinary work in, 349–350
    meteorology and, 349
    records and statistics in, 7–8
    regional, 184–217
    scales in, 4–5
    subfields of, 5–7
    sustainability and, 351–353
climographs
    for Accra, Ghana, *247*
    for Albuquerque, New Mexico, 187, *188*
    for Alice Springs, Australia, *247*
    for Antananarivo, Madagascar, *252*
    for Arequipa, Peru, *256*
    for Baghdad, Iraq, *189*
    for Barranquilla, Colombia, *244*
    for Beijing, China, *210*
    for Belize City, Belize, *236*
    for Berlin, Germany, *203*
    for Boise, Idaho, *191*
    for Brisbane, Australia, *249*
    for Broome, Australia, *248*
    for Buenos Aires, Argentina, *249*
    for Cairns, Australia, *243*
    for Cairo, Egypt, *246*
    for Calgary, Alberta, *191*
    for Capetown, South Africa, *250*
    for Caribou, Maine, *207*
    for Cedar City, Utah, *190*
    for Christchurch, New Zealand, *251*
    for Cold Bay, Alaska, *212*
    for Colombo, Sri Lanka, *236*
    for Columbus, Ohio, *205*
    for Dar es Salaam, Tanzania, *242*
    for Darwin, Australia, *243*
    for Des Moines, Iowa, *205*
    for Dhaka, Bangladesh, *238*
    for Doha, Qatar, *245*
    for Duluth, Minnesota, *207*
    for Elkins, West Virginia, *208*
    for Erie, Pennsylvania, *206*
    for Fairbanks, Alaska, *213*
    for Faro, Portugal, *200*
    for Hami, China, *187*
    for Hartford, Connecticut, *206*
    for Hilo, Hawaii, *235*
    for Hong Kong, China, *196*
    for Juba, Sudan, *241*
    for Kotzebue, Alaska, *212*

for Las Vegas, Nevada, *185*
for Little Rock, Arkansas, *194*
for Long Beach, California, *198*
for Lubbock, Texas, *190*
for Lusaka, Zambia, *242*
for Manaus, Brazil, *237*
for Manchester, England, *202*
for Manila, Philippines, *238*
for McMurdo Research Station, Antarctica, 254, *254*
for Mexico City, Mexico, *255*
for Miami, Florida, *240*
for Minsk, Belarus, *209*
for Mogadishu, Somalia, 245, *246*
for Moosonee, Canada, *214*
for Moscow, Russia, *209*
for Mumbai, India, *241*
for Murmansk, Russia, *212*
for New Delhi, India, *196*, 197
for Ottawa, Canada, *208*
for Perth, Australia, *250*
for Phoenix, Arizona, *185*
for Pretoria, South Africa, *252*
for Quito, Ecuador, *255*
for Raleigh-Durham, North Carolina, *194*
for Rapid City, South Dakota, *192*
for Reno, Nevada, *163*
for Reykjavik, Iceland, *203*
for Richmond, Virginia, *195*
for Rome, Italy, *200*
for Sacramento, California, *199*
for Santa Maria, California, *199*
for Santiago, Chile, *251*
for Seattle, Washington, *202*
for Singapore, Singapore, *235*
for South Pole, 254, *254*
for Spokane, Washington, *163*
for Tashkent, Uzbekistan, *192*
for Tegucigalpa, Honduras, *239*
for Thule Air Force Base, Greenland, *215*
for Tokyo, Japan, *195*
for Tucson, Arizona, *189*
for Tura, Russia, *214*
for Vladivostok, Russia, *210*
for Yellowknife, Canada, *213*
clock time, 83
clouds
    front types and, 165
    generation of, 35
    insolation and, 33
    as liquid aerosols, 11
    microphysics of formation of, 6
    in temperature variations, 35
cloud seeding, 7
cluster analysis, 170–171

coastal Kelvin waves, 65, 67
cold air damming, 178
cold air draining, 216
Cold Bay, Alaska
    climograph for, *212*
    water balance diagram for, *212*
cold-ENSO events. *See* La Niña
cold fronts, 165, *165*, 175
Cold Steppe (BSk) climates, 188–189, *189*, 247
Colombo, Sri Lanka
    climograph for, *236*
    water balance diagram for, *236*, 236–237
Colorado wave cyclones, 177
Columbus, Ohio
    climograph for, *205*
    water balance diagram for, *205*
Common Factor Analysis, 169, 321
Common Water, 60
components, 169
computerized climate models, 309–314
concentration-response curves, 299–300, *300*
condensation, 88, 104
    in atmosphere formation, 12
    at dewpoint temperature, 91
    lapse rate and, 96
    lifting level of, 98
conditional instability, 97
conduction, 89
confluence, 145, *145*
conservation of angular momentum, 130–131
constant absolute vorticity trajectory, *141*, 141–142
constant gases, 16–17
content analysis, 276
continental drift, 277, *277*
continentality, *33*, 33–35, 174
    in Antarctica, 232
    in Asia, 181–182, *182*
    definition of, 33
    latitude and, 49
    in North America, 174
    seasonality and, *50*, 50–51
continental polar (cP) air masses, 164, *164*
    in North America, 175
    North Atlantic Oscillation and, 325
continental tropical (cT) air masses, 164, *164*
    in Africa, 223
    tornado formation in, 175
continuity equation, 317–318
convection
    in cloud formation, 35
    continentality and, 34
    in deep ocean thermohaline currents, 58
    definition of, 18
    forced, 99

in general circulation, 127, *127*
on land vs. water, 132–133
in the troposphere, 79
in turbulent fluxes, 88–89
convection cells, 127, *127*
convective fluxes, 88–89
convective precipitation, 221
convergence, 144, *144. See also* South Pacific
    Convergence Zone (SPCZ)
horizontal, 221
velocity, 317–318, *318*
cooling degree-day (CDD), 343
coral reefs, 276
Coriolis effect (CE), *40,* 40–41
in Asian monsoons, 182
constant absolute vorticity trajectory and, 142
in convergence and divergence, 144
on cyclones and anticyclones, 43, 44
downwelling and, 57
in the free atmosphere, 80
friction and, 42
in general circulation, 127
low-level jet stream and, 177
northern vs. southern hemisphere, 222
in oceanic circulation, 46
on ocean surface currents, 54, 55
polar cells and, 129
on polar front jet stream, 146
South Pacific Convergence Zone and, 228
in surface wind systems, 129, 130
in tropical atmospheres, 220–221
in upper-level winds, 130–131, *131,* 132
vorticity and, 139
Coriolis parameter, 139, 197, 315
counter-radiation, 87, 263
counter-trade winds, *223,* 224–225
coupled atmosphere–ocean general circulation
    models, 311
crest, streamflow, 118–119
Crestview, Florida
extremes for, 4
normals for, 4
Cretaceous Period, *264,* 265
Criddle, W. D., 110
criteria pollutants, 299
crop heat unit (CHU), 343
Crop Moisture Index (CMI), 122, *123*
cross-cutting relationships, 271
Cryogenian Period, 263, *263*
cryosphere
applied climatology on, 333
climate change and, 73–75
definition of, 2
evidence of climatic change in, 271–274
feedbacks in, 74–75

hydroclimatology and, 6
interactions of with atmosphere, 3*t*
solid water in, 105
cumuliform clouds, 165, 166, 204, 234
cumulonimbus clouds, 165, 175
currents. *See* ocean currents
cyclogenesis
in Africa, 223
in Asia, 183–184
in Mediterranean climates, 201
mountains and, 143
polar lows and, 137
Rocky Mountains in, 177
Rossby waves in, 140–141
in tropical climates, 221, 222
cyclones, 43–44
airflow patterns in, 44, *44,* 45
definition of, 43
divergence and convergence in, 144, *144*
Gulf Coast, 179
in Humid Continental climates, 209
in North America, 175
pollution concentrations and, 298
troughs compared with, 138
vorticity in, 141

D

daily mean temperature, 7
Daily Synoptic Series, 340
dams, 73
Dar es Salaam, Tanzania
climograph for, *242*
water balance diagram for, *242*
Darwin, Australia, 63
climograph for, *243*
water balance diagram for, *243*
data collection, 8, *9,* 43
agencies in, 335–337
for applied climatology, 332–333
in general circulation models, 312, 314
on global warming, 295–296
inaccuracy in, 113–114
instrumentation in, *112,* 112–113, *113, 340,* 341–342
other variables in, 343–345
primary vs. secondary data in, 337
time of observation and, 342–343
data sets, 262
data sources, 335–346
daughter isotopes, 270–271
day length
Hipparchus on, 153
isotherms and, 49, *49*
latitude and, 29, 32–33
polar cells and, 136
revolution, rotation, and tilt in, 30–33, *31*

December solstice, 32
decreasing availability water balance model, 120
deep-water thermohaline conveyor model, *60*,
　　　60–61
deficits, water, 117, 158
　in Steppe climates, 184
　in Tropical Rain Forest climates, 237
deflection. *See* Coriolis effect (CE)
deforestation, 71–73
　Amazon basin, 231
　biosphere impacts of, 3*t*
　definition of, 71
　hydrologic cycle and, 106
degree-days, 343
dendroclimatology, 275
density, 36
　atmospheric, 18
　definition of, 18
　height and, 47–48
　hydrostatic equation and, 94–95
deposition, 73, 88, 104
　at dewpoint temperature, 91
　in Ice Cap climates, 215
　lapse rate and, 96
depressions, 39, 57
depression storage, 118
desert climates, 16, 157. *See also* Arid (B) climates;
　　　True Desert (BW) climates
desertification, 71–73, *73*
　in Africa, 223
　definition of, 72
　in South America, 231
Des Moines, Iowa
　climograph for, *205*, 207–208
　water balance diagram for, *205*
developing countries, weather data collection
　　　in, 8
dewpoint temperature (T_d), 91
　air temperature and, 110
　thermodynamic diagrams and, 98
　in tropical atmospheres, 221
dextral individuals, 274
Dhaka, Bangladesh
　climograph for, *238*
　water balance diagram for, *238*
diabatic heating, 316
diatomic nitrogen (N_2), 12–13, 13*t*, 20
diatomic oxygen (O_2), 12, 13, 13*t*, 20
diffluence, 145, *145*
diffuse radiation, 87
diffusion, 14
dimensionless quantities, 92
dimming, global, 297–298
dinosaurs, 264–265, 269–270, 297
direct radiation, 87

divergence, 144, *144*
　velocity, 318, *318*
diversity, 234
Dixie Alley, 175–176
Doha, Qatar
　climograph for, *245*
　water balance diagram for, *245*
doldrums, 129
downdrafts, 43
downscaling, 311, 332, 350
downwelling, 57
Drought Follows the Plow (Glantz), 223
droughts, 112
　deficits and, 117
　desertification and, 72–73
　from ENSO events, 66, 68
　indices of, 121–123, *122*, *123*, *124*
　in Latin America, 231
　in Mediterranean climates, 200–201
　solar output and, 281–282
　transpiration rates in, 92
dry climates, 156–157
dry deposition, 303
drylines, 175
dryness index (DI), 158
Duluth, Minnesota
　climograph for, *207*
　water balance diagram for, *207*
Dust Bowl, 282
dust devils, 5, 5*t*
Dust Veil Index, 70
dynamic climatology, 6, 337–345
dynamic instability, 320

E
Earth
　age of, 262
　axial precession of, *279*, 279–280
　axial tilt of, *29*, 29–30, *278*, 278–279
　circulation on a nonrotating, 127, *127*
　curvature of, path length and, 27–28, *28*
　eccentricity in orbit of, *278*, *278*
　gravity of, 12
　magnetic fields of, 21
　mass of, 11
　orbit changes of, 277–280
　origin of, 11–12
　radiation by, 87
　revolution of, 29, 30–33, *31*
　rotation of, 29
　uniqueness of, 12
Earth-ocean-atmosphere system
　carbon cycle in, 13–16
　in general circulation models, 311–312, *312*
　global hydrologic cycle in, 104

meteorology and, 4
temperature regulation in, 17–18
zones in, 2
Earth-Sun relationships, 29–33
   axial tilt, *29*, 29–30
   combined effect of revolution, rotation, and tilt,
      30–33, *31*
     negating, 33–34
     revolution, 29
East African low-level jet, 225
East Antarctic Ice Sheet, 75, *266*, 267
east-coast occlusions, 204
easterly waves, 221, *221*
eccentricity, 278, *278*
ecotones, 156, 188
eddy covariance, 94
eigenvalues, 171, 320–321
eigenvector analysis, 168–170, *169*, 320–321
Einstein, Albert, 11, 36
Ekman spiral, 46, *46*
   in Africa, 224
   downwelling and, 57
   geostrophic balance and, *57*, 57–58
   ocean surface currents and, 54, 55
   seasonality and, 51
   upwelling and, 55–56
Ekman, Walfrid, 46
El Azizia, Libya, temperatures in, 224
El Chichón, Mexico, eruption, 70*t*, 280
electromagnetic spectrum, 20
elevation
   pressure and, 38
   pressure gradient force and, 39–40
Elkins, West Virginia, 205–206, 210
   climograph for, *208*
   water balance diagram for, *208*
El Niño
   atmospheric consequences of, 65–66
   characteristics of, *64*, 64–66, *65*
   definition of, 61
   global effects of, 66–67, 67*t*
   historical data on, 63–64, 64*t*
   impacts of, 322, *323*
   Madden-Julian Oscillation compared with, 229
   in North American summer monsoon, 188
   precipitation and, 67, 68
   Rocky Mountains and, 177
   volcanic activity and, 280
El Niño-Southern Oscillation (ENSO) events, 61–69
   in Australia and Oceania, 226–227
   climatic change from, *282*, 282–283
   cryospheric effects of, 75
   effects on the U.S., 67–68
   frequency of extreme, 64, 69
   global effects of, *66*, 66–67 67*t*

global warming and, 68–69
historical observations of, 63–64, 64*t*
in Latin America, 231
Quasi-Biennial Oscillation in, 229–230
South Pacific Convergence Zone and, 228
Walker circulation in, *62*, 62–63
empirical approach, 305, 332, 350
empirical climatic classification systems, 159
Empirical Orthogonal Functions, 169
energy
   climate effects on, 334–335
   in the climate system, 80–90
   general circulation and, 126
   measuring radiant, 85
   momentum flux and, 99
   radiation balance and, 85–88
   Sun as source of, 80–84
   thermal and mechanical turbulence and Rich-
     ardson number and, 99–100
   thermodynamic equation for, 316–317
   turbulent fluxes and, 88–89
energy balance, 90
energy balance equation, 90, 107
energy balance model, 309
ENSO. *See* El Niño-Southern Oscillation (ENSO)
     events
environmental lapse rate (ELR), 18, 79, 96, *96*
environmental planning, 8, 262
Environmental Protection Agency (EPA), 299, 303
eons, 263
epochs, 265
equal availability water balance models, 120–121
equation of state, 36, 146, 318–319
   hydrostatic equation and, 94
   for moist air, 319
equation of time, 83–84, *85*, 86*t*
equator
   atmosphere thickness at, 19
   axial tilt and, 30
   day length at, 31, 33
   geopotential height at, 132, *132*, 147
   latitude of, 26
   ocean salinity at, 59
   seasonality and current near, 51
equatorial (E) air masses, 164, *164*
equatorial Kelvin waves, 64–65, 67
equatorial Rossby waves, 65
equilibrium, hydrostatic, 11
equilibrium line, 73–74
equinoxes
   autumnal, 32
   precession of, *279*, 279–280
   Sun path diagrams and, 82
   vernal, 32
eras, 265

Erie, Pennsylvania
  climograph for, *206*
  water balance diagram for, *206*
erratics, glacial, 272, *273*
eukaryotes, 13
Eurasian Humid Continental climate, 206
EuroFlux program, 345, *345*
Europe
  blocking anticyclones in, 181, *181*
  climatic setting of, 179–181
  deforestation in, 71
  Humid Continental climates in, 204
  Marine West Coast climate in, 201
  Mediterranean climate in, 197, 199, 201
  mountain ranges in, 180–181
  North Atlantic Oscillation and, 324–325
  ocean current effects on, 47, 180
  physiographic features of, *180*
  sea level changes and, *266*
  Tundra climates in, 213
eustatic sea levels, 233, 266
evaporation, 104
  altitude and, 48
  cooling from, 34–35
  desertification and, 73
  eddy covariance and, 94
  estimating, 110
  on land vs. water, 132–133
  latent heat in, 88
  radiant energy conversion in, 34
  in rain gauges, 114
  sources of, 6
  water vapor and, 16
evaporation pans, 92–93, *93*, 107–108
evapotranspiration (ET), 92, 104
  in global water balance equation, 107
  measuring, 92–94, *93*
  potential, 93
  in surface water balance, 107–110, *108, 109*
  surface water balance and, 110
  water flux via, 105–106
evolutionary bottlenecks, 71
evolution of hominids, 239–240
extinction, 72, 265, 269–270
extratropical northern hemisphere climates, 174–219
  Europe, 179–181
  regional climatology, 184–217
extremes, 3, 4
  regionalization and, 153
  temperature, 7–8
Eyjafjallajökull, Iceland, eruption, 71, 130, 281

F

Fagan, Brian, 335
faint young Sun paradox, 17–18, 263

Fairbanks, Alaska
  climograph for, *213*
  water balance diagram for, *213*
famines
  desertification and, 72–73
  from ENSO events, 67
Faro, Portugal
  climograph for, *200*
  water balance diagram for, *200*
Federal Weather Service, 336
feedback systems
  negative, 74, 294–295
  positive, 74–75, 223, 277, 291, 333, 351
fermentation, 13
Fick's law, 80, 99
field capacity, 117
Filchner Ice Shelf, 233
Findlater jet, 225
fine particles, 301, *302*
Finger Lakes Region, New York, 266
finite differencing, 311
fire potential, 122–123
first law of thermodynamics, 91, 95
Fischer-Porter rain gauges, 341, *341*
fish migrations, 67
floods, 214
  climate change and, 262
  in Latin America, 231
Florida, sea level changes and, 267
fluorocarbon 12, 16*t*
fluorocarbons, 290
flux, 13, 33
  energy balance and, 90
  Fick's law on, 80
  latent heat, 80
  local, of matter, 90–94, *91, 93*
  momentum, 99
  sensible heat, 79–80
  substrate heat, 89
  turbulent, 88–89, *89*
  of water, 105–106
foams, 20
Föehn winds, 181
fog, 179
  acid, 303
  advection, 200, 230
foraminifera, 274
forced convection, 99
forcing mechanisms, 174
  Antarctic, 233
  in arid climates, 184–185
  classification systems based on, 159, 161–167
  natural causes of climatic change, 276–283
forests
  boreal, 210

carbon dioxide sequestration in, 289
continentality and, 34–35
40–50 Day Oscillation, 228–229
fossil fuels
carbon cycle and, 14–16
carbon dioxide from, 287–288
climate effects of burning, 71
Kyoto Treaty on, 290, 304
U.S. consumption of, 291, *291*
fossils, 274–275
free atmosphere, 42, 79, *79,* 80
freezing, 104
French Revolution, 335
Freon, 20
frequencies
definition of, 4
of extremes, 8
friction
on cyclones and anticyclones, 43–44
free atmosphere layer, 80
in oceanic circulation, 46
wind and, 42–43
friction-free zone, 42
Frigid Zone, 153
frontal wave cyclones, 43, 193–195
fronts, 162–167
classification of, *165,* 165–167
cold, 165, *165*
definition of, 165
occluded, *166,* 166–167, 177
stationary, 165
warm, 165, *165*
frostpoint temperature, 91
fudge factors, 317
fumigation events, 302

G
Galileo, 153
gases, 11
compression of atmospheric, 18, 21
constant, 16–17
greenhouse, 17–18
ideal, 18–19
noble, 12
residence time of, 13
variable, 16–17, 16*t*
from volcanic activity, 69–70
general circulation, 126–150
in African climates, 223–225
definition of, 126
Hadley cells in, 127–128, 133–136
idealized, on a rotating planet, 127–132
land-water contrasts in, 132–133
midlatitude features in, 137
of a nonrotating Earth, 127, *127*

observed surface patterns in, 132–137
over mountainous terrain, 142–143, *143*
polar cells in, 128–129, 136–137
subtropical anticyclones in, 134–136, *135*
upper-level airflow in, 137–147
general circulation models (GCMs), 309–314
basic equations for, 314–320
data for, 312, 314
Earth-ocean-atmosphere system in, 311–312, *312*
types of, 310–311
weather forecasting models compared with, 319–320
genetic climatic classifications, 159, 161–167
geographic information systems (GIS), 114, 351
geography, 2
geological history, 262–266, *264,* 265*t*
evidence of climatic change in, 272–273
geology, 152
Geophysical Fluid Dynamics Laboratory, 310
geopotential height, 131–132, *132*
equatorial, 147
low-level jet stream and, 177
mean, in the northern hemisphere, 147, *148*
polar front jet stream and, 146
vorticity and, 141
geostrophic balance, 44
in the free atmosphere, 80
isohypses and, 144
in tropical atmospheres, 220–221
geostrophic flow, *57,* 57–58
geostrophic wind, 44, 197
Ghat mountains, 183
Gibson Desert, 244
glacial advance, 71, 73–74
glacial erratics, 272, *273*
glacial phases, 265, 295
glaciation, 17
axial tilt changes and, 278–279
Milankovitch cycles and, 278–280
glaciers, 73–74, 272–273
Glantz, Michael, 223
glass, specific heat of, 34*t*
global dimming, 297–298
Global Energy and Water Cycle Experiment (GEWEX), 346
global hydrologic cycle, 6, 104–125, *105*
surface water balance in, 107–119
Global Hydrology Center, 346
global warming, 286–297
adaptation to, 304–305
in Antarctica, 233
average annual temperature and, 8
carbon cycle and, 14–16, 15*t*
continued research on, 305
debate over, 294–297

deep-water conveyor and, 61
ENSO events and, 68–69
greenhouse effect in, 286–287
health impacts of, 335
mainstream perspective on, 294
mitigation of, 304
ocean fertilization and, 289
ozone layer damage and, 20–21
policymaking on, 350
prevention of, 304
skeptical perspective on, 294–297
urban heat island in, 292–294
global water balance equation, 107
Gobi Desert, 183, 184, 187, 188
Goddard Institute of Space Sciences, 310
Gondwanan Ice Age, *263*, 263–264, 277, 280
governmental policies, 8, 262
cap and trade, 304
climatology and, 350
gradient Richardson number (Ri), 99–100
gradient wind equation, 197–198
Grand Canyon, 271, *272*
granite, specific heat of, 34*t*
gravity
atmospheric stability and, 95
development of around the Earth, 12
force of on the atmosphere, 11
hydrostatic equation and, 94
vertical motion and, 43
gray smog, 302
grazing, in desertification, 72–73
Great Artesian Basin, 244
Great Lakes
climate effects on, 178
continentality around, 34
evaporation of, 4
formation of, 266
in Humid Continental climates, 204, 207–208
lake effect snows around, 48, *48*
Great Salt Lake, 267, 271
Great Sandy Desert, 244
Great Victoria Desert, 244
greenhouse effect, 16. *See also* global warming
faint young Sun paradox and, 17–18
fossil evidence of, 274
global temperature curve and, *288*
in global warming, 286–287
methanogens in, 263
pollution in, 49
in tropical rain forests, 234
greenhouse gases, 17, 287–292
carbon dioxide, 287–289
carbon monoxide, 290
emission trends in, 290–291
fluorocarbons, 290

in global warming, 68, 286–287
indirect effects of increasing, 291–292
methane, 289–290
nitrogen oxides, 290
nitrous oxide, 290
snow cover and, 75
water vapor, 287
wavelengths absorbed and emitted by, 287, 287*t*
greening initiatives, 350, 352
Greenland, 254
latitude of, 174
North Atlantic Oscillation and, 324
permanent ice in, 74
Viking colonization of, 268
Greenwich, England, 27
gridpoint models, 311, *312*
gridpoint technique, 114, 115–116
groundwater, 105
drought indices and, 122
percolation and, 107
growing degree-day (CDD), 343
growing season, 334
guard cells, 92, 120
Gulf Coast cyclones, 179
Gulf of Mexico
climate effects of, 174, 176–177
ENSO events and, 68
Loop current and, 179
Muller Weather Types for, 167, *167*
Gulf Stream, 47, 57
continental drift and, 277
deep-water thermohaline currents and, 60
European climate effects of, 180
in North American climate, 178–179
western-boundary intensification in, 58, *58*
gust fronts, 175
gymnosperms, 275
gyres, 143, 180

H
Hadley cells, 127–128
in Europe, 180
ITCZ and, *133*, 133–134, *134*
locations and strength of features in, 133–136
upper-level winds in, 130–131
Hadley Centre (England), 310
Hadley, George, 127, *127*
hailstorms, 4
half-life, 270
Hami, China
climograph for, 187, *187*
water balance diagram for, 187, *187*
Hancock County, Illinois, precipitation measurement in, 114–115, *115*
Harmattan winds, 224

Hartford, Connecticut
    climograph for, *206*
    water balance diagram for, *206*
Hawaiian high, *135,* 135–136, 179
    Africa and, 225
    Mediterranean climate and, 200–201
    pollution concentrations and, 298
hazard assessment, 8
hazard mitization, 262
heat capacity, 292
heat index, 35
heating degree-day (HDD), 343
helium, 12, 16
heterosphere, *20,* 22
Highland (H) climates, 215–217, 254–255
highland climatic classification, 155, 155*t*
Hilo, Hawaii
    climograph for, *235*
    water balance diagram for, *235,* 236–237
Himilayan Mountains
    in Asian monsoons, 182–183
    Highland climates in, 216
Hindu Kush, 183, 216
Hipparchus, 153
Historical Climatology Network, 346
historical data, 276
Hitler, Adolph, 326
hockey stick temperature curves, 296, *296*
Holdridge Life Zones Climatic Classification
    System, 159, *160, 161, 162*
Holdridge, L. R., 159
Holocene Epoch, 74, 266–267
Holocene Interglacial Phase, 74, 265, *265,* 268
homosphere, *20,* 22
Hong Kong, China
    climograph for, *196*
    water balance diagram for, *196*
horizontal convergence, 221
horse latitudes, 128
Hot Steppe (BSh) climates, 188, *189,* 246–247
Hourly Precipitation Data (HPE), 340, 340*t*
human bioclimatology, 6
human health and comfort, 335
humans. *See* anthropogenic changes
Humboldt Current, 56, *56,* 230
Humid Continental (Dfa, Dfb) climates, 204–210
Humid Continental Winter Dry (Dwa, Dwb)
    climates, 204
humidity, 35
    absolute, 91
    ocean surface currents and, 55
    in polar front jet stream, 146
    polar front jet stream and, 145–146
    Red Sea area, 16, 245
    relative, calculating, 90–91
    specific, 92
    thermodynamic diagrams and, 98
humidity index (HI), 158
Humid Subtropical (Cfa) climates, 191–195, 248
Humid Subtropical Winter Dry (Cwa) climates,
    195–197
humus, 289
Hurricane Katrina, 179
Hurricane Rita, 179
hurricanes
    Atlantic season of, 179, 262
    greenhouse gases and, 292
    historical data on, 276
    La Niña and, 68
    Quasi-Biennial Oscillation and, 230
    spatial-temporal relationships of, 5*t*
Hurricane Wilma, 179
hybrid climatic classification techniques, 171
hydrocarbons, 302–303
hydroclimatology, 6
hydrogen, 12
    nuclear fusion of, 17
    as variable gas, 16, 16*t*
hydrologic cycles, global, 6, 104–125. *See also*
    moisture; water
hydrosphere
    applied climatology on, 333–334
    in atmosphere evolution, 13
    atmosphere interactions with, 4
    definition of, 2
    global warming and, 61
    interactions of with atmosphere, 3*t*
    moisture movement through, 104
hydrostatic equation, 94–95, 315
hydrostatic equilibrium, 11, 43
hygrometers, 92
hygroscopic nuclei, 293
Hypsithermal, 268

I

Iberian Peninsula Steppe, 188
ice
    evidence of climatic change in, 273–274
    permanent, 74
    planetary, 73–75
    polar cells and, 128–129
ice ages, 17, 18, 263–264
    definition of, 74
    glacial phases in, 265
    interglacial phases in, 265
    Little, 268–269
    methanogens and, 263
    orbit eccentricity and, 278
    sea level changes and, *266,* 266–268, *267*
Ice Cap (EF) climates, 156, 215

ice caps, cryosphere impacts of melting, 3*t*
ice cores, 273–274, 287
ice dams, 214
icefish, 253
Iceland
    currents near, 47
    Marine West Coast climate in, 202–203
    North Atlantic Oscillation and, 324
    Tundra climates in, 213
    volcanic activity in, 70*t*, 71, 130
Icelandic low, 136, 180
ice shelves, 233
Idaho, baroclinic zone over, 144, *144*
ideal gases, 18–19, 36
ideal gas law, 36, 146, 318–319
India
    greenhouse gas emissions in, 290
    Mesothermal climates in, 191, 192
    Steppe climates in, 246
Indian Ocean
    currents to, 47
    ENSO events in, 66
    monsoons in, 182
    thermohaline currents in, 60, 61
    tropical cyclones in, 248
Indian Ocean high, *135,* 136, 223
indices
    of aerosols, 70–71
    drought, 121–123, *122, 123, 124*
    dryness, 158
    humidity, 158
    moisture, 157–158
indigenous people, 243
Indonesia
    downwelling around, 57
    rain forest destruction in, 72
Industrial Revolution, 14, 71, 287
industrial smog, 302
inertia, 41
inertial period, 41
infiltration, 107, 193
initialization, 319–320
insolation
    aerosol indices and, 70
    in Africa, 223
    albedo values and, 86–87, 87*t*
    circulation and, 35
    clouds and, 35
    continentality and, 34–35
    definition of, 17
    in Europe, 181
    evaporation and, 91
    ice and, 74
    in land vs. water, 132–133
    latitude and, *27,* 27–28, *28*

    pollution concentrations and, 299
    potential evapotranspiration and, 107
    rotation, tilt, and revolution effects on, 32
    surface temperatures and, 17–18
    wavelengths in, 19–20, *20*
instrumentation, *112,* 112–113, *113, 340,* 341–342
insurance, 335
interception, 118
interdisciplinary work, 349–350
interglacial phases, 74, 265
    Milankovitch cycles and, 278–280
    orbit eccentricity and, 278
    rapid climate change and, 351
Intergovernmental Panel on Climate Change, 296,
    350
internal energy, 316
international date line, 83, 174
Intertropical Convergence Zone (ITCZ), 129,
    133–134
    in Africa, 222–223
    in Australia and Oceania, 226
    in general circulation, 137
    global view of, *133*
    low pressure from, 222
    migration of, 133–134, *134*
    North American weather effects of, 190
    in Tropical Rain Forest climates, 235, 237
    in Tropical Savanna climates, 240–241
inverse square law, 278
inverted troughs, 222
Iowa, soil moisture storage in, 117
irrigation, 73
isobars
    definition of, 38, *38*
    mean surface, in semipermanent surface
        pressure cells, *135*
isohyetal method, 114, 116
isohyets, 114, 116
isohypses, 144, 145
isostatic rebound, 267–268
isothermal layers, 19
isotherms, 49–51
    definition of, 49
    in July, 49, *49*
isotopes, 270–271
ITCZ. *See* Inter*t*ropical Convergence Zone (ITCZ)
iterative models, 310

J
Japan
    cyclones in, 183–184
    Humid Subtropical (Cfa) climates in, 192
    Meteorological Agency, 310
    mountain effects in, 183
    oroshi in, 217

jet streak, 145
jet streams
    definition of, 67
    ENSO events and, 67–68, *68*
    Findlater, 225
    in North American climate, 176–177
    Pacific-North American pattern and, 326–328,
        327
    snowfall and, 74–75
    Somali, 225
    transverse wind shear and, 139–140, *140*
    Turkana, 225
Juba, Sudan
    climograph for, *241*
    water balance diagram for, *241*
June solstice, 30, 31–32

K

Kalahari Desert, 224, 244
    Steppe climates in, 246
Kalkstein, Laurence, 168
katabatic winds, 215, 216–217, 253–254
Katmai, Alaska, eruption, 70*t*
Keeling, Charles, 15, 15*t*
Keeling curve, 15, 15*t*
Keetch-Byram Drought Index, 122–123, *124*
Kelvin temperature scale, 36, 90
Kelvin waves
    coastal, 65, 67
    equatorial, 64–65, 67
Kilauea, Hawaii, volcanic activity in, 69
kinetic energy, 36, 89
    geopotential height and, 131–132
    thermodynamic energy equation and,
        316–317
Köppen Climatic Classification System, 153–154,
        154
    air mass classification vs., 164
    in Asia, 182
    data in, 159
    first-order divisions in, 155, 155*t*
    major climatic types in, 156*t*
    modified, 154–157
    second-order subdivisions in, 155–156, 155*t*
    third-order subdivisions in, 155*t*, 156
    Thornthwaite system compared with, 158–159,
        160
Köppen, Vladimir, 153–154
Kotzebue, Alaska
    climograph for, *212*
    water balance diagram for, *212*
Krakatau easterlies, 229
Krakatau, Indonesia, eruption, 70*t*, 71, 229
krypton, 16
K-T extinction, 265, 269–270, 297

Kuala Lumpur, Malaysia, climatic classification of,
        152
Kyoto Treaty (1997), 290, 304

L

Labrador current, 179
Lake Bonneville, 271
lake effect snows, 48, *48*, 178, 208
Lake Vostok, 233
Laki Fissure, Iceland, eruption, 71
Lamb Weather Types, 167–168, 171
laminar layer, 79, *79*, 80
La Nada, 61, 62, *62. See also* El Niño-Southern
        Oscillation (ENSO) events
    atmospheric consequences of, 65–66
    climatic change from, 282–283
    ocean response to, 63, 64, *64*
    Walker circulation and, 63
land covers, 6
landforms, change in, 277, *277*
Landsberg, Helmut, 162
land/sea breezes, 5, 5*t*
land use, 71–73
land-water contrasts, 132–133, 178–179
La Niña, 61, 62. *See also* El Niño-Southern
        Oscillation (ENSO) events
    in Australia and Oceania, 227, *227*
    characteristics of, 66
    climatic change from, 282–283
    global effects of, 67
    in Latin America, 231
    Madden-Julian Oscillation and, 229
    ocean response to, 63, 64, *64*, 66, *66*
    Pacific Decadal Oscillation and, 69
    precipitation and, 68
lapse rate, 95–97
    environmental, 18, 96, *96*
    saturated adiabatic, 96, *97*
    unsaturated adiabatic, 95–96, *97*, 177
Las Vegas, Nevada
    climate of, 184
    climograph for, *186*
latent energy, 34, 72, 73
latent heat flux (Q_E), 80, 88–89, *89*, 90
latent heat of vaporization, 107
Latin America. *See South* America
latitude, 26–29, *27*, *28*
    atmospheric circulation and, 35
    centrifugal force and, 41–42
    climate effects of, 28–29
    combined effects of revolution, rotation, and tilt
        on, 30–33
    Coriolis effect and, *40*, 40–41
    in Europe, 179–180
    horse, 128

in North America, 174
oceanic circulation and, 46
pressure and, 5
rotational speed and, 41
laws of motion, 11, 46
lead, 301
leeward areas
continentality in, 34
potential vorticity in, 143, *143*
precipitation in, 48
Rocky Mountain, 177
in Tropical Rain Forest climates, 236
Legates, David, 113–114
legislation, on air quality, 299
level of nondivergence, 141
Libyan Erg dune, 243
lifting condensation level, 98
lightning, causes of, 6
liquid aerosols, 11
lithification, 288–289
lithosphere
applied climatology on, 334
in atmosphere evolution, 13
definition of, 2
evidence of climatic change in,
271–274
hydroclimatology and, 6
interactions of with atmosphere, 3*t*
water percolation through, 106–107
Little Climatic Optimum (LCO), 268, 296
Little Ice Age (LIA), 268–269, 295, 296
Little Rock, Arkansas
climograph for, *194*
water balance diagram for, *194*
livestock, methane from, 290
Livezey, Robert, 327
loadings, 169, *169*, 170, *170*, 321
local apparent time, 84
local atmosphere
assessing stability in, 97–99
moisture in, 90–94, *91*, *93*
Local Climatological Data (LCD), 337, 338*t*, 340
local features, *48*, 48–49
local scale, 5
local vertical, *40*, 40–41
Loch Lomond Stadial, 268
loess, 272
London smog event (1952), 302
Long Beach, California
climograph for, *198*
water balance diagram for, *198*
longitude, 26–27, *27*
international date line and, 174
in North America, 174
rotation of the Earth and, 29

Long-term Ecological Research (LTER) Program,
345, *345*
longwave radiation, 78
from the earth, 87
low-level jet stream and, *176*, 176–177
long waves, 138
Loop current, 179
Lorenz, Edward, 320
low-level jet (LLJ) stream, *176*, 176–177
East African, 225
Lubbock, Texas
climograph for, *190*
water balance diagram for, *190*
Lusaka, Zambia
climograph for, *242*
water balance diagram for, *242*
lysimeters, 93, *93*, 107–108

M
Madagascar, drought in, 66
Madden-Julian Oscillation (MJO), 228–229, 327
magma, 280
magnetic fields, 21
Manaus, Brazil
climograph for, *237*
water balance diagram for, *237*
Manchester, England
climograph for, *202*
water balance diagram for, *202*
Manilla, Philippines
climograph for, *238*
water balance diagram for, *238*
March equinox, 32
Marine Dry Winter (Cwb) climates, 252
Marine West Coast (Cfb, Cfc) climates, 197–204,
250–253
Marine West Coast Cool Summer (Cfc) climates,
201–204
Marine West Coast Warm Summer (Cfb) climates,
201–204, 250–251
Maritime Continent, 57, 61, 225–230
maritime effects, 34
latitude and, 49
ocean surface currents and, 55
seasonality and, *50*, 51
upwelling in, 46–47
maritime polar (mP) air masses, 164, *164*, 180
maritime tropical (mT) air masses, 164, *164*
in Africa, 223
in North America, 174
Martha's Vineyard, formation of, 266
mass, 11, 36
height and, 47–48
of the troposphere, 79
Massachusetts Institute of Technology, 36

mass conservation equation, 317–318

mass spectrometry, 270

Mather, John, 120

matter

    general circulation of, 126

    local flux of, 90–94, *91, 93*

Mauna Loa, Hawaii, 15

Maunder Minimum, 281–282

maximum/minimum thermometers, 342

Max Planck Institute for Meteorology (Germany), 310

McMurdo Research Station, Antarctica, climograph for, 254, *254*

mean free path, 22

mechanical turbulence, 79, 99–100

Medieval Warm Period, 268

Mediterranean (Csa, Csb) climates, 197–201, 248–250

Mediterranean Hot Summer (Csa) climates, 198, 249

Mediterranean Warm Summer (Csa) climates, 249

Mediterranean Warm Summer (Csb) climates, 198

meltemi, 216

melting, 88

mercury

    in barometers, 37, *38*

    specific heat of, 34*t*

meridians, 26. *See also* longitude

    principal, 83, *84*

meridional flow, 137–138, *138*

    constant absolute vorticity trajectory and, 142

    vorticity in, 140

mesopause, *20,* 21

mesoscale, 5

Mesoscale Model version 5 (MM5) model, 311

mesosphere, *20,* 21–22

Mesothermal (C) climates

    in Australia and Oceania, 226

    Humid Subtropical, 191–197, 248

    Marine West Coast, 201–204, 250–253

    Mediterranean, 197–201, 248–250

    northern hemisphere, 191–204

    southern hemisphere, 247–253

mesothermal climatic classification, 155, 155*t,* 156, 158*t*

Mesozoic Era, *264,* 264–265

meteorologists, 3, 4

meteorology

    climatology and, 349

    climatology compared with, 4

    definition of, 3

    physical climatology and, 6

meteors, 280

methane (CH_4), 12, *290*

    in global warming, 289–290

    from methanogens, 18

    as variable gas, 16, 16*t*

methane hydrates, 290

methanogens, 18, 263, 289

Mexico City, Mexico

    climograph for, *255*

    water balance diagram for, *255*

Miami, Florida

    climograph for, *240*

    water balance diagram for, *240*

microfossils, 274–275

micrometers, 19–20

microscale, 5, 78

microthermal climatic classification, 155, 155*t,* 156, 182, 204

Microthermal Midlatitude (D) climates, 204–212

    Humid Continental, 204–210

    Ice Cap, 215

    Subarctic, 210–212

Middle East, 184

midlatitude cold climatic classification, 155, 155*t*

Midlatitude Cold Desert (BWk) climates, 185–188, *190,* 242

midlatitude mild climatic classification, 155, 155*t*

midlatitudes

    climatic classification of, 155, 156

    cyclones in, 5, 5*t*

    northern hemisphere, 175

    prevailing winds in, 35

    as source regions, 163

    surface features in, 137

    upper-level winds in, 131–132, *132*

midlatitude wave cyclones, 43

    fronts in, *165,* 165–166

    in Marine West Coast climates, 203

    Mediterranean climates and, 201

    in North America, 175

    pollution concentrations and, 298

    Rocky Mountains and, 177

midlatitude westerlies

    in Africa, 223

    definition of, 130

    in North America, 175

midoceanic ridges, 69

Milankovitch cycles, 277

Milankovitch, Milutin, 277

millibars (mb), 38, 90

Minsk, Belarus

    climograph for, *209*

    water balance diagram for, *209*

Mississippian Period, *263,* 263–264

Mississippi River, 106

    formation of, 267

    saltwater intrusion in, 118

    soil moisture storage around, 117

    volcanic activity and, 281

mistral winds, 216

mitigation, 304
mixing ratio, 319
MJO. *See* Madden-Julian Oscillation (MJO)
MM5 (Mesoscale Model version 5) model, 311
Mogadishu, Somalia
    climograph for, 245, *246*
    water balance diagram for, *246*
Mo, Kingtse, 327
moisture
    in air mass classification, 163–164, 164*t*
    anomalies, 226
    atmospheric statics and, 94–99
    in Australia and Oceania, 226
    in Desert climates, 185, 187–188
    in Humid Continental climates, 207–210
    in Humid Subtropical climates, 193–195, 248
    in Ice Cap climates, 254
    local atmosphere, 90–94, *91, 93*
    in Marine West Coast climates, 203–204, 251–253
    in Mediterranean climates, 200–201, 250
    in Polar climates, 253
    in Steppe climates, 190–191, 247
    in Subarctic climates, 211–212
    in Subtropical Dry Winter climates, 196–197
    surface boundary layer, 92
    in Tropical Monsoon climates, 238–239
    in Tropical Rain Forest climates, 234–237
    in Tropical Savanna climates, 240–241
    in True Desert climates, 245–246
    in Tundra climates, 215
moisture conservation equation, 317
moisture index (MI), 157–158
momentum, 6
    of air, 43
    conservation of angular, 130–131
    flux in, 99
    general circulation of, 126
    in oceanic circulation, 46
monsoons
    Asian, 182
    definition of, 182
    effects of in Asia, 182–183
    summer, 184
    west African monsoon system, 246
Moon, formation of, 12
Moosonee, Canada
    climograph for, *214*
    water balance diagram for, *214*
moraines, 272
Moscow, Russia
    climograph for, *209*
    water balance diagram for, *209*
motion
    Navier-Stokes equations of, 42–43, 314–316
    Newton's laws of, 11, 46

primitive equations of, 316
    second law of, 11, 129
mountains
    airflow over, 142–143, *143*
    in Asian climate, 183
    in Asian monsoons, 182–183
    boundary-layer climatology and, 6
    in European climate, 180–181
    formation of, 277
    glacier retreat in, 75
    in North American climate, 177–178
mountain winds, 216
Mount Pinatubo, Philippines, eruption, 70*t*, 71, 281
Mount Saint Helens, Washington, eruption, 71
Mount Toba, Sumatra, eruption, 71
mudslides, 67, 112, 231
Mullen, George, 18
Muller Weather Types, *167,* 167–168, 171
Mumbai, India
    climograph for, *241*
    water balance diagram for, 241, *241*
Murmansk, Russia, 47
    climograph for, *212*
    ocean current effects on, 180
    water balance diagram for, *212*

N

Namib Desert, 244
National Aeronautics and Space Administration (NASA), 310, 346
National Air Quality Standards (NQAS), 299
National Center for Atmospheric Research, 311, 337
National Centers for Environmental Prediction, 337
National Climate Program Act of 1978, 336
National Climatic Data Center (NCDC), 336
National Data Buoy Center, 345–346
National Environmental Satellite, Data, and Information Service (NESDIS), 336
National Oceanic and Atmospheric Administration (NOAA), 118–119, 336, *336,* 340
National Snow and Ice Data Center, 345
National Weather Service, 3, 43, 336
    evapotranspiration measurement by, 92–93, *93*
    water balance models used by, 118–119
Native Americans, migration of, 266
natural systems, climate impacts on, 333–334
Navier-Stokes equations of motion, 42–43, 314–316
nebula, 12, *12*
negative feedback systems, 74, 294–295
negative vorticity, 139
neon, 12, 16

net radiation (Q*), 85–88, *87,* 107
    in Budyko Climatic Classification, 159
    potential evapotranspiration and, 109
neutral atmosphere, 95
    potential temperature in, 98–99, *99*
    Richardson number in, 100
Nevada, anticyclones in, 178
New Delhi, India
    climograph for, *196,* 197
    water balance diagram for, *196*
New Orleans, Louisiana
    zone mean time in, 83
Newton, Isaac, 11, 46
Newtons (N), 38
New York, New York, ocean current effects
        on, 47
New Zealand
    Marine West Coast climates in, 251
    Mesothermal climates in, 248
nitric acid (NHO$_3$), 303
nitric oxide (NO), 299
nitrogen
    as constant gas, 16
    diatomic (N$_2$), 12–13, 13*t*, 20
nitrogen dioxide (NO$_2$), 299
nitrogen oxides (NO$_x$), 290, 302
nitrous oxide (N$_2$O), 16, 16*t*, 290
noble gases, 12
noise, 260
nonattainment zones, 299
nonthreshold pollutants, 299–300, *300*
nor'easters, 178–179
normals
    annual climatic, 262
    definition of, 3–4
    regionalization and, 153
    temperature, 7
North America
    air motion over, 147
    boreal forest in, 210
    climatic setting of, 174–179
    deforestation in, 71
    Gulf of Mexico effects on, 174, 176–177
    Highland climates in, 216
    Humid Continental climates in, 204
    Humid Subtropical (Cfa) climates in, 192–193
    Marine West Coast climate in, 201
    Mediterranean climate in, 197, 198, 200–201
    mountain ranges in, 177–178
    teleconnections over, 326–328, *327*
    temperature range in, 51
    Tundra climates in, 213
North American Steppes, 184, 188–191
North American summer monsoon, 187–188
North Atlantic Deep Water, 60

North Atlantic Drift, 47, 57
    continental drift and, 277
    European climate effects of, 180
    Ice Cap climates and, 215
    in Marine West Coast climates, 201
North Atlantic Oscillation (NAO), 322, 324–325,
        325
northeast trade winds, 62
    in Africa, 224
    easterly waves in, 221, *221*
    in surface wind systems, 129, 130
North Equatorial Current, 57, 62–63, 180
northern hemisphere
    atmospheric behavior in, 222
    Coriolis effect in, *40,* 41
    cyclone and anticyclone airflow in, 43–44, *44, 45*
    day length in, 30–32, *31*
    Ekman spiral in, 46, *46*
    extratropical climates in, 174–219
    January isotherms in, 49–50, *50*
    July isotherms in, 49, *49*
    mean geopotential height in, 147, *148*
    ocean surface currents in, 54, 55
northern hemisphere climates, 174–219
    Europe, 179–181
    regional, 184–217
northern lights, 21, *21*
North Pacific high, *135,* 135–136
North Pacific Intermediate Water, 60
North Pole
    geopotential height at, 132, *132*
    latitude of, 26
    rotation viewed from, 29
    surface wind systems in, 130
North Sea, 201
North Star, 30
Norway, ocean current effects on, 47
nuclear autumn, 297
nuclear fusion, 12, 17
nuclear winter, 297

O

oases, 243
occluded fronts, *166,* 166–167, 177
occlusions, 204
ocean currents
    in circulation, 54–61
    deep ocean thermohaline, 58–61
    in European climate, 180
    in North American climate, 178–179
    salinity inequalities and, 46–47
    seasonality and, *50,* 51
    surface, 54–58, *55, 56*
    upwelling, downwelling, and mass advection in,
        55–58

Oceania
  climatic setting of, 225–230
  ENSO events in, 226–227
  general characteristics of, 225–226
  Madden-Julian Oscillation in, 228–229
  physiographic features of, *225*
  Quasi-Biennial Oscillation in, 229–230
oceanic general circulation models, 310
oceans. *See also* continentality
  advection in, 34
  as carbon reservoirs, 13–14, *14,* 16, 289
  circulation in, *46,* 46–47, 54–61
  in climate system, *33,* 33–35
  convection in, 34
  pressure in, 36–37
  a reservoirs, 104
  salinity in, 58–61
  surface temperatures of, 8
  upwelling in, 46
  water transparency in, 34
Oimekon, Siberia, 211
Oort cloud, 12
optical effects, 6
orbit changes, 277–280
Ordovician Period, 263, *263*
Orion Nebula, *12*
orogeny, 277
orographic effects, 48
  in Asian monsoons, 182–183
  precipitation measurement and, 116
  in tropical climates, 221
  in Tropical Rain Forest climates, 236
oroshi, 217
Ottawa, Canada
  climograph for, *208*
  water balance diagram for, *208*
Outback, 244
outflow boundaries, 175
outgassing, 12, 13, 69
oxygen
  deep-water conveyor movement of, 61
  diatomic (O_2), 12, 13, 13*t,* 20
  historical changes in levels of, 263
  methanogens and, 18
  rain forests and, 72
oxygen-18, 273–275
Oymyakon, Siberia, temperature range in, 211
ozone (O_3)
  photochemical smog and, 302–303, *303*
  Polar climates and, 253
  production and dissociation of, 20–21
  as variable gas, 16, 16*t*
ozone (O_3) layer, 13
  hole in, 253
  processes in, *19,* 19–20

P
Pacific Decadal Oscillation (PDO)
  ENSO events and, 69
  teleconnections in, 322, *323, 324*
Pacific-North American (PNA) pattern, 326–328,
    327
Pacific Ocean
  downwelling in, 57
  extratropical teleconnections in, 322, *323, 324*
  monsoons in, 182
  as source area, 59
  Southern Oscillation Index and, 63
  thermohaline currents in, 59, *59,* 60, 61
  trade winds, 61
  tropical cyclones in, 248
Pacific Subarctic Water, 60
paleoclimatogists, 268
paleoclimatology, 6, 268. *See also* climatic change
    and variability
  secondary sources on, 346
paleotempestology, 269
Paleozoic Era, 263, *263*
Palmer Drought Severity Index (PDSI), 121–122,
    122
Palmer Hydrological Drought Index, 122
palynology, 275
pan adjustment coefficients, 93
Pangaea, 264, 280
parallelism, *29,* 29–30
parallels, 26. *See also* latitude
Paramenides, 153
parameterizations, 317
parent isotopes, 270–271
particulates, 70, 301, *302*
Pascals (Pa), 38, 90
path length, 27–28, *28,* 81
PE. *See* potential evapotranspiration (PE)
pendulum day, 41
Penman, H. L., 109
Penman-Monteith equation, 109–110
Pennsylvanian Period, *263,* 263–264
percolation, 106–107
perihelion, 29, 222
  orbit eccentricity and, 278
  precession and, *279,* 279–280
periodic variability, 260, *261*
periods, 263, *263*
permafrost, 211–212, 214
permeability, 117
Permian Period, *263,* 263–264
peroxyacetyl nitrates, 303
perspiration, 92
Perth, Australia
  climograph for, *250*
  water balance diagram for, *250*

Peru
    ENSO events in, 63
    maritime effects on, 56
Peru Current, 56, *56*, 230
PGF. *See* pressure gradient force (PGF)
Phanerozoic Eon, 263, *263*
Philippines
    cyclones in, 184
    downwelling around, 57
    mountain effects in, 183
    volcanic activity in, 70*t*, 71
Phoenix, Arizona
    climate of, 185
    climograph for, *186*
photochemical smog, 299, 302–303
photodissociation, 20, 302
photosynthesis, 81
    Amazon basin in, 231
    in atmosphere formation, 13
    carbon dioxide in, 289
    ozone layer and, 20
    in rain forests, 72
physical climatology, 6, 346
phytoplankton, 253
Pineapple Express, 177
plane of the ecliptic, *29*, 29–30
planetary albedo, 86
planetary boundary layer, 80
planetary scale, 5, 78
    of climatic classification, 153
    waves of motion, 127
planetary vorticity (*f*), 139
planetary wind systems, 129–132
planetesimals, 12
Pleistocene Epoch, 74, 273
plumbism, 301
pluvial lakes, 267
PM$_{2.5}$ particulates, 301, *302*
PM$_{10}$ particulates, 301
polar cells, 128–129
    locations and strength of features in, 136–137
    upper-level winds in, 130–131
Polar (E) climates, 212–215
    Ice Cap, 156, 215, 253–254
    Tundra, 156, 212–215, 253
polar climatic classification, 155, 155*t*, 156
polar easterlies, 130, 326
polar front jet stream, 67–68, *145*, 145–146
    Himilayas and, 183
    Pacific-North American pattern and, 326–328, *327*
    seasonality of, 147
polar highs, 129
polar ice caps, cryosphere impacts of melting, 3*t*
Polaris, 30
polar jet stream, 74–75

polar regions
    atmosphere thickness at, 19
    molecular kinetic energy at, 132
    water vapor in, 16
pollen, 275
pollutants, classification of air, 299–304
pollution
    air quality legislation and, 299
    atmospheric, 297–299
    atmospheric factors affecting, 298–299
    classification of air pollutants, 299–304
    climate influences of, 49
    in global dimming, 297–298
    smog, 293
    volcanic, 70
porosity, 117
positive feedback systems, cryospheric, 74–75, 223
    greenhouse gases and, 291
    mountain formation and, 277
    natural system impacts and, 333
    rapid climate change and, 351
positive vorticity, 139
potential evapotranspiration (PE), 93
    global annual average of, 107, *109*
    Holdridge method of calculating, 159
    measuring, 107–108
    moisture index and, 157–158
    Penman-Monteith equation for, 109–110
    seasonality and, 107, 112
    in surface water balance, 107–110, *108*, *109*
    Thornthwaite, 108–109, 158
    unadjusted, 108–109
    U.S. annual average of, 107, *108*
    in water balance diagrams, 121
potential temperature, 98–99, *99*
potential vorticity, 143, *143*, 277
precession, *279*, 279–280
precipitable water, 104
precipitation
    annual average throughout the world, *111*
    annual U.S., *111*
    in Antarctica, 233
    in anticyclones, 45
    in Asian monsoons, 182–183
    in Australia and Oceania, 226–227, *227*
    cloud thickness and, 165
    convective, 221
    cooling by, 35
    ENSO events and, 65–66, 67, 68
    instrumentation in measuring, 111–112, *112*, *113*,
        341, *341*, *342*
    ITCZ in, 137
    measuring, *112*, 112–113
    in rain forests, 72
    spatial analysis of, 114–116, *115*

in surface water balance, 110–116
thermohaline currents and, 58–59
topography and, 48
undercatch of, 113–114, 341
urban heat islands and, 294
water vapor levels and, 16
pressure, 36–38, *37*
centrifugal force and, 41–42
cyclones and anticyclones and, 43–45
in deep ocean thermohaline currents, 58
definition of, 36
high, 38
isobars, 38, *38*
latitudinal, 5
low, 38
mean surface, in semipermanent surface
pressure cells, *135,* 135–136
notations for, 37–38
Southern Oscillation Index of, 63
pressure gradient force (PGF), *39,* 39–40, 42
in the free atmosphere, 80
in surface wind systems, 129
vertical motion and, 43
Pretoria, South Africa
climograph for, *252*
water balance diagram for, *252*
prevention of global warming, 304
primary climate impacts, 333
primary data, 337
primary pollutants, 300–302
prime meridian, 27
primitive equations of motion, 316
principal components analysis (PCA), 169, 321
principal meridians, 83, *84*
principle of superposition, 271
Principle of Uniformitarianism, 269
prokaryotes, 13
Proterozoic Eon, 263, *263*
prototype atmospheric circulation patterns, 171
proxy evidence, 6, 269, 276, 286
Pyrenees, 188

Q
QBO. See Quasi-Biennial Oscillation (QBO)
quantitative analysis, 168–171
cluster, 170–171
eigenvector, 168–170, *169*
hybrid techniques for, 171
quantitative methods, 309–330
atmospheric teleconnections and, 321–328
basic equations in, 314–320
computerized climate models, 309–314
statistical techniques, 320–321
Quarternary Period, *264,* 265
Quasi-Biennial Oscillation (QBO), 229–230, 283

Quito, Ecuador
climograph for, *255*
water balance diagram for, *255*

R
radiant energy, 34, 90
radiation
atmospheric window for, 287, 287*t*
balance of, 85–88, *87*
counter-, 87
diffuse, 87
direct, 87
longwave, 78
net, 85–88, *87*
path length of, 27–28, *28*
shortwave, 20
ultraviolet, 13, 20–21
wavelengths of, 19–20, *20,* 78
radiation balance equation, 87–88
radioactive decay, 270
radiometric dating, 270–271
radiosondes, 96, 269
rain forests. *See* tropical rain forests
rain gauges, *112,* 112–113, 341, *341, 342*
rain machine, 72, 231
rain shadow effects, 183, 184, 185
Atacama Desert, 244
in Steppe climates, 188
Raleigh-Durham, North Carolina
climograph for, *194*
water balance diagram for, *194*
random variabiity, 260, *261*
Rapid City, South Dakota
climograph for, *192*
water balance diagram for, *192*
reanalysis data set, 337
records, 7–8, *9*
recreation, 334
recycling, 351–352
Red Sea, 16, 245
reflection, radiation balance and, 85–86. *See also*
albedo
reforestation, 289
refraction, 33
Regional Climate Centers, 336, *337*
regional climate models (RCMs), 311
regional climatology, 6, 184–217
regionalization of climate, 153, 168, *168*
regression lines, *169,* 169–170
relative dating, 271
relative humidity, 90–91, 92
around Asian mountains, 183
wet bulb depression and, 110
relative vorticity, 139
relativity, theory of, 11, 36

relic landforms, 271
Renewable Resource Data Center, 346
Reno, Nevada
    climograph for, *163*
    forcing mechanisms in, 161
research, on global warming, 305
reservoirs, 13, 73
residence time, 13
    of carbon, 14
    of constant gases, 16
    of variable gases, 16
residuals, 170
resistance, potential evapotranspiration and, 109
resolution, 310
revolution of the Earth, 29, 30–33, *31*
Reykjavik, Iceland
    climograph for, 203, *203*
    water balance diagram for, *203*
Richardson number, 99–100
Richmond, Virginia
    climograph for, *195*
    water balance diagram for, *195*
ridges, 137–138
    constant absolute vorticity trajectory and, 141–142
    convergence and divergence around, 144
    in energy balance, 139
    ENSO events and, 68, *68*
    mountains and, 277
    negative vorticity in, 139, *140*
    northern vs. southern hemisphere, 222
    over North America, 147
    Rocky Mountains and, 177
    in tropical atmospheres, 221
riming, 254
risk management industry, 335
River Forecast Centers, *118*, 118–119
rock striations, 272, *272*
Rocky Mountains, 177–178, 216
Rogers, Jeff, 327
Rome, Italy
    climograph for, *200*
    ocean current effects on, 47
    water balance diagram for, *200*
Ronne Ice Shelf, 233
Rossby, Carl, 138, 141
Rossby waves, 138–139
    action centers in, 142
    air mass transfer by, 164
    constant absolute vorticity trajectory and, *141,* 141–142
    cyclogenesis in, 140–141
    diffluence and confluence in, 145, *145*
    divergence and convergence in, 144, *144*
    equatorial, 65
    mean patterns of, 146–147

northern vs. southern hemisphere, 222
    Pacific-North American teleconnection and, 326–327, *327*
    Rocky Mountains and, 177
    in tropical atmospheres, 220, *221*
    vorticity of, 139–141, *140*
Ross Ice Shelf, 233
Ross Sea, 233
rotation of the Earth, 29
    Coriolis effect and, *40,* 40–41
    vorticity and, 139
roughness layer, *79,* 79–80
roughness length, 100
Rub-al-Khali, 224
runoff, 106, 118–119
    in Africa, 223
    global distribution of, *106*
rural areas, weather data collection in, 8

S
Sacramento, California
    climograph for, *199*
    water balance diagram for, *199*
Sagan, Carl, 18
Sahara Desert, 72–73, *73,* 223, 224
    Arid climate in, 242–244
    Steppe climate around, 246
Sahel, Africa
    desertification in, 72–73, *73,* 223
    Steppe climates in, 246
salinity
    ocean currents and, 46–47
    thermohaline currents and, 58–61
saltwater intrusion, 118, 291
sand, specific heat of, 34*t*
sand dunes, 243
San Diego, California
    forcing mechanisms in, 159, 161
    Sun angle in, 27
San Francisco, California
    continentality effects on, 33–34, 35
    maritime effect in, 35–36
Santa Ana winds, 178
Santa Maria, California
    climograph for, *199*
    water balance diagram for, *199*
Santa Maria, Guatemala, eruption, 70*t*
Santiago, Chile
    climograph for, *251*
    water balance diagram for, *251*
Santorre, Santorio, 153
Sargasso Sea, 128
satellite monitoring and recording, 8, 86
    future of, 350–351
    global warming skeptics on, 295–296

saturated adiabatic lapse rate, 96, *97*

saturation vapor pressure (es), 90, *91,* 104, 145

scales, in climatology, 4–5

scattering, 81, 85–86

scores, 169, *169,* 170, *170*

scree plots, 171

sea level

    in Antarctica, 233

    El Niño and, *64,* 64–65

    eustatic, 233

    ice ages and, *266,* 266–268, *267*

    ice sheet melting and, 75, *266,* 267

    La Niña and, 66, *66*

    pressure at, 36, 38

seasonality

    axial tilt changes and, 279

    carbon dioxide cycles and, 15

    Earth-Sun relationships in, 29–33

    maritime effect and, 55

    in North America, 175

    North Atlantic Oscillation and, 324–325

    northern vs. southern hemisphere, 222

    in the polar front jet stream, 147

    potential evapotranspiration and, 107, 112

    prevailing winds and, 35–36

    Rossby waves and, 138

    of surface pressure features, 136

    temperature variation and, *49,* 49–51, *50*

    tropical, 26

sea surface temperatures (SSTs), El Niño and,

    61–69

Seattle, Washington

    climograph for, *202*

    water balance diagram for, *202*

secondary circulations, 126–127, 137–147

    anticyclones, 166

    cyclones, 137, 166

    midlatitude wave cyclones as, 166

    northern vs. southern hemisphere, 222

    Rossby waves, 138–147

secondary climate impacts, 333

secondary data, 337–345

secondary pollutants, 301, 302–304

second law of motion, 11, 126, 129

second law of thermodynamics, 11, 35, 39

sedimentary rock, 13, 14, 271, 288–289

semiarid regions, 156, 157

    desertification of, 72–73

    northern hemisphere, 184

semipermanent high-pressure cells, *135,* 135–136

sensible energy, 34, 72, 73

sensible heat flux (Q_H), 79–80, 88, *89*

    energy balance and, 90

    potential evapotranspiration and, 109

Seoul, South Korea, climatic classification of, 152

September equinox, 32

severe weather, 335. *See also* cyclogenesis

    in Europe, 181

    in North America, 175–176, *176*

shearing stress, 80

Sheridan, Scott, 168

ship logs, weather data in, 63, 276

shortwave radiation, 20, 78

    global budget of, 86, *87*

    greenhouse effect and, 287

    in low-level jet stream formation, 177

    radiation balance and, 87–88

Siberia

    low temperatures in, 136

    ocean current effects on, 47

    temperature extremes in, 181–182, 211

Siberian high, 136, 182

siderite ($FeCO_3$), 18

Sierra Madre Occidental, 230, *230*

Sierra Madre Oriental, 230, *230*

Sierra Nevada Mountains, 161, *163,* 177

    Highland climates in, 216

    Steppe climate around, 184

Simpson Desert, 244

Singapore, Singapore

    climograph for, *235*

    water balance diagram for, *235*

sinistral individuals, 274

sinks, carbon dioxide, 14, 288–289

Skew T-Log P diagrams, *97,* 97–98, *98*

slab models, 311

slash and burn agriculture, 71–72

smog, 293, 299

    industrial, 302

    photochemical, 299, 302–303

snow and snowfall. *See also* precipitation

    in Antarctica, 254

    average annual global, *114*

    average annual U.S., 112–113, *113*

    historical data on, 276

    in Humid Continental climates, 209–210

    lake effect, 48, *48*

    in Marine West Coast climates, 204

    measuring, 112–113, *113*

    polar cells and, 128–129

    whiteouts, 233

snowball earth hypothesis, 263

societal systems, climate impacts on, 334–335

soil

    as carbon reservoir, 14

    in climatic classification, 154

    moisture storage in, *116,* 116–117

    permeability of, 117

    porosity of, 117

    in tropical rain forests, 71–72

soil heat flux plates, 89

soil moisture recharge, 117, 121, 223

soil moisture storage capacity, 116–117

soil moisture withdrawal, 117, 121

solar declination, 27–28, 29, 33

    in Africa, 224–225

    axial tilt changes and, 279

    ITCZ and, 134

    revolution, rotation, and tilt in, 30

    solar noon and, 83

    Sun path diagrams and, 81–82, *82*

solar noon, 32–33, 82–83, 127

solar output, 17, *281,* 281–282

solar system, formation of, 12

solar time, 82–84

solar wind, 12

solid aerosols, 11, 70

solstices

    precession and, *279,* 279–280

    summer, 30, 31–32, 81

    winter, 32, 81

Somali jet, 225

source areas, 59

source regions, 163, 176, 180

South America

    Arid climates in, 242

    climatic setting of, *230,* 230–231

    ENSO events in, 63, 66

    Humid Subtropical climates in, 248

    rain forest destruction in, 72

    temperature range in, 51

    upwelling around, *56*

South Atlantic high, *135,* 136, 223

southeast trade winds, 62

    in Africa, 222–223

    in surface wind systems, 129, 130

South Equatorial Current, 62–63

southern hemisphere

    atmospheric behavior in, 222

    Coriolis effect in, *40,* 41

    day length in, 32

    January isotherms in, *50,* 50–51

    July isotherms in, 49, *49*

    ocean surface currents in, 54, 55

southern hemisphere climates

    Africa, 222–225

    Antarctica, 232–233

    Australia and Oceania, 225–230

    Latin America, 230–231

    regional, 233–255

    tropical vs. extratropical, 220–222

southern lights, 21

Southern Oscillation, 62, 69, 226. *See also* El Niño-Southern Oscillation (ENSO) events

    climatic change from, 282–283

impacts of, 322, *323*

    South Pole temperatures and, 233

Southern Oscillation Index (SOI), 63, *282*

South Pacific Convergence Zone (SPCZ), 227–228, *228*

South Pacific high, *135,* 136

South Pacific subtropical anticyclone, 56

South Pole, 232, *232*

    climograph for, 254, *254*

    continental drift and, 277

    day length at, *31,* 32

    geopotential height at, 132, *132*

    latitude of, 26

    surface wind systems in, 130

southwesterly monsoons, 182, 225

spatial analysis methods, 114–116, *115*

spatial downscaling, 311

spatial resolution, 310

spatial scales, 4–5, 5*t,* 78

Spatial Synoptic Classification, 168

SPCZ. *See* South Pacific Convergence Zone (SPCZ)

Spearfish, South Dakota, temperature change in, 177

specific heat, 34, 34*t,* 132, 316–317

specific humidity, 92, 254

specific volume, 315

spectral models, 311–312, *313*

Spokane, Washington

    climograph for, *163*

    forcing mechanisms in, 161

spores, 275

squall lines, 175

stability

    assessing local atmosphere, 97–99

    atmospheric, 95–97

    hydrostatic equation and, 94–95

    moisture movement and, 94

stable atmosphere, 55

    pollution concentrations in, 298–299

    potential temperature in, 98–99, *99*

stable stratification, 58, 167

standard nonrecording rain gauges, 341

star formation, 12, 17

State Climatology Program, 336

statics, atmospheric, 94–99

stationary fronts, 165

statistical techniques, 320–321

statistics, 7–8, *9*

step changes, 260, *261*

Steppe (BS) climates, 157, 184, 188–191

St. Lawrence Seaway, 267

St. Louis, Missouri, continentality effects on, 33–34, 35

stomates, 92, 120

Stommel, Henry, 60

Stommel model, *60,* 60–61

Storm Data, 340
storms. *See* cyclogenesis
Strait of Otranto, 216
stratification, stable, 58, *58*
stratiform clouds, 165, 203–204
stratopause, *20,* 21
stratosphere, *19,* 19–21, 22, 70
streamline analysis, 220
Styrofoam, 20
Subarctic (Dfc, Dwc, Dwd) climates, 210–212
subduction zones, 280
sublimation, 73, 88, 291
subpolar lows, 129, 130, 136–137
substrate heat flux (Q_G), 89, *89,* 109
subtropical anticyclones
    in Africa, 223–224
    in arid climates, 184–185
    as forcing mechanisms, 159, 161
    in general circulation, 134–136, *135*
    Hadley cells and, 128
    location and strength of, 134–136, *135*
    in Mediterranean climates, 197–198, *198,*
        200–201
    migration of, 134–135
    ocean salinity and, 59
    ocean surface currents and, 54
    pollution concentrations and, 298
    in surface wind systems, 129–130
Subtropical Desert (BWh) climates, 242
Subtropical Dry Winter (Cwa, Cwb) climates,
    195–197
Subtropical Hot Desert (BWh) climates, 185
subtropical jet stream, 67, 131
sulfur, 12
sulfur dioxide (SO_2)
    in industrial smog, 302
    volcanic activity and, 69, 71
sulfuric acid (H_2SO_4), 303
sulfur oxides, 300–301
summer monsoons, 184
    Subtropical Dry Winter climates and, 195
    in Tokyo, 193
summer solstice, 30, 31–32, 81
summer thermal efficiency concentration, 158, 159*t*
Sun, 12
    angle of, climate and, 174
    angle of, latitude and, *27,* 27–28, *28,* 32–33
    changes in output from, 17, 263, *281,* 281–282
    Earth-Sun relationships and, 29–33
    energy from, 80–84
    faint young Sun paradox and, 17–18
    ozone layer and, 19–20
    radiation emission spectrum of, 87
    wavelengths from, 19–20, *20*
Sun path diagrams, *81,* 81–82, *82*

sunspots, *281,* 281–282
surface airways observations, 340, 340*t*
surface boundary layer (SBL), 80
surface currents, ocean, 54–58, *55, 56*
surface tension, 117
surface water balance, 107–119
    in Africa, 223
    applied climatology and, 333
    deficits in, 117
    diagrams on, 121, *121*
    drought indices and, 121–123, *122, 123, 124*
    example of, 119–120, 119*t*
    Great Lakes in, 178
    model types for, 120–121
    potential evapotranspiration in, 107–110, *108, 109*
    precipitation in, 110–116
    snowfall in, 112–113, *113*
    soil moisture storage in, 116–117
    surplus and runoff in, 117–119
    in Tropical Rain Forest climates, 236–237
surface water balance models, 309
surface weather maps, 38, *38*
surface wind systems, 129–130
surplus water, 117–119, 158
sustainability, 351–353
swamp models, 311
swamps, drainage of, 73
synoptic climatology, 6, 321, 337–345
synoptic scale, 5

T
taiga, 210–211, 212
Taiwan
    cyclones in, 184
    Humid Subtropical (Cfa) climates in, 192
Tallahassee, Florida, thermodynamic diagram for,
    97
Tambora, Indonesia, eruption, 70*t,* 71, 268
Tashkent, Uzbekistan, water balance diagram for,
    192
Tasmania, 251
Tegucigalpa, Honduras, 238
    climograph for, *239*
    water balance diagram for, *239*
teleconnections, 142, 321–328
    extratropical Atlantic, 322–326, *325*
    extratropical Pacific, 322, *323, 324*
    over North America, 326–328, *327*
    teleconnections between, 327–328
Temperate Zone, 153
temperature
    absolute zero, 36, 87, 287, 316–317
    in air mass classification, 163–164, 164*t*
    altitude and, 47–48
    average, 7, 8

clouds and, 35
daily mean, 7
degree-days, 343
in Desert climates, 185
dewpoint, 91, 98, 110, 221
Earth's revolution and, 29
faint young Sun paradox, 17–18
fossil evidence on, 274–275
in fronts, 165
frostpoint, 91
geological history of changes in, 262–266
glacial advances and, 74
global mean annual average, *269*
greenhouse effect and, 15–16
hockey stick curves on, 296, *296*
in Humid Continental climates, 204–206
in Humid Subtropical climates, 192–193, 248
in Ice Cap climates, 254
Kelvin scale for, 36
lapse rate of, 95–97
in Marine West Coast climates, 201–203, 251
maximum–minimum calculation of, 7
in Mediterranean climates, 198–200, 249–250
mesopause, 21
mesosphere, 21
normals of, 7
in Polar climates, 253
potential, 98–99, *99*
pressure and, 36
records and statistics on, 7–8
regionalization of, 168, *168*
saturation vapor pressure and, 90–91, *91*
seasonality and, *49,* 49–51, *50*
sea surface, 61–69
solar output and, 281–282
in Steppe climates, 188–190, 246–247
in Subarctic climates, 211
in Subtropical Dry Winter climates, 196
surface, 17–18
in Tropical Monsoon climates, 237–238
in Tropical Rain Forest climates, 234
in Tropical Savanna climates, 240
tropopause, 19
in True Desert climates, 245
in Tundra climates, 214
urbanization and, 8
water, 35
temperature inversions, 19
in Africa, 223
in low-level jet stream, 177
net radiation and, 85
stratospheric, 20–21
thermosphere, 22
temporal downscaling, 311
temporal resolution, 310

temporal scale
in climatology, 4–5
records and statistics and, 7–8
spatial relationships with, 4–5, *5t*
temporal trends, 260
Terjung, Werner, 350
Tertiary Period, *264,* 265
The Little Ice Age (Fagan), 335
thermal conductivity, 292
thermal efficiency index, 158
thermal lows, 182
thermal turbulence, 79, 99–100
thermoclines, 66, *66,* 231
thermodynamic diagrams, *97,* 97–98, *98*
thermodynamic energy equation, 316–317
thermodynamic processes, 18–19
thermodynamics
first law of, 91, 95
second law of, 11, 35, 39, 126
thermohaline circulation, 47
deep ocean, 58–61
rapid climate change and, 351
thermometers, 153, 342
thermosphere, *20,* 22
Thiessen polygon method, 114, 115, 116
30–60 Day Oscillation, 228–229
Thompson, David, 325
Thornthwaite Climate Classification System, 154,
157–159, *158t*
Köppen system compared with, 158–159, *160*
Thornthwaite, C. W., 107, 108, 112, 120, 154, 157
Thornthwaite potential evapotranspiration, 108–109
threshold pollutants, 299–300, *300*
Thule Air Force Base, Greenland
climograph for, *215*
water balance diagram for, *215*
thunderstorms, spatial-temporal relationships of,
5t. See also severe weather
Tibetan low, 136, 182, 235
Tibetan Plateau, 183
Tien Shan mountains, 216
Tierra del Fuego, 253
time of day, record keeping and, 8, 342–343
time of observation bias, 342
time steps, 310
time zones, 29, *83,* 83–84, *84*
tipping bucket rain gauges, 341, *342*
Titan, 263
Tokyo, Japan
climograph for, *195*
water balance diagram for, *195*
topography, 47–48
of air mass source regions, 163
circulation and, 127
in North America, 175, *175*

precipitation measurement and, 116
in Tropical Rain Forest climates, 235–236
Tornado Alley, 175–176, *176*
tornadoes
acceleration in, 11
centrifugal force in, 42
as cyclones, 43
low-level jet stream in, 177
in North America, 175–176, *176*
Rocky Mountains and, 177
spatial-temporal relationships of, 5*t*
Toronto, Canada, continentality in, 34
Torricelli, Evangelista, 37, 153
Torrid Zone, 153
trade surges, 237
trade winds
Arctic Oscillation and, 326
counter-, *223*, 224–225
El Niño and, 61
ITCZ and, 133–134, *134*
La Niña and, 66, *66*
in Latin America, 231
migration of, 134
northeast, 62, 129, 130, 221, *221*, 224
southeast, 62, 120, 130, 222–223
in surface wind systems, 129
in Tropical Rain Forest climates, 235
Walker circulation and, 64
transition layer, 79, *79*, 80
transpiration, 72, 92, 104
transportation, 334
transverse wind shear, 139–140, *140*
treeline, 156, 212, 276
tree rings, 275
Triple Bottom Line (TBL) concept, 352
Tropical (A) climates
in Australia and Oceania, 225–226
extratropical vs., 220–222
southern hemisphere, 233–241
Tropical Monsoon, 237–239
Tropical Rain Forest, 234–237
Tropical Savanna, 239–241
tropical climatic classification, 155, 155*t*
tropical cyclones, 43
in Asia, 183–184
Atlantic Ocean, 248
greenhouse gases and, 292
Indian Ocean, 248
in North America, 175
Pacific Ocean, 248
Quasi-Biennial Oscillation in, 229–230
typhoons, 236
Tropical Monsoon (Am) climates, 237–239
Tropical-Northern Hemisphere (TNH) teleconnection, *327*, 327–328

Tropical Rain Forest (Af) climates, 234–237
tropical rain forests
deforestation of, 71, 230, 231
diversity in, 72
soils in, 71–72
in Tropical Monsoon climates, 238–239
Tropical Savanna (Aw) climates, 237, 239–241
Tropic of Cancer, 30, 33
axial tilt changes and, 279
ITCZ and, 133–134, *134*
Tropic of Capricorn, 32, 33
axial tilt changes and, 279
ITCZ and, 133–134, *134*
tropics
definition of, 33
net energy surplus in, 132
seasonality in, 26
tropopause, 19, *19*, 132
troposphere, 18–19, *19*
circumpolar vortex and, 253
cyclones and anticyclones and, 43–44
Hadley cells and, 127–128
near-surface, *79*, 79–80
properties of, 78–79
temperature and altitude in, 47
troughs, 138
constant absolute vorticity trajectory and, 142
convergence and divergence around, 144
in energy balance, 139
ENSO events and, 67–68, *68*
inverted, 222
mountains and, 277
over North America, 147
in tropical atmospheres, 221–222
vorticity in, 140
True Desert (BW) climates, 157
in Asia, 183
in North America, 185
northern hemisphere, 184, 185
southern hemisphere, 242–246
Tuareg, 243
Tucson, Arizona
climograph for, *189*
water balance diagram for, *189*
Tundra (ET) climates, 156, 212–215, 253
Tura, Russia
climograph for, *214*
water balance diagram for, *214*
turbulence, 79
mechanical, 79, 99–100
thermal, 79, 99–100
turbulent flux of latent heat, 80, 88–89, *89*, 90
turbulent flux of sensible heat, 79–80, 88, *89,* 90
Turc method, 110
Turkana jet stream, 225

Twain, Mark, 35
typhoons, 236
Typhoon Tip (1979), 236

U

ultraviolet (UV) radiation, 13, 20–21, 253
undercatch, 113–114, 341
United States
    air quality in, 303–304
    air quality legislation in, 299
    annual potential evapotranspiration in, 107, *108*
    average annual precipitation in, *111*
    average annual snowfall in, *113*
    ENSO effects on, 67–68
    fossil fuel consumption in, 291, *291*
    greenhouse gas emissions in, 290–291
United States Weather Bureau, 336
universe, origin of, 11–12
unsaturated adiabatic lapse rate (UALR), 95–96, *97*,
        177
unstable atmosphere, 55
    atmospheric stability and, 95
    pollution concentrations in, 298–299
    potential temperature in, 98–99, *99*
    Richardson number in, 99–100
updrafts, 43
upper-level winds, 130–132, *131, 132*
    constant absolute vorticity trajectory and, *141,*
        141–142
    divergence and convergence in, 144, *144*
    vorticity in, 139–141
upwelling, 46, 55–56, *56*
    in La Niña, 66, *66*
    seasonality and, 51
urban canyons, *293*, 293–294
urban heat islands, 8, 48–49, 292–294
urbanization
    boundary-layer climatology and, 6
    desertification and, 73
    global warming skeptics and, 295
    greenhouse effect and, 292–294
    ice sheet melting and, 75, 268
    weather data collection and, 8
U.S. Department of Energy, 346
U.S. Energy Information Administration, 290–291
U.S. Forest Service, 122–123
Utah
    anticyclones in, 178
    baroclinic zone over, 144, *144*

V

Vancouver, British Columbia, Sun angle in, 27
vaporization, latent heat of, 107
vapor pressure *(e)*, 90, 109
variability. *See* climatic change and variability

variability changes, 260, *261*
variable gases, 16–17, 16*t*
varves, 6, 274
veering, 43–44
velocity, Navier-Stokes equation for, 314–315
velocity convergence, 317–318, *318*
velocity divergence, 318, *318*
Verkhoyansk, Siberia, 211
vernal equinox, 32
vertical flux of horizontal momentum, 80
vertical motion, 43, 315
vertical wind shear, 99
vertical zonation, 215, 254
Vietnam, cyclones in, 183–184
Vinson Massif, Antarctica, 232, *232*
visualization, 351
Vladivostok, Russia
    climograph for, *210*
    water balance diagram for, *210*
vog, 70
volcanic activity
    aerosol indices and, 70–71
    atmosphere composition and, 12, 69
    climate effects of, 69–71, 280–281
    general effects of, 69–70
    lithosphere impacts of, 3*t*
    major eruptions, 70*t*, 71
Volcanic Explosivity Index, 70–71
vorticity, 139–141, *140*
    absolute, 141–142, 166
    jet stream, 147
    negative, 139
    potential, 143, *143*, 277
Vostok, Antarctica, coldest temperature in, 232
Vostok ice core, 274
Vostok Research Station Ice, 233

W

Walker, Gilbert, 324
Walker circulation, *62*, 62–63
Wallace, John, 325
warm-ENSO events. *See* El Niño
warm fronts, 165, *165*
warm sectors, *166*, 166–167
Washington, DC
    continentality effects on, 33–34
    maritime effect in, 35–36
    thermodynamic diagram for, *97*
water
    altitude and, 48
    deficits, 117
    depression storage, 118
    fresh, 105
    global distribution of, 104, *105*
    interception, 118

latent heat flux in, 88–89
precipitable, 104
runoff, 106, *106*, 118–119
in the solar system, 12
specific heat of, 34*t*
surface tension of, 117
surface water balance, 107–119
surplus, 117–119
transparency of, 132
water balance. *See* surface water balance
water balance diagrams, 121
for Accra, Ghana, *247*
for Albuquerque, New Mexico, 187, *188*
for Alice Springs, Australia, *247*
for Antananarivo, Madagascar, *252*
for Arequipa, Peru, *256*
for Baghdad, Iraq, *189*
for Barranquilla, Colombia, *244*
for Baton Rouge, Louisiana, *193*
for Beijing, China, *210*
for Belize City, Belize, *236*
for Berlin, Germany, *203*
for Boise, Idaho, *191*
for Brisbane, Australia, *249*
for Broome, Australia, *248*
for Buenos Aires, Argentina, *249*
for Cairns, Australia, *243*
for Cairo, Egypt, *246*
for Calgary, Alberta, *191*
for Capetown, South Africa, *250*
for Caribou, Maine, *207*
for Cedar City, Utah, *190*
for Christchurch, New Zealand, *251*
for Cold Bay, Alaska, *212*
for Colombo, Sri Lanka, *236*
for Columbus, Ohio, *205*
for Dallas, Texas, 121, *121*
for Dar es Salaam, Tanzania, *242*
for Darwin, Australia, *243*
for Des Moines, Iowa, *205*
for Dhaka, Bangladesh, *238*
for Doha, Qatar, *245*
for Duluth, Minnesota, *207*
for Elkins, West Virginia, *208*
for Erie, Pennsylvania, *206*
for Fairbanks, Alaska, *213*
for Faro, Portugal, *200*
for Hami, China, *187*
for Hartford, Connecticut, *206*
for Hilo, Hawaii, *235*
for Hong Kong, China, *196*
for Juba, Sudan, *241*
for Kotzebue, Alaska, *212*
for Las Vegas, Nevada, *185*
for Little Rock, Arkansas, *194*

for Long Beach, California, *198*
for Lubbock, Texas, *190*
for Lusaka, Zambia, *242*
for Manaus, Brazil, *237*
for Manchester, England, *202*
for Manila, Philippines, *238*
for Mexico City, Mexico, *255*
for Miami, Florida, *240*
for Minsk, Belarus, *209*
for Mogadishu, Somalia, *246*
for Moosonee, Canada, *214*
for Moscow, Russia, *209*
for Mumbai, India, 241, *241*
for Murmansk, Russia, *212*
for New Delhi, India, *196*
for Ottawa, Canada, *208*
for Perth, Australia, *250*
for Phoenix, Arizona, *185*
for Pretoria, South Africa, *252*
for Quito, Ecuador, *255*
for Raleigh-Durham, North Carolina, *194*
for Rapid City, South Dakota, *192*
for Reykjavik, Iceland, *203*
for Richmond, Virginia, *195*
for Rome, Italy, *200*
for Sacramento, California, *199*
for Santa Maria, California, *199*
for Santiago, Chile, *251*
for Seattle, Washington, *202*
for Singapore, Singapore, *235*
for Tashkent, Uzbekistan, *192*
for Tegucigalpa, Honduras, *239*
for Thule Air Force Base, Greenland, *215*
for Tokyo, Japan, *195*
for Tucson, Arizona, *189*
for Tura, Russia, *214*
for Vladivostok, Russia, *210*
for Yellowknife, Canada, *213*
Watertown, New York, continentality in, 34
water vapor
continentality and, 34–35
deposition of, 73
faint young Sun paradox and, 17
as greenhouse gas, 287
riming, 254
sublimation of, 73
as variable gas, 16, 16*t*
wavelengths, 19–20, 78, 138. *See also* radiation
weather
accuracy of forecasting, 3, 43
data collection, 8, *9*
definition of, 3
normals, 3–4
regionalization of, 153
troposphere in development of, 18

weather forecasting models, 319–320
weather sphere. *See* troposphere
weather stations, 8, 43
Weddell Sea, 233
west African monsoon system, 246
West Antarctic Ice Sheet, 75, *266,* 267
west-coast occlusions, 204
western-boundary intensification, 58, *58*
wet bulb depression, 110
wet deposition, 303
wetted perimeter, 114
whiteouts, 233
Willmont, Cort, 108
wilting point, 117
windbreaks, 7
wind channeling, 216
wind-chill factor, 254
winds
    altitude and, 100
    Antarctic, 232
    atmosphere stability and, 99–100, *100*
    backing, 44
    Beaufort scale for, 344–345, 344*t*
    belts of, 5
    bora, 216
    centrifugal force and, 41–42
    Chinook, 177
    in climate control, 38–43
    Coriolis effect and, *40,* 40–41
    data collection on, 343–345, 344*t*
    definition of, 38
    Föehn, 181
    friction and, 42–43
    general circulation of, 126–150
    geostrophic, 44, 197
    Harmattan, 224
    katabatic, 215, 216–217, 253–254
    latent heat flux and, 89
    mistral, 216
    mountain effects on, 177–178
    ocean currents and, 47
    planetary boundary layer and direction of,
        80, *80*
    planetary systems of, 129–132
    pollution concentrations and, 298
    precipitation measurement and, 114
    pressure and generation of, 36–38
    pressure gradient force and, *39,* 39–40
    prevailing, 35
    secondary circulations, 126–127
    solar, 12
    surface systems of, 129–130
    surface vs. upper-level, 43
    trade, 61, 62
    upper-level, 130–132, *131, 132*
    in urban canyons, *293,* 293–294
    veering of, 43–44
    in water advection, 106
wind shears
    shearing stress and, 80
    transverse, 139–140, *140*
    vertical, 99
windward areas
    continentality in, 34
    precipitation in, 48
    Sierra Nevada Mountains, 177
    in Tropical Rain Forest climates, 236
winter solstice, 32, 81
Wisconsin Glacial Phase, 74, 233, *265,* 265–266
within-group variability, 152–153, 171
World Climate Research Programme, 346
World Data Center for Paleoclimatology, 346
World Meteorological Organization (WMO),
    340–341, 350
Wyoming, drought index for, 122

X
xenon, 16

Y
Yarnal, Brent, 6
"Year without a Summer" (1816), 71, 268, 280
Yellowknife, Canada
    climograph for, *213*
    water balance diagram for, *213*
Yosemite Valley, formation of, 266
Younger Dryas, 268

Z
zenith angle, 81, *81*
zonal flow, 137, *138*
    constant absolute vorticity trajectory and, 142
    Pacific-North American pattern and,
        326–328, *327*
    transverse wind shear in, 139–140, *140*
    vorticity in, 140
zone mean time, 83, 84
zooplankton, 253

Photo Credits

Chapter Openers, © Eyewire, Inc.

Part 1 Opener (page 1): Courtesy of NASA

Figure 2.1/Color Plate 2.1: Courtesy of NASA, ESA, M. Robberto (Space Telescope Science Institute/ESA) and the Hubble Space Telescope Orion Treasury Project Team

Figure 2.7/Color Plate 2.7: © Roman Krochuk/ShutterStock, Inc.

Part 2 Opener (page 24): © Dmitry Naumov/ShutterStock, Inc.

Figure 4.3/Color Plate 4.3: Courtesy of NASA Goddard Space Flight Center

Figure 4.9/Color Plate 4.9: Courtesy of NASA/JPL-Caltech

Figure 5.8: Courtesy of CIA

Figure 5.10: Courtesy of Anthony Ayiomamitis

Figure 5.16: Courtesy of NWS/NOAA

Figure 5.17: Courtesy of Ernest Clawson

Figure 6.8: Courtesy of High Sierra Electronics, Inc.

Figure 6.9: Courtesy of Alan Sim

Figure 7.5a: Courtesy of Global Hydrology and Climate Center/Marshall Space Flight Center/NASA

Figure 7.5b/Color Plate 7.5b: Courtesy of Global Hydrology Resource Center/NASA

Part 3 Opener (page 151): © Ulises Sepúlveda Déniz/ShutterStock, Inc.

Figure 10.18: © vega bogaerts/ShutterStock, Inc.

Part 4 Opener (page 259): © Leigh Prather/ShutterStock, Inc.

Figure 11.6/Color Plate 11.6: © Chelsea/ShutterStock, Inc.

Figure 11.7: Courtesy of RG Johnsson/NPS

Figure 11.12: Courtesy of D. Roddy (U.S. Geological Survey), Lunar and Planetary Institute/NASA

Part 5 Opener (page 331): © TebNad/ShutterStock, Inc.

Figure 14.3: Courtesy of Robert V. Rohli

Figure 14.4: Courtesy of Sean Linehan, NOS, NGS/NOAA